Gold

Cardiff University, United Kingdom
4–6 July 2011

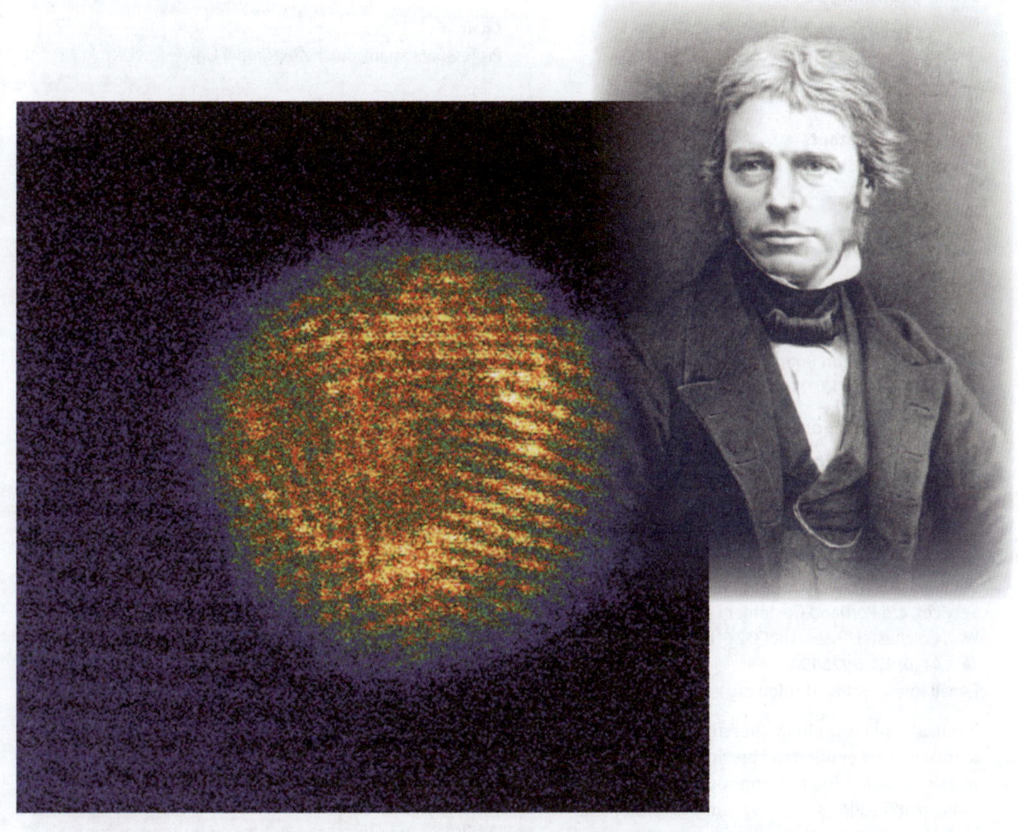

FARADAY DISCUSSIONS
Volume 152, 2011

RSC Publishing

The Faraday Division of the Royal Society of Chemistry, previously the Faraday Society, founded in 1903 to promote the study of sciences lying between Chemistry, Physics and Biology.

EDITORIAL STAFF

Editor
Philip Earis

Deputy editor
Jane Hordern

Senior publishing editor
Nicola Nugent

Publishing editors
Jonathan Counsell, Anna Watson

Publishing assistants
Hannah Porter, Claire Sissen

Publisher
Niamh O' Connor

Faraday Discussions (Print ISSN 1359-6640, Electronic ISSN 1364-5498) is published 4 times a year by the Royal Society of Chemistry, Thomas Graham House, Science Park, Milton Road, Cambridge, UK CB4 0WF. Volume 152 ISBN-13: 978 1 84973 2376

2011 annual subscription price: print+electronic £669, US $1,247; electronic only £602, US $1,122. Customers in Canada will be subject to a surcharge to cover GST. Customers in the EU subscribing to the electronic version only will be charged VAT. All orders, with cheques made payable to the Royal Society of Chemistry, should be sent to RSC Distribution Services, c/o Portland Customer Services, Commerce Way, Colchester, Essex, UK CO2 8HP.
Tel +44 (0) 1206 226050;
E-mail sales@rscdistribution.org

If you take an institutional subscription to any RSC journal you are entitled to free, site-wide web access to that journal. You can arrange access *via* Internet Protocol (IP) address at www.rsc.org/ip. Customers should make payments by cheque in sterling payable on a UK clearing bank or in US dollars payable on a US clearing bank. Periodicals postage is paid at Rahway, NJ and at additional mailing offices. Airfreight and mailing in the USA by Mercury Airfreight International Ltd., 365 Blair Road, Avenel, NJ 07001, USA.

US Postmaster: send address changes to *Faraday Discussions*, c/o Mercury Airfreight International Ltd., 365 Blair Road, Avenel, NJ 07001. All despatches outside the UK by Consolidated Airfreight.

PRINTED IN THE UK

Faraday Discussions documents a long-established series of *Faraday Discussion* meetings which provide a unique international forum for the exchange of views and newly acquired results in developing areas of physical chemistry, biophysical chemistry and chemical physics.

SCIENTIFIC COMMITTEE, Volume 152

Chair
Professor Graham Hutchings (Cardiff University, UK)

Professor Mike Bowker (Cardiff University, UK)
Dr Barry Murrer (Johnson Matthey, UK)
Professor Robert Schlögl (Fritz-Haber-Institut der Max-Planck-Gesellschaft, Germany)
Professor Rutger van Santen (Eindhoven University of Technology, The Netherlands)

FARADAY STANDING COMMITTEE ON CONFERENCES

Chair
D E Heard (Leeds, UK)

W A Brown (UCL, UK)
I Hamley (Reading, UK)
J Hirst (Nottingham, UK)
A Mount (Edinburgh, UK)

Gold

Faraday Discussions

www.rsc.org/faraday_d

A General Discussion on Gold was held at Cardiff University, Cardiff, United Kingdom on 4th, 5th and 6th July 2011.

RSC Publishing is a not-for-profit publisher and a division of the Royal Society of Chemistry. Any surplus made is used to support charitable activities aimed at advancing the chemical sciences. Full details are available from www.rsc.org

CONTENTS

ISSN 1359-6640; ISBN 978-1-84973-237-6

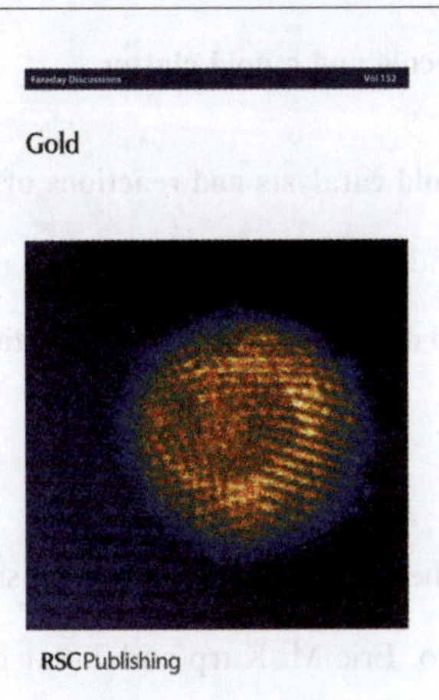

Cover
See Willock *et al.*, *Faraday Discuss.*, 2011, **152**, 135–151.

The image shows a high resolution STEM image of a Au/Pd core shell particle on an oxide support. The diameter of the particle is around 4 nm.

Image reproduced by permission of Dr David Willock from *Faraday Discuss.*, 2011, **152**, 135.

PREFACE

INTRODUCTORY LECTURE

PAPERS AND DISCUSSIONS

CONCLUDING REMARKS

ADDITIONAL INFORMATION

Preface

Graham J. Hutchings

Received 8th September 2011, Accepted 9th September 2011
DOI: 10.1039/c1fd90029c

Faraday Discussion FD152 has the very bold title of "Gold". As single word titles go, gold is perhaps one of the easiest to use. Admittedly I had initially proposed a much longer title but was readily persuaded by the President of the Faraday Division that "Gold" just said it all. Everyone knows that gold is a beautiful lustrous yellow metal much renowned for its value. Of course at the time of writing gold has reached remarkably high market values and is clearly still viewed as a major haven in times of financial uncertainty. Until very recently gold was also renowned as being the most noble metal and consequently compared to the other elements in the periodic table it was always considered to have a very limited interest for chemists. However, when gold is prepared in a nano-particulate form it can become highly reactive. During the past twenty five years the chemistry, and in particular the physical chemistry, of gold nanoparticles has seen a marked ascendency in the chemical literature. The realisation that gold nanoparticles were active as heterogeneous catalysts was mainly due to the pioneering work of Masatake Haruta who showed that small gold nanoparticles 2–5nm in diameter when supported on an oxide support could oxidise carbon monoxide at temperatures as low as $-70\ ^\circ$C. This unprecedented catalytic activity was discovered in the mid 1980's and this has fired the imagination of scientists worldwide, all trying to answer the question of how can gold be so active at the nanoscale. At present several hundred articles and patents appear annually concerned with gold nanoparticles, many of these on catalysis by gold. It was therefore apparent that the subject of catalysis by gold was ready for a Faraday Discussion and this is the context in which the discussion meeting was convened.

The physical chemistry associated with gold catalysis was explored in four sessions:

- Gold Catalysis & Materials Science
- Theoretical Insights on Gold Catalysis
- Gold Catalysis at the Gas–Solid Interface
- Gold Catalysis and Enhanced Selectivity

Each of these had a mix of invited speakers and speakers selected from an extensive number of submitted papers.

A key feature of Faraday Discussions is the careful selection of both the opening and closing speakers as these can set the scene in many ways. In respect of gold catalysis I have noted that the field of gold catalysis owes its great popularity to the seminal studies of Masatake Haruta; it was therefore wholly appropriate that he opened the Faraday Discussion on gold as the RSC Spiers Medal Lecturer. The choice of closing speaker is also crucial at any Faraday Discussion. In this case we wanted someone who had no previous experience or real exposure to gold catalysis, in this way the intention was that there was the potential for radically new insights into the field of gold catalysis that could inspire future lines of investigation. We were pleased that Martyn Poliakoff agreed to do this difficult task and you will read the interesting suggestions he has been able to make concerning the future of this field. Gold Catalysis is a field that is changing rapidly as new techniques and theoretical methods are able to delve even deeper into the complexity of this subject. We can anticipate further discussion meetings in the future as catalysis by gold will continue to fascinate many scientists.

Graham J Hutchings
Cardiff University
7th September 2011

Spiers Memorial Lecture
Role of perimeter interfaces in catalysis by gold nanoparticles

Masatake Haruta*

Received 24th August 2011, Accepted 31st August 2011
DOI: 10.1039/c1fd00107h

Gold can be deposited as nanoparticles (NPs) of 2 to 5 nm in diameter on a variety of materials such as metal oxides and carbides, carbons, organic polymers and exhibits surprisingly high catalytic activities for many reactions in both gas and liquid phases. The mechanisms for the genesis of catalysis by gold NPs is discussed based on real powder catalysts and model single crystal catalysts for two simple reactions, low-temperature oxidation of CO in which gold NPs catalysts are exceptionally active and for dihydrogen dissociation in which gold NPs catalysts are still poorly active. For both the two reactions, it has been revealed that reactions take place at perimeter interfaces around gold NPs.

1. Introduction

About 7 thousand years ago human beings recognized the existence of gold on the earth. Gold can exist as metallic species in the nature because thermodynamically oxide formation does not spontaneously take place. The melting point of gold is 1336 K, noticeably lower than those of Pd (1825 K) and Pt (2045 K). Gold could be melted even in a primitive rock furnace, allowing the production of larger ingots. Its softness enabled hand crafting to make jewelley and decorations for religious ceremonies and festivals. In the course of time gold has been regarded as the most precious metal because it shines brilliantly and eternally. This feature attracted the governors of communities as a symbol of eternal youth, beauty, and power.

The eternal brightness of gold allured many rulers or governors to artificially produce it from cheap elements such as lead, tin, copper, iron, mercury and sulfur.[1] The alchemy started in Egypt and then grew in Greece and was succeeded by Islam at around 10th century. This attempt was continued and intensified in west Europe until the dawn of modern science. Although alchemy was over in vain, it has brought about modern chemistry which currently produces a variety of functional materials such as organic polymers, food additives, medicines, pesticides, and semiconductors. In the classical chemistry gold was treated as an inert material. The major industrial use is very thin wires for electrical bonding utilizing high electrical conductivity and chemical inertness of gold. The sole chemistry of gold might be wine red colour of colloids of stained glass windows in cathedrals and churches. It was Michael Faraday that initiated scientific approaches to small colloidal gold particles, which presented the fourth state of materials in addition to gas, liquid, and solid phases.

As shown in Fig. 1, the size of Au colloids prepared by Michael Faraday in 1850s was estimated to be around 30 nm,[2] which is an optimum size for surface plasmon resonance absorption giving a colour of wine red.[3] Then Zsigmondy studied

Department of Applied Chemistry, Graduate School of Urban Environmental Sciences, Tokyo Metropolitan University, 1-1 Minami-osawa, Hachioji, Tokyo 192-0397, Japan. E-mail: haruta-masatake@center.tmu.ac.jp

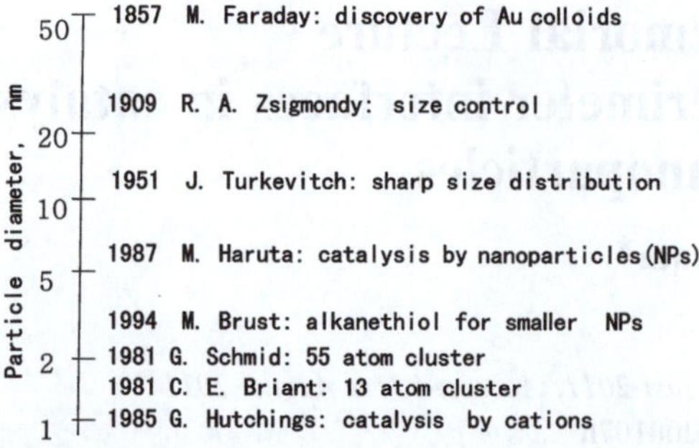

Fig. 1 An overlook on the major contributions to gold chemistry in view of size.

reducing agents to control the size of gold colloids.[4] Turkevitch prepared monodisperse colloids by choosing suitable stabilizers.[5] Haruta found that gold was catalytically very active for CO oxidation when dispersed on base metal oxides as nanoparticles (NPs) below 5 nm.[6] Brust succeeded in preparing tiny Au NPs smaller than 2 nm by using alkanethiol as a stabilizer.[7] Schmid[8] and Briant[9] synthesized 55 atoms and 13 atoms clusters bonded with organic ligands, respectively. Hutchings reported that cationic gold could be theoretically the most active catalyst for the hydrochlorination of ethylene[10] and later experimentally confirmed.[11]

2. Catalysis by gold leading to green chemistry

Japanese industries have made tremendous efforts to save energies since 1973, the year of the first oil crisis. In particular, chemical industry was a top runner in energy saving and succeeded in reducing the standard unit energy consumption by 50% within ten years.[12] However, since 1983 the unit energy consumption has remained almost steady, implying that further energy saving is hard to achieve by technological improvements alone. In Japan, chemical industry consumes the largest fraction of energies, 34% of the total industrial consumption. Because the situation of chemical industry can be assumed to be similar in Europe and USA, if technological "innovation" in chemical industry happens, the impact would be very great all over the world.

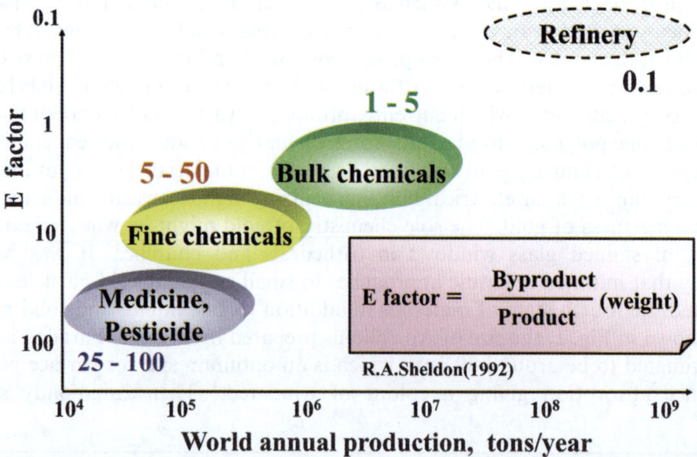

Fig. 2 E-factors for the production of chemicals as a function of annual production (Produced from the data given in ref. 13).

A key question is whether there is still a room for technological innovation in chemical industry. Fig. 2 shows E-factors which are defined by Sheldon as the weight of products divided by the weight of byproducts.[13] The E-factors for producing bulk chemicals, which are produced in a scale of ten million tons per year all over the world, are above 1 meaning that byproducts are inevitably produced more than the products we want. In fine chemicals production, byproducts are formed about one order of magnitude more than products. The production of medicines, pesticides, and high-purity compounds for semiconductor industry needs a number of reaction steps and is accompanied by the formation of a huge amount of byproducts exceeding 50 times as that of main products. A major reason why we are producing unwanted byproducts much more than what we want is that we have to adopt stoichiometric reactions which inevitably cause byproduct formation.

A typical example is the production of propylene oxide (PO). Currently about half of PO is produced by chlorohydrin process which is based on the following reactions. To produce one mol of PO one mol of HCl and a half mol of $CaCl_2$ are byproduced. If we could replace these stoichiometric reactions with simple gas phase reactions by the creation of new catalysts, the contribution to environmental maintenance would be significantly large. Gold catalysts appear to have potential capability in this direction of technological innovation.

< Classical industrial process: stoichiometric reaction accompanied by byproducts. >

$$CH_3CH{=}CH_2 + Cl_2 + H_2O \rightarrow CH_3CH(OH){-}CH_2Cl + HCl$$

$$CH_3CH(OH){-}CH_2Cl + \tfrac{1}{2}Ca(OH)_2 \rightarrow CH_3CH(O)CH_2 + \tfrac{1}{2}CaCl_2$$

< Future target: green and sustainable chemistry to produce only what we need >

$$CH_3CH{=}CH_2 + \tfrac{1}{2}O_2 \rightarrow CH_3CH(O)CH_2$$

World market sales of catalysts amount about 9 billion US dollar/year.[14] Chemical industry occupies about half of the market through polymerization and chemicals syntheses. The rest of half is shared equally by petroleum refinery and by automotive exhaust gas treatments. In principle, polymerization and fine chemicals syntheses use homogeneous catalysts, while bulk chemicals production, petroleum refinery, automotive exhaust treatments, utilize heterogeneous catalysts. Among heterogeneous catalysts, metal catalysts which can proceed both oxidation and reduction are most widely used.

As shown in Fig. 3, there are only 12 elements that are applicable to metal catalysts. The history started at the very beginning of the 20th century with 3d metals, Fe, Co, Ni, and Cu, which contributed to the development of coal chemical industries by using synthetic gas composed of CO and H_2. In the middle of 20th century 4d metals, Ru, Rh, Pd, and Ag catalysts were developed to establish new chemical processes using petroleum, unsaturated hydrocarbons as raw materials. Among 5d metals, osmium is excluded from catalytic metals because its oxide is toxic. Iridium has limited applications but is used, in particular, for hydrazine rocket fuel ignition. Platinum is an all-round player and useful not only for organic syntheses and hydrocracking of petroleum feed stock but also for automotive exhaust gas treatments and fuel cells. The only exception is gold, which was considered to be too noble to work as a catalyst.

Until now, gold catalysts have been investigated for a number of different reactions in gas and liquid phases, usually at temperatures below 473 K. However, major ruling factors which define the catalytic activity and selectivity of gold have not yet

Group Valence Orbital	VIII			IB
	8	9	10	11
3d	Fe NH₃	Co gasoline	Ni fats	Cu CH₃OH
4d	Ru cyclohexene	Rh hydroformy-lation	Pd coupling	Ag ethylene oxide
5d	Os Toxic	Ir hydrazene combustion	Pt cracking, exahust gas	Au

Fig. 3 Catalytic metals in the periodic table and their representative applications.

been well summarized. In this article, by focusing on two simple reactions, CO oxidation and dihydrogen dissociation, the active states of gold and probable mechanism for the genesis of catalysis by gold will be discussed.

3. Active states of gold that catalyze CO oxidation at room temperature

Many hypotheses have been presented so far concerning catalytically active states of gold for CO oxidation. Fig. 4 shows turn over frequencies (TOFs), rates per surface metal atom, as a function of gold sizes (diameter, height, or infringes). Two extremes, atomically dissociated gold cations and unsupported metallic gold in a shape of sponge or tube are less active by one or two orders of magnitude than supported gold clusters, thin films, and NPs. It is obvious that there is an optimum size range from 0.5 to 5 nm for gold to exhibit TOF at around 1 s⁻¹, which means one molecule of CO reacts with O_2 to form CO_2 per one surface metal atom per 1 s. This value is really high in comparison with other reactions.

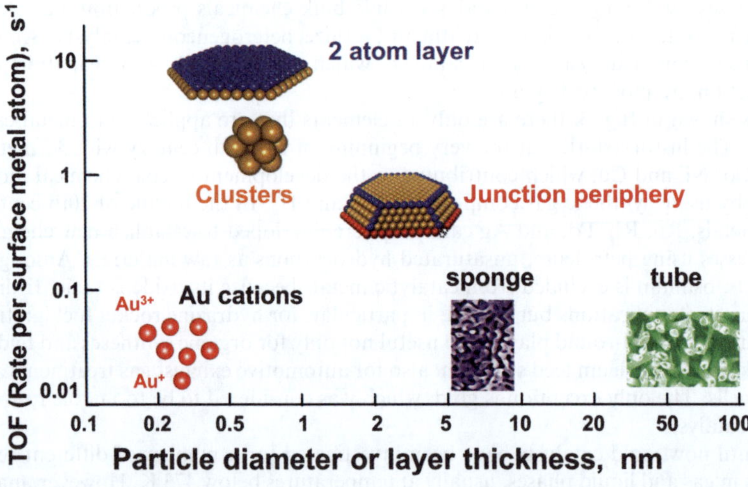

Fig. 4 Turn over frequency of CO oxidation at room temperature for various states of gold.

3.1 Unsupported gold sponges and polymer-embedded gold microtubes

It was a surprise that unsupported gold sponges and gold tubes which were larger than 5 nm exhibited catalytic activity at room temperature. Gold sponges with ligament size of 5 nm[15] and 50 nm[16] were prepared by leaching Ag from Ag–Au alloys. The surfaces of gold sponge were inevitably contaminated by Ag_2O, which markedly enhanced the catalytic activity of gold.[17] Microtubes of gold with diameter of 100 nm embedded within the pores of polycarbonate template membranes showed high catalytic activity at 300 K in a CO and O_2 gas stream saturated with water vapour.[18] In the co-presence of aqueous alkali solution activity is further improved, which coincides well with the fact that gold in bulk is far more active than bulk Pt in the electrochemical oxidation of CO in alkali solution.[19]

3.2 Cationic gold

Gates has made an intensive study of gold supported on MgO,[20] La_2O_3,[21] CeO_2,[22] and zeolite NaY[23] by means of XANES, EXAFS and has obtained quantitative correlations between the fractions of cationic gold and TOF for CO oxidation. Over basic metal oxide (MgO, La_2O_3, and NaY zeolite) supports, cationic gold species are stabilized to some extent and they show catalytic activity higher than that of metallic gold. Trivalent gold cation, Au^{3+} is more active but less stable than monovalent Au^+ during reaction. The TOF of cationic gold is not high and in the level of 0.01 s^{-1}, which was one or two orders of magnitude lower than those of the most active Au/TiO_2 and Au/Fe_2O_3. Over CeO_2 support, gold clusters of about 15 atoms were more active than cationic gold species.[22]

The catalytic behavior of gold cations is considered to be different from that of metallic gold NPs in terms of apparent activation energy (very high, 138 kJ mol^{-1} for cations[22]), reaction orders, and influence of moisture. It is interesting to investigate whether these catalysts are enhanced by moisture or not. If they behave like base metal oxides such as Co_3O_4 and NiO, they would be deactivated by moisture.

Flyzani-Stephanopoulos reported for the first time that catalytic activity of Au/CeO_2 for water gas shift reaction did not change after leaching metallic gold particles by aqueous solution of NaCN.[24] It should be noticed that the activity of their Au/CeO_2 catalysts prepared by coprecipitation was lower than that of commercial $Cu/ZnO/Al_2O_3$ catalysts, whereas our Au/CeO_2 catalysts were more active than the commercial copper based catalysts.[25] They assumed that all the cationic gold species were present as isolated species in the matrices of the support. On the other hand, as shown in Fig. 5, Oyama has recently reported that NaCN leaching forms sodium gold cyanide and this crystalline compound shows catalytic activity for hydrogenation of propylene in a mixture of H_2 and O_2.[26,27] Accordingly, experiments carried out by using gold catalysts leached with an aqueous solution of NaCN may not allow proper discussion on the active species of gold. As far as CO oxidation concerns, metallic gold before leaching is more active than oxidic gold.[28]

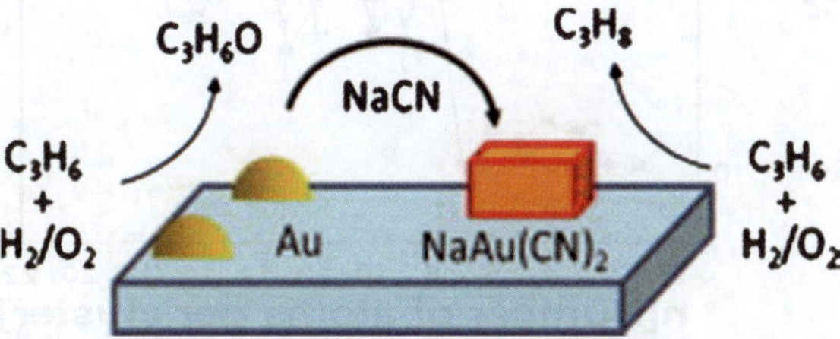

Fig. 5 Schematic representation of leaching supported gold catalysts with aqueous NaCN to form Na Au(CN)₂.[26,27]

3.3 Gold clusters smaller than 2 nm

The third hypothesis is that gold clusters having electronic structures different from that of bulk are responsible for low temperature CO oxidation. Almost 50 years ago Kubo predicted that small particles should behave differently from bulk.[29] As shown in Fig. 6, metal clusters smaller than 2 nm show discrete energy band structure similar to those of semiconductors. It is also reasonable to assume that the support can affect the electronic structure of tiny metal clusters much more than those of larger NPs.

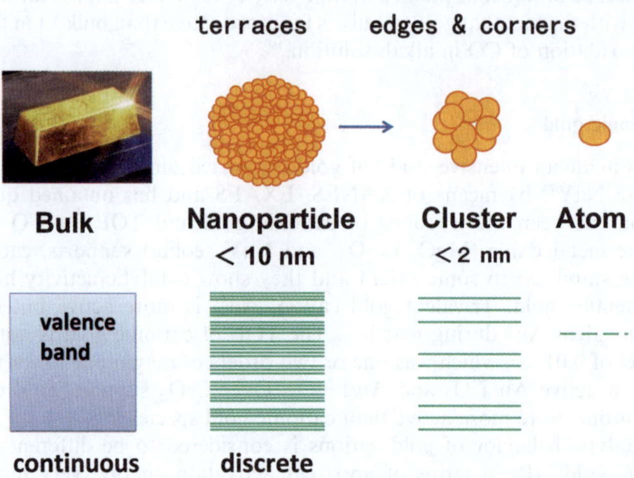

Fig. 6 Major surface states and valence band structures of bulk, nanoparticles, clusters, and atoms.

Fig. 7 shows that more than 7 atoms, at least 8 atoms are necessary for the genesis of catalytic activity of gold for CO oxidation.[30] This is because stable structure can be formed only with 8 atoms or more. In addition to the bottom up approach, top

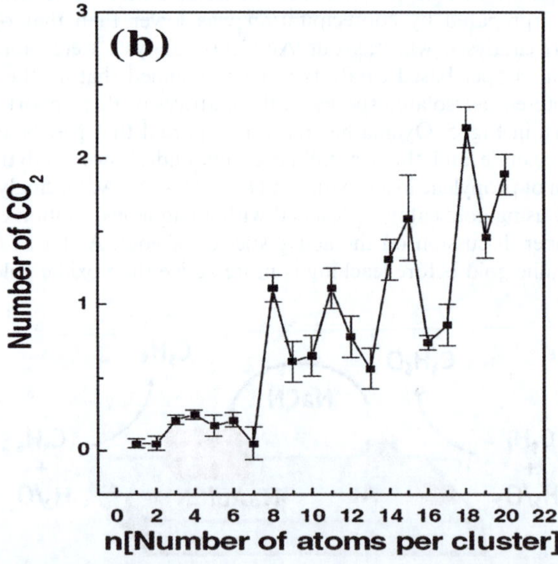

Fig. 7 Catalytic activity of gold clusters deposited on MgO for CO oxidation as a function of the number of atoms.[30]

down approach to the electronic structure change has been made.[31] Since the band structure of valence band of glassy carbon is simple, it is advantageous to deposit gold on glassy carbon to investigate the gold components of electronic structure. As shown in Fig. 8, d-band spin-orbital splitting tends to decrease at around 100 atoms. Eight to 100 atoms is the range of clusters which provide electronic structures different from those of atoms and of bulk. This size range of gold clusters may lead to novel catalysis by gold clusters in the near future.

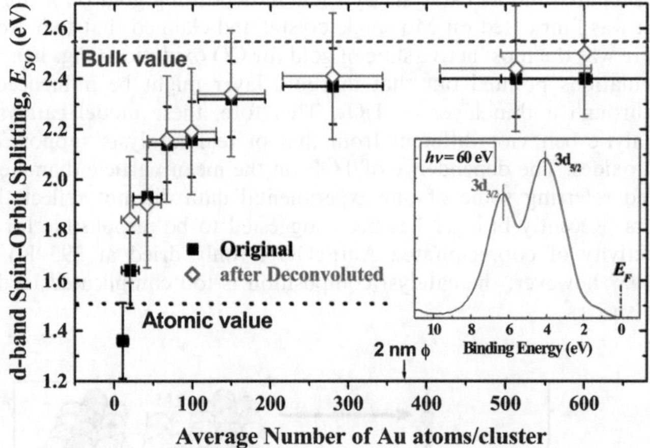

Fig. 8   d-band spin-orbital splitting of gold clusters deposited on glassy carbon as a function of the average number of atoms.[31]

Over alkaline earth metal hydroxides, $Be(OH)_2$ and $Mg(OH)_2$, and $La(OH)_3$ as supports, gold exhibits very high catalytic activity for CO oxidation even at 200 K, but only when it is smaller than 1.5 nm in diameter. The requirement to gold size is very strict and the active catalysts suddenly die when the size of gold exceeds 1.5 nm, usually after about 4 months. Based on HR-TEM, EXAFS, X-ray scattering results, it is assumed that only 13 atoms clusters are responsible

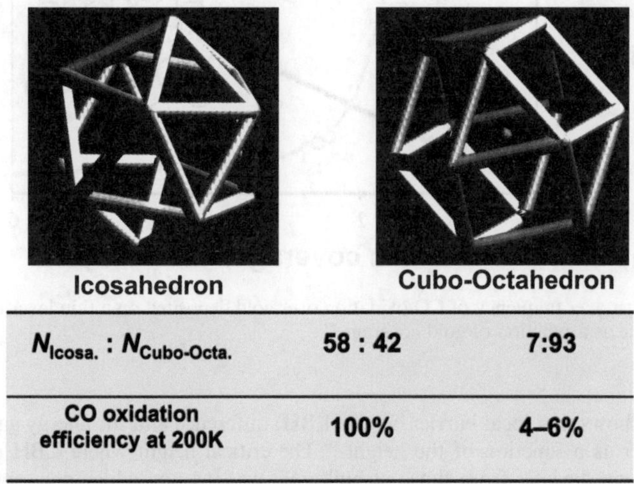

	Icosahedron	Cubo-Octahedron
$N_{Icosa.} : N_{Cubo-Octa.}$	58 : 42	7:93
CO oxidation efficiency at 200K	100%	4–6%

Fig. 9   CO oxidation at 200 K over gold clusters supported on $Mg(OH)_2$ by deposition-precipitation method.[32] The number of gold atoms and three dimensional structures were estimated by X-ray scattering combined with Debye functional analysis.

for CO oxidation at 200 K. The Debye Functional Analysis of X-ray scattering[32] for differently prepared $Au/Mg(OH)_2$ suggests that the catalytically active structure is not cubo-octahedron but is icosahedron (Fig. 9). If the shape of gold was not bulky sphere but raft-like layer owing to the strong interaction with $Mg(OH)_2$, other flat structure might be more probable than round icosahedron.

3.4 Bi-layer of gold

Goodman prepared model catalysts by vacuum-depositing gold on a thin layer of TiO_2 which was fabricated on Mo single crystal and claimed that two atoms layer raft structure was the most active state of gold for CO oxidation (Fig. 10).[33,34] Theoretical calculations pointed out that the gold layer might be influenced by Mo substrate through a thin layer of TiO_2. Therefore, these model catalysts might exhibit catalytic behavior different from that of real catalysts supported on the bulk metal oxides. The dependence of TOF on the mean particle diameter of gold they showed referring some of our experimental data did not reflect the whole data of ours. Recently bi-layer has been suggested to be responsible for the high catalytic activity of coprecipitated Au/FeOOH (only dried at 393 K) at room temperature,[35] however, the catalyst composition is too complicated to define the active gold species.

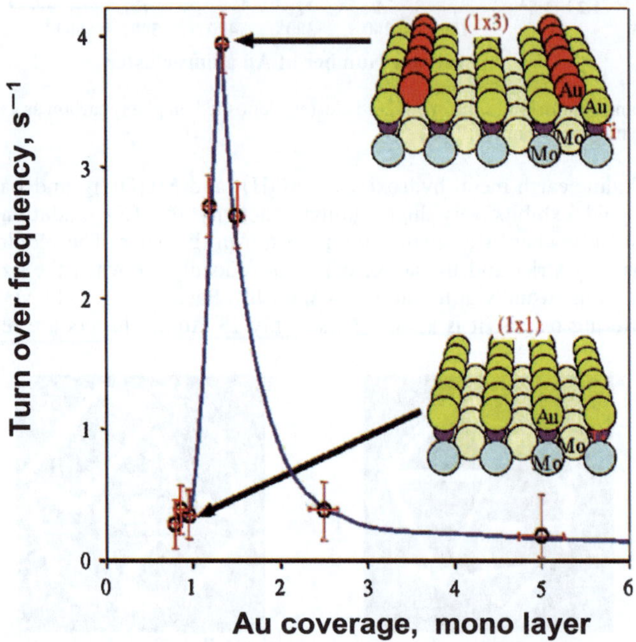

Fig. 10 Turn over frequency of CO oxidation over gold deposited on a thin layer of TiO_2 on Mo substrate as a function of gold coverage.[34]

Fig. 11 shows that local barrier height (LBH) difference and the energy gap of each gold cluster as a function of the height.[36] The critical height where LBH difference and band gap deviates from those of bulk gold was 0.4 nm, which corresponds to 2 atom layer. One atom layer patches cannot be assumed to be catalytically active for CO oxidation, because Au/Fe_2O_3 and Au/Co_3O_4 prepared by coprecipitation, which contained a lot of single layer patches in addition to gold NPs, were less active than the catalysts prepared by deposition-precipitation which were almost free from single-layer patches.[37,38] It is hard to understand why there was a big difference between 2-atoms and 3-atoms thickness although the electronic structure was not appreciably different. It would be interesting to investigate whether mono-layer patches exhibit unique catalysis for reactions other than CO oxidation. On the other hand, it is likely that bi-layer of gold combined with O_2 activation at periphery sites constitutes an ideal nano-structure of gold catalysts with a minimum amount of gold.

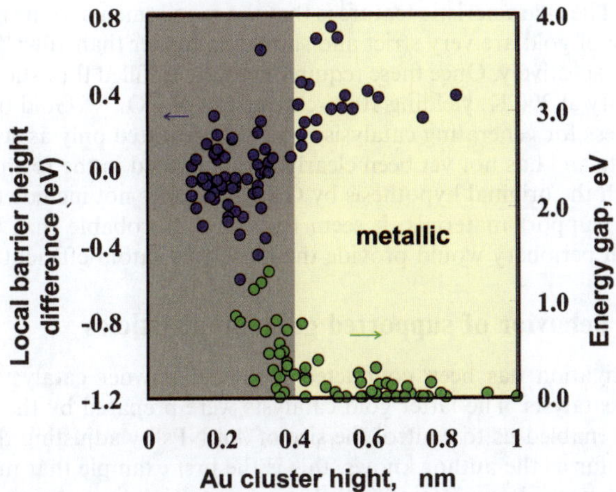

Fig. 11 Local barrier height difference and energy gap as a function of the height of gold clusters deposited on single crystal of rutile TiO_2 (110).[36]

3.5 Gold nanoparticles of 2–5 nm

Fig. 12 displays five active states of gold in the axes of four factors which make gold catalytically active for low-temperature CO oxidation. Gold NPs supported on semiconductive and/or reducible metal oxides such as TiO_2, MnO_2, Fe_2O_3, Co_3O_4, NiO, ZnO, ZrO_2, and CeO_2 exhibit very high activity at temperatures below 273 K in the abscence of

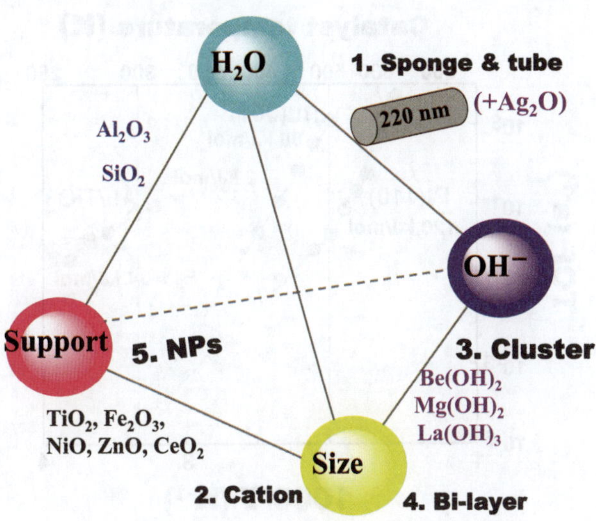

Fig. 12 Four conditions to make gold active as a heterogeneous catalyst.

moisture. Gold NPs supported on insulating and non-reducible metal oxides such as Al_2O_3, SiO_2 and metal carbides such as TiC are moderately active at room temperature and require a significant amount of water above 100 ppm.

Unsupported gold sponges show moderate catalytic activity at room temperature owing to decoration with Ag_2O contaminants.[15,16] Polymer embedded microtubes are active in the copresence of aqueous alkali solution.[18] Gold cations are also moderately active exhibiting higher activity over basic metal oxide supports than over neutral metal oxides. Gold clusters supported on the hydroxides of alkaline earth metals (Be and Mg) and lanthanum are extremely active at a temperature as low as 200 K. The characteristic feature is that the requirements to the metal loading and to the size of gold are very strict and should be higher than 10wt% and smaller than 1.5 nm, respectively. Once these requirements are fulfilled they show the highest catalytic activity at 200 K, yielding 100% conversion of CO.[32,39] Gold bi-layer as the critical thickness for generating catalysis has been presented only as surface science model catalysts and has not yet been clearly demonstrated in the form of real catalysts. Although the original hypothesis by Goodman does not include the contribution from the support materials, it seem to be more probable that bi-layer gold combined with periphery would provide the most gold- atom-efficient catalysts.

4.   Kinetic behavior of supported gold nanoparticles

Kinetic investigation has been conducted with real powder catalysts and model single crystal catalysts. The latter gold catalysts were prepared by the cathodic arc plasma which enabled us to control the size of Au NPs by adjusting the condenser capacity.[40] As far as the author knows, this is the first example that makes the size control of Au possible maintaining the same metal loading. At one monolayer coverage, with a decrease in the condenser capacity to 360μF, the mean diameter of gold NPs was reduced down to 1.3 nm, which roughly corresponds to 55 atoms.

A characteristic feature of supported gold catalysts prepared by wet methods is shown in Fig. 13 that there is a transition in reaction mechanism at around 333 K.[41] This feature has been reproduced by model single crystal catalysts in their Arrhenius plots.[42] The apparent activation energy is very small, 2–3 kJ mol^{-1} at temperatures above 333 K, while at lower temperatures in the range of 26–34 kJ mol^{-1}. These apparent activation energies for CO oxidation are appreciably lower than those reported for single crystal surfaces of other noble metals, 80–120 kJ mol^{-1}.[43]

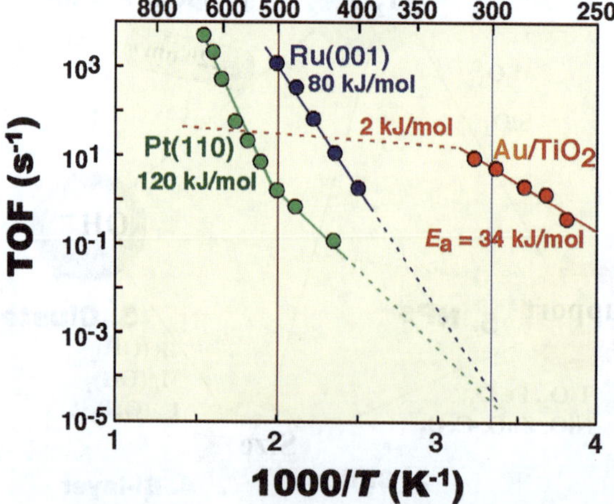

Fig. 13   Arrhenius plots for CO oxidation over noble metal catalysts.

The second feature is a big difference between wet powder catalysts and dry model catalysts was that the model catalysts were not active at all under dry conditions.[42] The model gold catalysts need a trace amount of water to exhibit catalytic activity because they are almost completely dry, whereas powder gold catalysts prepared by the wet methods contain a certain amount of water or OH^- groups. The third feature is that the effect of water is different depending on the reaction temperature. The oxidation of CO at temperatures above 333 K does not need water, whereas at room temperature the reaction needs a trace amount of water.

At 400 K, the rate per surface Au atoms is independent on particle diameter while the rate per perimeter Au atoms linearly decreases with a decrease in the mean diameter. This means that the rate of CO oxidation over one catalyst specimen increases linearly with the number of atoms at the surfaces. Accordingly it is very likely that the reaction takes place at the gold surfaces. The Arrhenius plots for (111) and (100) planes of single gold crystals and gold NPs supported on TiO_2 are similar to each other with small apparent activation energies of 3 to 5 kJ mol^{-1}. These results strongly support a mechanism that at temperatures above 333 K CO oxidation takes on the gold surfaces.

At 300 K the rate per perimeter interfaces is independent on the mean diameter of gold NPs, while the rate per surface atoms sharply increases with a decrease in the particle size. This means that the rate is in proportion to the number of perimeter gold atoms and at room temperature CO oxidation takes place not on the gold surfaces but on the perimeter sites.

The rate dependency on the partial pressure of CO is the first order up to 4 Torr or 0.5 vol% and then it becomes 0th order. The behavior is the same at 400 and 300 K. Zeroth order at higher partial pressure means that CO is adsorbed on the gold surfaces nearly to saturation. The rate dependency on the partial pressure of O_2 also shows similar tendencies. Above 10 Torr or 1 vol% of O_2 the rate is 0th order irrespective of reaction temperature. Under lower partial pressure the reaction orders are 0.5 at 400 K and 1 at 300 K. The above kinetic behaviour suggests the Langumuir-Hinshelwood mechanism in which the rate determining step is the reaction of CO adsorbed on the surfaces of gold NPs with oxygen adsorbed at the periphery sites around gold NPs.

5. Role of perimeter interfaces in supported gold nanoparticles

In the periphery hypothesis, CO is considered to adsorb on the gold surfaces, most likely at edges and corners and oxygen molecule is activated at the perimeter interfaces.[30,41,43,44] Fig. 14 depicts the fractions of edges and corners, and of perimeter interfaces changes with a decrease in diameter.[45]

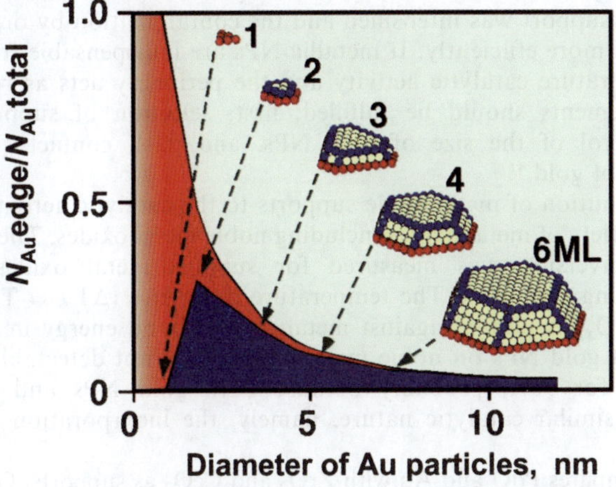

Fig. 14   Fractions of edge & corner atoms and of perimeter atoms as a function of the diameter of hemispherical gold nanoparticles.[45]

As can be seen from TEM images in Fig. 15, a conventional gold catalyst prepared by the impregnation method contains large gold NPs of 30–100 nm, which are spherical in shape and as large as the particles of the metal oxide support. The interaction with the support materials appears to be very weak. With this structure gold is poorly active as a catalyst. In contrast, when gold catalysts were prepared by the coprecipitation and deposition-precipitation methods, gold were deposited as hemispherical NPs being attached to each particle of the support at their flat planes. In particular, the deposition-precipitation method hardly forms raft-structured gold unlike the coprecipitation method, therefore presenting better defined catalysts. The majority of gold NPs are really hemispherical as schematically shown in Fig. 14.

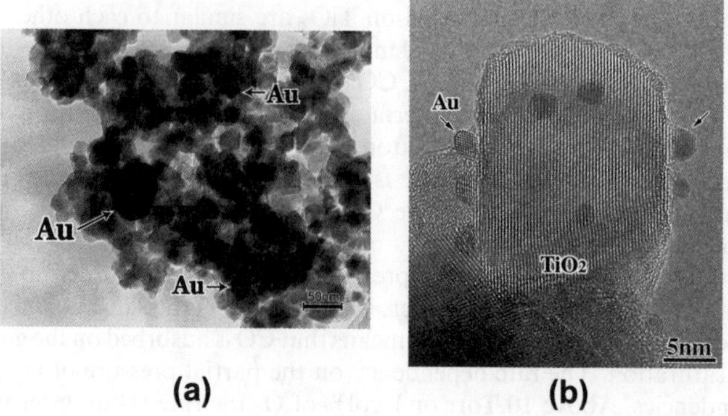

Fig. 15 TEM images of Au/TiO$_2$ prepared by (a) the impregnation method and (b) the deposition-precipitation methods followed by calcination in air at 673 K. Note that the support material is the same and Degussa TiO$_2$, p-25.

The possibility of bi-layers and cationic gold as indispensably important state of gold can be excluded because without those species (deposition-precipitation) the catalytic activity is already as high as that of coprecipitated samples. Strong evidence is shown in Fig.16 that Au colloids immobilized by metal oxide supports can exhibit high activity and calcination at higher temperatures results in higher catalytic activity.[46,47] This is probably because the contact between gold NPs and the support was intensified and the contamination by organic ligands was removed more efficiently. If metallic NPs are indispensable for the genesis of low-temperature catalytic activity and the periphery acts as reaction sites, three requirements should be fulfilled; first, selection of support material, second, control of the size of gold NPs, and, last, contact structure and morphology of gold.[41]

The contribution of metal oxide supports to the catalytic activity was evaluated for a variety of metal oxides including noble metal oxides. The temperature for 50% conversion was measured for support metal oxides alone and after depositing gold NPs. The temperature difference ($\Delta T_{1/2} = T_{1/2}$ of MO$_x$ − $T_{1/2}$ of Au/MO$_x$) is plotted against metal-oxygen bond energy in Fig. 17. The deposition of gold NPs on noble metal oxides does not detectably change the temperature for 50%, probably because both gold NPs and noble metal oxides have similar catalytic nature, namely, the incorporation of oxygen is very slow.

Fig. 18 compares PdO and Au with ZrO$_2$ and CeO$_2$ as supports. Gold is always more active than PdO and Pd. The temperature difference is larger than 100 K between PdO/TiO$_2$ and Au/TiO$_2$. Depending on catalytic metals the selection of

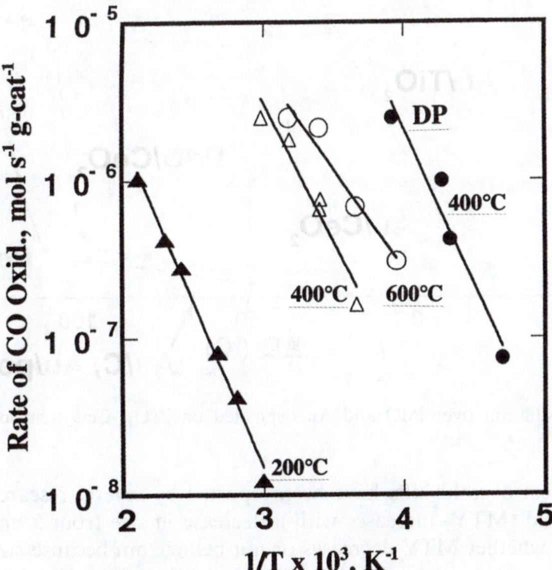

Fig. 16 Catalytic activity for CO oxidation of Au/TiO$_2$ prepared by colloid immobilization followed by calcination in air at different temperatures.[47]

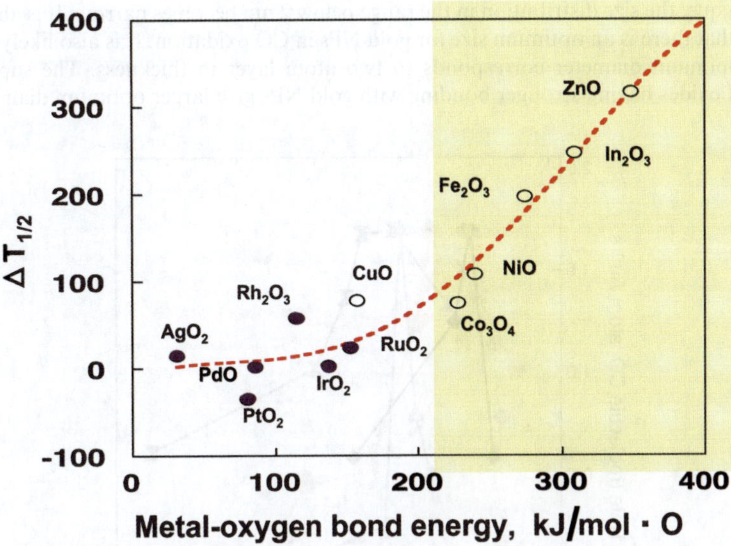

Fig. 17 Difference in temperature for 50% conversion in CO oxidation between metal oxide supports and gold nanoparticles supported on them.

support metal oxides changes. Palladium oxide prefers CeO$_2$ to TiO$_2$ whereas gold prefers TiO$_2$. It should be noted that gold NPs deposited on activated carbon and organic polymers did not exhibit catalytic activity in gas phase up to 373 K but they are active for glucose oxidation in liquid water. This means that supports which cannot activate molecular oxygen do not make gold NPs catalytically active in gas phase.

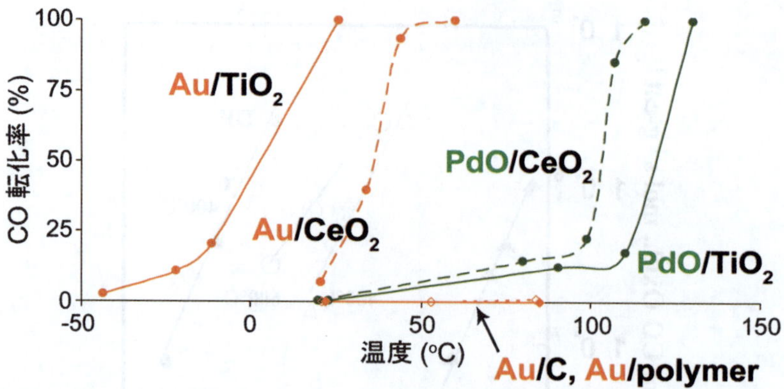

Fig. 18 CO oxidation over PdO and Au deposited on ZrO₂, CeO₂, carbon, and organic polymer.

The size effect of gold NPs have been reported by several research groups.[48–51] Metal time yield (MTY) increases with a decrease in size from 5 nm, however, it was uncertain whether MTY decreases or not below 2nm because size distribution of gold NPs were too wide to definitely discuss the size effect. Fig. 19 shows several examples of experimental data and, in principle, volcano-like correlations.[48] Recently we have dispersed organic gold clusters having a fixed number of gold atoms, such as $Au_{11}(PPh_3)_8Cl_3$, $Au_{55}(PPh_3)_{12}Cl_{16}$, $Au_{101}(PPh_3)_{21}Cl_5$ over metal oxide supports and removed organic ligands by means of oxygen plasma followed by calcination in air preventing original gold clusters from coagulation. With this technique, the size distribution in the range below 2 nm becomes narrow. It is almost sure that there is an optimum size for gold NPs in CO oxidation. It is also likely that the optimum diameter corresponds to two atom layer in thickness. The support metal oxides having stronger bonding with gold NPs give larger optimum diameter.

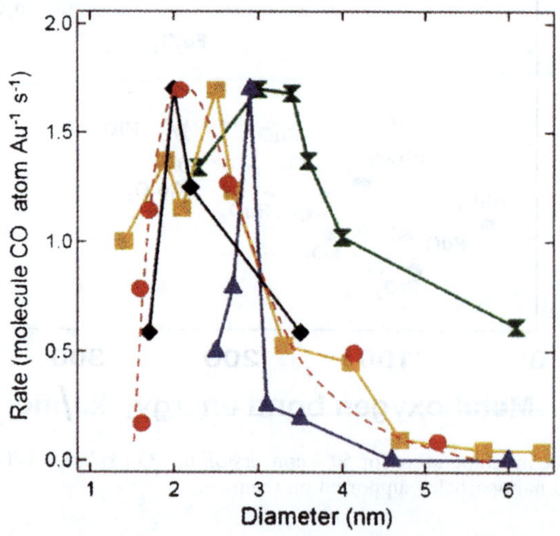

Fig. 19 Metal Time Yield (MTY) as a function of mean diameters of gold NPs deposited on TiO₂.[48]

The contact interface structure between gold NPs and metal oxide supports may affect the reactivity of periphery. High Angle Angular Dark Field-Scanning Transmission Electron Microscopy (HAADF-STEM) which enabled us to determine the position of each element (Fig. 20) showed that gold NPs grew at (111) plane, which is the most densely packed plane with spacing of 0.235 nm, on (110) plane of TiO_2 with a lattice spacing of 0.325 nm. The contact interface had a spacing of 0.33 nm, which suggests that oxygen is present at the interfaces as depicted in the model structure.

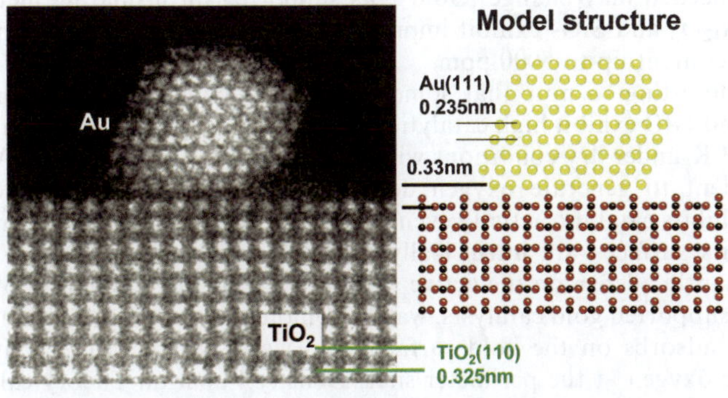

Fig. 20 HAADF-STEM image of the contact interface of a gold nanoparticle deposited on TiO_2 single crystallites and schematic atomic configuration.

6.   Promoting effect of moisture

It is beneficial that the catalytic activity of supported gold NPs and clusters is markedly enhanced by moisture in CO oxidation at room temperature. By using a special fixed bed flow-type catalytic reactor, which was constructed based on the ultraclean technology of semiconductor industry, the effect of moisture was investigated in a wide range of 0.08 to 6,000 ppm (Fig. 21).[52] Over gold NPs deposited on reducible metal oxides prepared by the wet methods, water

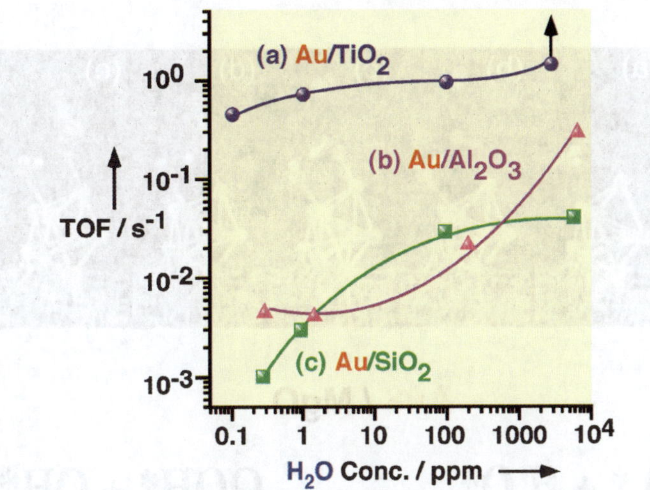

Fig. 21 TOF for CO oxidation at 273 K over Au/TiO$_2$, Au/Al$_2$O$_3$, Au/SiO$_2$ as a function of moisture content in the reactant feed.[52]

promotes CO oxidation but is not indispensable. As previously described, over dry model gold catalysts for surface scientific study, water is indispensable at temperatures below 333 K. These results imply that at least a trace amount of water is necessary for CO oxidation at room temperature. Depending on the capability of dioxygen activation of metal oxide supports, the minimum amount of water needed may change. Gold NPs supported on insulating metal oxides such as Al_2O_3 and SiO_2 exhibit improved catalytic activity with an increase in moisture content up to 5000 ppm.

It is interesting to note that some base metal oxides, in particular, MnO_2, Co_3O_4, and NiO exhibit high catalytic activity for CO oxidation at temperatures below 273 K under dry conditions with moisture less than 0.1 ppm.[53] After mild pretreatment to generate oxygen defect sites[53] or by morphology control,[54] Co_3O_4 exhibits catalytic activity even at 200 K without gold deposition, suggesting that its surfaces are intrinsically active at low temperatures. Water may predominantly adsorb on the active sites to prohibit the adsorption of CO and O_2. Over supported gold catalysts, water promotes the reaction because CO preferentially adsorbs on the gold surfaces and water enhances the activation of molecular oxygen at the perimeter sites. Density Function Theory calculations present a scheme in Fig. 22 that water molecule and oxygen molecule adsorb on gold cluster and then react with each other to form hydroperoxide and OH^-.[55] The hydroperoxo species react with CO to form CO_2, being accompanied by water recovery.

$$H_2O + O_2 \rightarrow OOH^* + OH^*$$

Tanaka has recently proposed that reaction of CO with OH^- anion is promoted by a positive feed back cycle given by H_2O,[56] which can explain why gold NPs are not active in acidic environments (Fig. 23). Concerning oxygen species, electron paramagnetic resonance measurements for Au/Al_2O_3 and Au/TiO_2 revealed that O_2^- was formed during reaction.[57] Because O_2^- was not detected over Al_2O_3 and TiO_2 alone, it is likely that the activation of molecular oxygen took place on either the surfaces of gold NPs or on the perimeter interfaces.

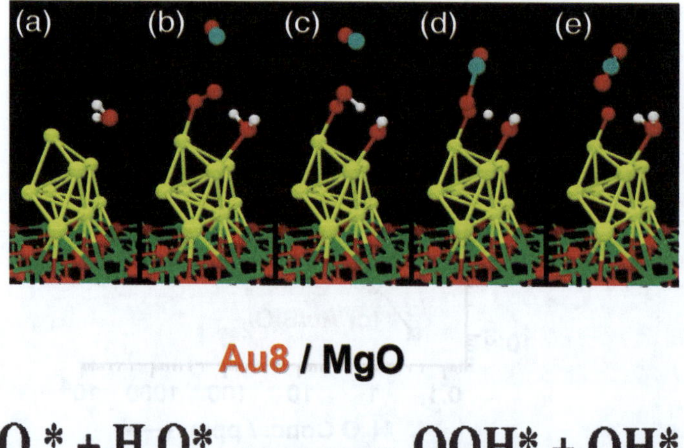

$$O_2^* + H_2O^* \rightleftharpoons OOH^* + OH^*$$

Fig. 22 Schematic representation for CO oxidation over Au_8/MgO promoted by moisture.[55]

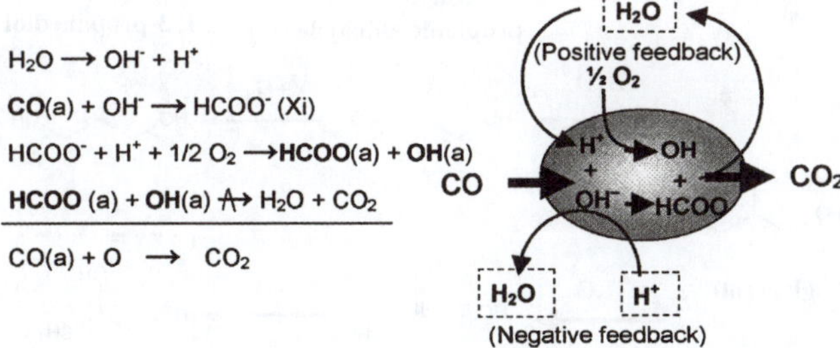

Fig. 23 Reaction of CO with OH⁻ promoted by the positive feedback cycle of water.[56]

Based on catalytic activity differences of gold NPs deposited on different kind of materials and on kinetic and spectroscopic investigation, a probable reaction route at room temperature has been schematically shown in Fig. 24. The rate determining step is the reaction of CO adsorbed on gold surfaces and oxygen adsorbed at the periphery. At temperatures above 333K oxygen molecules can be activated at the gold NPs surfaces.[58]

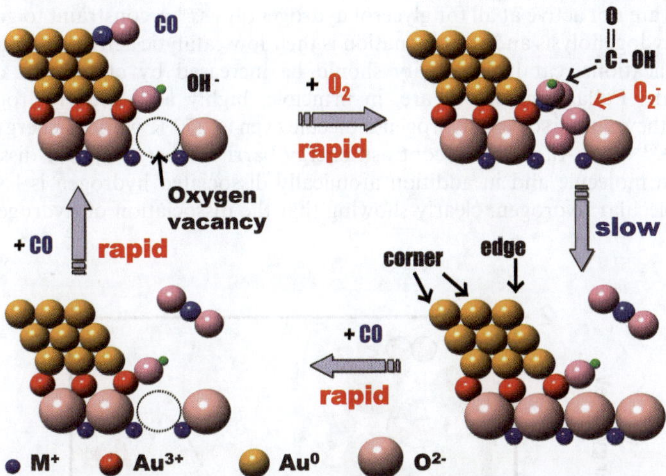

Fig. 24 Probable mechanism for CO oxidation over supported gold catalysts at room temperature.

7. Hydrogen dissociation below 473 K: another perimeter mechanism

Petrochemical industries currently use alkenes as raw materials and accordingly depend more on oxidation than on hydrogenation, because oxygen-containing organic compounds are useful for the production of polymers and fine chemicals. Biomass resources, on the other hand, are composed of carbohydrates so that they need dehydration and hydrogenation more than oxidation. A typical example is hydrogenolysis in water solvent of glycerol, which is produced in a large amount as a byproduct of biodiesel fuels (Fig. 25). There are two major routes in the transformation of glycerol. One is to produce, through the formation of 3-hydroxy propionic aldehyde, 1, 3 propane diol (PD), which has already been commercialized by Du-Pont[59] for polymer production. Another route is, through acetol, to produce 1, 2 PD, called as propylene glycol, which is used as human-friendly anti-freezer and

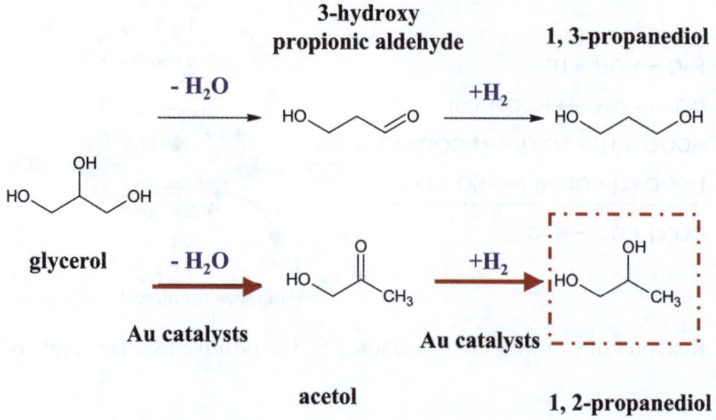

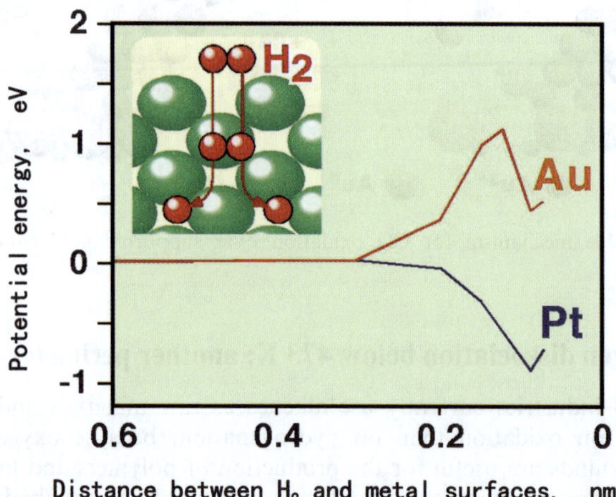

Fig. 25 Hydrogenolysis of glycerol: an example for biomass chemistry.

as raw materials for producing polyurethane polymers. This route has also been planned for commercialyzation.[59]

Supported gold catalysts exhibit selectivity to 1, 2 PD above 90% and among them Au/Al_2O_3 and Au/ZrO_2 are the most active. To the contrary, supported platinum catalysts are not active at all for glycerol hydrogenolysis.[60] A constraint to gold catalysts in hydogenolysis and hydrogenation is their low catalytic activity. For commercial applications, catalytic activity should be increased by about two order of magnitude. Palladium and Pt are, in principle, highly active for hydrogenation because they can dissociate hydrogen molecule even at 273 K without energy barrier (Fig. 26).[61] Over Au (111), in contrast, energy barrier is very high for dissociating hydrogen molecule and in addition atomically dissociated hydrogen is less stable than molecular hydrogen, clearly showing that the dissociation of hydrogen mole-

Fig. 26 Dissociation of hydrogen molecule over Pt (111) and Au (111) surfaces: Density function theory calculation.[61]

cule is unfavorable at room temperature. In fact, over the terraces of (111) and (110) and steps of (311) of gold no dissociation of hydrogen molecule takes place neither at room temperature nor at 473 K.[40]

Fujitani *et al.* have recently studied H_2–D_2 exchange reaction at 423 K over model Au/TiO_2 (rutile single crystal) catalysts having the same gold loading of one monolayer but different mean diameters of gold NPs.[40] Fig. 27 shows that the rate of HD formation per each catalyst specimen at 423 K markedly increases with a decrease in the mean diameter of gold NPs. The HD formation rate per the total number of gold atoms at the periphery sites is almost constant irrespective of the mean diameters of gold NPs, meaning that the rate of hydrogen dissociation in proportion to the perimeter distance. This fact strongly indicates that hydrogen molecule dissociates at the periphery around gold NPs. The dissociation of H_2 at the periphery is also indirectly supported by the experimental results that hydrogenation reaction takes place more rapidly over Au/TiO_2, Au/ZrO_2, and Au/CeO_2 than over gold NPs on other metal oxide supports. It is very likely that hydrogen molecule is heterolytically cleaved into a proton bonded to a support oxygen and a hydride bonded to gold.[62]

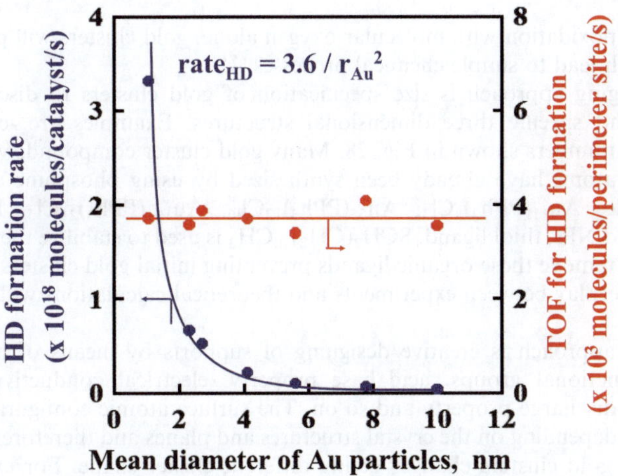

Fig. 27 Rate of H_2 and D_2 exchange reaction over model Au/TiO_2 (110) catalysts as a function of the mean diameter of gold particles.[40]

8. Future prospects

The number of published papers dealing with Au catalysts (homogeneous and heterogeneous) has dramatically been increasing year by year. When the present author began to work on gold catalysts in 1982, the number of papers was less than 40 per year. It now reaches 1,000. The first wave might be ignited by the low temperature oxidation of CO at 200 K. The second wave might be driven by selective oxidation, in particular, propylene epoxidation in gas phase[63] and glucose oxidation in liquid water.[64] The third wave is coming, probably driven by clusters and by hydrogenation.

Taken into account of the potential capability of gold catalysts in room temperature oxidation, promotion by water, and unique selectivity, a series of gold catalysts will become principal players in green and sustainable chemistry. In addition to gold NPs of 2–5 nm, tiny clusters having a specific number of atoms may show unique catalysis for a specific reaction which was considered to hardly take place over conventional catalysts. Of course, selection of support materials is also crucial since it defines the bonding strength with gold clusters and changes the three dimensional structures of gold clusters much more than gold NPs. As typically shown by direct

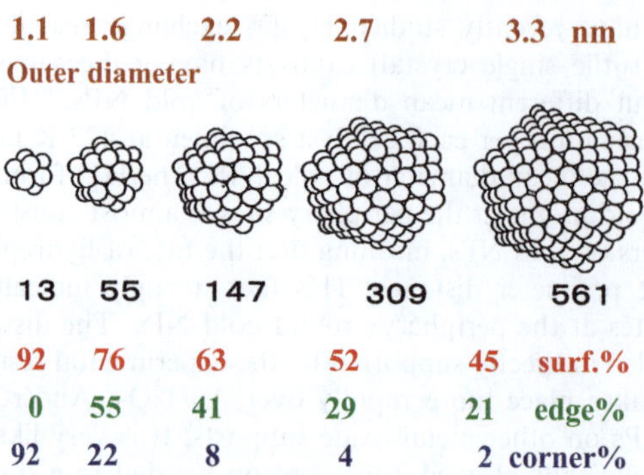

Fig. 28   Magic numbers, outer diameters, edge and corner sites for icosahedrons.

propylene epoxidation with molecular oxygen alone, gold clusters will provide new routes which lead to simple chemical processes.[65,66]

An emerging approach is size specification of gold clusters to discover magic numbers and specific three dimensional structures. Examples are icosahedrons with magic numbers shown in Fig. 28. Many gold cluster compounds with defined number of atoms have already been synthesized by using phosphine as a ligand, for example, $Au_{11}(PPh_3)_8Cl_3$, $Au_{55}(PPh_3)_{12}Cl_{16}$, $Au_{101}(PPh_3)_{21}Cl_5$. For monodisperse gold NPs, thiol ligand, $SCH_2(CH_2)_{10}CH_3$ is used to stabilize them. It is critical how to remove these organic ligands preventing initial gold clusters from coagulation. Interplay between experiments and theoretical calculations will work most efficiently.

Another approach is creative designing of supports by means of morphology control, functional groups, acid–base property, electrical conductivity, oxygen absorption-discharge property and so on. The surface atomic configuration markedly differs depending on the crystal structures and planes and therefore the contact strength of gold clusters changes giving different cluster shape. For example, the (110) plane of Co_3O_4 exposes catalytically active Co^{3+} ions, whereas (111) and (100) planes exposes only Co^{2+} ions.[54,67]

A serious constraint to metal cluster catalysts might be chemical and thermal stability during catalytic reactions. Concerning chemical stability, Gibbs energies for the formation of metal oxides are -130, -168, -22.4, and $+53.3$ kJ/mol $-O_2$ for PdO, PtO_2, Ag_2O, and Au_2O_3, respectively. Only gold does not spontaneously form oxides. In this consequence, gold clusters are advantageous over the clusters of other noble metals. As for thermal stability, gold is inferior to other noble metals because its melting point is 1336 K, lower than those of Pd and Pt by 492 K and 710 K. The melting point of gold clusters was markedly lowered down with a decrease in diameter; 649 K at 3.2 nm, 495 K at 2.3 nm, and 592 K at 2.0, 1.5, and 1.2 nm.[68] It appeared that the melting point did not monotonously go down but leveled off at around 592 K. Molecular-dynamic simulations employing embedded-atom interactions resulted in similar decrease in melting point of gold clusters; 760 K at 459 atoms (3.0 nm), 625 K at 146 atoms (2.0 nm), 550 K at 75 atoms (1.5 nm in diameter).[69] In fact, even after propylene epoxidation at 473 K for 17 h, gold clusters smaller than 2 nm were almost remained as they had been before reaction. If the bonding strength of gold clusters with the support is strong enough as in the case of CeO_3,[70,71] gold clusters can be stable enough during reaction as far as reaction temperature is below 473 K. It is a good news that most of reactions in gas and liquid phases can proceed at temperatures below 473 K over supported gold catalysts.

References

1 E. J. Holmyard, *Alchemy*, Penguin Books, 1957.
2 P. P. Edwards and J. M. Thomas, *Angew. Chem., Int. Ed.*, 2007, **46**, 2.
3 M. Faraday, *Philos. Trans. R. Soc. London*, 1857, **147**, 145.
4 R. Zsigmondy, in '*Colloids and Ultramicroscope*', J. Alexander, transl., Wiley, New York, 1909.
5 J. Turkevitch, P. C. Stevenson and J. Hillier, *Discuss. Faraday Soc.*, 1951, **11**, 55.
6 M. Haruta, T. Kobayashi, H. Sano and N. Yamada, *Chem. Lett.*, 1987, 405.
7 M. Brust, M. Walker, D. Behell, D. J. Schiffrin and R. Whyman, *J. Chem. Soc., Chem. Commun.*, 1994, 801.
8 G. Schmid, R. Boese, R. Pfeil, F. Bandermann, S. Mayer, G. H. M. Calis and J. W. A. van der Velden, *Chem. Ber.*, 1981, **114**, 3634.
9 C. E. Briant, B. R. C. Theobald, J. W. White, L. K. Bell and D. M. P. Mingos, *J. Chem. Soc., Chem. Commun.*, 1981, 201.
10 G. J. Hutchings, *J. Catal.*, 1985, **96**, 292.
11 B. Nkosi, N. J. Coville and G. J. Hutchings, *Appl. Catal.*, 1988, **43**, 33.
12 *Handbook of Energy & Economic Statistics in Japan*, The Energy Data and Modelling Center, IEEJ, 2009, p. 61.
13 R. A. Sheldon, *Chem. Ind. (London, U.K.)*, 1992, 903.
14 M. McCoy, *Chem. Eng. News*, 1999, **77**(38), 17.
15 C. X. Xu, J. X. Su, X. H. Xu, P. P. Liu, H. J. Zhao, F. Tian and Y. Ding, *J. Am. Chem. Soc.*, 2007, **129**, 42.
16 V. Zielasek, B. Jürgens, C. Schulz, J. Biener, M. M. Biener, A. V. Hamza and M. Bäumer, *Angew. Chem., Int. Ed.*, 2006, **45**, 8241.
17 Y. Iizuka, T. Miyamae, T. Miura, M. Okumura, M. Date and M. Haruta, *J. Catal.*, 2009, **262**, 280.
18 M. A. Sanchez-Castillo, C. Couto, W. B. Kim and J. A. Dumesic, *Angew. Chem., Int. Ed.*, 2004, **43**, 1140.
19 H. Kita, H. Nakajima and K. Hayashi, *J. Electroanal. Chem.*, 1985, **190**, 141.
20 J. Guzman and B. C. Gates, *J. Am. Chem. Soc.*, 2004, **126**, 2672.
21 J. C. Fierro-Gonzalez, V. A. Bhirud and B. C. Gates, *Chem. Commun.*, 2005, 5275.
22 V. Aguilar-Guerrero and B. C. Gates, *J. Catal.*, 2008, **260**, 351.
23 J. C. Fierro-Gonzalez and B. C. Gates, *J. Phys. Chem. B*, 2004, **108**, 16999.
24 Q. Fu, H. Saltsburg and M. Flytzani-Stephanopoulos, *Science*, 2003, **301**, 935.
25 H. Sakurai, T. Akita, S. Tsubota, M. Kiuchi and M. Haruta, *Appl. Catal., A*, 2005, **291**, 179.
26 S. T. Oyama, J. Gaudet, W. Zhang, D. S. Su and K. K. Bando, *ChemCatChem*, 2010, **2**, 1582.
27 J. Gaudet, K. K. Bando, Z. X. Song, T. Fujitani, W. Zhang, D. S. Su and S. T. Oyama, *J. Catal.*, 2011, **280**, 40.
28 W. Deng, C. Carpenter, N. Yi and M. Flytzani-Stephanopoulos, *Top. Catal.*, 2007, **44**, 199.
29 R. Kubo, *J. Phys. Soc. Jpn.*, 1962, **17**, 975.
30 U. Landman, B. Yoon, C. Zhang, U. Heiz and M. Arenz, *Top. Catal.*, 2007, **44**, 145.
31 A. Visikovskiy, H. Matsumoto, K. Mitsuhara, T. Nakada, T. Akita and Y. Kido, *Phys. Rev. B: Condens. Matter Mater. Phys.*, 2011, **83**, 165428.
32 D. A. H. Cunningham, W. Vogel, H. Kageyama, S. Tsubota and M. Haruta, *J. Catal.*, 1998, **177**, 1.
33 M. Valden, X. Lai and D. W. Goodman, *Science*, 1998, **281**, 1647.
34 M. S. Chen and D. W. Goodman, *Science*, 2004, **306**, 252.
35 A. A. Herzing, C. J. Kiely, A. F. Carley, P. Landon and G. J. Hutching, *Science*, 2008, **321**, 1331.
36 Y. Maeda, M. Okumura, S. Tsubota, M. Kohyama and M. Haruta, *Appl. Surf. Sci.*, 2004, **222**, 409.
37 M. Haruta, N. Yamada, T. Kobayashi and S. Iijima, *J. Catal.*, 1989, **115**, 301.
38 M. Haruta, S. Tsubota, T. Kobayashi, H. Kageyama, M. J. Genet and B. Delmon, *J. Catal.*, 1993, **144**, 175.
39 T. Takei, I. Okuda, K. K. Bando, T. Akita and M. Haruta, *Chem. Phys. Lett.*, 2010, **493**, 207.
40 T. Fujitani, I. Nakamura, T. Akita, M. Okumura and M. Haruta, *Angew. Chem., Int. Ed.*, 2009, **48**, 9515.
41 M. Haruta, *Chem. Rec.*, 2003, **3**, 75.
42 T. Fujitani and I. Nakamura, *Angew. Che. Int. Ed.*, in press.
43 C. H. F. Peden, Surface Science of Catalysis. *In Situ Probes and Reaction Kinetics*, ACS Symp. Ser. 482, D. J. Dwyer, F. M. Hoffmann, Washington, DC, 1992, p.143.

44 I. X. Green, W. Tang, M. Neurock and J. T. Yates Jr., *Science*, 2011, **333**, 736.
45 P. M. Mavrikakis, P. Stoltze and J. K. Norskov, *Catal. Lett.*, 2000, **64**, 101.
46 Y. Liu, C.-J. Jia, J. Yamasaki, O. Terasaki and F. Schuth, *Angew. Chem. Int. Ed.*, 2010, **49**.
47 S. Tsubota, T. Nakamura, K. Tanaka and M. Haruta, *Catal. Lett.*, 1998, **56**, 131.
48 I. Laoufi, M.-C. Saint-Lager, R. Lazzari, J. Jupille, O. Rabach, S. Garaudèe, G. Cabailh, P. Dolle, H. Cruguel and A. Bally, *J. Phys. Chem. C*, 2011, **115**, 4673.
49 Sh. K. Shaikhutdinov, R. Meyer, M. Naschitzki, M. Bäumer and H.-J. Freund, *Catal. Lett.*, 2003, **86**, 211.
50 Y. Tai, W. Yamaguchi, K. Tajiri and H. Kageyama, *Appl. Catal., A*, 2009, **364**, 143.
51 R. Zanella, S. Giorgio, C.-H. Shin, C. R. Henry and C. Louis, *J. Catal. A*, 2004, **222**, 357.
52 M. Date, M. Okumura, S. Tsubota and M. Haruta, *Angew. Chem., Int. Ed.*, 2004, **43**, 2129.
53 Y. Yu, T. Takei, H. Ohashi, H. He, X. Zhang and M. Haruta, *J. Catal.*, 2009, **267**, 121.
54 X. Xie, Y. Li, Z.-Q. Liu, M. Haruta and W. Shen, *Nature*, 2009, **458**, 746.
55 A. Bongiorno and U. Landman, *Phys. Rev. Lett.*, 2005, **95**, 106102.
56 K. Tanaka, *Catal. Today*, 2010, **154**, 105.
57 M. Okumura, J. M. Coronado, J. Soria, M. Haruta and J. C. Conesa, *J. Catal.*, 2001, **203**, 168.
58 S. K. Ritter, *Chem. Eng. News*, May 31, 2004, 31; B. K. Min and C. M. Friend, *Chem. Rev.*, 2007, **107**, 2709.
59 M. McCoy, *Chem. Eng. News*, March 26, 2007, 9.
60 S. Akita, H. Okatsu, T. Takei, M. Haruta, unpublished data.
61 B. Hammer and J. K. Norskov, *Nature*, 1995, **376**, 238.
62 R. Juarez, S. F. Parker, P. Concepcion, A. Corma and H. Garcia, *Chem. Sci.*, 2010, **1**, 731.
63 T. Hayashi, K. Tanaka and M. Haruta, *J. Catal.*, 1998, **178**, 566.
64 L. Prati and M. Rossi, *J. Catal.*, 1998, **176**, 552.
65 S. Lee, L. M. Molina, M. L. Lopez, J. A. Alonso, B. Hammer, B. Lee, S. Seifert, R. E. Winans, J. W. Elam, M. J. Pellin and S. Vajda, *Angew. Chem., Int. Ed.*, 2009, **48**, 1467.
66 J. Huang, T. Akita, J. Faye, T. Fujitani, T. Takei and M. Haruta, *Angew. Chem., Int. Ed.*, 2009, **48**, 7862.
67 J. A. Van Bokhoven, *ChemCatChem*, 2009, **1**, 363.
68 T. Castro, R. Reifenberger, E. Choi and R. P. Andres, *Phys. Rev. B: Condens. Matter*, 1990, **42**, 8548.
69 C. L. Cleveland, W. D. Luedtke and U. Landman, *Phys. Rev. B: Condens. Matter*, 1999, **60**, 5065.
70 J. A. Farmer and C. T. Campbell, *Science*, 2010, **329**, 933.
71 S. C. Parker and C. T. Parker, *Top. Catal.*, 2007, **44**, 3.

Gold nanoparticle-polymer/biopolymer complexes for protein sensing†

Daniel F. Moyano,[a] Subinoy Rana,[a] Uwe H. F. Bunz[b] and Vincent M. Rotello[*a]

Received 19th February 2011, Accepted 2nd March 2011
DOI: 10.1039/c1fd00024a

Nanoparticle-based sensor arrays have been used to distinguish a wide range of biomolecular targets through pattern recognition. Such biosensors require *selective* receptors that generate a unique response pattern for each analyte. The tunable surface properties of gold nanoparticles make these systems excellent candidates for the recognition process. Likewise, the metallic core makes these particles fluorescence superquenchers, facilitating transduction of the binding event. In this report we analyze the role of gold nanoparticles as receptors in differentiating a diversity of important human proteins, and the role of the polymer/biopolymer fluorescent probes for transducing the binding event. A structure–activity relationship analysis of both the probes and the nanoparticles is presented, providing direction for the engineering of future sensor systems.

Introduction

Nature has generated diverse mechanisms for sensing molecular systems and triggering appropriate responses. The majority of biological recognition processes occur via *specific* interactions such as antibody–antigen interactions.[1] However, sensory processes such as taste and smell use an array of cross-reactive receptors that use differential binding for analyte identification. These receptors bind to their analytes through interactions that are *selective* rather than *specific*.[2] Such array-based sensing platforms can be trained to create a response fingerprint for each analyte, enabling their detection and differentiation.[3] The potential of this method has been demonstrated in numerous biomacromolecule sensing including peptides,[4] proteins,[5] amino acids,[6,7] sugars,[8] bacteria,[9] and even mammalian cells.[10,11] Likewise, a number of synthetic biomolecular platforms including porphyrin,[12] oligopeptide functionalized resins,[13] polymers,[14] have been employed to create these sensor arrays.

Nanoparticles (NPs) provide a versatile scaffold for array-based sensing of biomolecules.[15] Gold NPs (AuNPs) offer many advantages in terms of recognition and transduction of binding events.[16] The surface of AuNPs can be decorated easily with various ligands of interest. Hence, physicochemical properties including charge, hydrogen-bonding ability, hydrophobicity/hydrophilicity and surface topology, can easily be modulated on the AuNP surface.[17] AuNPs can be readily fabricated in sizes comparable with biomacromolecules, facilitating high affinity interactions.[18] In addition, the extraordinary stability and high resistance to exchange by amines (*e.g.* lysine residues)[19] provide stable platforms for recognition. Finally, the strong

[a]Department of Chemistry, University of Massachusetts, 710 North Pleasant Street, Amherst, Massachusetts, 01003, USA. E-mail: rotello@chem.umass.edu
[b]School of Chemistry and Biochemistry, Georgia Institute of Technology, 770 State Street, Atlanta, Georgia, 30332, USA

† Electronic supplementary information (ESI) available. See DOI: 10.1039/c1fd00024a

---

quenching ability of AuNPs[20] provides facile transduction of the recognition process, making AuNPs attractive candidates for application in differential sensing.[21]

In our research, we have used arrays of AuNPs featuring different ligand structures at their surface to detect and differentiate a diversity of analytes. As shown

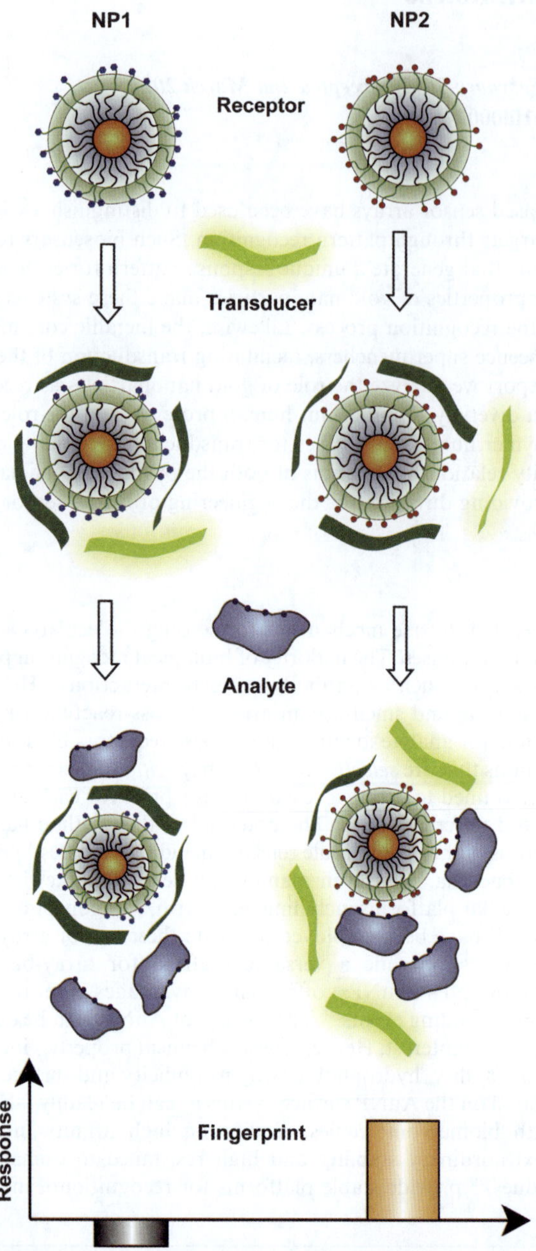

Fig. 1 Schematic illustration of NP/transducer sensor-array. Each nanoparticle and fluorescent probes are mixed followed by incubation with analytes. Depending on the nanoparticle/probe/analyte competitive binding, "lighting up" or further quenching of fluorescence can be observed. Such fluorescence response patterns are distinct and characteristics of each analyte, allowing differentiation.

in Fig. 1, the sensor unit is formed by adding a fluorescent probe to a supramolecu-
larly complimentary NP, turning off the fluorescence. When analytes are incubated
with the NP-probe complexes, there is competitive binding between analyte and NP-
probe complexes. Selective displacement of the probe from the particle by the analytes restores fluorescence, transducing the binding event in a "turn on" fashion. The
NPs are chosen to possess different affinities for dissimilar analytes depending on the
NP structure and analyte surfaces. As a result, each analyte generates a unique
signature that allows differentiation and recognition.

Our initial biosensing studies involved proteins identification employing AuNPs
as the receptors and conjugated polymers as the transducer.[14] Our strategy relied
on the displacement of the fluorescent probe from the supramolecular complexes
between cationic AuNPs and anionic polymers, generating distinct fluorescence
response patterns. This system used six NPs, and was able to properly identify seven
different proteins in buffer solution. Later studies showed that analogous arrays
could differentiate cell state on the basis of cell surface properties.[22] However, aggregation problems of the polymers restricted their applicability in complex matrices
such as serum. To overcome this issue, we used green fluorescent protein (GFP)
as the fluorophore, replacing the polymer transducer with a biopolymer analog.
Not only that, the biocompatibility of the probe allowed us to use the system
without affecting the target protein conformations. These GFP-AuNP sensor arrays
were capable of differentiating bio-relevant changes (500 nM changes in 1 mM overall protein content) in human serum protein.[23]

Our efforts have concurrently been directed towards enhancing the stability and
biocompatibility of synthetic polymers. Recently, we synthesized a new polymer
specifically designed to avoid aggregation by introducing oligo(ethylene glycol)
chains designed to prevent aggregation. This polymer was used with three nanoparticles to successfully differentiate twelve bacteria including three strains of the same
species.[24]

In this report, we analyze the structure–activity relationship of the particle–poly-
mer/biopolymer sensing system, providing insight into how to enhance sensor efficiency. This study uses six AuNPs with variations in hydrophobicity, hydrogen
bonding ability and aromatic recognition unit (Fig. 2a). We analyzed the effect of
structural variation and relative molar emissivity of the transducers using three fluorescent polymers (PPE1, PPE2, PPE3) and GFP as shown in Fig. 2b.

Experimental section

Materials

The ligands (Fig. 2) were synthesized according to previous reports[25] starting with
11-bromoundecanol, triphenylmethanethiol, tetra(ethylene glycol) and the corresponding tertiary amines, each purchased from Sigma Aldrich. NP gold cores
were synthesized according to literature procedure[26] using 1-pentanethiol as the
stabilizer, and purified by successive precipitation/redispersion cycles of the NPs
in ethanol and acetone.

Place exchange reactions[27] were performed to functionalize the surface of the NPs
using the synthesized ligands, and the resulting cationic gold nanoparticles were
purified by dialysis using a 10,000 MWCO membrane to remove the 1-pentane-
thiol/disulfide and the excess of ligand. TEM (core size), NMR (presence of free
ligand), DLS (hydrodynamic size) and LDI-MS (purity) techniques were used to
characterize the NPs.

Three different fluorescent polymers were synthesized and characterized according to literature procedures.[28] GFP was expressed according to reported methodology,[29] starting from cultures of a glycerol stock of GFP in BL21(DE3) that was
grown overnight in culture media at 37 °C. These initial cultures were then added
to a Fernbach flask that contained culture media and the mixture was shaken until

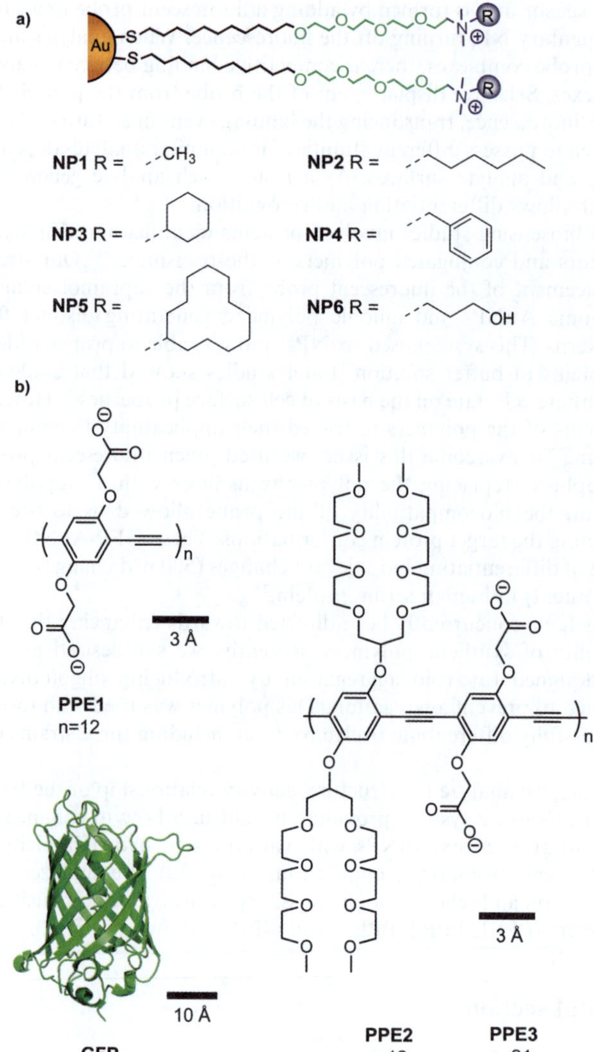

Fig. 2 a) Structure of the nanoparticles and ligands used in the current study, featuring an aliphatic interior, a tetra(ethylene glycol) layer and the head group responsible for analyte recognition. b) Molecular structures and size of the different fluorescent probes including the conjugated polymers (PPE1, PPE2, PPE3) and GFP.

the optical density at 600 nm was ~0.6. GFP formation was then induced by adding isopropyl-β-D-thiogalactopyranoside, shaking at 28 °C for three hours. The solution was centrifuged and the cells were resuspended, lysed and then the protein was purified using HisPur Cobalt columns and dialysis.

Sensor optimization

Nanoparticle titrations were performed to find the optimal nanoparticle/probe ratio for sensing purposes, measuring the change in fluorescence intensities after the addition of cationic nanoparticles (0–200 nM in 5mM sodium phosphate buffer pH 7.4). The measurement wavelengths were 465nm for the polymers (λ_{ex} 430nm) and 510nm for GFP (λ_{ex} 495nm). All the values were normalized using a neutral non-interacting nanoparticle as control (gold core functionalized with a ligand lacking the amine head group) to subtract the absorption effect of gold NP due to its optical properties. The curve fitting was done with Origin 8.0 using a previous reported algorithm[30] to find the physicochemical parameters of the binding process, namely binding constant (K_s) and binding sites per nanoparticle (n). The ultimate nanoparticle/probe ratio was chosen to balance sensitivity and dynamic range.

Sensing studies

Fluorescence of the samples was measured in a M5 SpectraMax Molecular Devices plate reader instrument using 96 well black plates. A stock solution of the each fluorescent probe was prepared in 5mM sodium phosphate buffer pH 7.4 with a concentration that gave $\sim$1000 a.u. of emission. This solution was then divided and nanoparticles added. The conjugates were added to microwell plates. After 15 min of incubation initial fluorescence was recorded, and 10 μL of a solution of the analyte proteins (with absorbance of 0.005 at 280 nm, Table 1) was added. After 15 min additional incubation the final fluorescence intensity was determined.

For data analysis, the change in fluorescence was plotted against each descriptor to get the fingerprint patterns for all the analytes. The obtained data were analyzed by Linear Discriminant Analysis (LDA) applying Mahalonobis clustering analysis, using the software Systat 11.[31] Canonical score plots were obtained in order to parameterize each case, using the nanoparticle as predictors. A total of 12 different proteins were analyzed using 6 replicates in each case.

Results and discussions

Comparison of the fluorescent probes

Emissivity is a key determinate of fluorescence sensor design, dictating fluorophore concentrations required for sensing.[32] As shown in Fig. 3, the fluorescence intensity of PPE1, PPE2 and GFP are similar at 50 nM. However, PPE3 shows a drastic increase in emission. This difference in molar emissivity can be explained by polymer structure and length. The extended conjugation of the aromatic rings of PPE3 provides the possibility of multiple fluorescent units along the chain of the polymer.[33] Additionally, PPE3 carries more charges, potentially inhibiting self-quenching. For the below studies we employed concentrations of the probes that produced similar emission intensity (PPE1: 50nM, PPE2: 50nM, PPE3: 5nM, GFP: 150nM).

Table 1 Analytes properties and concentrations used in the study

Protein	M.W. (kDa)	Isoelectric Point	Charge at pH 7.4	ε_{280} (M^{-1} cm^{-1})	Conc. (nM) $A_{280} = 0.005$
Human Serum Albumin	69.4	5.2	Negative	38E + 03	132
Cytochrome c	12.3	10.7	Positive	23E + 03	216
Acid Phosphatase	110	5.2	Negative	26E + 04	19
Alkaline Phosphatase	140	5.7	Negative	63E + 03	80
Hemoglobin	64.5	6.8	Neutral	13E + 04	40
Lysozyme	14.4	11.0	Positive	38E + 03	132
Ribonuclease A	13.7	9.4	Positive	10E + 03	500
Myoglobin	17.0	7.2	Neutral	14E + 03	359
Fibrinogen	340	5.5	Negative	51E + 04	10
α-Antitrypsin	52	4.6	Negative	28E + 03	181
Transferrin	80	$\sim$6.0	Neutral	11E + 04	45
Immunoglobulin G	150	$\sim$8.0	Positive	21E + 04	24

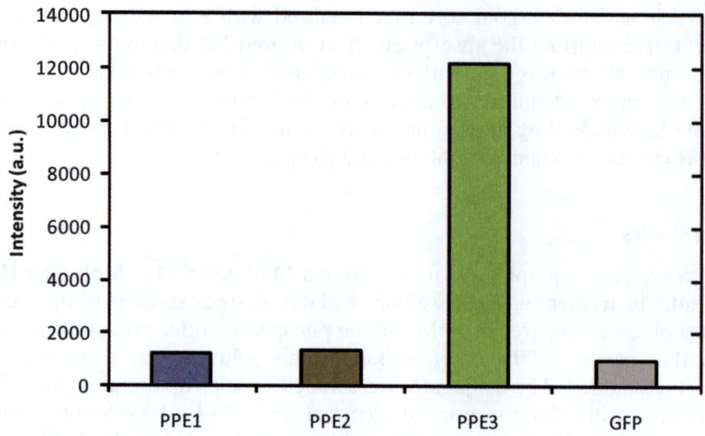

Fig. 3 Relative molar emissivity of different fluorescent probes used in the study with a concentration of 50 nM.

Next, we assessed the complexation between the fluorescent probes and the cationic NPs using fluorescence titrations. The fluorescence of GFP and the polymers were quenched significantly for all six AuNPs (Fig. 4 and SI†). The complex stability constants (K_S) and association stoichiometries (n) obtained through nonlinear least-squares curve-fitting analysis are tabulated in Fig. 4. The variation in complex stabilities and the binding stoichiometry demonstrate the significant effect of AuNP head groups in the NP-probe affinity. Differences in the molecular structure and properties of the transducers also play a key role in the complex supramolecular interactions, as reflected in the parameters.

From the binding curves it is apparent that PPE3 binds to the NPs with higher affinity and lower stoichiometries than the other polymers or GFP due to the increased contact area and size of PPE3. In addition to the lower amount of polymer required, the optimum NP concentration ranged from 1–7 nM for PPE3 and 5–25 nM for the other probes, demonstrating that PPE3 allows a more parsimonious use of materials.

Finally, we investigated the ability of the sensor arrays to differentiate analytes. Twelve proteins with diverse structural features including molecular weight and isoelectric point (pI) were used as the target analytes (Table 1). In our protocol we used UV absorbance as a means of decoupling protein concentrations from sensor response. We used standard absorbance ($A_{280} = 0.005$) for the sensor studies,

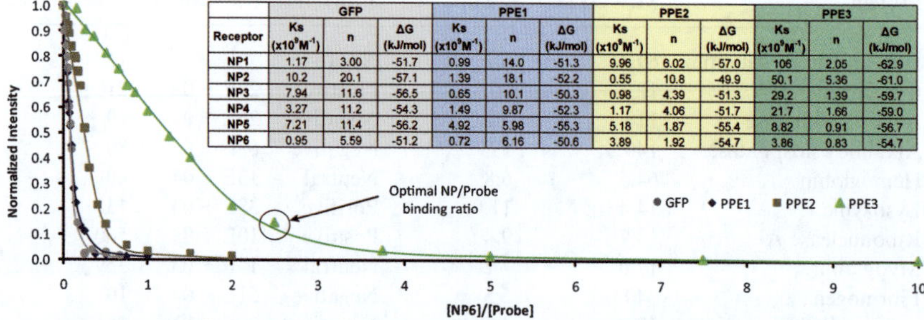

Receptor	GFP			PPE1			PPE2			PPE3		
	K_S ($\times 10^9$M^{-1})	n	ΔG (kJ/mol)	K_S ($\times 10^9$M^{-1})	n	ΔG (kJ/mol)	K_S ($\times 10^9$M^{-1})	n	ΔG (kJ/mol)	K_S ($\times 10^9$M^{-1})	n	ΔG (kJ/mol)
NP1	1.17	3.00	-51.7	0.99	14.0	-51.3	9.96	6.02	-57.0	106	2.05	-62.9
NP2	10.2	20.1	-57.1	1.39	18.1	-52.2	0.55	10.8	-49.9	50.1	5.36	-61.0
NP3	7.94	11.6	-56.5	0.65	10.1	-50.3	0.98	4.39	-51.3	29.2	1.39	-59.7
NP4	3.27	11.2	-54.3	1.49	9.87	-52.3	1.17	4.06	-51.7	21.7	1.66	-59.0
NP5	7.21	11.4	-56.2	4.92	5.98	-55.3	5.18	1.87	-55.4	8.82	0.91	-56.7
NP6	0.95	5.59	-51.2	0.72	6.16	-50.6	3.89	1.92	-54.7	3.86	0.83	-54.7

Fig. 4 Titration curves of the different probes with NP6. The circle indicates the point of optimal NP/probe binding ratio chosen. The table shows the thermodynamic parameters calculated for NPs against the fluorescent probes.

diluting stock solutions to obtain the desired levels. These protein solutions were added to the fluorescence-quenched supramolecular complexes using NP/probe ratios obtained from the titrations (Fig. 4). The changes in fluorescence (Figure S5†) upon addition of protein generate a pattern characteristic of the individual proteins. We used LDA to quantitatively differentiate the fluorescence response patterns. LDA is a statistical technique that maximizes the ratio of between-class variance to within-class variance, allowing response patterns to be differentiated. The 72 training cases (12 proteins × 6 replicates) were efficiently clustered into 12 separate groups with 95% confidence ellipses using LDA. Distinction of the groups was reflected in the overall classification accuracy obtained from Jack-knifed classification matrix.[34]

Distinct canonical score plots were obtained from LDA analysis for each transducer and the most significant scores are plotted in Fig. 5. It is observed that the PPE1 and PPE2 show considerable differences in classification abilities in spite of their similar relative emissivities and size. In fact, PPE2 possesses the lowest identification efficiency among all the transducers. However, PPE3 features greater capability of differentiation, possibly due to higher binding affinity and molar emissivity. Although the classification accuracy of PPE3 is similar to PPE1, it possesses added advantages of biocompatibility and lack of aggregation in complex media. It is

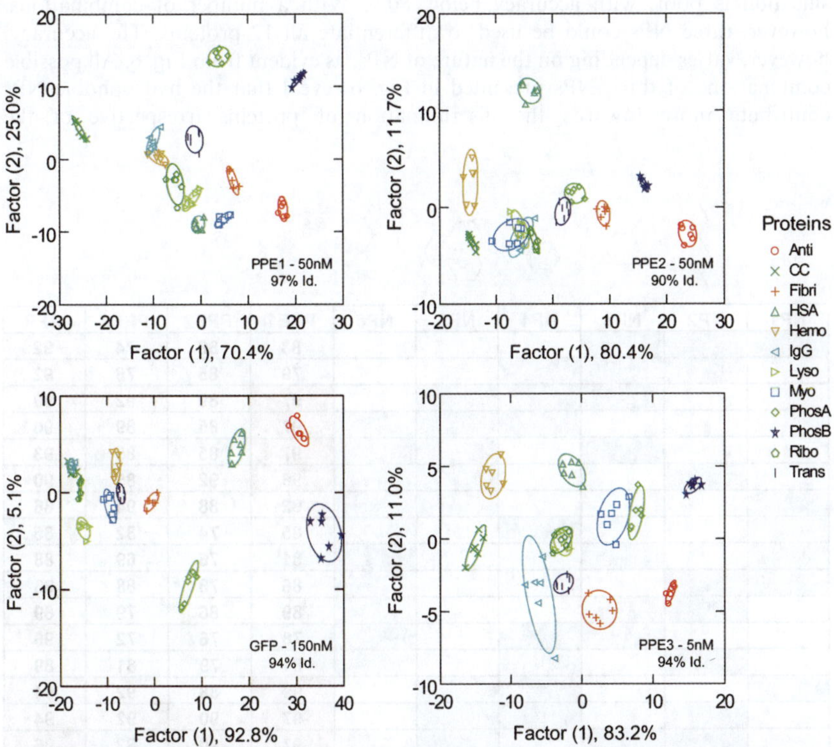

Fig. 5 Canonical score plots for the fluorescence patterns of the different probes as obtained from LDA use of six NP-probe complexes against twelve analyte proteins (analytes added in a concentration 10–500 nM, with absorbance of 0.005 at 280 nm). The 95% confidence ellipses for the each analyte, the concentration of the probe used and the percentage of identification accuracy are shown on the plots. Acronyms used: Anti = α-Antitrypsin, CC = Cytochrome c, Fibri = Fibrinogen, HSA = Human Serum Albumin, Hemo = Hemoglobin, IgG = Immunoglobulin G, Lyso = Lysozyme, Myo = Myoglobin, PhosA = Phosphatase acidic, PhosB = Phosphatase basic, Ribo = Ribonuclease A, Trans = Transferrin.

worth noting that PPE3 needs 10-fold less polymer concentration that the other two polymers, indicating much higher sensitivity.

In the present study, we achieved high efficiency of identification of the analyte proteins using GFP as well as with PPE1 and PPE3. Compared to the other transducers, GFP shows higher dispersion of clustering on LDA analysis suggesting greater possibility of differentiating multiple analytes at the same time. However, GFP-based sensing required substantially more fluorophore and nanoparticle compared to PPE3.

Nanoparticle surface functionality: structure–activity relationships in sensing

One of the goals for array-based sensor design is maximal differentiation with the minimal number of sensing elements. The head groups of the ligands used in this study were designed to exhibit different physicochemical properties such as hydrophobicity, hydrogen-bonding sites, and π-stacking systems. The purpose of this structural diversity was to obtain clues about which interactions best generated *selectivity* in analyte binding. Insight into the nature of the multivalent interactions involved in the NP–protein interactions would help us in designing appropriate NP surfaces with enhanced selectivity.[35,36]

We used a Jackknifed classification matrix to determine which particles were most effective at analyte differentiation. When using only two NPs as predictors the classification is poor, with accuracy below 80%. With a number of combinations, however, three NPs could be used to differentiate all 12 proteins. The accuracy, however, varies depending on the nature of NPs, as evident from Fig. 6. All possible combinations of three NPs presented in Fig. 6 reveal that the hydrophobic NPs contribute more towards the discrimination of proteins irrespective of the

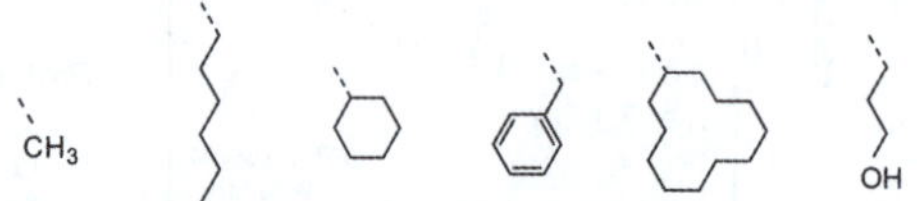

NP1	NP2	NP3	NP4	NP5	NP6	PPE1	PPE2	PPE3	GFP
■	■	■				83	86	74	92
	■	■	■			79	85	78	92
		■	■	■		97	89	92	90
			■	■	■	97	85	89	90
		■		■	■	97	85	85	93
■				■	■	96	92	85	90
■		■	■			92	88	94	86
■	■		■			85	74	82	85
■	■			■		81	76	69	88
■		■		■		86	78	88	83
	■	■			■	89	86	79	89
	■			■	■	78	76	72	90
		■	■		■	81	79	81	89
■			■		■	99	88	92	94
■		■			■	97	90	92	94
■	■				■	97	90	82	96
	■		■	■		97	82	93	93
■			■	■		97	89	92	94
	■		■		■	74	76	72	93
	■	■		■		86	85	81	88

Fig. 6 Jackknifed classification matrix for identification using selected combinations with three nanoparticles as predictors in LDA analysis. Combinations with hydrophobic changes on nanoparticle functionality show better identification accuracy.

transducer. In general, the combinations containing NP4 and NP6 show poorer differentiation grouping (*e.g.* NP2–NP4–NP6; NP1–NP4–NP6 combinations). Interestingly, NP5 appears to be highly effective at differentiating analytes, as all highly efficient sets contained this NP.

Conclusions

In this study we demonstrate the versatility of AuNPs in array-based sensing, identifying twelve different proteins using a variety of polymer and biopolymer transducers. Of the transducers, PPE3 shows greatest sensitivity due to its higher intrinsic molar emissivity. The other polymers were equally effective in differentiation, and possess their respective advantages for sensing applications.

From our studies, we also determined some general guidelines for nanoparticle design. First, we found that hydrophobic particles were better able to discriminate among analyte proteins. Second, structure is important, as the cyclododecyl-terminated NP5 was a very effective partner in protein differentiation. Taken together, these studies demonstrate that through tuning of receptor and transducer structure highly effective array based sensors for proteins can be produced.

Acknowledgements

The research was supported by the U.S. Department of Energy, Office of Basic Energy Sciences, Division of Materials Sciences and Engineering under Award DE-FG02-04ER46141, the National Science Foundation (NSF) Center for Hierarchical Manufacturing at the University of Massachusetts (NSEC, DMI-0531171), and the NIH (GM077173 and AI073425).

References

1 D. Raftos and R. L. Raison, *Immunol. Cell Biol.*, 2008, **86**, 479–481.
2 K. J. Albert, N. S. Lewis, C. L. Schauer, G. A. Sotzing, S. E. Stitzel, T. P. Vaid and D. R. Walt, *Chem. Rev.*, 2000, **100**, 2595–2626.
3 A. P. Umali and E. V. Anslyn, *Curr. Opin. Chem. Biol.*, 2010, **14**, 685–692.
4 T. Zhang, N. Y. Edwards, M. Bonizzoni and E. V. Anslyn, *J. Am. Chem. Soc.*, 2009, **131**, 11976–11984.
5 O. R. Miranda, C.-C. You, R. Phillips, I.-K. Kim, P. S. Ghosh, U. H. F. Bunz and V. M. Rotello, *J. Am. Chem. Soc.*, 2007, **129**, 9856–9857.
6 J. F. Folmer-Andersen, M. Kitamura and E. V. Anslyn, *J. Am. Chem. Soc.*, 2006, **128**, 5652–5653.
7 A. Buryak and K. Severin, *J. Am. Chem. Soc.*, 2005, **127**, 3700–3701.
8 J. W. Lee, J. S. Lee and Y. T. Chang, *Angew. Chem., Int. Ed.*, 2006, **45**, 6485–6487.
9 A. Duarte, A. Chworos, S. F. Flagan, G. Hanrahan and G. C. Bazan, *J. Am. Chem. Soc.*, 2010, **132**, 12562–12564.
10 A. Bajaj, S. Rana, O. R. Miranda, J. C. Yawe, D. J. Jerry, U. H. F. Bunz and V. M. Rotello, *Chem. Sci.*, 2010, **1**, 134–138.
11 K. El-Boubbou, D. C. Zhu, C. Vasileiou, B. Borhan, D. Prosperi, W. Li and X. F. Huang, *J. Am. Chem. Soc.*, 2010, **132**, 4490–4499.
12 L. Baldini, A. J. Wilson, J. Hong and A. D. Hamilton, *J. Am. Chem. Soc.*, 2004, **126**, 5656–5657.
13 A. T. Wright and E. V. Anslyn, *Chem. Soc. Rev.*, 2006, **35**, 14–28.
14 C. C. You, O. R. Miranda, B. Gider, P. S. Ghosh, I.-B. Kim, B. Erdogan, S. A. Krovi, U. H. F. Bunz and V. M. Rotello, *Nat. Nanotechnol.*, 2007, **2**, 318–323.
15 M. Di Marco, S. Shamsuddin, K. A. Razak, A. A. Aziz, C. Devaux, E. Borghi, L. Levy and C. Sadun, *Int. J. Nanomed.*, 2010, **5**, 37–49.
16 S. Rana, Y. Yeh and V. M. Rotello, *Curr. Opin. Chem. Biol.*, 2010, **14**, 828–834.
17 D. F. Moyano and V. M. Rotello, *Langmuir*, 2011, DOI: 10.1021/la2004535.
18 I. Lynch and K. A. Dawson, *Nano Today*, 2008, **3**, 40–47.
19 S. Nath, S. K. Ghosh, S. Kundu, S. Praharaj, S. Panigrahi and T. Pal, *J. Nanopart. Res.*, 2006, **8**, 111–116.

20 C. Fan, S. Wang, G. C. Bazan, K. W. Plaxco and A. J. Heeger, *Proc. Natl. Acad. Sci. U. S. A.*, 2003, **100**, 6297–6301.

21 O. R. Miranda, B. Creran and V. M. Rotello, *Curr. Opin. Chem. Biol.*, 2010, **14**, 7138–736.

22 A. Bajaj, O. R. Miranda, I.-B. Kim, R. L. Phillips, D. J. Jerry, U. H. F. Bunz and V. M. Rotello, *Proc. Natl. Acad. Sci. U. S. A.*, 2009, **106**, 10912–10916.

23 M. De, S. Rana, H. Akpinar, O. R. Miranda, R. R. Arvizo, U. H. F. Bunz and V. M. Rotello, *Nat. Chem.*, 2009, **1**, 461–465.

24 R. L. Phillips, O. R. Miranda, C. C. You, V. M. Rotello and U. H. F. Bunz, *Angew. Chem., Int. Ed.*, 2008, **47**, 2590–2594.

25 Z.-J. Zhu, P. S. Ghosh, O. R. Miranda, R. W. Vachet and V. M. Rotello, *J. Am. Chem. Soc.*, 2008, **130**, 14139–14143.

26 M. Brust, M. Walker, D. Bethell, D. J. Schiffrin and R. Whyman, *J. Chem. Soc., Chem. Commun.*, 1994, 801–802.

27 A. C. Templeton, W. P. Wuelfing and R. W. Murray, *Acc. Chem. Res.*, 2000, **33**, 27–36.

28 I.-B. Kim, R. Phillips and U. H. F. Bunz, *Macromolecules*, 2007, **40**, 5290–5293.

29 M. De, S. Rana and V. M. Rotello, *Macromol. Biosci.*, 2009, **9**, 174–178.

30 R. L. Phillips, O. R. Miranda, D. E. Mortenson, C. Subramani, V. M. Rotello and U. H. F. Bunz, *Soft Matter*, 2009, **5**, 607–612.

31 SYSTAT11.0, *SystatSoftware*, Richmond, CA 94804, USA, 2004.

32 J. Fu, G. Li, Y. Qiin and W. J. Freeman, *Sens. Actuators, B*, 2007, **125**, 489–497.

33 I. Filatov and S. Larsson, *Chem. Phys.*, 2002, **3**, 575–591.

34 P. C. Jurs, G. A. Bakken and H. E. McClelland, *Chem. Rev.*, 2000, **100**, 2649–2678.

35 J. R. Roan, *Phys. Rev. Lett.*, 2006, **96**, 248301.

36 J. Wang, S. Tian, R. A. Petros, M. E. Napier and J. M. Desimone, *J. Am. Chem. Soc.*, 2010, **132**, 11206–11313.

The active site behaviour of electrochemically synthesised gold nanomaterials†

Blake J. Plowman, Anthony P. O'Mullane* and Suresh K. Bhargava*

Received 11th February 2011, Accepted 3rd March 2011
DOI: 10.1039/c1fd00017a

Even though gold is the noblest of metals, a weak chemisorber and is regarded as being quite inert, it demonstrates significant electrocatalytic activity in its nanostructured form. It is demonstrated here that nanostructured and even evaporated thin films of gold are covered with active sites which are responsible for such activity. The identification of these sites is demonstrated with conventional electrochemical techniques such as cyclic voltammetry as well as a large amplitude Fourier transformed alternating current (FT-ac) method under acidic and alkaline conditions. The latter technique is beneficial in determining if an electrode process is either Faradaic or capacitive in nature. The observed behaviour is analogous to that observed for activated gold electrodes whose surfaces have been severely disrupted by cathodic polarisation in the hydrogen evolution region. It is shown that significant electrochemical oxidation responses occur at discrete potential values well below that for the formation of the compact monolayer oxide of bulk gold and are attributed to the facile oxidation of surface active sites. Several electrocatalytic reactions are explored in which the onset potential is determined by the presence of such sites on the surface. Significantly, the facile oxidation of active sites is used to drive the electroless deposition of metals such as platinum, palladium and silver from their aqueous salts on the surface of gold nanostructures. The resultant surface decoration of gold with secondary metal nanoparticles not only indicates regions on the surface which are rich in active sites but also provides a method to form interesting bimetallic surfaces.

Introduction

The gas and liquid phase heterogeneous catalytic behaviour of unsupported and supported gold has received significant attention since the discovery that gold when transformed from an often regarded inert bulk material to the nanostructured form demonstrates significant activity.[1–6] This is attributed to a variety of factors including size, shape, crystallographic orientation and interactions with an underlying support. Recently, nanosized gold has also attracted widespread attention in the field of electrochemistry with widespread applications as electrochemical sensors[7,8] and also as an effective electrocatalyst for fuel cell relevant reactions such as oxygen reduction and small organic molecule oxidation.[9–20]

However, it is interesting to note that gold is often regarded as a model electrode material for electrochemical studies. This is due to the well characterised behaviour of this metal in the bulk form in aqueous solution which exhibits a well defined double layer region that is considered to exhibit purely capacitive behaviour and

---

School of Applied Sciences, RMIT University, GPO Box 2476V, VIC, 3001, Australia. E-mail: anthony.omullane@rmit.edu.au; suresh.bhargava@rmit.edu.au

† Electronic supplementary information (ESI) available: See DOI: 10.1039/c1fd00017a

---

a monolayer oxide formation/removal process. Significantly however, the majority of electrocatalytic reactions that occur at a gold surface in aqueous solution are well within the double layer region with few reactions catalysed by the main oxide formation process.[14,21] Recently, it has been shown for gold and other metals including platinum, palladium, silver and copper that significant oxidation processes can occur within the double layer region of these metals which can be detected by the cyclic voltammetry technique once the electrode material has been sufficiently activated. This type of surface activation includes polarisation in the hydrogen evolution region,[22,23] multilayered oxide growth followed by removal,[24] repetitive potential cycling,[9] thermal treatments followed by rapid quenching[25] and sonochemical[26] methods. In all cases energy is inserted into the metal lattice which often results in the disruption of the outer layers of the metal to create an active surface. The creation of active sites, which are believed to consist of low lattice stabilised gold atoms, or clusters of atoms, are more readily oxidised compared to the bulk material where the atoms are fully lattice stabilised. This more facile oxidation of active gold results in significant electrochemical responses recorded in the double layer region. It is the creation of these low coordinated active sites which has been postulated to mediate electrocatalytic reactions of gold in aqueous solution.[21,27]

A further significant area of research has been the electrochemical synthesis of gold in the nanostructured form. This approach is advantageous in that it eliminates the need for organic species which are used as capping agents in chemical synthesis, and allows the bare metal to be investigated. This alleviates any problems associated with the influence that the capping species may have on the electrocatalytic reaction of interest or treatments that would be required for its removal which may also affect the surface chemistry of the metal. Using appropriate electrochemical protocols a wide variety of structures including flowers,[28] spikes,[29] rods,[30] pyramids,[31] spheres[32] and dendrites[33] can be produced which have been shown to possess good electrocatalytic and surface enhanced Raman scattering properties often attributed to the creation of active sites or hot spots. Even though the solution based electrochemical behaviour of nanoparticles has been well documented in terms of the observation of discrete single electron transfer processes[34-36] the standard electrochemical behaviour of immobilised nanostructures of gold on substrates in aqueous solution is often overlooked. This can offer a wealth of information regarding the effectiveness of such materials as an electrocatalyst and give an insight into what drives electrocatalytic processes at nanostructured gold.[14] In this work we investigate the electrochemical behaviour of gold that is often employed in applied studies in the form of a conventional commercial bulk electrode, as an evaporated thin film, and in the nanostructured form such as well dispersed hierarchical structures and a continuous porous honeycomb network where the latter is an extension of our previous communication.[37] This analysis is done with conventional electrochemical techniques such as dc cyclic voltammetry but also with large amplitude Fourier transformed ac (FT-ac) voltammetry which can effectively discriminate between capacitive and Faradaic processes occurring at an electrode surface.[38,39] We demonstrate that active sites exist on gold in all the forms investigated whose coverage depends on the method of synthesis. Significantly, we demonstrate how these active sites allow for the spontaneous and discrete electroless deposition of metals such as silver, platinum and palladium in the form of nanoparticles on their surfaces.

Experimental

Chemicals

Solutions of $KAuBr_4$, H_2SO_4, NaOH (Ajax Finechem), ethanol (Merck), $AgNO_3$, K_2PtCl_4 (Aldrich), $Ni(NO_3)_2$, $Pd(NO_3)_2$ (BDH) were used as received and made up with deionized water (resistivity of 18.2 $M\Omega$ cm) purified by use of a Milli-Q reagent deioniser (Millipore). High purity oxygen (5.0 coregas) was purged into the relevant solution for at least 20 minutes, otherwise solutions were degassed with nitrogen for at least 10 minutes.

Electrode materials

Gold and platinum electrodes (BAS) of diameter 1.6 mm were used. Gold thin films were deposited by a Balzers™ electron beam evaporator. The layer composed of 1500Å of Au and 100Å of Ti. The films were deposited sequentially by electron evaporation process onto the bare AT-cut quartz substrates. The purpose of the Ti layer is to assist with the adhesion of the Au layer to the substrate surface. Indium tin oxide (ITO) (Prazisions Glas and Optik GmbH) coated glass with a sheet resistance of 10 Ω sq^{-1} as quoted by the manufacturer was used as the substrate for gold nano-material electrodeposition.

Materials characterisation

SEM measurements were performed on a FEI Nova SEM instrument (Nova 200). Prior to SEM imaging, samples were thoroughly rinsed with Milli-Q water and dried under a flow of nitrogen. X-ray diffraction data were obtained with a Bruker AX 8: Discover with General Area Detector Diffraction System (GADDS). X-ray photoelectron spectroscopy (XPS) characterization of samples was done using a Thermo K-Alpha instrument at a pressure better than 1×10^{-9} Torr. The core level binding energies (BEs) were aligned with the adventitious C 1s binding energy of 285 eV.

Electrochemical measurements

Cyclic voltammetric experiments were conducted at (20 ± 2) °C with a CH Instruments (CHI 760C) electrochemical analyser in an electrochemical cell (BAS) that allowed reproducible positioning of the working, reference, and counter electrodes and a nitrogen inlet tube. ITO and e-beam evaporated Au films when used as the working electrode were sonicated and washed in acetone and methanol respectively followed by drying in a stream of nitrogen gas prior to use. The gold or platinum (BAS) electrode was polished with an aqueous 0.3 μm alumina slurry on a polishing cloth (Microcloth, Buehler), sonicated in water and dried with nitrogen. The reference electrode was Ag/AgCl (aqueous 3 M KCl). For electrodeposition experiments an inert graphite rod (3 mm diameter, Johnson Matthey Ultra "F" purity grade) was used as the counter electrode to prevent contamination by any electrolysis products.[40] All electrochemical experiments were commenced after degassing the electrolyte solutions with nitrogen for at least 10 min prior to any measurement.

FT-ac measurements

A description of the FT voltammetric instrumentation used in this study is available elsewhere.[41] Sine waves of frequencies $f = 21.29$ Hz and amplitudes of $\Delta E = 100$ mV were employed as the ac perturbation. DC voltammetric experiments were also carried out with this instrumentation by using a zero amplitude perturbation to compare with results obtained using the CH Instruments potentiostat.

Results and discussion

Electrochemical characterisation of unactivated and activated gold electrodes

Illustrated in Figure 1 is a typical cyclic voltammogram for a polycrystalline commercial gold electrode in 1 M H_2SO_4. It can be seen that there is an extensive double layer region from the beginning of the sweep until the onset of compact monolayer oxide formation (Au_2O_3) at *ca.* 1.20 V. On the reverse sweep this oxide

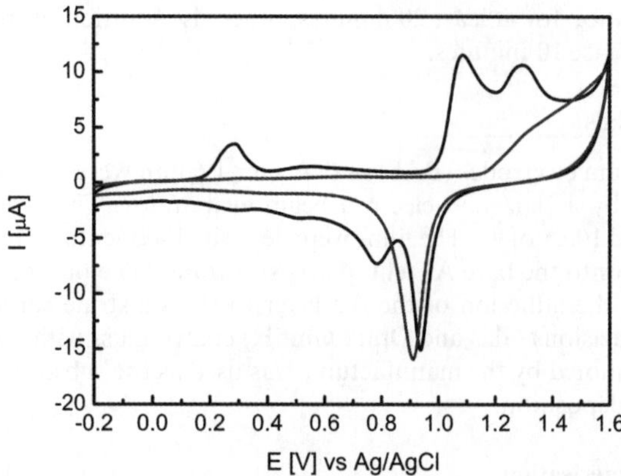

Fig. 1 CVs recorded in 1 M H_2SO_4 at a sweep rate of 100 mV s^{-1} for an untreated polycrystalline gold electrode (grey) and activated by polarisation at -1.5 V for 10 min in 1 M H_2SO_4 (black).

is removed in a sharper reduction process with a peak potential at 0.91 V where the hysteresis observed is due to a place exchange process which occurs during oxide formation.[24] If this electrode is activated by polarisation at a potential where hydrogen gas is evolved from the surface for 10 minutes and transferred to fresh H_2SO_4 solution then distinctly different voltammetry is observed. The double layer region now shows distinct oxidation processes at *ca.* 0.29, 0.55 and 1.08 V. These processes show a degree of reversibility with the associated reduction peaks observed at 0.26, 0.50 and 0.78 V respectively.

It has been postulated by Burke that polarisation of gold in the hydrogen evolution region disrupts the outer layer of the metal through a process akin to hydrogen embrittlement.[42] This creates low co-ordinated gold clusters which are more readily oxidised than the bulk metal which begins at 1.20 V. Recently it has been demonstrated that not only gold[43] but also silver[44] and lead[45] nanoparticles show evidence of size dependent electro-oxidation where the oxidation potential occurs at less positive potentials as the size of the nanoparticles decreases. This may also be the origin of such processes occurring at different potentials in that the cluster size of atoms generated are highly likely to be polydisperse in nature. The possibility of impurities being electrodeposited and subsequently stripped from the surface is discounted as this type of behaviour has also been reported for thermally treated electrodes that have been rapidly quenched under nitrogen[25] and oxygen[9] atmosphere and more recently by gold nanoparticles that have been sonochemically activated[26] where the possibility of depositing metallic impurities is minimal. It also raises the question as to whether calculating the surface area of gold using the charge associated with the reduction of the monolayer oxide as described by Woods[46] and used extensively in this area is valid for such highly active surfaces.

It is interesting to note that a significant number of electrocatalytic reactions occur on gold in acidic medium at discrete electrode potentials which coincide with the potentials at which these active site responses are observed. Illustrated in Figure 2 are three electrocatalytic reactions which occur at different regions within the double layer region of gold. It can be seen that hydrogen peroxide reduction (Figure 2b) begins at *ca.* 0.29 V, hydrazine oxidation (Figure 2c) at *ca.* 0.63 V and hydrogen peroxide oxidation (Figure 2d) at *ca.* 1.00 V which are at the potentials in which the activated gold electrode was shown to be oxidised (Figure 2a). It has been shown previously that several electrocatalytic reactions on gold in acidic solution are grouped at these distinct potentials in which the generation of surface active sites controls the reactions of interest.[21]

    This journal is © The Royal Society of Chemistry 2011

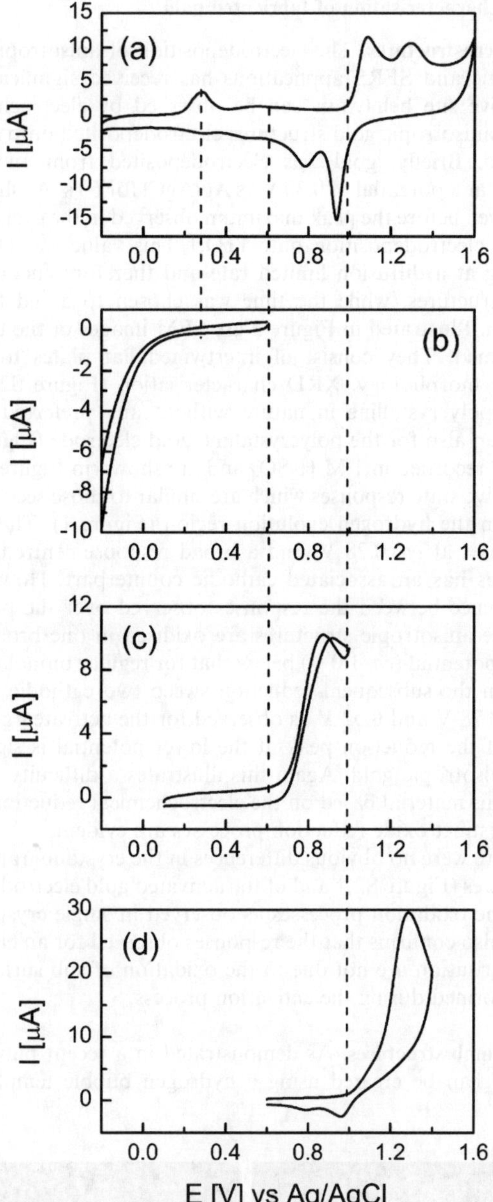

Fig. 2 CVs recorded in 1 M H_2SO_4 for (a) an activated gold electrode at a sweep rate of 100 mV s^{-1} and an unactivated gold electrode in (b) containing 20 mM H_2O_2 at a sweep rate of 20 mV s^{-1}, (c) containing 50 mM N_2H_4 at 50 mV s^{-1} and (d) containing 20 mM H_2O_2 at a sweep rate of 20 mV s^{-1}.

It has been reported that activating a gold surface may enhance electrocatalytic activity by promoting the formation of active sites. However, given that electrocatalytic reactions are mediated by the above mentioned processes at an unmodified gold electrode, suggests that active sites are inherently present on the surface. To investigate this phenomenon a variety of gold surfaces were investigated to see if active sites were indeed present and if their coverage was dependent on the method of fabrication.

Electrochemical characterisation of fabricated gold

Anisotropic microstructures. The electrodeposition of anisotropic gold structures for electrocatalytic and SERS applications has received significant attention. To determine if active site behaviour can be observed by electrochemical methods, "hedgehog" like anisotropic gold structures electrodeposited onto an ITO substrate were investigated. Briefly, gold was electrodeposited from an aqueous 5 mM $KAuBr_4$ solution at a potential of 0.30 V vs Ag/AgCl for 60 s. As shown in Figure S1 this potential is well before the peak maximum observed in the cyclic voltammogram recorded for Au electrodeposition onto ITO.† This value was chosen to prevent growth occurring at a diffusion limited rate and therefore encourage the growth of anisotropic structures, while the time was chosen to avoid the formation of a continuous film. Illustrated in Figure 3 are SEM images of the type of structures that can be formed. They consist of intertwined flat plates to give an overall "hedgehog" type morphology. XRD characterisation (Figure S2) confirmed that this material is polycrystalline in nature without any preferred crystallographic orientation as seen also for the polycrystalline gold electrode (Figure S2).†

CVs were then recorded in 1 M H_2SO_4 and are shown in Figure 4. It can be seen that there are active state responses which are similar to those seen for the gold electrode polarised in the hydrogen evolution region (Figure 1). There are active site oxidation responses at *ca.* 0.28 V, and a broad response centred at 0.78 V. Each of these processes has an associated cathodic counterpart. However, there is an interesting difference between the responses observed over the potential range of 1.0 to 1.6 V. The anisotropic structures are oxidised in one broad process which commences at a potential (*ca.* 1.0 V) below that for regular monolayer oxide formation (1.20 V). On the subsequent reduction sweep two cathodic processes can be observed at *ca.* 0.78 V and 0.52 V as observed for the activated gold electrode but the magnitude of the reduction peak at the lower potential is significantly higher in the case of anisotropic gold. Again this illustrates a difficulty in estimating the surface area of this material based on the electrochemical reduction of a monolayer oxide, as three distinct oxide reduction processes are evident.

Given that there were no obvious differences in the crystallographic nature of the gold nanostructures (Figure S2)† and of the activated gold electrode, the presence of specific gold plane oxidation processes as observed in single crystal studies can be discounted.[16] It also confirms that the responses observed for an electrode activated by cathodic polarisation are not due to the oxidation of sub surface hydrides that may have been formed during the activation process.

Porous honeycomb structures. As demonstrated in a recent publication[37] porous honeycomb gold can be created using a hydrogen bubble templating method in

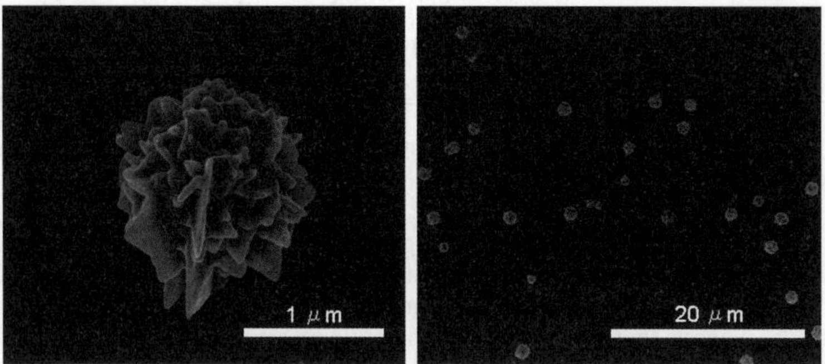

Fig. 3 SEM images of anisotropic gold structures deposited for 60 s at 0.3 V on an ITO substrate.

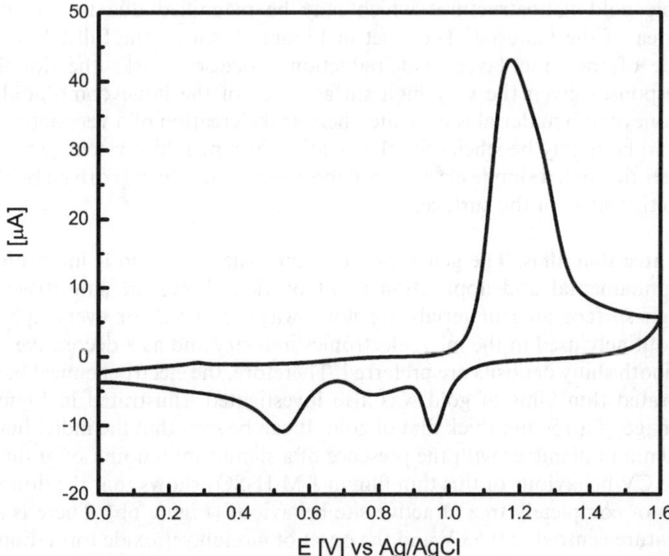

Fig. 4 CV for anisotropic Au structures (as in Figure 3) on ITO in 1 M H_2SO_4 recorded at a sweep rate of 50 mV s^{-1}.

which vigorous hydrogen gas evolution during the course of the electrodeposition process allows metal to be deposited around the evolving hydrogen bubbles to create a highly porous structure. In this case a honeycomb gold film was created from a 0.1 M $KAuBr_4$ + 1.5 M H_2SO_4 solution using a current density of 2 A cm^{-2} for 30 s. A typical SEM image of the type of structure is shown in Figure 5a with the inset showing a higher magnification image demonstrating the highly dendritic nature of the internal wall structure. Given this type of nanostructure and the fact that vigorous hydrogen evolution occurred during the synthesis, which is analogous to the activation method described for creating the type of surface characterised in Figure 1, it is not surprising that significant active site responses are observed in the double layer region. It is interesting to note the responses at potentials below 0.20 V (Figure 5b) which were not observed for the activated gold electrode or the

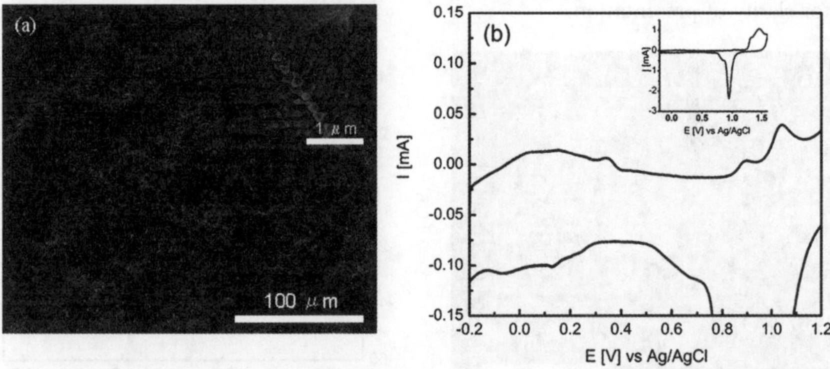

Fig. 5 SEM image showing the honeycomb film formed through the dynamic hydrogen bubble technique (a), where the dendritic nature of the pore walls is seen in the inset image. CV of the honeycomb film recorded at 50 mV s^{-1} in 1 M H_2SO_4 (b), showing a magnified view along with the overall response (inset).

anisotropic gold nanostructures which may be magnified due to the increased surface area of the material. The inset in Figure 5b shows the full CV where the magnitude of the monolayer oxide/reduction processes masks the double layer region responses given the very high surface area of the honeycomb gold.[37] This porous honeycomb material is presented here as the creation of a very active surface is shown to be highly beneficial for the creation of bimetallic systems which is discussed later through a simple electroless deposition route which is driven by the presence of active sites on the surface.

Evaporated thin films. The generation of nanostructured gold is interesting from both a fundamental and application point of view, however, nanostructured or rough high surface area materials are not always required for every application. Gold is routinely used in the microelectronics industry and as a decorative material where smooth shiny deposits are preferred. Therefore, the electrochemical behaviour of evaporated thin films of gold was also investigated. Illustrated in Figure 6a is a SEM image of a 150 nm thick film of gold. It can be seen that the individual grains are *ca.* 50 nm in diameter with the presence of a significant amount of grain boundaries. The CV behaviour of this thin film in 1 M H_2SO_4 shows that the double layer region is not completely free of active site behaviour (Figure 6b). There is a broad anodic feature centred at 0.53 V and the onset of monolayer oxide formation occurs at *ca.* 1.00 V as seen in the case of an activated gold electrode and the anisotropic gold nanostructures. Therefore, this data suggests that even relatively smooth evaporated films of gold have an inherent activity which may dictate the electrocatalytic performance of gold.

Electrochemical characterization of porous honeycomb gold using large amplitude FT-ac voltammetry

The possibility that the responses observed in the double layer are purely due to capacitive effects associated with specific anion adsorption can also be ruled out. Recently, Bond *et al.* have developed the large amplitude Fourier Transformed ac voltammetry technique which offers the unique advantage of being able to effectively discriminate between capacitive and Faradaic processes in the higher ac harmonic responses occurring at the electrode surface.[39,47–50] Also, analysis of the shape of the higher harmonic responses allows the reversible nature of the Faradaic process to be investigated. Illustrated in Figure 7 are the dc and fundamental to fourth harmonic ac responses recorded for honeycomb porous gold in 1 M H_2SO_4. The data is split into the forward (a1 to e1) and reverse (a2 to e2) sweeps for clarity of presentation.

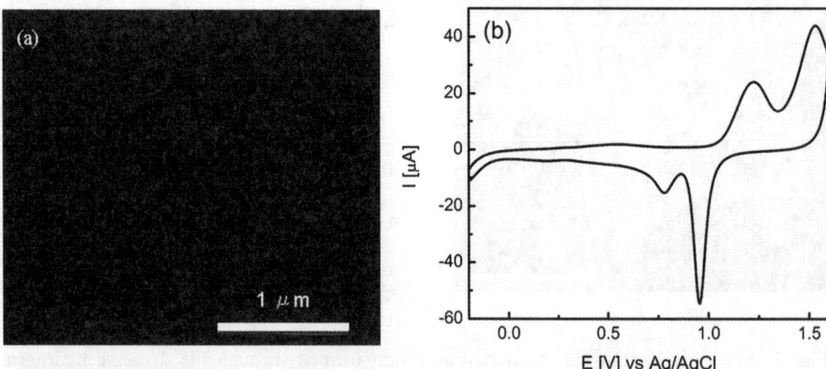

Fig. 6 An SEM image of an evaporated gold film (a) and a CV of the gold film recorded at 50 mV s^{-1} in 1 M H_2SO_4 (b).

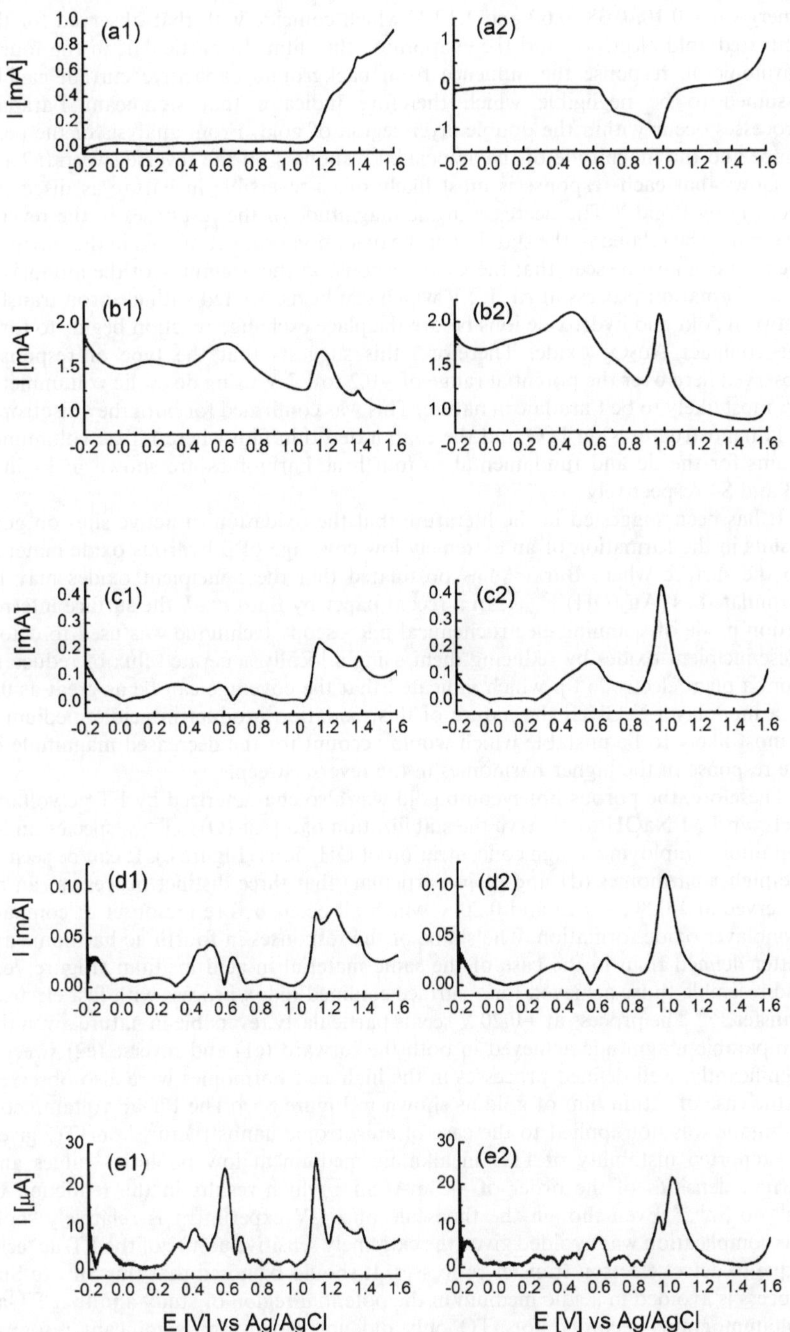

Fig. 7 Large amplitude Fourier transformed ac cyclic voltammograms obtained for the dc (a) and fundamental to fourth harmonics (b–e) for honeycomb porous gold in 1 M H_2SO_4. Conditions employed: $f = 21.16$ Hz, $\Delta E = 100$ mV and $\nu = 67.06$ mV s^{-1}.

In the dc component (a1) very little evidence is seen for active site behaviour as the response is masked by the magnitude of the monolayer oxide formation response as discussed previously. However, from analysis of the fundamental (b1), second (c1), third (d1) and fourth (e1) ac harmonics it can be seen that distinct processes begin to emerge at −0.10, 0.38, 0.60 and 1.12 V which coincide with that observed for the activated gold electrode and the evaporated thin film. In particular, in the fourth harmonic ac response the influence from background capacitive current can be assumed to be negligible which therefore indicates that significant Faradaic processes occur within the double layer region of gold. From analysis of the peak shape and the magnitude of the associated responses on the reverse sweep (a2–e2) it shows that each response is most likely quasi-reversible in nature as discussed recently by Bond.[51] The decrease in the magnitude of the responses in the reverse sweep may be related to the stability of the oxidation product of gold in this environment. It can also be seen that there is a response at the beginning of the monolayer oxide formation process at *ca.* 1.2 V which can be associated with electron transfer between gold and hydroxide ions before the place exchange reaction begins to form the compact Au_2O_3 oxide. Therefore, this suggests that the type of responses observed here over the potential range of −0.2 to 1.2 V using dc cyclic voltammetry are most likely to be Faradaic in nature. This was confirmed for both the anisotropic gold nanostructures on ITO and the evaporated thin film where FT-ac voltammograms for the dc and fundamental to fourth ac harmonics are shown in Figures S3 and S4 respectively.

It has been suggested in the literature that the oxidation of active sites on gold results in the formation of an extremely low coverage of β-hydrous oxide material on the surface where Burke[24] has postulated that these incipient oxides may be formulated as $[Au_2(OH)_9]^{3-}{}_{ads}$. In a recent paper by Bard *et al.* the surface interrogation mode of scanning electrochemical microscopy technique was used to detect these incipient oxides by reducing them using a locally generated flux of reductant from a microelectrode tip which indicated that the coverage can be as great as 0.2 of a monolayer.[52] Given the nature of this oxidation product in acidic medium it is most likely to be unstable which would account for the decreased magnitude of the response in the higher harmonics in the reverse sweep.

Therefore, the porous honeycomb gold was also characterized by FT-ac voltammetry in 1 M NaOH to observe the stabilization of a $[Au_2(OH)_9]^{3-}{}_{ads}$ species under conditions employing a high concentration of OH^- ions (Figure 8). It can be seen in the higher harmonics (d1 and e1 in particular) that three distinct processes can be observed at −0.80, −0.15 and 0.20 V which all occur before the onset of compact monolayer oxide formation. The shape of the responses in fourth ac harmonic are better defined than in the case of the same material in acid medium (Figure 7e1) and resemble those achieved for a surface confined redox process with fast electron transfer.[39,48] The process at −0.20 V seems particularly reversible in nature given the comparable magnitude achieved in both the forward (e1) and reverse (e2) sweeps. Significantly, well defined processes in the higher ac harmonics were also observed in the case of a thin film of gold as shown in Figure S5.† The FT-ac voltammetry technique was not applied to the case of anisotropic nanostructures on ITO given the reported instability of ITO in alkaline medium at low potential values and current densities of the order of −2 mA cm^{-2} which results in the reduction of Sn^{IV} to Sn^{II}.[53] Even though the timescale of a CV experiment is relatively short this complication was avoided given the extremely sensitive nature of the FT-ac technique to detect electron transfer processes. It should be noted that this Sn^{IV} to Sn^{II} process is avoided in acidic medium in the potential region of study and that FT-ac voltammograms recorded for ITO only did not shown any significant response above the noise level of the instrument.

Electrocatalytic reactions at porous honeycomb gold

Illustrated in Figure 9a is the electrooxidation of hydrogen peroxide at porous honeycomb gold in 1 M H_2SO_4 solution. It can be seen that the onset potential is at *ca.* 0.66 V which is 0.34 V less positive than that observed on the polycrystalline electrode (Figure 2d). This shift in onset potential is generally regarded as being an

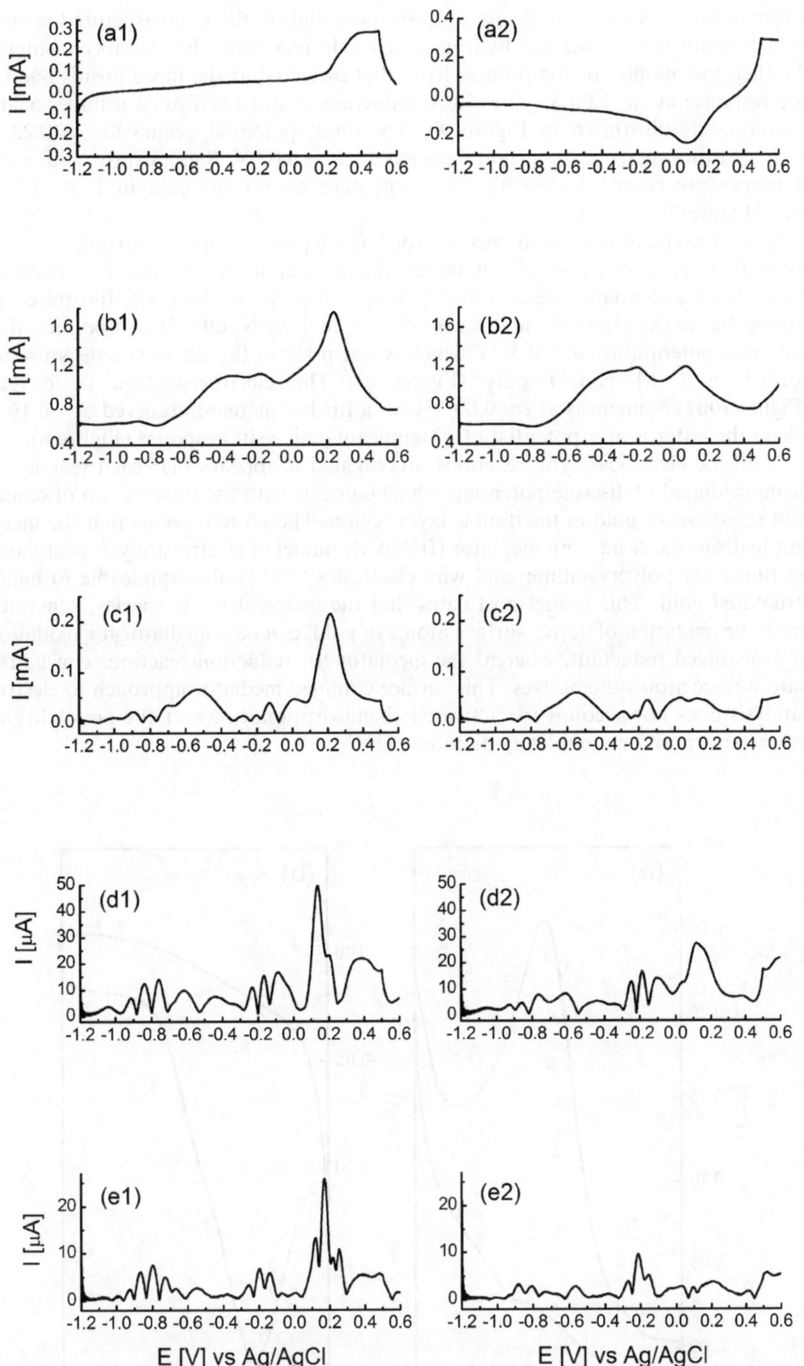

Fig. 8 Large amplitude Fourier transformed ac cyclic voltammograms obtained for the dc (a) and fundamental to fourth harmonics (b–e) for honeycomb porous gold in 1 M NaOH. Conditions employed: $f = 21.16$ Hz, $\Delta E = 100$ mV and $v = 63.33$ mV s^{-1}.

indication of an electrocatalytic effect. It is interesting to note however that the reaction commences at a potential where an active state response was recorded with the FT-ac voltammetry technique at 0.60 V (Figure 7). This suggests that the presence of a significant active state response at lower potential on the nanostructured porous gold is more favourable for hydrogen peroxide oxidation, by possibly changing the thermodynamics of the process from that observed at the more positive active site response at *ca.* 1.00 V. The electroreduction of dissolved oxygen under acidic conditions is illustrated in Figure 9b. The onset potential occurs at *ca.* 0.22 V with a further increase in current detected at *ca.* −0.02 V. Significantly, active site responses are recorded close to these potentials on porous gold in 1 M H_2SO_4 only (Figure 7).

A similar type of behaviour was recorded at the porous honeycomb gold for electrocatalytic reactions carried out under alkaline conditions in that the reactions occurred at potentials where active sites responses were observed. Illustrated in Figure 10a is the electro-oxidation of ethanol in 1 M NaOH. It can be seen that the onset potential is *ca.* −0.38 V which is just prior to the active site response recorded in 1 M NaOH only (Figure 8). The electroreduction of oxygen (Figure 10b) commences at *ca.* 0.00 V with a further increase observed at −0.16 V where the latter is at a potential of a significant active site response (Figure 8).

In all the electrocatalytic reactions investigated it appears that each reaction is being mediated at discrete potentials which coincide with the observation of oxidation responses of gold in the double layer region. Therefore it seems that the incipient hydrous oxide adatom mediator (IHOAM) model of electrocatalysis postulated by Burke for polycrystalline gold wire electrodes[14,25,54] is also applicable to nanostructured gold. This model postulates that the incipient oxide species, generated upon the oxidation of active surface atoms of gold, can be a mediator for oxidation of a dissolved reductant, whereas the mediator for reduction reactions can be the bare surface atoms themselves. This surface confined mediator approach to electrocatalysis does not account for activated chemisorption, however the possibility of the latter cannot be completely discounted.[21,55]

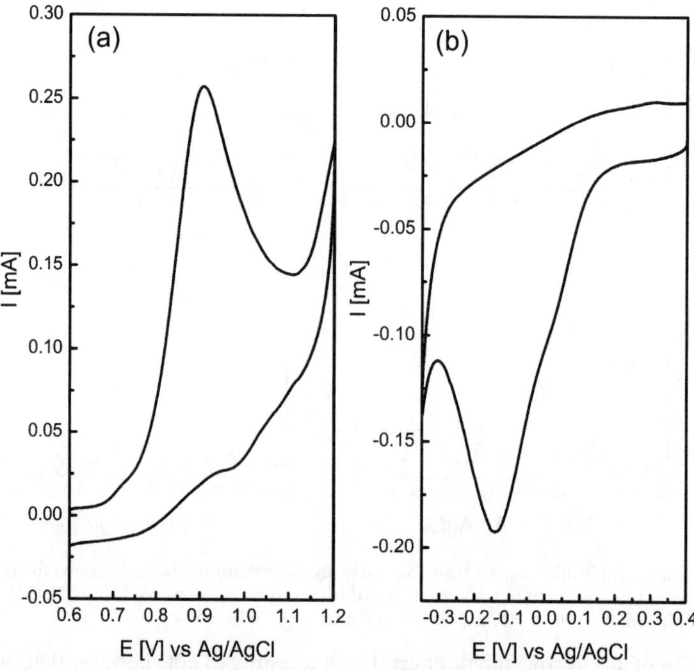

Fig. 9 CVs recorded at porous honeycomb gold at a sweep rate of 20 mV s^{-1} in 1 M H_2SO_4 solution containing (a) 20 mM H_2O_2 and (b) saturated with oxygen.

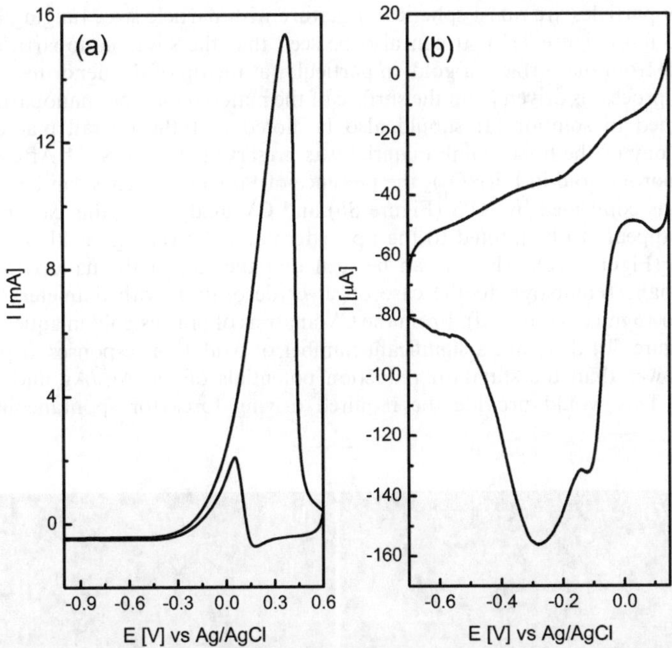

Fig. 10 CVs recorded at porous honeycomb gold in 1 M NaOH solution containing (a) 1 M ethanol obtained at a sweep rate of 50 mV s^{-1} and (b) saturated with oxygen obtained at a sweep rate of 20 mV s^{-1}.

Decoration of active gold with metal nanoparticles

A recent method to create bimetallic nanoparticles and surfaces is the galvanic replacement method which has resulted in an array of porous, hollow, dendritic and high surface area materials with compositions such as Ag/Au, Ag/Pt, Co/Pt, and Fe/Au.[56–61] When this approach is used, the thermodynamic driving force for the reaction is the difference in the standard electrode reduction potentials between the sacrificial metal/metal ion couple and the solution based metal/metal ion couple. For the case of gold its decoration using this approach is problematic given the high standard reduction potential of the Au^{3+}/Au couple (1.50 V vs SHE). Even if a chloride media was employed the standard reduction potential of AuCl$_4^-$/Au is still high at a value of 1.002 V vs SHE). In this work the active state responses observed in the double layer region of gold are assumed to be due to the more facile oxidation of gold in an active state than the bulk material. Therefore if active gold is more readily oxidised then it should allow for spontaneous electroless deposition of metal species to occur that have a standard reduction potential greater than the potential at which these active sites are oxidised. A porous honeycomb gold sample was then immersed in separate 1 mM aqueous solutions of Ni(NO$_3$)$_2$, AgNO$_3$, Pd(NO$_3$)$_2$ and K$_2$PtCl$_4$ where the standard reduction potentials of Ni^{2+}/Ni, Ag$^+$/Ag, Pd^{2+}/Pd and PtCl$_4^{2-}$/Pt are −0.257 V, 0.799, 0.915 and 0.758 V vs SHE respectively. It can be seen that these standard reduction potentials are significantly lower than that for the Au^{3+}/Au couple and therefore the spontaneous replacement of gold is not expected. However, upon immersion of honeycomb porous gold in AgNO$_3$ and Pd(NO$_3$)$_2$ significant decoration of the surface with nanoparticles can be observed (Figure 11).

For the case of silver, surface decoration was confirmed by XPS (Figure S6),† and CV analysis and SEM images are shown in Figures 11a and b. It can be seen that the dendritic nanostructures are well decorated with silver nanoparticles (Figure 11a).

The nanoparticles are quasi spherical in nature with particle sizes ranging between 15 to 40 nm (Figure 11b). It can also be seen that the silver nanoparticles grow outwards from the surface of gold, in particular at the tip of the dendrites, showing that this process is driven from the surface of the material and that nanoparticles are not formed in solution. It should also be noted that the overall macroporous morphology of the honeycomb material was preserved (Figure S7).† After immersion of porous gold in $Pd(NO_3)_2$ the presence of Pd nanoparticles can be observed which was confirmed by XPS (Figure S6) and CV analysis. In this case however, growth appears to be limited to the tips, edges and protruding backbone of each dendrite (Figure 11c). Also, it can be seen that the size of the nanoparticles are much smaller compared to the case of silver decoration with diameters ranging from 5 to 15 nm (Figure 11d). From the CV analysis of porous gold in aqueous solution (Figure 5b) there are a significant number of oxidation responses at potential values lower than the standard reduction potentials of the Ag^+/Ag and Pd^{2+}/Pd couples. This would provide the required driving force for spontaneous metal

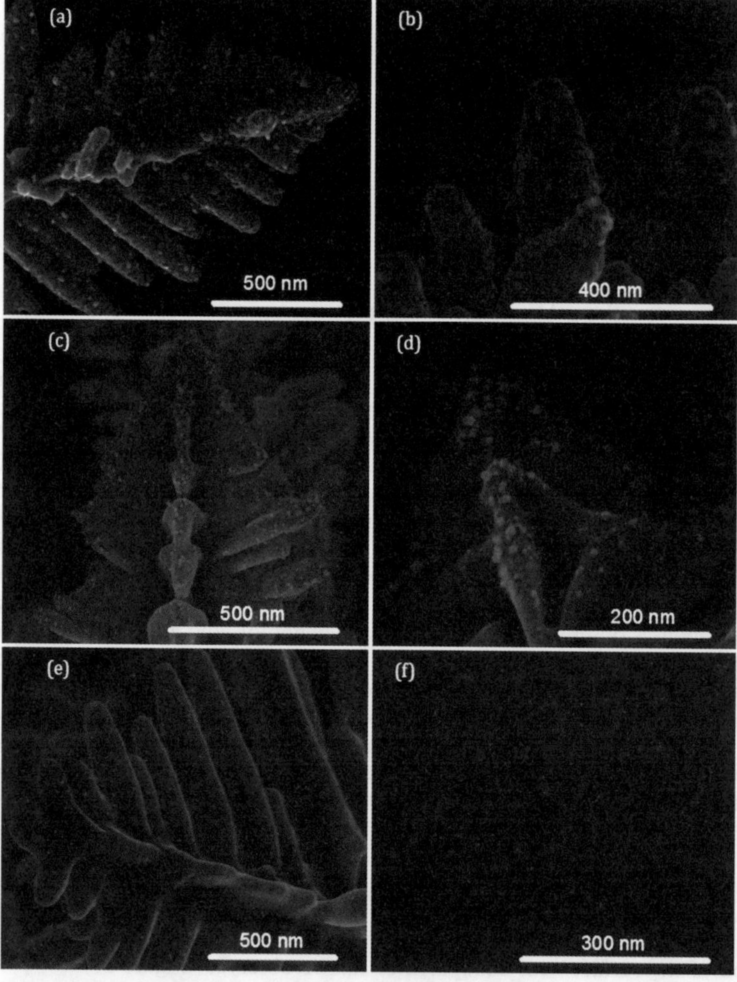

Fig. 11 SEM images showing the decoration of honeycomb structures after 5 min immersion in 1 mM solutions of $AgNO_3$ (a, b) $Pd(NO_3)_2$ (c, d) and K_2PtCl_6 (e) along with an evaporated gold substrate after 5 min immersion in 1 mM $Pd(NO_3)_2$.

deposition to occur whereby the facile oxidation of active gold provides the electrons for reduction of the Ag^+ and Pd^{2+} ions. To provide further evidence for this mechanism it was noted that when $Ni(NO_3)_2$ solution was employed there was no observation of surface decoration with Ni nanoparticles by SEM or that could be detected with XPS and CV analysis. The Ni^{2+}/Ni couple has a standard reduction potential of -0.257 V vs SHE which is in the hydrogen evolution region on gold where no active site responses can be observed and it is highly unlikely that gold could be oxidised at potentials lower than this. It should be noted that the synthesis of the porous gold material requires vigorous hydrogen evolution and so the possibility of adsorbed hydrogen acting as a reductant was investigated. This mechanism can be discounted as an evaporated gold film immersed in $Pd(NO_3)_2$ showed the presence of palladium nanoparticles as illustrated in Figure 11f. The particles have a diameter of less than 10 nm as in the case of porous gold but interestingly do not show preference for decoration of the grain boundaries suggesting that active sites are present all over the surface.

Interestingly, when a 1 mM solution of K_2PtCl_4 was used, surface decoration with Pt nanoparticles could not be observed by SEM (Figure 11e). However, XPS analysis confirmed the presence of metallic platinum even though the signal was quite weak (Figure S6c).† A CV was obtained in 1 M H_2SO_4 (Figure 12a) and it can be seen that there is a characteristic hydrogen adsorption/desorption region over a potential range of -0.22 to -0.14 V associated with metallic platinum. When the hydrogen evolution reaction is carried out a significant shift in onset potential and magnitude of the response is observed compared to the unmodified porous gold which further confirms the presence of platinum (Figure S8).† A continuous thin layer of platinum on the surface is discounted as distinct voltammetry from gold can be seen. Indeed the surface coverage with platinum appears to be quite low when the CV responses associated with platinum and gold are compared which explains the weak XPS signal. Taking the electrochemically active surface area of both gold and platinum indicates a 5% coverage of the surface with platinum. The surface area of gold was calculated using the charge associated with the monolayer oxide reduction peak[46] at 0.93 V and a value of 400 μC cm^{-2}. The contribution from the reduction of incipient oxides formed on the forward sweep at potentials lower than 0.93 V on active gold is not taken into account here. It should be noted that these responses are significantly less than the monolayer oxide reduction peak unlike the case of anisotropic structures on gold which demonstrated three significant reduction peaks. The surface area of platinum was calculated using the charge associated with the desorption of hydrogen[62] and a value of 210 μC cm^{-2}. The calculated metal coverage is slightly higher than the *ca.* 1% of an unactivated surface that has been estimated by Burke to be active for electrocatalytic reactions. This is most likely due to the method of synthesis which introduces a higher amount of metastable surface sites. This percentage coverage with platinum suggests that perhaps extremely small nanoparticles are produced that cannot be observed at the resolution achievable by the field emission SEM instrument used in this study. A high resolution TEM study will be undertaken to confirm this hypothesis. It can also be seen that the double layer region is now absent of active site responses from 0.20 to 1.10 V (inset Figure 12a) where significant responses were observed for the unmodified surface (Figure 5b) illustrating the role of these sites in the electroless deposition process.

The CV responses for the Au/Pd and Au/Ag systems are shown in Figures 12b and c respectively. For Au/Pd a characteristic hydrogen adsorption/desorption region can be observed, indicating the presence of metallic palladium, and similarly metallic platinum is observed in the case of Au/Pt (Figure 12a and inset), where the former is shifted to more negative potentials compared to the latter. There is some evidence of active site behaviour, with a process observed at 0.52 V, which suggests that not all active sites are decorated with palladium. The Au/Ag material was characterised by CV in 1 M NaOH to avoid the electrodissolution of Ag that occurs in acidic media.

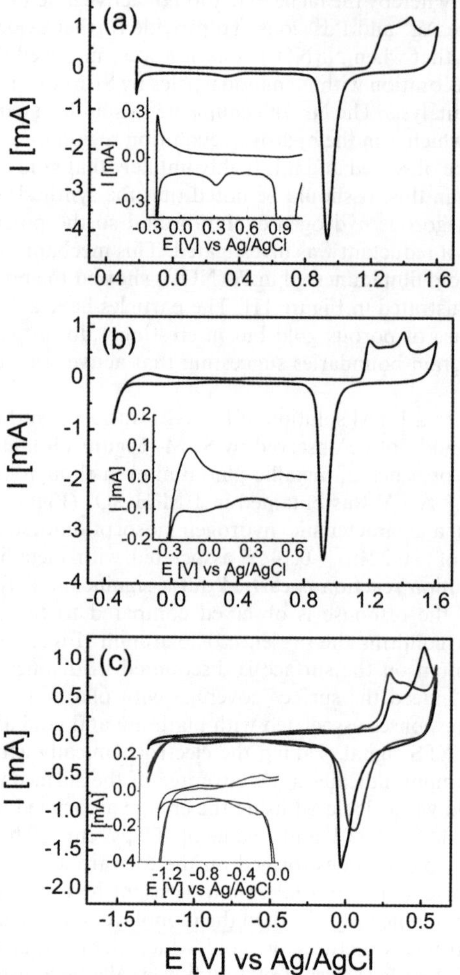

Fig. 12 CVs recorded at porous honeycomb gold at a sweep rate of 50 mV s^{-1} decorated with (a) platinum, (b) palladium and (c) silver. The electrolyte employed was 1 M H_2SO_4 (a and b) and 1 M NaOH (c) where the CV is also shown for porous gold only (grey).

It can be seen in Figure 12c that a distinct oxidation process at 0.50 V can be observed which is indicative of oxide formation on silver.[63] On the reverse sweep the large cathodic peak centred at -0.03 V is at a slightly lower potential than the unmodified gold surface due to the added contribution from silver oxide reduction which occurs over a similar potential region.[63] It can also be seen that the increase in current observed from -1.0 to -1.2 V on gold only, due to the evolution of hydrogen, is nearly completely absent once the surface is decorated with silver (inset of Figure 12c). Given that the surface is only partly decorated with silver nanoparticles (Figure 11a) and that there is a significant amount of gold still available for the hydrogen evolution reaction suggests that the silver nanoparticles block the active sites on gold which are responsible for this electrocatalytic reaction.

This method of chemically interrogating the surface of gold is in a way analogous to the recent work of Scholz who elegantly demonstrated that active sites on gold are attacked by Fenton's reagent where asperities, assumed to be the origin of the active site, are dissolved through a reaction with hydroxyl radicals.[27] The activity of a gold

     This journal is © The Royal Society of Chemistry 2011

electrode was shown to decrease with exposure time to Fenton's reagent for the reduction of dissolved oxygen and also the hydrogen evolution reaction. We believe that our approach of decorating gold with other metals in fact indicates the location of regions with higher densities of active sites on gold surfaces. In the case of the evaporated gold film the coverage was quite uniform, however, the honeycomb porous gold showed highly preferred deposition at the tips and edges of the dendritic nanostructures within the internal wall structure, in particular when palladium was used. The extent of coverage and size of nanoparticles may also be indicative of the reactive nature of the active site in question as the electroless deposition process will be governed by both thermodynamic and kinetic effects. It is also a facile route for the creation of bimetallic surfaces which have received a lot of interest as electrocatalysts due to beneficial synergistic effects that can enhance both the current response and prolong the use of the catalyst by negating poisoning effects with carbonaceous species.[64-67] Recent work has focused on biofuels such as ethanol as a fuel source, but its electro-oxidation is difficult due to significant poisoning of

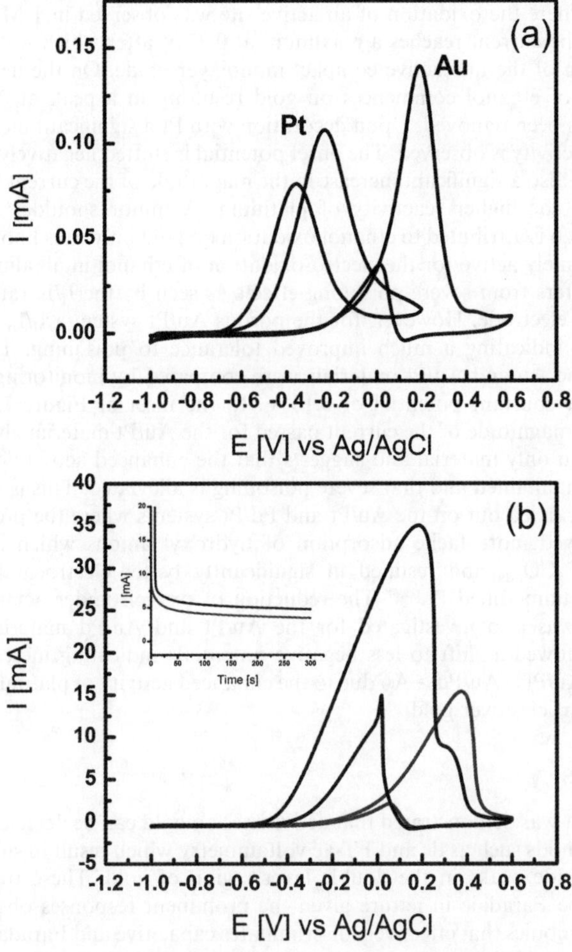

Fig. 13 CVs recorded at a sweep rate of 50 mV s^{-1} in 1 M NaOH containing 1 M ethanol at (a) Pt and Au electrodes and (b) porous gold (grey) and porous gold decorated with platinum (black). The inset shows the current time response when the potential of the relevant electrode was held at 0.18 V.

the electrocatalyst surface.[65] It has been identified that the Au/Pt system is particularly effective in destabilizing adsorbed poisons such as CO by a bi-functional mechanism.[64,65]

The electrocatalytic performance of the bimetallic Au/Pt system was then evaluated for ethanol oxidation in 1 M NaOH which shows significantly different behaviour at gold and platinum electrodes and therefore allows for any synergistic effects of a Au/Pt electrode to be investigated. For a platinum electrode the oxidation of ethanol commences on the positive sweep at −0.60 V and reaches a maximum (I_f) at −0.23 V (Figure 13a). On the negative sweep an oxidation peak is also observed (I_b) which is due to the oxidation of poisoning carbonaceous species, most likely adsorbed CO_{ads}, formed during the positive sweep. It is established that the ethanol oxidation reaction proceeds through the formation of extensive poisoning species such as CO_{ads} on the Pt group of metals. The I_f/I_b ratio observed for Pt is 1.37 and is often used to compare the tolerance of metallic catalysts to poisoning.[68] The electro-oxidation of ethanol on a gold electrode occurs at more positive potentials, demonstrating that it is a less effective catalyst than platinum for this particular reaction. A CV for the oxidation of ethanol on honeycomb porous gold is shown in Figure 13b. The onset for the reaction occurs at *ca.* −0.25 V which is again at a potential where the oxidation of an active site was observed in 1 M NaOH only (Figure 8). The current reaches a maximum at 0.37 V after which it decays due to the formation of the unreactive compact monolayer oxide. On the negative sweep re-oxidation of ethanol commences on gold resulting in a peak at 0.061 V once this oxide has been removed. Upon decoration with Pt a significant increase in electrocatalytic activity is observed. The onset potential is shifted negatively by 0.28 V to −0.53 V with also a significant increase in the magnitude of the current which can be attributed to the higher reactivity of platinum. A minor shoulder at 0.31 V is observed which is attributed to ethanol oxidation on gold only. It is known that platinum is extremely active for the electro-oxidation of ethanol in alkaline conditions but often suffers from severe poisoning effects as seen by the I_f/I_b ratio of 1.37 at the platinum electrode. However, for the porous Au/Pt system a I_f/I_b ratio of 2.30 was attained indicating a much improved tolerance to poisoning. Therefore the stability of the Au and Au/Pt materials were compared by monitoring the current response at a constant potential of 0.18 V. In the inset of Figure 13b it can be seen that the magnitude of the current passed for the Au/Pt material always exceeds that of the Au only material and suggests that the enhanced activity of decorative platinum is maintained and that severe poisoning is alleviated. This is in agreement with studies carried out on the Au/Pt and Pd/Pt systems where the presence of Au and Pd allowed more facile adsorption of hydroxyl anions which inhibited the formation of CO_{ads} and resulted in significantly better electrocatalytic activity than that of unmodified Pt.[65,68] The reduction of oxygen under acidic conditions (Figure S9) was also investigated for the Au/Pt and Au/Pd materials which in both cases showed a shift to less negative potentials indicating increased activity in the order Au/Pt > Au/Pd > Au due to the enhanced activity of platinum and palladium nanoparticles over gold.

Conclusions

In this work it was demonstrated that active sites on gold can be detected by electrochemical methods such as dc and FT-ac voltammetry which result in significant pre-monolayer oxide peaks in the double layer region of gold. These transitions are assumed to be Faradaic in nature given the prominent responses observed in the higher ac harmonics that effectively discriminates capacitive and Faradaic processes. It is also suggested that active sites are not only present on severely activated gold electrodes but also on electrodeposited nanostructured gold and even evaporated thin films. The electrocatalytic behaviour of gold was investigated and several reactions were found to proceed at a potential where distinct active site oxidation processes were observed in the absence of the analyte under study. This is in agreement with the IHOAM model of electrocatalysis which was previously studied at planar polycrystalline gold electrodes. Significantly, it was found that advantage could be taken of the more facile oxidation of gold in an active state whereby the surface could be decorated with platinum, palladium and silver which is thermodynamically forbidden at bulk fully lattice stabilised gold. This method allows for the formation of interesting bimetallic surfaces such as Au/Pt which was shown to be particularly beneficial for the ethanol oxidation reaction under alkaline conditions. It is also speculated that this type of surface decoration may indicate regions of high density of active sites on surfaces and may be applicable to many other metals and interesting applications.

Acknowledgements

The authors acknowledge Prof. A. M. Bond and Dr. C.-Y. Lee for the provision of instrumentation and assistance with large amplitude FT-ac experiments. Financial support from the Australian Research Council is also gratefully acknowledged.

References

1 M. Haruta, N. Yamada, T. Kobayashi and S. Iijima, *J. Catal.*, 1989, **115**, 301.
2 J. Zeng, Q. Zhang, J. Chen and Y. Xia, *Nano Lett.*, 2010, **10**, 30.
3 T. A. Baker, X. Liu and C. M. Friend, *Phys. Chem. Chem. Phys.*, 2011, **13**, 34.
4 A. Wittstock, J. Biener and M. Baumer, *Phys. Chem. Chem. Phys.*, 2010, **12**, 12919.
5 G. J. Hutchings, *Catal. Today*, 2005, **100**, 55.
6 P. A. Sermon, G. C. Bond and P. B. Wells, *J. Chem. Soc., Faraday Trans. 1*, 1979, **75**, 385.
7 Y. Li, H. J. Schluesener and S. Xu, *Gold Bull.*, 2010, **43**, 29.
8 J. M. Pingarron, P. Yanez-Sedeno and A. Gonzalez-Cortes, *Electrochim. Acta*, 2008, **53**, 5848.
9 J. H. Shim, J. Kim, C. Lee and Y. Lee, *J. Phys. Chem. C*, 2011, **115**, 305.
10 W. Chen and S. Chen, *Angew. Chem., Int. Ed.*, 2009, **48**, 4386.
11 J. Hernandez, J. Solla-Gullon, E. Herrero, A. Aldaz and J. M. Feliu, *J. Phys. Chem. C*, 2007, **111**, 14078.
12 M. S. El-Deab, T. Sotomura and T. Ohsaka, *Electrochim. Acta*, 2006, **52**, 1792.
13 Z. Borkowska, A. Tymosiak-Zielinska and G. Shul, *Electrochim. Acta*, 2004, **49**, 1209.
14 L. D. Burke, *Gold Bull.*, 2004, **37**, 125.
15 Y. Chen, W. Schuhmann and A. W. Hassel, *Electrochem. Commun.*, 2009, **11**, 2036.
16 R. R. Adzic, S. Strbac and N. Anastasijevic, *Mater. Chem. Phys.*, 1989, **22**, 349.
17 X. Han, D. Wang, J. Huang, D. Liu and T. You, *J. Colloid Interface Sci.*, 2011, **354**, 577.
18 D. H. Nagaraju and V. Lakshminarayanan, *J. Phys. Chem. C*, 2009, **113**, 14922.
19 B. K. Jena and C. R. Raj, *J. Phys. Chem. C*, 2007, **111**, 15146.
20 J. Hernandez, J. Solla-Gullon, E. Herrero, A. Aldaz and J. M. Feliu, *Electrochim. Acta*, 2006, **52**, 1662.
21 L. D. Burke and P. F. Nugent, *Gold Bull.*, 1998, **31**, 39.
22 L. D. Burke and A. P. O'Mullane, *J. Solid State Electrochem.*, 2000, **4**, 285.
23 V. Díaz, S. Real, E. Téliz, C. F. Zinola and M. E. Martins, *Int. J. Hydrogen Energy*, 2009, **34**, 3519.
24 L. D. Burke and P. F. Nugent, *Gold Bull.*, 1997, **30**, 43.
25 L. D. Burke, L. M. Hurley, V. E. Lodge and M. B. Mooney, *J. Solid State Electrochem.*, 2001, **5**, 250.
26 Y.-H. Lee, G. Kim, M. Joe, J.-H. Jang, J. Kim, K.-R. Lee and Y.-U. Kwon, *Chem. Commun.*, 2010, **46**, 5656.
27 A. M. Nowicka, U. Hasse, G. Sievers, M. Donten, Z. Stojek, S. Fletcher and F. Scholz, *Angew. Chem., Int. Ed.*, 2010, **49**, 3006.
28 S. Guo, L. Wang and E. Wang, *Chem. Commun.*, 2007, 3163.
29 B. Plowman, S. J. Ippolito, V. Bansal, Y. M. Sabri, A. P. O'Mullane and S. K. Bhargava, *Chem. Commun.*, 2009, 5039.
30 F. Gao, M. S. El-Deab and T. Ohsaka, *Indian J. Chem., A.*, 2005, **44A**, 932.
31 Y. Tian, H. Liu, G. Zhao and T. Tatsuma, *J. Phys. Chem. B*, 2006, **110**, 23478.
32 L. Komsiyska and G. Staikov, *Electrochim. Acta*, 2008, **54**, 168.
33 W. Ye, J. Yan, Q. Ye and F. Zhou, *J. Phys. Chem. C*, 2010, **114**, 15617.
34 J. F. Hicks, D. T. Miles and R. W. Murray, *J. Am. Chem. Soc.*, 2002, **124**, 13322.

35 R. W. Murray, *Chem. Rev.*, 2008, **108**, 2688.
36 R. Sardar, A. M. Funston, P. Mulvaney and R. W. Murray, *Langmuir*, 2009, **25**, 13840.
37 B. J. Plowman, A. P. O'Mullane, P. Selvakannan and S. K. Bhargava, *Chem. Commun.*, 2010, **46**, 9182.
38 J. Zhang, S.-X. Guo, A. M. Bond and F. Marken, *Anal. Chem.*, 2004, **76**, 3619.
39 S. Guo, J. Zhang, D. M. Elton and A. M. Bond, *Anal. Chem.*, 2004, **76**, 166.
40 L. D. Burke, A. P. O'Mullane, V. E. Lodge and M. B. Mooney, *J. Solid State Electrochem.*, 2001, **5**, 319.
41 A. M. Bond, N. W. Duffy, S. X. Guo, J. Zhang and D. Elton, *Anal. Chem.*, 2005, **77**, 186A.
42 P. D. Cobden, B. E. Nieuwenhuys, V. V. Gorodetskii and V. N. Parmon, *Platinum Metals Review*, 1998, **42**, 141.
43 K. Z. Brainina, L. G. Galperin, Y. V. Vikulova, N. Y. Stozhko, A. M. Murzakaev, O. R. Timoshenkova and Y. A. Kotov, *J. Solid State Electrochem.*, 2011, **15**, 1049.
44 O. S. Ivanova and F. P. Zamborini, *J. Am. Chem. Soc.*, 2010, **132**, 70.
45 L. A. Hutton, M. E. Newton, P. R. Unwin and J. V. MacPherson, *Anal. Chem.*, 2011, **83**, 735.
46 D. A. J. Rand and R. Woods, *J. Electroanal. Chem.*, 1971, **31**, 29.
47 J. Zhang, S.-X. Guo and A. M. Bond, *Anal. Chem.*, 2007, **79**, 2276.
48 J. Zhang and A. M. Bond, *J. Electroanal. Chem.*, 2007, **600**, 23.
49 C.-Y. Lee and A. M. Bond, *Langmuir*, 2010, **26**, 16155.
50 A. P. O'Mullane, J. Zhang, A. Brajter-Toth and A. M. Bond, *Anal. Chem.*, 2008, **80**, 4614.
51 B. Lertanantawong, A. P. O'Mullane, W. Surareungchai, M. Somasundrum, L. D. Burke and A. M. Bond, *Langmuir*, 2008, **24**, 2856.
52 J. Rodriguez-Lopez, M. A. Alpuche-Aviles and A. J. Bard, *J. Am. Chem. Soc.*, 2008, **130**, 16985.
53 E. Matveeva, *J. Electrochem. Soc.*, 2005, **152**, H138.
54 L. D. Burke, J. M. Moran and P. F. Nugent, *J. Solid State Electrochem.*, 2003, **7**, 529.
55 L. D. Burke, J. K. Casey, J. A. Morrissey and M. M. Murphy, *Bull. Electrochem.*, 1991, **7**, 506.
56 A. Pearson, A. P. O'Mullane, V. Bansal and S. K. Bhargava, *Chem. Commun.*, 2010, **46**, 731.
57 D. Zhao, Y.-H. Wang, B. Yan and B.-Q. Xu, *J. Phys. Chem. C*, 2009, **113**, 1242.
58 Q.-S. Chen, S.-G. Sun, Z.-Y. Zhou, Y.-X. Chen and S.-B. Deng, *Phys. Chem. Chem. Phys.*, 2008, **10**, 3645.
59 L. Au, X. Lu and Y. Xia, *Adv. Mater.*, 2008, **20**, 2517.
60 J. Chen, B. Wiley, J. McLellan, Y. Xiong, Z. Y. Li and Y. Xia, *Nano Lett.*, 2005, **5**, 2058.
61 I. Najdovski, A. P. O'Mullane and S. K. Bhargava, *Electrochem. Commun.*, 2010, **12**, 1535.
62 R. Woods, in *Electroanalytical Chemistry*, ed. A. J. Bard, Dekker, New York, 1976.
63 V. Bansal, V. Li, A. P. O'Mullane and S. K. Bhargava, *CrystEngComm*, 2010, **12**, 4280.
64 W. Tang, S. Jayaraman, T. F. Jaramillo, G. D. Stucky and E. W. McFarland, *J. Phys. Chem. C*, 2009, **113**, 5014.
65 O. A. Hazzazi, G. A. Attard, P. B. Wells, F. J. Vidal-Iglesias and M. Casadesus, *J. Electroanal. Chem.*, 2009, **625**, 123.
66 J. Zeng, J. Yang, J. Y. Lee and W. Zhou, *J. Phys. Chem. B*, 2006, **110**, 24606.
67 N. Fujiwara, K. A. Friedrich and U. Stimming, *J. Electroanal. Chem.*, 1999, **472**, 120.
68 S. S. Mahapatra, A. Dutta and J. Datta, *Electrochim. Acta*, 2010, **55**, 9097.

   This journal is © The Royal Society of Chemistry 2011

Aberration corrected analytical electron microscopy studies of sol-immobilized Au + Pd, Au{Pd} and Pd{Au} catalysts used for benzyl alcohol oxidation and hydrogen peroxide production

R. C. Tiruvalam,[a] J. C. Pritchard,[b] N. Dimitratos,[b]
J. A. Lopez-Sanchez,[b] J. K. Edwards,[b] A. F. Carley,[b]
G. J. Hutchings[b] and C. J. Kiely*[a]

Received 15th February 2011, Accepted 16th March 2011
DOI: 10.1039/c1fd00020a

In this study, a systematic series of AuPd bimetallic particles were prepared by colloidal synthesis methods, in order to gain better control over the particle size distribution and structure. Particles having random alloy structures, as well as 'designer' particles with Pd-shell/Au-core and Au-shell/Pd-core morphologies, have been prepared and immobilized on both activated carbon and TiO_2 supports. Aberration corrected analytical electron microscopy (ACEAM) has been extensively used to characterize these sol-immobilized materials. In particular, state-of-the-art z-contrast STEM-HAADF imaging and STEM-XEDS spectrum imaging has been employed. These techniques have provided invaluable new (and often unexpected) information on the atomic structure, elemental distribution within particles, and compositional variations between particles for these controlled catalyst preparations. In addition, we have been able to compare their differing thermal stability, sintering and wetting behaviors on activated carbon and TiO_2 supports. These sol immobilized materials have also been compared as catalysts for (i) benzyl alcohol oxidation and (ii) the direct production of H_2O_2 in an attempt to elucidate the optimum particle morphology/support combination for each reaction.

Introduction

There is a growing interest in using supported bimetallic AuPd alloy nanoparticles as catalysts for a variety of reactions, including the direct synthesis of H_2O_2 from H_2 and O_2,[1] epoxidation of alkenes,[2,3] oxidation of alcohols[4] and polyols,[5] and the oxidation of toluene.[6] In our first reports on this topic, the AuPd alloy nanoparticles were prepared *via* an impregnation route using $HAuCl_4$ and $PdCl_2$ as precursors. After calcination at 400 °C, this generated AuPd alloy particles having a broad particle size distribution, with the larger particles being Au-rich and the smaller ones being Pd-rich.[7] Interestingly we found that on oxide supports the AuPd particles tended to adopt a Pd-rich shell/Au-rich core structure, whereas on activated carbon they always remained as random homogenous AuPd alloys.[8]

[a]*Department of Materials Science and Engineering, Lehigh University, 5 East Packer Avenue, Bethlehem, PA, 10185-3195, USA. E-mail: chk5@lehigh.edu*
[b]*Cardiff Catalysis Institute, School of Chemistry, Cardiff University, Main Building, Park Place, Cardiff, CF10 3AT, UK*

While the impregnation method has the distinct advantage of being a relatively simple method for generating the catalyst, it offers little control over the particle size distribution and chemical composition of the individual Au–Pd nanoparticles produced after calcination. In order to gain better control over these parameters, we have investigated the sol-immobilization process for producing Au–Pd nanoparticles.[9] Random Au + Pd alloys can be produced in sol form by reducing Au and Pd salts using $NaBH_4$ in the presence of PVA (poly vinyl alcohol), which acts as a ligand controlling the growth and agglomeration of the formed nanoparticles. These colloidal nanoparticles can then be readily immobilized onto supports such as carbon and TiO_2.

In our preliminary study,[9] it was found that Au + Pd alloy sol-immobilized catalysts dried at 120 °C were the more active (as compared to their impregnated counterparts) on both carbon and TiO_2 supports for the direct oxidation of benzyl alcohol. Even though thermogravimetric analysis of these sol-immobilized samples suggested that the PVA ligands were still attached to the alloy nanoparticles up to ~300 °C, this did not seem to detrimentally affect their catalytic activity. Additionally, the catalyst samples dried at 120 °C showed almost no loss of activity upon re-use. In order to check the performance of these catalysts upon ligand removal, the catalysts were calcined at 400 °C. However, it was found that this calcination process led to a significant drop in their catalytic activity.[9]

For the direct hydrogenation of oxygen to form hydrogen peroxide, the sol-immobilized Au + Pd mixed alloy catalysts dried at 120 °C were more active than the corresponding impregnated catalysts. They showed higher initial activity for the formation of hydrogen peroxide, but were also very active for the further hydrogenation of the formed hydrogen peroxide into water, making their overall selectivity for the production of H_2O_2 poorer than the impregnated catalysts.[9] For the sol-immobilised materials to be utilized in the direct hydrogen peroxide synthesis, they must be somehow "engineered" to give a better selectivity towards hydrogen peroxide.

"Core-shell" morphology nanoparticles comprising Au and Pd can also be prepared by modifying the sequence of metal addition during colloid preparation. Instead of adding Au and Pd salts simultaneously in the parent solution and reducing them together, it is also possible to reduce the metal salts sequentially. This in principle gives us access to three different combinations of Au and Pd bimetallic nanoparticles; namely, (i) mixed Au–Pd alloy (denoted here as Au + Pd), Au_{shell}–Pd_{core} (denoted Au{Pd}) and Pd_{shell}–Au_{core} (denoted Pd{Au}).

Ever since Turkevich *et.al.*,[10] demonstrated the possibility of forming core shell Au_{shell}–Pd_{core} particles by successively reducing Pd and Au ions to metals in solution, AuPd nanoparticles of different morphologies; (*e.g.*, random alloys,[9,11–13] core-shell morphologies,[14–22] tri-layered structures[23,24] and cluster in cluster structures[25,26]) have been prepared. Some of these materials have also been tested as catalysts for various reactions.[27] However, generating core-shell particles *via* colloidal routes is not always as straightforward as one might first imagine. For example, Toshima *et.al*[28] tried to prepare Au_{core}–Pd_{shell} particles by sequentially reducing Pd^{2+} and Au^{3+} ions, using poly N-vinyl-2-pyrrolidone (PVP) as the stabilizing reagent. After the first reduction step, monometallic, PVP protected Pd nanoparticles which were only a few nanometers in size, were observed. However, upon further reduction of Au^{3+} in the presence of the Pd 'seeds', phase separated connected clusters of Au–Pd were observed instead of core-shell structures. In contrast, reduction of Pd^{2+} ions in a solution containing PVP protected Au 'seeds' led to the formation of Pd_{shell}–Au_{core} nanoparticles as expected. Ferrer *et al.*,[23,24] performed the successive reduction of Pd^{2+} and Au^{3+} ions using ethylene glycol (commonly known as the polyol method) to form bimetallic nanoparticles. In their study, it was found that particles greater than 5 nm in size adopted a three-layered onion like structure with a Pd core, Au-rich intermediate layer and Pd-rich outer layer. In contrast, particles less than 5 nm in size were thought to be alloyed with no visible segregation. These examples serve to show that solution synthesis of core-shell nanoparticles is a complex phenomenon that yields particle structures and morphologies that are highly dependent on the preparation route. Several reasons have been proposed to resolve the dichotomy between the desired and observed nanoparticle structures; including, (i) differences in rates and order of metal ion reduction,[16,17,29,30] (ii) dynamical dissolution of the formed metallic clusters[16,23,29] and (iii) the complex thermodynamics involved in the phase stabilities of nanoparticles.[31] The latter factor in particular is fascinating and unexpected, as Shirinyan and Wautelet[31] have shown that phase separation can occur under certain conditions at the nanoscale, even in systems that are completely miscible in bulk alloy form, which is the case for the Au–Pd system.[32]

There is a strong motivation for wanting to explore core-shell configurations of Au–Pd bimetallic particles in catalysis as they may well behave quite differently to monometallic Au- or Pd-catalysts or even randomly mixed AuPd alloys. Two primary reasons have been proposed to explain why core-shell morphologies could show subtle differences in catalytic behaviour.[33–37] Firstly there is the *strain effect*, whereby the average bond lengths in the shell metal are modified from its native state due to the strain induced by a lattice mismatch with the underlying material. Secondly, there is the so called *ligand effect*, in which the surface electronic structure is modified due to the heterometallic bonding interactions between the core and shell. This latter terminology should not be confused with the organic surfactants or "ligands" used as stabilizers in colloidal syntheses. The final catalytic activity is probably a complex interplay of these two factors that could in principle act co-operatively or competitively. Physical parameters such as core size, shell thickness, surface oxidation, and the degree of layer intermixing can potentially have a strong influence on the strain and ligand effects displayed by the core-shell particle. Hence it is critical that advanced characterization techniques and protocols are developed that allow us to carefully monitor these parameters in core-shell colloidal particles.

In this contribution, we extend our preliminary work[9,38] on sol immobilized Au + Pd catalysts for benzyl alcohol oxidation and hydrogen peroxide production. We have prepared a systematic set of Au + Pd, Au{Pd} and Pd{Au} colloidal particles and immobilized them on activated C and TiO_2 supports. These bimetallic nanoparticles have been characterized by state-of-the-art aberration corrected analytical electron techniques in both the sol and immobilized forms. We have also studied the structural evolution of the Au + Pd, Au{Pd} and Pd{Au} particles as a function of increasing calcination temperature by STEM and XPS, in order to compare and contrast their differing behaviours on activated C and TiO_2 supports. This set of sol immobilized materials have also been compared as catalysts for (i) benzyl alcohol oxidation and (ii) the direct production of H_2O_2 in an attempt to elucidate the optimum particle morphology/support combination for each reaction.

Experimental

Colloid preparation

Three distinct Au–Pd sols were prepared which differed primarily in the sequence of metal salt addition and reduction. Aqueous solutions of $PdCl_2$ (Johnson Matthey) and $HAuCl_4.3H_2O$ (Johnson Matthey) of the desired concentration were first prepared. Fresh aqueous solutions of poly vinyl alcohol (PVA) (1 wt% aqueous solution, Aldrich, MW = 10000, 80% hydrolyzed) and of 0.1 M $NaBH_4$ were also prepared.

(a) **Au + Pd sol.** The $PdCl_2$ and $HAuCl_4$ stock solutions were mixed in the desired ratio and the required amount of a PVA solution (1 wt%) was added (PVA/(Au + Pd) (wt/wt) = 1.2); the freshly prepared solution of $NaBH_4$ ($NaBH_4$/(Au + Pd) (mol/mol) = 5) was then added to the solution to form a dark-brown sol.

(b) Pd{Au} sol. Firstly, the required amount of a PVA solution (1 wt%) was added (PVA/(Au + Pd) (wt/wt) = 1.2) to an aqueous $HAuCl_4$ solution of the desired concentration; a freshly prepared solution of $NaBH_4$ (0.1 M, $NaBH_4$/Au (mol/mol) = 5) was then added to form a red sol. The solution was then stirred for 30 minutes to allow the complete reduction of all the Au^{3+} present in the solution. Then, the required amount of the stock aqueous $PdCl_2$ solution was added, followed by the desired amount of $NaBH_4$ ($NaBH_4$/Pd (mol/mol) = 5), to produce a dark brown sol. The solution was stirred for a further 30 minutes.

(c) Au{Pd} sol. To an aqueous $PdCl_2$ solution of the desired concentration the required amount of a PVA solution (1 wt%) was added (PVA/(Au + Pd) (wt/wt) = 1.2); a 0.1 M freshly prepared solution of $NaBH_4$ ($NaBH_4$/Pd (mol/mol) = 5) was then added to form a brown sol. The solution was stirred for 30 minutes to allow the complete reduction of all the Pd^{2+} present in the solution. Then, the desired amount of the stock $HAuCl_4$ aqueous solution was added, followed by the desired amount of $NaBH_4$ ($NaBH_4$/Au (mol/mol) = 5), until a dark brown sol was obtained. The solution was stirred for a further 30 minutes.

Colloid immobilization on C and TiO_2

Six distinct catalyst systems were prepared utilizing the three morphology combinations (*i.e.*, the Au + Pd alloy and Pd{Au} and Au{Pd}) onto two different supports *i.e.*, C and TiO_2). The precise support materials used were: TiO_2 (Degussa P25) and amorphous carbon (Darco G60, Aldrich). Immobilization of the sol is accomplished by adding activated carbon or TiO_2 (acidified at pH 1 by sulfuric acid) after 30 min of sol-generation under vigorous stirring conditions. The total meal loading was maintained at 1% wt. In all three catalyst systems, the Au:Pd weight ratio was maintained at 1:1, so that the materials could be directly compared with AuPd catalysts previously prepared by impregnation.[5–7] After 2 h the slurry was filtered, the catalyst washed thoroughly with distilled water (neutral mother liquors) and dried at 120 °C for 16 h. The sets of dried catalysts were also calcined in the 200–400 °C temperature range under static air for 3 h.

Catalyst characterization by HREM/STEM

Samples of sol-immobilized catalysts were prepared for TEM/STEM analysis by dry dispersing the catalyst powder onto a holey carbon TEM grid. In the case of the starting Au + Pd, Au{Pd} and Pd{Au}-PVA sols, a drop of the colloidal sol was deposited, and then allowed to evaporate, onto a 300-mesh copper TEM grid covered with an ultrathin continuous C film. High resolution electron microscopy (HREM) and high-angle annular dark field (HAADF) imaging experiments were carried out using a 200kV JEOL 2200FS transmission electron microscope equipped with a CEOS aberration corrector. All the STEM-HAADF images were treated with a light low pass filter using a 3×3 kernel to decrease the high frequency noise. The JEOL 2200FS TEM/STEM was equipped with a Thermo Scientific Inc Si(Li) detector for X-ray energy dispersive spectroscopy (XEDS) analysis. Point spectra were acquired with a total acquisition time of 120 s. Spectrum images were acquired using a pixel dwell time of 800 ms. Multi-variate statistical analysis (MSA) of the XEDS data cubes was carried out utilizing the MSA plug-in for Digital Micrograph.

X-ray photoelectron spectroscopy

XPS analysis was carried out on a Kratos Axis Ultra DLD spectrometer employing a monochromatic Al Kα X-ray source (75–150 W) and analyser pass energies of 160 eV (for survey scans) or 40 eV (for detailed scans). Samples were mounted using double-sided adhesive tape and binding energies referenced to the C(1s) binding energy of adventitious carbon contamination which was taken to be 284.7 eV.

This journal is © The Royal Society of Chemistry 2011

The ratio of Au:Pd has been calculated by measuring the intensity ratio of the Pd $(3d_{3/2})$ to the $Au(4d_{5/2})$ peaks and making use of elemental sensitivity factors.

Catalyst testing

Benzyl alcohol oxidation was carried out in a stirred reactor (100 mL, Parr Instruments). The vessel was charged with benzyl alcohol (40 mL) and catalyst (0.05 g). The autoclave was then purged 5 times with oxygen leaving the vessel at 10 bar gauge. The stirrer was set at 1500 rpm and the reaction mixture was raised to the required temperature; the reaction time was started as soon as the required reaction temperature was reached. Samples from the reactor were taken periodically, *via* a sampling system. For the analysis of the products a GC-MS and GC (a Varian star 3400 cx with a 30m CP-Wax 52 CB column) were employed, and the products were identified by comparison with known standards. For the quantification of the amounts of reactants consumed and products generated, an external calibration method was used.

Hydrogen peroxide synthesis was performed using a stainless steel autoclave (Parr Instruments) with a nominal volume of 50 ml and a maximum working pressure of 14 MPa. The autoclave was equipped with an overhead stirrer (0–2000 rpm) and provision for measurement of temperature and pressure. Typically, the autoclave was charged with the catalyst (0.01 g unless otherwise stated), solvent (5.6 g MeOH and 2.9 g H_2O), purged three times with $5\%H_2/CO_2$ (3 MPa) and then filled with 5% H_2/CO_2 and 25% O_2/CO_2 to give a hydrogen to oxygen ratio of 1:2 at a total pressure of 3.7 MPa. Stirring (1200 rpm unless otherwise stated) was commenced on reaching the desired temperature (2 °C), and all experiments were carried out for 30 min unless otherwise stated. The H_2O_2 yield was determined by titration of aliquots of the final filtered solution with acidified $Ce(SO_4)_2$ (7×10^{-3} mol l^{-1}). The $Ce(SO_4)_2$ solutions were standardized against $(NH_4)_2Fe(SO_4)_2.6H_2O$ using ferroin as indicator. Hydrogen peroxide hydrogenation was carried out in an identical manner, but in this case the 25% O_2/CO_2 was not added so that hydrogen peroxide could not be synthesized; instead part of the water was replaced by H_2O_2 (50 vol%) to give a hydrogen peroxide concentration of 4 wt%.

Results

Structural characterization

Starting colloid materials. STEM-HAADF imaging was employed to evaluate the particle size distribution of the starting colloids deposited onto continuous carbon films. The mean particle sizes of the Au + Pd, Au{Pd} and Pd{Au} starting sols were all found to be rather similar at 2.9 nm, 3.0 nm and 2.6 nm respectively (Table 1). An advantage of the aberration corrected STEM-HAADF imaging technique is that it can also be used to monitor the spatial distribution of Au and Pd within the individual particles. Those regions within the nanoparticle with a higher local concentration of Au appear brighter than the Pd-rich areas by virtue of the higher atomic number of Au. As an example, STEM-HAADF images of a pure Au nanoparticle and a similarly sized Au + Pd (1:1wt%) random alloy nanoparticle are shown in Figure 1(a) and (b) respectively. The intensity of the pure Au nanoparticle (Fig. 1(a)) appears lowest at its edge and reaches a maximum at its centre, due to the gradual systematic change in thickness from the periphery to the centre of the particle. The Au + Pd nanoparticle (Fig. 1(b)), shows a similar trend of intensity variation across its diameter, but superimposed on top of this is an additional characteristic speckle type contrast arising from the random distribution of heavier Au atoms and lighter Pd atoms within the particle. The variation of intensity is observed much more clearly when these same images have been colour coded with a temperature gradient as shown in Figs. 1(c) and (d). The "speckled" intensity variations within the alloy nanoparticle are not a result of thickness variation alone and can

Table 1 Summary of Au–Pd particle size data for the three starting sols and six sol-immobilized catalysts dried at 120 °C for 8 h, and selected samples calcined at 400 °C for 3 h

Particle/Catalyst Description	Median Size/nm	Mean Size/nm
Au + Pd colloid	3.2	2.9
Au + Pd/C (120 °C dried)	3.3	3.7
Au + Pd/TiO$_2$ (120 °C dried)	3.5	3.9
Au + Pd/C (400 °C calcined)	—	36
Au + Pd/TiO$_2$ (400 °C calcined)	7.1	7.5
Au@Pd colloid	2.7	3.0
Au@Pd/C (120 °C dried)	3.2	3.1
Au@Pd/TiO$_2$ (120 °C dried)	4.6	4.6
Pd@Au colloid	2.6	2.6
Pd@Au/C (120 °C dried)	3.9	4.3
Pd@Au/TiO$_2$ (120 °C dried)	4.4	4.6

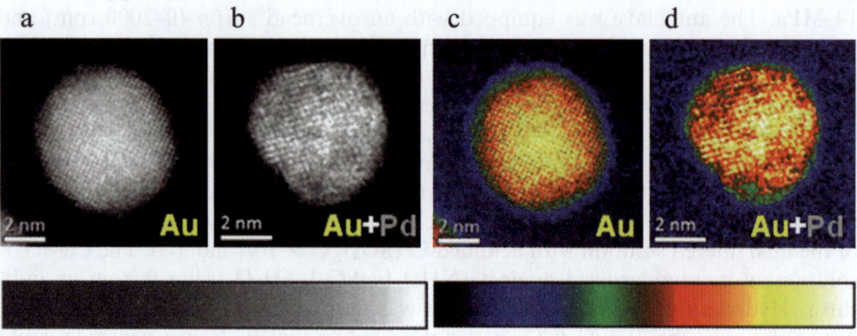

Fig. 1 A comparison of STEM-HAADF images of: **(a)** pure Au and **(b)** Au + Pd alloy nanoparticles; **(c and d)** thermal colour palette applied to images (a) and (b) in order to compare the pixel intensity variations within the nanoparticle. As compared to pure Au, the Au + Pd alloys particles shows an additional 'speckle' Z-contrast superimposed on the contrast from thickness variations.

be therefore be taken as an indication of the random mixing of Au and Pd atoms in the structure.

Representative HAADF images of Au + Pd, Au{Pd} and Pd{Au} colloidal sols drop cast onto a continuous carbon TEM grid are shown in Figure 2. The as-prepared Au + Pd sol (Figure 2(a) and (b)) consisted of a mixture of icosahedral (I$_h$), decahedral (D$_h$) and cub-octahedral (co) particles, with sizes in the 1–5nm range. All of the particles greater than ~1.5nm in size showed speckle type intensity variations indicative of random alloying between Au and Pd. For example, a 2nm cub-octahedral Au + Pd mixed alloy particle is highlighted in Fig. 2(a) (*insert*), where the brighter columns are Au-rich. Several nanometer scale clusters of atoms are also present in the colloidal sols (indicated by circles). These are possibly homogeneously nucleated structures that form at a later stage during the reduction of metal salts. It was not possible to obtain a detailed structural or chemical analysis of the clusters as they were very unstable under electron beam irradiation.

STEM-HAADF images of the Au{Pd} colloids supported on a continuous carbon film are shown in Figs. 2(c), (d) and (e). Particles with several distinctive morphologies were noted, including icosahedral (I$_h$), decahedral (D$_h$), cuboctahedral (co), dodecahedral (DD$_h$) and singly twinned particles (STP). There is a characteristic Z-contrast effect noticeable in Fig. 2(c), where the center of the particle has a considerably lower intensity than the periphery, suggesting either a Pd$_{core}$–Au$_{shell}$

     This journal is © The Royal Society of Chemistry 2011

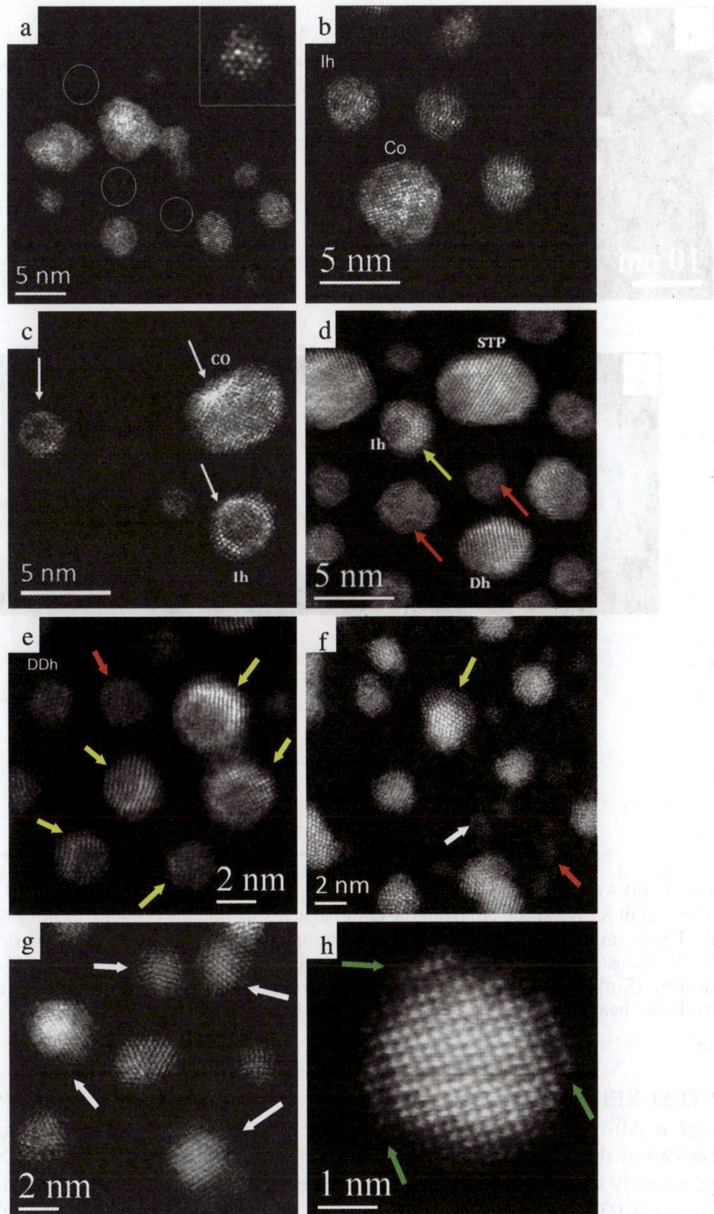

Fig. 2 STEM HAADF images of the starting colloids dispersed onto a continuous carbon thin film: (a, b) Au + Pd colloids; (c, d, e) Au{Pd} colloids; and (f, g, h) Pd{Au} colloids.

nanostructure or a central void in the particle. In order to verify our Pd_{core}–Au_{shell} interpretation of the contrast, a detailed STEM-XEDS spectrum image analysis of such a particle is presented in Figure 3. An entire XEDS spectrum was acquired at each pixel point of ADF image shown in Fig. 3(a). The summed XEDS spectrum from the entire particle is shown in Fig. 3(b) and confirms that both Au and Pd are present. The spatial distributions of Au and Pd as extracted from the spectrum image data-cube[8] are shown in Figs. 3(c) and (d) respectively, and a composite RGB image colour overlay map is presented in Fig. 3(e). It is evident from the

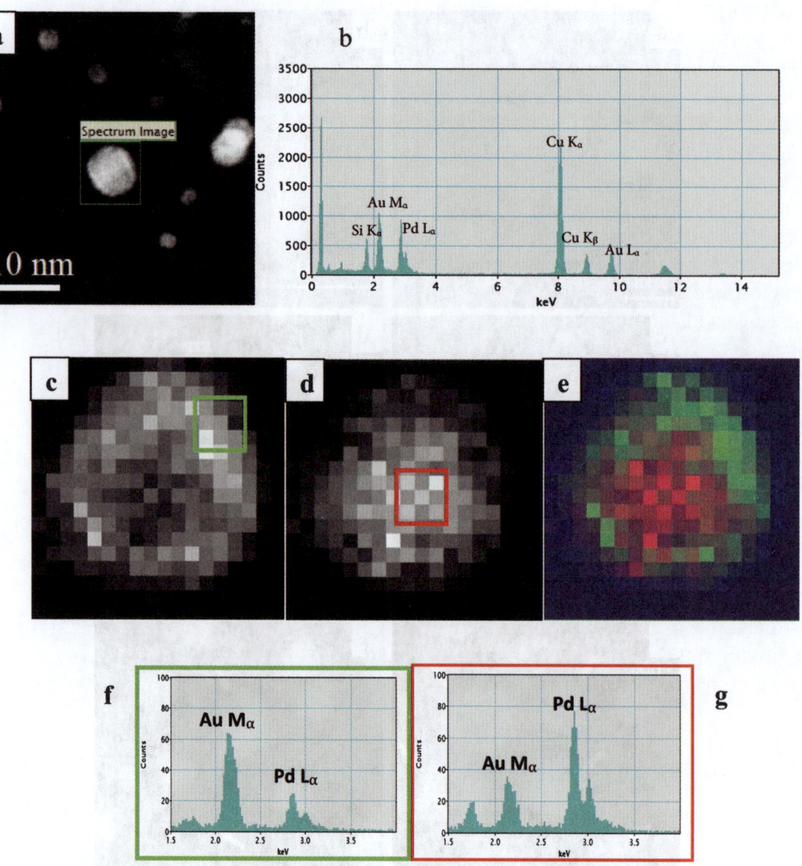

Fig. 3 **(a)** An ADF image of an Au{Pd} nanoparticle selected for STEM-XEDS mapping; **(b)** The overall XEDS spectrum from the nanoparticle showing the presence of both Au and Pd; **(c, d).** The Au and Pd elemental maps of the nanoparticle obtained after MSA processing: **(e)** An RGB reconstruction showing the distribution of Au (green) and Pd (red)-the blue channel is neutral; **(f, g)** the cumulative XEDS spectra from 3 × 3 pixels (indicated by squares in (c & d)) from the center and periphery of the particle respectively.

STEM-XEDS map that the particle has a core-shell structure with a Pd-rich core and a Au-rich shell. Summed spectra extracted from the centre and periphery regions of the nanoparticle are shown in Figs. 3(g) and (h). The STEM-XEDS analysis clearly shows the centre of the particle to be Au-rich and the shell to be Pd-rich.

The Au{Pd} sol particles in Figs. 2(c), (d) and (e) show a mixture of structures; namely, (i) Pd cores with complete Au coverage (indicated by white arrows), (ii) particles with off-centre Pd cores and incomplete Au coverage (yellow arrows), and (iii) Pd particles with no Au coverage (red arrows). Almost all of the particles greater than 5 nm in diameter were seen to have a complete Au coverage. From analysis of numerous such micrographs it was found that the Pd core size varied between 2 and 5nm. The sizes of the Au-shell thickness can also vary between 0.5–1.0 nm, as seen in Figure 2 (e). Shell thicknesses smaller than 0.5 nm are difficult to image in the continuous carbon supported sols. The variation and non-uniformity in Au shell coverage of the Pd nanoparticle is probably due to insufficient shell material (in this case Au) being introduced during the particle preparation as will be discussed later.

STEM-HAADF images of the Pd{Au} sol are shown in Figs. 2(f), (g) and (h). Once again, a mixture of icosahedral, decahedral, dodecahedral, cub-octahedral and singly twinned structures were found. Several nanometre scale metallic clusters were also observed (Figure 2(f)). As expected, many particles in these HAADF images show the opposite sense of Z-contrast to that noted for the Au{Pd} particles describe above; namely, bright (Au-rich) cores and fainter (Pd-rich) shells (as indicated by white arrows in Figs. 2(f) and (g)). In this sample, the Au_{core}–Pd_{shell} configurations dominate, with the Au cores varying between 2 and 4 nm in diameter. The Pd shells were also seen to vary from 2–5 atomic layers in thickness. Fig. 2(h) shows the STEM-HAADF image of a core-shell nanoparticle with a bi-layer coverage of Pd. It is interesting to note that there are some abnormally bright atoms observed in the Pd shell (green arrows), indicating incorporation of Au atoms into the Pd shell region. Core-shell structures are clearly seen in particles as small as 4 nm in total diameter in this Pd{Au} sample. However, there were still some particles, arrowed red, that were devoid of a core-shell structure.

Sol-immobilized catalysts dried at 120 °C. Particle size distribution data derived from STEM-HAADF imaging studies of the three starting sols and six sol-immobilized catalysts dried at 120 °C are summarized in Table 1. The mean particle sizes of the starting sols are 2.9 nm, 3.0 nm and 2.6 nm for Au + Pd, Au{Pd} and Pd{Au} sols respectively. On the TiO_2 supports they show a modest size increase to 3.7 nm, 3.2 nm and 4.3 nm respectively after the samples were dried for 16 hours at 120 °C. On carbon supports the corresponding mean particle sizes after drying were 3.9 nm, 4.4 nm and 4.6 nm. Hence we conclude that, the drying procedure does not result in significant particle coarsening on either support. XPS data collected on these samples has been described thoroughly in our previous publications.[9,38] Briefly, XPS analysis confirmed that the Pd:Au molar ratio is close to the expected nominal 2:1 value for all samples (Table 2). The sol-immobilized bimetallic catalysts dried at 120 °C all contain Au in the metallic state and Pd predominantly in the metallic state.

Representative images of the 120 °C dried Au + Pd particles immobilized on both C and TiO_2 supports are shown in Figure 4. On the activated-C supports, the Au + Pd particles have quite rounded morphologies and seem to be relatively non-wetting (Figs 4(a, b)). In comparison, on the TiO_2 supports (Figs. 4(c), (d)) the Au + Pd particles have a greater tendency to be faceted and form an extended flat interface with the underlying crystalline TiO_2 surface. A similar wetting behavior trend was also observed for the Au{Pd} and Pd{Au} nanoparticle variants immobilized on C and TiO_2. This intimate wetting behavior of Au–Pd on TiO_2 implies that the PVA ligands have been locally displaced from the particle/support interface region, which may in turn lead to enhanced adhesion of the particle to the support. It also increases the number of particle-support 'periphery' sites which are thought to important for certain reactions.[39,40]

Representative STEM-HAADF images of individual Au + Pd, Au{Pd} and Pd {Au} nanoparticles immobilized on both carbon and TiO_2 supports and dried at 120 °C are shown in Figures 5, 6 and 7 respectively. The Au + Pd particles (Figure 5) exhibit characteristic speckle contrast indicating that they remain as random alloys

Table 2 XPS derived elemental atomic ratios for Pd{Au} catalysts supported on C and TiO_2. Samples were prepared by sol-immobilisation, dried at 120 °C, and calcined at 400 °C

Sample	Catalyst pre-treatment	XPS derived corrected Pd: Au Molar ratio
Pd{Au}/C	Dried 120 °C/Air	2.2
Pd{Au}/C	Calcined 400 °C/Air	5.7
Pd{Au}/TiO_2	Dried 120 °C/Air	1.6
Pd{Au}/TiO_2	Calcined 400 °C/Air	1.5

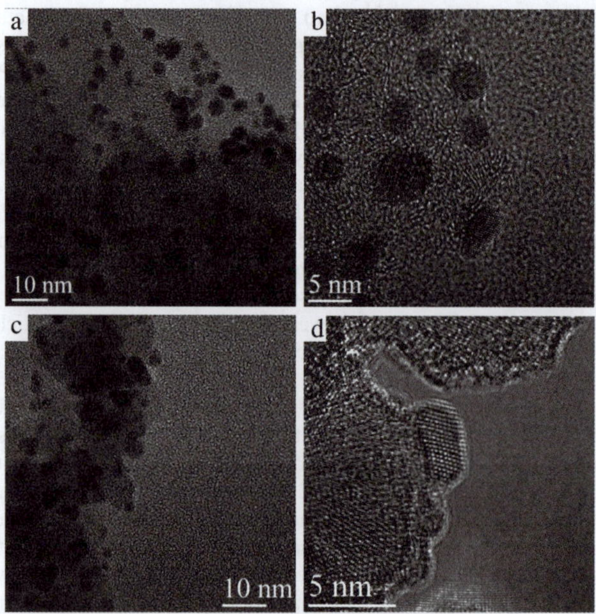

Fig. 4 Representative HREM micrographs of the sol-immobilized Au + Pd nanoparticles on **(a, b)** activated C and **(c, d)** TiO$_2$ supports, after drying at 120 °C for 16 h.

on both supports (Figure 5), however the smallest nm-scale clusters noted in the starting colloid (circled in Fig. 2(a)) are now absent having been consumed into the larger particles by either coarsening or Ostwald ripening during the drying

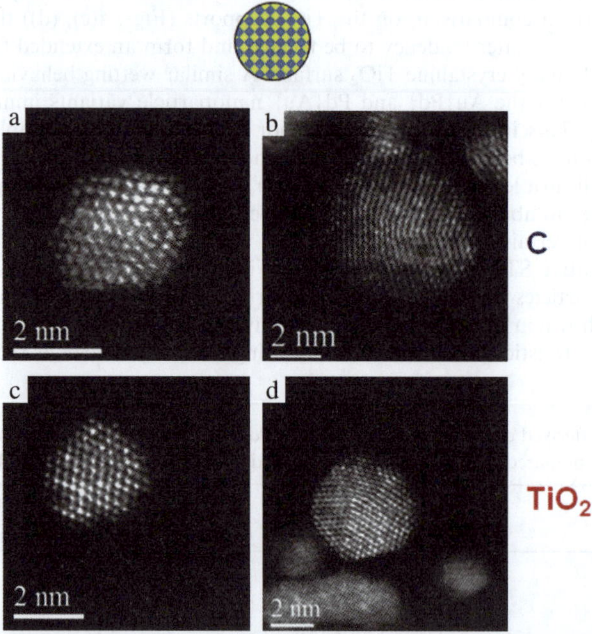

Fig. 5 Representative STEM-HAADF images of individual sol-immobilized Au + Pd nanoparticles on **(a)** and **(b)** activated C and **(c)** and **(d)** TiO$_2$ supports after drying at 120 °C for 16 h.

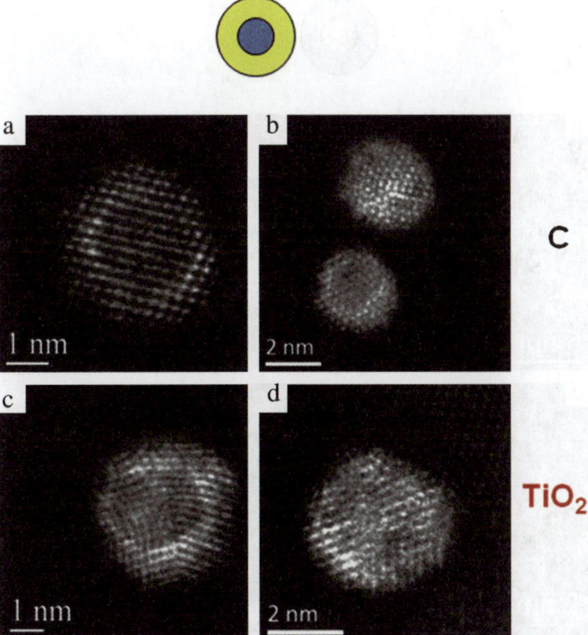

Fig. 6 Representative STEM-HAADF images of individual sol-immobilized Au{Pd} nanoparticles on (a) and (b) activated C and (c) and (d) TiO_2 supports after drying at 120 °C for 16 h.

process. Many Au{Pd} particles above $\sim$3nm in size immobilized on activated C and TiO_2 supports still showed clear evidence of Pd_{core}–Au_{shell} configurations (Figure 6), indicating their robustness to the relatively low temperature drying process. It is interesting to note that in some such particles (Figs. 6(b) and (c)) the Pd cores also show a distinct facet structure. The Au outer layer on Au{Pd} particles were always $\sim$1nm or less in thickness. Au shells were typically not visible on particles below 3nm in diameter, and once again the initial population of nanometer scale clusters was missing, presumably as they had been absorbed into the larger particles during drying. The third nanoparticle variant, Pd{Au}, were also seen to largely retain their Au_{core}–Pd_{shell} configurations when immobilized on activated C and TiO_2 supports (Figure 7). All of the larger diameter particles (>5 nm) showed core-shell configurations and compared to their Au{Pd} counter parts, their shells were thicker ($\geq$ 2nm). Once again the internal Au/Pd interfaces in the Au_{core}–Pd_{shell} particles were sometimes observed to be highly facetted (*e.g.* Fig. 7(d)). In these Pd{Au} samples, most particles smaller than $\sim$3 nm appeared to be monometallic, suggesting that they are possibly "seed" Au particles that have not, for reasons to be discussed later, received any significant Pd shell coverage during preparation.

Sol-immobilized catalysts calcined at 200 °C. The effect of a 3h calcination treatment at 200 °C on the Pd{Au} catalyst immobilized on C and TiO_2 supports was also analysed by STEM-HAADF imaging (see Figure 8) and an interesting support dependant phenomenon was observed. The Au_{core}–Pd_{shell} nanoparticle structure was generally well preserved on the activated C support (Figs 8(a), (b) and (c)). However, the onset of sintering of neighbouring Pd{Au} particles was also commonly observed, as shown in Figs. 8(b) and (c), suggesting that the particle/ support interaction was not very strong. By way of contrast, the Pd{Au} particles on the TiO_2 support, showed no strong evidence of sintering, (Figs. 8(d), (e) and (f)), but their Au_{core}–Pd_{shell} structure had been compromised instead. Presumably

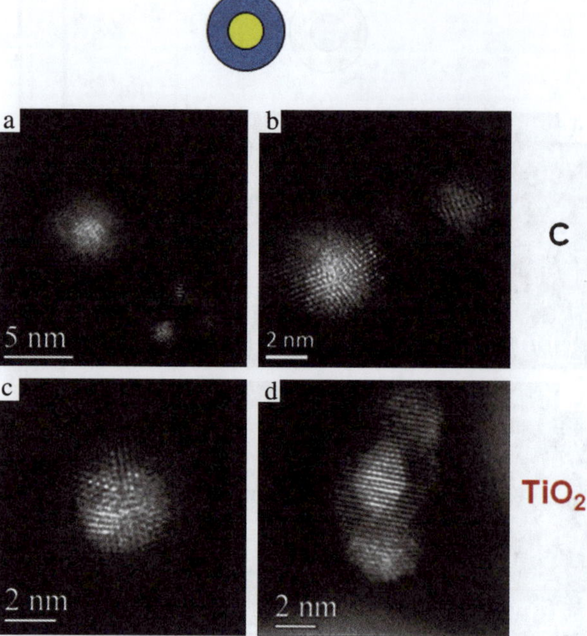

Fig. 7 Representative STEM-HAADF images of individual sol-immobilized Pd{Au} nano-particles on (a) and (b) activated C and (c) and (d) TiO$_2$ supports after drying at 120 °C for 16 h.

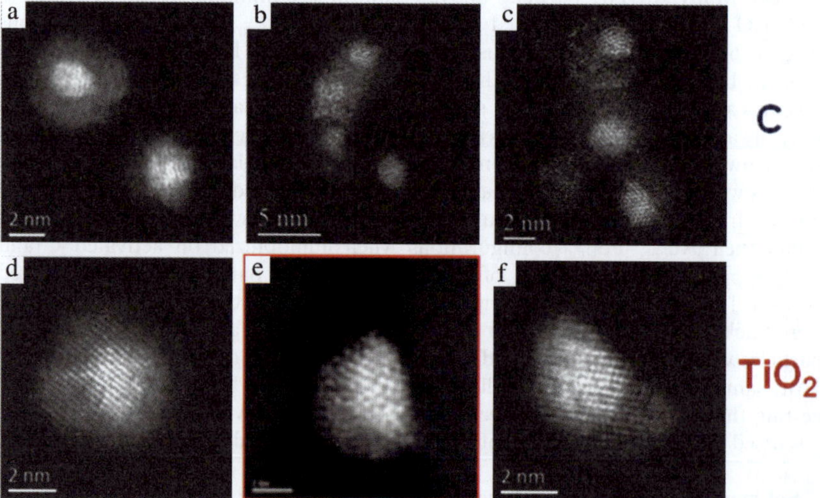

Fig. 8 Representative STEM-HAADF images of individual sol-immobilized Pd{Au} nano-particles on (a, b, c) activated C and (d, e, f) TiO$_2$ supports after calcination at 200 °C for 3 h.

their better resistance to sintering is associated with the fact that the metal particles had formed extended flat interfaces with the TiO$_2$ support, which improved their overall adhesion and thermal stability. It is also fascinating to note that in those particles where a core-shell was still visible, it was not uncommon to find that the Au component had diffused towards, and attached itself preferentially to the TiO$_2$

    This journal is © The Royal Society of Chemistry 2011

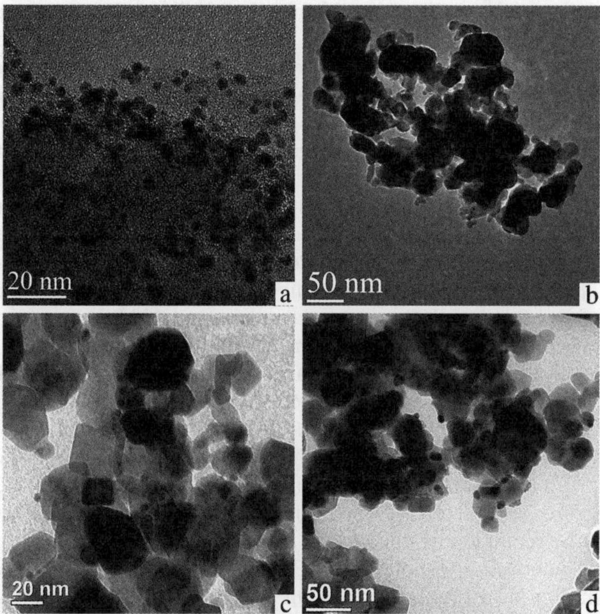

Fig. 9 Representative bright field TEM micrographs of sol-immobilized Au + Pd nano-particles on activated C **(a)** before and **(b)** after calcination at 400 °C for 3 h. Corresponding images of sol-immobilized Au + Pd nanoparticles on TiO_2 **(c)** before and **(d)** after calcination at 400 °C for 3 h.

support, while the Pd component was displaced outwards forming a hemispherical cap (Figs. 8(e) and (f)).

Sol-immobilized catalysts calcined at 400 °C. Calcination at 400 °C for 3h was found to be extremely detrimental to the all the C supported catalysts irrespective of the Au–Pd morphology variant. As an example, Figs. 9(a) and (b) show bright field TEM images of the Au + Pd/C catalyst before and after calcination at 400 °C. The mean size of the Au + Pd particles increases to ~35 nm (up from ~4 nm) after calcination, accompanied by a large loss of activated C-support mass. The gross Au + Pd particle sintering is driven by both the removal of the protective PVA ligands above 300 °C and the loss of activated C surface area due to support burn-off. Analysis of XEDS point spectra showed no significant change in the bulk Au:Pd ratio within these metal particles before and after the 400 °C calcination treatment. Complementary XPS analysis however, showed that the Pd:Au molar ratio increased from 2:1 to 5.7 (Table 2) which suggests a Pd enrichment is occurring on the surface of these highly sintered bimetallic nanoparticles on C at 400 °C.

The Au + Pd, Au{Pd} and Pd{Au} catalysts immobilized on TiO_2 responded very differently to the 400 °C calcination treatment as compared to their C-supported counterparts. Bright field images of the Au + Pd/TiO_2 sample before and after calcination at 400 °C are shown in Figs. 9(c) and (d) respectively. The mean Au + Pd particle size is seen to undergo a much more modest increase from 3.6 nm to 7.2 nm after heat treatment. The strong wetting interaction on TiO_2 clearly imparts good adhesion of the metal particles to the support, even in the absence of the PVA ligands. The Au–Pd catalysts immobilized on TiO_2 were also analysed by HAADF imaging to monitor any changes to the metal nanoparticle morphology upon high temperature calcination. Representative STEM-HAADF images from the Au + Pd/TiO_2, Pd{Au}/TiO_2, and Au{Pd}/TiO_2 catalysts are shown in Figs. 10(a), (b) and (c) respectively. For the 400 °C calcined Au + Pd/TiO_2 samples (Fig. 10(a)), all the metal particles remained as random AuPd alloys. However, no core-shell structures were retained at all in the 400 °C dried Au{Pd}/TiO$_2$ and Pd{Au}/TiO$_2$ materials (Figs. 10(b) and (c)). Instead, three new distinct new particle morphologies had emerged: namely; (i) bright particles with extended 1–10 nm faint tails (arrowed blue), (ii) isolated faint particles (arrowed red) and (iii) isolated bright particles (arrowed yellow). Localized XEDS measurements from these three distinct types of particles are presented in Figure 11. For those particles exhibiting extended tails (Fig. 11(a)), the faint tail feature was found to be very Pd-rich, whereas the brighter end of the structure was Au-rich. The tailed particles must therefore be a snapshot of ongoing phase separation of the Au and Pd components from the Pd$_{core}$–Au$_{shell}$ and Au$_{core}$–Pd$_{shell}$ nanoparticles upon high temperature calcination. XPS analysis of the Au{Pd}/TiO$_2$ and Pd{Au}/TiO$_2$ samples at calcined at 400 °C shows a corresponding transformation of Pd^0 to Pd^{2+}, which may be also be correlated to the appearance of numerous phase separating 'tail-like' particles noted in Figs. 10 and 11 for the Au {Pd} and Pd{Au} variants. This unexpected phase separation may possibly be driven by the desire of the Pd component to form PdO at higher calcination temperatures. Point XEDS measurements from the isolated minority faint particles (Fig. 11 (b)) showed them to be composed of Pd-only, whereas the brighter particles without tail-like structures (Fig 11(c)) were identified to be Au–rich alloys. These minority particles may be remnants of uncoated Au and Pd particles present in the starting sols. It is also interesting to note that XPS measurements (Table 2) show that the 400 °C calcination measurement treatment did not seem to significantly affect the overall Pd:Au molar surface ratio for these catalysts immobilized on TiO$_2$.

Catalyst testing

Benzyl alcohol oxidation. The aerobic oxidation of benzyl alcohol has been investigated using these catalyst samples, as it is often used as a model reaction for alcohol oxidation.[9,38] This reaction can yield a range of products, such as

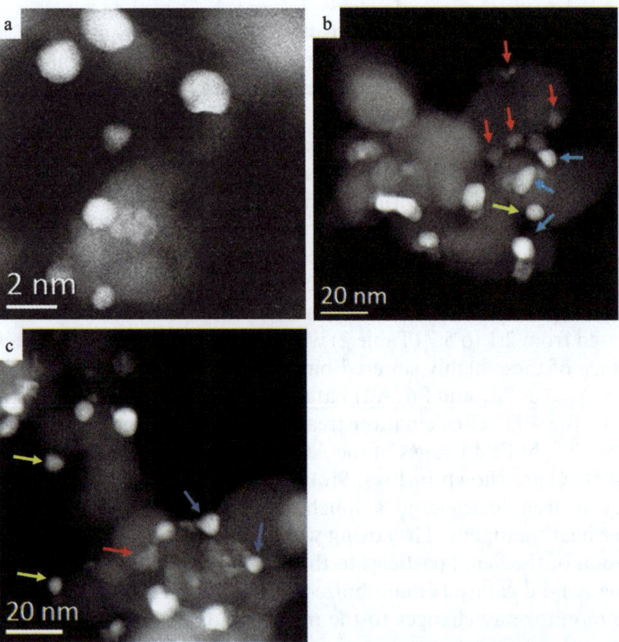

Fig. 10 Representative STEM-HAADF images of the **(a)** Au + Pd/TiO$_2$, (b) Pd{Au}/TiO$_2$ and (c) Au{Pd}/TiO$_2$ catalysts after calcination at 400 °C for 3 h.

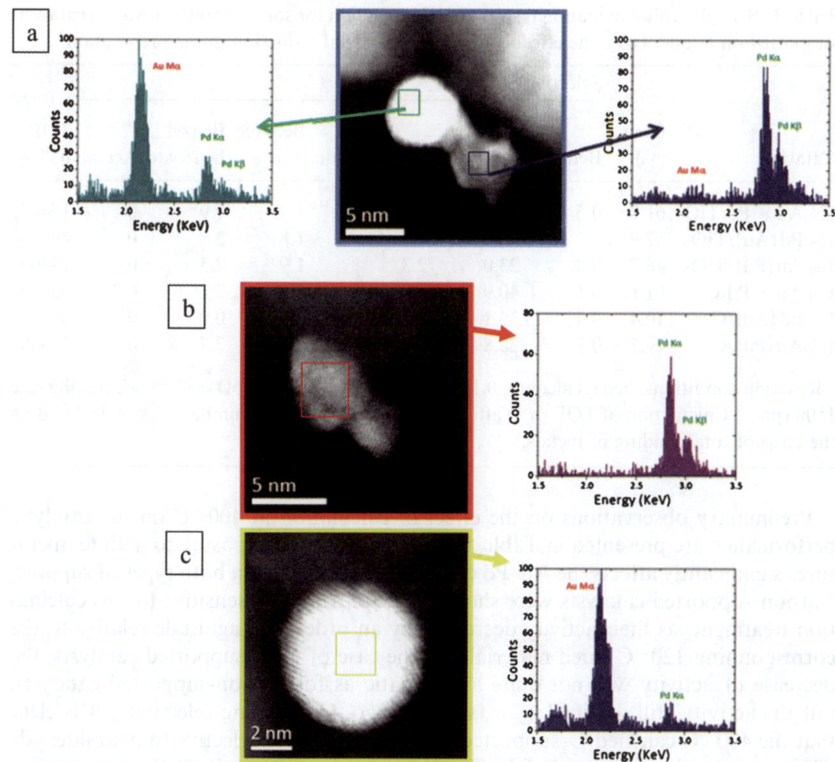

Fig. 11 STEM-XEDS point spectra of the three particle types highlighted in Fig. 10(b); **(a)** 'tailed' particles (indicated by blue arrows); **(b)** 'fainter' particles (indicated by red arrows); and **(c)** 'brighter' particles (indicated by yellow arrows).

benzeldehyde, benzoic acid, toluene and benzyl benzoate, so the design of a metal supported catalyst that can facilitate the formation of a specific product with high selectivity is highly desirable. The aerobic liquid phase oxidation of benzyl alcohol was investigated using our systematic series of Au–Pd/C and Au–Pd/TiO$_2$ catalysts and we observed that the order of metal addition and reduction during initial sol-formation affects not only the activity, but also the selectivity. In the case of Au–Pd nanoparticles immobilized on TiO$_2$, dried at 120 °C, the highest activity (Table 3) (calculated at initial rate, 0.5 h) was the highest for Pd{Au}/TiO$_2$, with the following ranking order of activity:- Pd{Au}/TiO$_2$ > Au{Pd}/TiO$_2$ > Pd + Au/TiO$_2$. A similar comparison at an iso-conversion level of 50% (Table 4) indicates that all the 120 °C dried TiO$_2$ supported catalysts have similar selectivities to benzaldehyde and toluene production. In the case of the corresponding 120 °C dried carbon-supported Au–Pd catalysts, the nanoparticle morphology had a significant effect on the initial rate, with Pd{Au}/C showing the highest activity among the carbon-supported catalysts, although they all showed similar activity after a 2 h reaction time (Table 3). The order of activity displayed was: Pd {Au}/C > Au + Pd/C > Au{Pd}/C. At iso-conversion, the selectivities to benzaldehyde were all similar; however, a higher selectivity to benzaldehyde was found for the Au{Pd}/C variant of the catalyst (Table 4). In a more general comparison between the two types of support for the sol-immobilization process, the activated C supported materials dried at 120 °C were consistently more active than the corresponding TiO$_2$-materials, but the latter were more selective to benzaldehyde production (Tables 3 and 4).

Table 3 Benzyl alcohol oxidation results after 2 h reaction for sol immobilized Au–Pd catalysts supported on C and TiO$_2$: the effect of the order of metal reduction in the preparation[a]

Catalyst	Conv. (%)	Selectivity (%)						TOF/ h^{-1}[b]
		Benzene	Toluene	Benzaldehyde	Benzoic acid	Benzyl benzoate	Acetal	
1% Au + Pd/TiO$_2$	61.2	0.5	26.7	69.2	1.7	1.9	0	15400
1% Pd{Au}/TiO$_2$	57.8	0.2	18.1	77.1	1.8	2.8	0	19300
1% Au{Pd}/TiO$_2$	48.7	0.4	23.0	72.3	1.9	2.5	0	17400
1% Au + Pd/C	81.1	0.5	40.9	55.0	1.3	2.1	0.2	35400
1% Pd{Au}/C	70.4	0.4	35.1	62.6	2.0	0.0	0	41900
1% Au{Pd}/C	79.2	0.7	28.8	65.0	2.7	2.7	0	24300

[a] Reaction conditions: benzyl alcohol, 0.1g of catalyst, T = 120 °C, pO$_2$ = 150 psi, stirring rate 1500 rpm. [b] Calculation of TOF (h^{-1}) after 0.5 h of reaction. TOF numbers were calculated on the basis of total loading of metals.

Preliminary observations on the effect of calcination at 400 °C on the catalytic performance are presented in Table 5. It is clear that the exposure to high temperatures significantly affects the Au–Pd catalytic performance on both types of support. Carbon-supported catalysts were shown to be particularly sensitive to this calcination treatment, as their activity decreased by an order of magnitude relative to the corresponding 120 °C dried materials. In the case of TiO$_2$-supported catalysts, the decrease of activity was not quite as dramatic as for carbon-supported catalysts, but the activity still declined by a factor of two. Concerning selectivity, it is clear that the 400 °C calcined C–supported materials are more selective to benzaldehyde when compared at iso-conversion (Table 6). Taking into account the larger mean particle size of the calcined materials, it is tempting to ascribe the improved selectivity to benzaldehyde to the very substantial particle size increase noted above for the C-supported materials.

Direct production of H$_2$O$_2$. The catalytic performance of the sol immobilized Au–Pd catalysts dried at 120 °C was also investigated for the direct synthesis of H$_2$O$_2$ as shown in Table 7. Very high initial productivities were observed with these sol-immobilized materials, especially for the C-supported catalysts, where the productivity values were in the 1300–1640 mol$_{\text{H}_2\text{O}_2}$/kg$_{\text{cat}}$/h range, which are amongst the highest values reported to date. Carbon-supported catalysts were more active by a factor of 5 when compared with their corresponding TiO$_2$-supported counterparts. For the

Table 4 Benzyl alcohol oxidation results at iso-conversion for sol-immobilized Au–Pd catalysts supported on C and TiO$_2$: the effect of the order of metal reduction in the preparation [a]

Catalyst	Conv. (%)	Selectivity (%)					
		Benzene	Toluene	Benzaldehyde	Benzoic acid	Benzyl benzoate	Acetal
1% Au + Pd/TiO$_2$	50	0.5	26.9	69.1	1.5	2.0	0
1% Pd{Au}/TiO$_2$	50	0.2	23.0	72.8	1.5	2.5	0
1% Au{Pd}/TiO$_2$	50	0.4	23.0	72.2	1.9	2.5	0
1% Au + Pd/C	50	0.3	44.0	51.3	1.7	2.7	0.1
1% Pd{Au}/C	50	0.3	41.6	54.6	1.6	1.8	0.1
1% Au{Pd}/C	50	0.6	38.9	57.1	1.6	1.7	0.1

[a] Reaction conditions: benzyl alcohol, 0.1g of catalyst, T = 120 °C, pO$_2$ = 150 psi.

Table 5 Benzyl alcohol oxidation results after 2 h reaction for sol immobilized Au–Pd catalysts supported on C and TiO$_2$ (calcined at 400 °C in air): the effect of the order of metal reduction in the preparation [a]

Catalyst	Conv. (%)	Selectivity (%)						TOF/ h^{-1} [b]
		Benzene	Toluene	Benzaldehyde	Benzoic acid	Benzyl benzoate	Acetal	
1% Au + Pd/TiO$_2$	22.7	0.8	22.2	69.7	1.9	5.5	0	3940
1% Pd{Au}/TiO$_2$	32.8	0.6	21.3	71.6	1.9	4.6	0	8650
1% Au{Pd}/TiO$_2$	33.5	0.4	25.9	68.7	1.3	3.7	0	8780
1% Au + Pd/C	6.7	5.5	2.4	78.7	3.6	9.8	0	2490
1% Pd{Au}/C	6.0	5.2	2.9	75.1	4.3	12.6	0	2430
1% Au{Pd}/C	7.2	1.7	4.7	79.1	3.6	10.8	0	3410

[a] Reaction conditions: benzyl alcohol, 0.1g of catalyst, T = 120 °C, pO$_2$ = 150 psi, stirring rate 1500 rpm. [b] Calculation of TOF (h^{-1}) after 0.5 h of reaction. TOF numbers were calculated on the basis of total loading of metals.

carbon supported materials the order of ranking for activity found was:- Au + Pd/C > Pd{Au}/C ≈ Au{Pd}/C, while in the case of the TiO$_2$ supported catalysts, the following order was observed:- Pd + Au/TiO$_2$ > Au{Pd}/TiO$_2$ > Pd{Au}/TiO$_2$. These results show that in terms of the sol-preparation method, the most active nanoparticle variant for direct H$_2$O$_2$ production were the (Au + Pd) materials created by simultaneous addition and reduction of both metal precursors, in which the Au and Pd are randomly alloyed. In contrast, both the Au{Pd}and Pd{Au} core-shell variants resulted in a lower catalytic activity. A time-on-line analysis (Table 7) indicates that the rate of hydrogen peroxide production declines with increasing reaction time, probably due to the consecutive hydrogenation reaction of hydrogen peroxide, which leads to water formation, and results in a drop in the hydrogen peroxide production rate.[1,9] Therefore, we have also examined the H$_2$O$_2$ hydrogenation activity of these catalyst materials (see Table 8). It is evident that very high hydrogenation activities were obtained across the entire range of the TiO$_2$ and C-supported catalysts. The initial hydrogenation activities (after 0.5 min of reaction) are in the 13000–19700 mol$_{H_2O_2}$/kg$_{cat}$/h range, suggesting that the H$_2$O$_2$ generated will be rapidly hydrogenated to H$_2$O. Comparing the hydrogenation activities at 30 min, the most active nanoparticle type for hydrogenation was Pd{Au}/C for C-supported materials, while for the TiO$_2$ supported materials the Pd{Au}/TiO$_2$ and Au + Pd/TiO$_2$ variants presented rather similar hydrogenation activities. The effect of calcination on the Pd{Au} materials was also studied to assess how it affected the synthesis of H$_2$O$_2$ synthesis (Table 9). A substantial decrease in both H$_2$O$_2$ productivity and hydrogenation activity was observed for the calcined materials on both supports. This implies that the metal particle growth, which is substantially greater for the Pd{Au}/C catalyst in comparison to the Pd{Au}/TiO$_2$ catalyst, does not facilitate H$_2$O$_2$ formation or enhance hydrogenation activity. It is clear that the calcination temperature needs to be optimised in order to maintain the H$_2$O$_2$ productivity while simultaneously stifling the undesired H$_2$O$_2$ hydrogenation reaction. We will present the results of a detailed catalyst optimization study in a follow-on publication.

Discussion

Structural comparison of 'as-prepared' and immobilized Au–Pd nanoparticles

The sol-immobilization technique clearly affords a much tighter control over the Au–Pd particle size distribution that can be achieved by conventional impregnation

Table 6 Benzyl alcohol oxidation results at iso-conversion for sol immobilized Au–Pd catalysts supported on C and TiO_2 (calcined at 400 °C in air): the effect of the order of metal reduction in the preparation[a]

Catalyst	Catalyst pre-treatment	Conv. (%)	Selectivity (%)					
			Benzene	Toluene	Benzaldehyde	Benzoic acid	Benzyl benzoate	Acetal
1% Au + Pd/C	dried	5	0.2	44.2	54.9	0.6	0.1	0
1% Pd{Au}/C	dried	5	0.3	40.1	56.8	1.0	1.8	0
1% Au{Pd}/C	dried	5	0.6	35.2	59.5	1.4	3.3	0
1% Au + Pd/C	Calcined 400 °C/Air	5	6.1	3.6	76.1	3.5	10.6	0
1% Pd{Au}/C	Calcined 400 °C/Air	5	6.8	3.4	73.7	4.1	12.0	0
1% Au{Pd}/C	Calcined 400 °C/Air	5	2.2	4.3	85.1	2.1	6.3	0
1% Au + Pd/TiO$_2$	dried	20	0.4	21.3	75.3	1.0	2.0	0
1% Pd{Au}/TiO$_2$	dried	20	0.3	33.1	63.8	0.9	1.9	0
1% Au{Pd}/TiO$_2$	dried	20	0.4	28.8	66.9	0.9	3.0	0
1% Au + Pd/TiO$_2$	Calcined 400 °C/Air	20	0.8	22.3	69.8	1.8	5.4	0
1% Pd{Au}/TiO$_2$	Calcined 400 °C/Air	20	0.5	23.8	71.0	1.3	3.3	0
1% Au{Pd}/TiO$_2$	Calcined 400 °C/Air	20	0.5	28.3	66.9	1.1	3.2	0

[a] Reaction conditions: benzyl alcohol, 0.1g of catalyst, T = 120 °C, pO_2 = 150 psi.

Table 7 Direct synthesis of H_2O_2 using sol-immobilized Au–Pd on C and TiO$_2$ catalysts dried at 120 °C[a]

Catalyst	Productivity (mol H_2O_2/kg$_{cat}$/h)		
	30 min	2 min	30 s
Au + Pd/TiO$_2$	31	159	254
Pd{Au}/TiO$_2$	25	117	231
Au{Pd}/TiO$_2$	29	117	238
Au + Pd/C	158	720	1303
Pd{Au}/C	143	706	1384
Au{Pd}/C	142	929	1641

[a] Reaction conditions:10 mg catalyst in 5.6 g methanol, 2.9 g water solvent, 420 psi 5%H_2/CO_2 + 160 psi 25%O_2/CO_2, productivity calculated after 2 and 30 minutes reaction with stirring 1200 rpm at 2 °C.

methods. Comparison of the "as-prepared" sols and their immobilized counterparts on C and TiO$_2$ reveals the absence of nm-scale clusters in the samples dried at 120 °C. The loss of these ultra-small clusters by sintering explains the slight increase in mean particle size observed after drying at 120 °C. Furthermore, aberration corrected HAADF imaging provides direct evidence that the morphologies of the immobilized nanoparticles produced are largely as intended *i.e.*, co-reduction of the metal salts leads to the formation of Au + Pd random alloy particles, whereas successive reduction of the metal salts generates either Au{Pd} or Pd{Au} morphologies. Indeed, particles as small as 4 nm have been shown to exhibit core-shell morphologies in the Au{Pd}/C and Pd{Au}/C systems.

It is evident from STEM-HAADF imaging experiments on the Au{Pd} and Pd{Au} core-shell structures, that both the core size and shell thickness vary from particle-to-particle and is not uniform as one might have hoped. In the Pd{Au} system, particles less than 3 nm in size were found to be Au-rich, whereas the larger particles were Au$_{core}$–Pd$_{shell}$ type in which the shell thickness varied from as little as a monolayer up to 8 atomic layers thick in some cases. In comparison, the Pd$_{core}$–Au$_{shell}$ particles in the Au{Pd} system often had incomplete shells and off-centre cores. The cause of such heterogeneities in the core-shell structures could in principle arise from (i) differences in the reduction rates of the two salts (as Au^{3+} and Pd^{2+} ions have different reduction potentials), (ii) a lack of shell material to sufficiently cover all of the seed particles, and (iii) subtle variations in other processing parameters.

Table 8 Hydrogenation of H_2O_2 using carbon and TiO$_2$-supported Au–Pd catalysts prepared by sol-immobilisation.[a]

Catalyst	Hydrogenation activity (mol H_2O_2/kg$_{cat}$/h)		
	30 min	2 min	30 s
Au + Pd/TiO$_2$	384	6030	19300
Pd{Au}/TiO$_2$	371	4120	14600
Au{Pd}/TiO$_2$	331	4660	15600
Au + Pd/C	318	4560	17200
Pd{Au}/C	654	4120	13700
Au{Pd}/C	370	5640	19700

[a] Reaction conditions: same as for the H_2O_2 synthesis, but in the absence of 160 psi 25%O_2/CO_2 and with 4 wt% H_2O_2 added to the solvent.

Table 9 Direct synthesis of H_2O_2 using C and TiO_2-supported Pd{Au} catalysts prepared by sol-immobilisation: effect of calcination[a]

Catalyst	Catalyst pre-treatment	Productivity (mol H_2O_2/kg$_{cat}$/h)	Hydrogenation activity (mol H_2O_2/kg$_{cat}$/h)
Pd{Au}/TiO$_2$	Dried 120 °C/Air	25	371
	Calcined 400 °C/Air	4	275
Pd{Au}/C	Dried 120 °C/Air	143	654
	Calcined 400 °C/Air	4	252

[a] Reaction conditions:10 mg catalyst in 5.6 g methanol 2.9 g water solvent, 420 psi 5%H_2/CO_2 + 160 psi 25%O_2/CO_2, 30 minutes reaction with stirring 1200 rpm at 2 °C.

In the Pd{Au} colloid preparation, the molar concentration of Pd introduced was nearly twice that of Au (since the Au:Pd weight ratio was kept at 1:1). Hence, the primary cause for any heterogeneities in the Pd{Au} system could relate to non-ideal processing parameters which causes a few particles to develop extraordinarily thick shells, which is necessarily compensated by other cores having insufficient shell material available to obtain complete coverage. On the other hand, for the Au {Pd} system the origin of any insufficient shell coverage is thought to be due to the low Au molar percentage chosen for the preparation. As the Pd:Au ratio was initially fixed at 1:1 wt% (*i.e.*, 1.8:1 mol%) in all cases, the amount Au introduced may not be sufficient for a complete coverage of all the generated Pd seed particles during their shell growth phase.

Composition variations in the core-shell samples

Estimates of the total weight fraction of the shell element theoretically required to obtain complete coverage of the core material with an integer number of shell layers are shown in Tables 10 and 11. These tables have been derived using the Mackay icosahedal model[41] for the Au{Pd} (Table 10) and the Pd{Au} (Table 11) systems. The calculations show that a 4.3 nm Au core (*i.e.*, a core containing 1415 Au atoms) requires only 20 wt% of Pd to generate a complete monolayer shell. By way of contrast, to cover a 4.1 nm Pd core (*i.e.*, a core composed of 1415 Pd atoms) with a monolayer of Au, requires over 45 wt% of Au. In the current study, the wt% of Au was fixed at 50 wt% which would not be sufficient for the complete coverage of sub-3 nm Pd seeds during the preparation of Au{Pd} nanoparticles. Hence, many smaller Pd particles were seen with little or just a partial shell coverage in the Au{Pd} sols.

From such simple geometric calculations and the starting composition constraints used in this study, a Pd coverage of a monolayer and above may be expected for all Au core sizes greater than 1.44 nm in the Pd{Au} system, whereas only Pd core sizes greater than 3.6 nm can be expected to have a monolayer coverage of Au in the Au {Pd} system. Increasing the weight fraction of Au is therefore an obvious remedial step for future preparations of Au{Pd} particles in order to ensure more uniform coverage of the Pd cores. In the case of Pd{Au} samples, numerous Au cores were found to have Pd shells that were several atomic layers thick. For example, Pd{Au} nanoparticles supported on TiO$_2$ and C with shells thicker than 4–5 atomic layers are shown in Figure 7. According to Table 10, a 2nm Au core with a coverage 4–5 atomic shells of Pd, requires more than 90 wt% of Pd. Many particles in the Pd {Au}/C and Pd{Au}/TiO$_2$ catalysts have shells that are 4–5 atomic layers thick (*i.e.*, exceeding their nominal Pd quota), so a significant population of Au cores with little or no Pd coverage are to be expected. Such heterogeneities in the shell thickness, which imply very significant variations in the particle composition from the nominal value, should be addressed by optimization of the processing parameters. For

Table 10 The weight percentage of Pd required for the complete coverage of Au cores of different sizes with integer numbers of Pd atomic layers. The core-shell combinations possible with 1:1 wt% Au:Pd preparation as used in this study are indicated in italics

#Pd shells	1 atom shell	2 atom shells	3 atom shells	4 atom shells	5 atom shells
Au core size/nm	Wt % Pd required for complete coverage				
0.86	64.2	85.1	92.7	95.9	97.5
1.44	*48.2*	72.0	83.6	89.8	93.2
2.02	*38.0*	61.0	74.6	82.7	87.8
2.59	*31.2*	52.5	66.5	75.9	82.2
3.17	*26.4*	*45.8*	59.7	69.6	76.6
3.74	*22.8*	*40.6*	53.9	64.0	71.5
4.32	*20.1*	*36.3*	*49.1*	59.0	66.7
4.9	*18.0*	*32.9*	*45.0*	54.7	62.5
5.47	*16.2*	*30.0*	*41.4*	50.9	58.6
6.05	*14.8*	*27.6*	*38.4*	*47.5*	55.1

example, tighter control of the initial core size may aid in creating particles with more uniform shell coverage, and hence tighter composition control. It should be noted that the use of Tables 10 and 11 presents a simple way to estimate the compositions of individual core-shell particles where the cores and shell dimensions have been experimentally measured by HAADF imaging. While it is a reasonable approximation for icosahedral (and cuboctahedral) structures, care must be taken when analyzing decahedral structures which have significantly different core to shell ratio requirements.

Such compositional inhomogeneities from particle-to-particle are not just restricted to core-shell systems. Even Au + Pd mixed alloy particles prepared by colloidal routes can show a significant compositional variation between particles of different sizes, despite the expected efficient intimate mixing of the metal salts achieved during the co-reduction process. In a recent study,[42] we have found that for Au + Pd preparations with a nominal ratio of Au:Pd = 1:7 mol%, the actual ratio of Au:Pd in the particles varied systematically according to the size of the particles. As determined from XEDS analysis, the smaller particles ($\sim$2–3 nm) were consistently Au-rich, the medium sized particles (5–7 nm) contained roughly equal amounts of Au and Pd whereas while the larger particles (10–12 nm) were all Pd-rich. The reason for such a systematic variation of nanoparticle composition with size is not clearly understood; however, the

Table 11 The weight percentage of Au required for the complete coverage of Pd cores of different sizes in integer atomic layers. The core-shell combinations possible with 1:1 wt% Au:Pd preparation as used in this study are indicated in **bold italics**

# Au shells	1 atom shell	2 atom shells	3 atom shells	4 atom shells	5 atom shells
Pd core size/nm	Wt % Au required for complete coverage				
0.82	85.3	94.9	97.6	98.7	99.2
1.37	75.1	89.3	94.3	96.6	97.8
1.92	66.5	83.5	90.5	93.9	95.9
2.47	59.5	78.2	86.6	91.1	93.7
3.01	53.7	73.3	82.8	88.1	91.4
3.56	***49.0***	68.9	79.1	85.2	89.0
4.11	***45.0***	64.9	75.8	82.4	86.7
4.66	***41.5***	61.4	72.6	79.6	84.4
5.21	***38.6***	58.1	69.6	77.0	82.1
5.75	***36.0***	55.2	66.9	74.6	79.9

difference in the rate of reduction of Au and Pd in a mixed solution could contribute to such compositional inhomogeneities.[16]

Thermal stability of nanoparticles

This study also illustrates that post-preparative heat treatment procedures can drastically affect the morphology and oxidation state of the metallic components in the bimetallic nanoparticles. XPS and HAADF imaging studies suggest that oxidation of the Pd component is minimal during the 120 °C drying stage of the immobilized catalysts, irrespective of the Au–Pd variant or the support. When calcined at 200 °C, the Au–Pd particles supported on C retained their initial sol morphologies but showed the onset of particle sintering. In contrast, a calcination temperature of 200 °C did not cause excessive Au–Pd particle sintering on the TiO_2 supports, but was found to be detrimental to the integrity of the desired core–shell structures in the Au{Pd}/TiO_2 and Pd{Au}/TiO_2 catalysts.

When the catalysts were calcined at 400 °C, the structure of the nanoparticles on both C and TiO_2 underwent massive structural changes. In case of the C-supported Au + Pd, Pd{Au} and Au{Pd} catalysts, the calcination treatment resulted in the burning away of the carbon support accompanied by severe sintering of the metal nanoparticles. On TiO_2 supports, the Au + Pd variant retained its random alloy structure and did not phase separate. However, the Pd{Au} and Au{Pd} variants on TiO_2 showed the co-existence of (i) Au-rich particles associated with long PdO tails, (ii) Au-rich particles and (iii) PdO particles. The mere existence of the latter two morphologies implies that a population of Pd and Au monometallic particles existed alongside the desired core-shell nanoparticles in both the Au{Pd} and Pd {Au} starting sol preparations. Hence we conclude that developing alternative low temperature methods of ligand removal are highly desirable in order to maintain the size distribution and designer morphology of the starting sol.

Comments on catalyst activity and structure

It is clear that the supported dried nanoparticles are highly effective as catalysts for the oxidation of benzyl alcohol when considering the initial rates when expressed as the TOF at 30 min reaction time (Table 3). In particular, the Pd{Au}/TiO_2 and Pd {Au}/C catalysts show the highest activities. However, the selectivities to the desired product is relatively low and the amount of toluene by-product, formed by a disproportionation reaction of benzyl alcohol,[43] is significant, particularly for the carbon-supported nanoparticles. This indicates that although the oxidation pathway is available by which benzaldehyde is formed directly, the disproportionation reaction in which two molecules of benzyl alcohol react to form benzaldehyde, toluene and water is equally facile. This is not observed with this reaction for supported Au–Pd alloy nanoparticles prepared by an impregnation method.[4] It is possible that the PVA ligands that remain on the catalyst surface may be having a marked effect on catalyst performance and may be blocking or masking the active centers for the oxidation reaction. Calcination of the catalysts, which is known to remove PVA ligands from the surface of nanoparticles,[44] does not however improve the selectivity. It is noted by XPS that the nanoparticles prepared using the sol-immobilization method have Pd-enriched surfaces and this may be facilitating the formation of toluene.

With respect to H_2O_2 synthesis, we had previously proposed that small nanoparticles could be preferred for this reaction.[7] Indeed, the carbon-supported dried nanoparticles are highly effective as catalysts for this reaction, displaying much higher activity than that observed with catalysts prepared by impregnation.[7] However, this is not the case for the TiO_2-supported nanoparticles, which are about half as active as AuPd catalysts prepared by impregnation.[7] Unfortunately, the striking feature of the sol-immobilized catalysts is that they display exceptionally high activities for the hydrogenation of H_2O_2, hence making them rather unsuitable for this

     This journal is © The Royal Society of Chemistry 2011

reaction. It is therefore possible that ultra-small AuPd nanoparticles may not be the optimal catalyst structure for this demanding synthesis. They may, however, be suitable catalysts for other less demanding hydrogenation reactions, and this possibility is something for a future investigation.

Acknowledgements

GJH would like to thank the EPSRC, Solvay and the Leverhulme Trust for Financial support. CJK would like to acknowledge the financial support of the NSF (grants nos. DMI-0304180 and DMI-0457602)

References

1 J. K. Edwards, B. Solsona, E. Ntainjua N, A. F. Carley, A. A. Herzing, C. J. Kiely and G. J. Hutchings, *Science*, 2009, **323**, 1037.
2 M. D. Hughes, Y. J. Xu, P. Jenkins, P. McMorn, P. Landon, D. I. Enache, A. F. Carley, G. A. Attard, G. J. Hutchings, F. King, E. H. Stitt, P. Johnston, K. Griffin and C. J. Kiely, *Nature*, 2005, **437**, 1132.
3 J. Huang, T. Akita, J. Faye, T. Fujitani, T. Takei and M. Haruta, *Angew. Chem., Int. Ed.*, 2009, **48**, 7862.
4 D. I. Enache, J. Edwards, P. Landon, B. Solsana-Espriu, A. F. Carley, A. Herzing, M. Watanabe, C. J. Kiely, D. W. Knight and G. J. Hutchings, *Science*, 2006, **311**, 362.
5 L. Prati and M. Rossi, *J. Catal.*, 1998, **176**, 552.
6 L. Kesavan, R. Tiruvalam, M. I. Bin Saiman, D. I. Enache, R. Jenkins, N. Dimitratos, J. A. Lopez-Sanchez, S. H. Taylor, D. W. Knight, C. J. Kiely and G. J. Hutchings, *Science*, 2011, **351**, 195.
7 J. K. Edwards, A. F. Carley, A. A. Herzing, C. J. Kiely and G. J. Hutchings, *Faraday Discuss.*, 2008, **138**, 225.
8 A. A. Herzing, M. Watanabe, C. J. Kiely, J. Edwards, M. Conte, Z. R. Tang and G. J. Hutchings, *Faraday Discuss.*, 2008, **138**, 337.
9 J. A. Lopez-Sanchez, N. Dimitratos, P. Miedziak, E. Ntainjua, J. K. Edwards, D. Morgan, A. F. Carley, R. C. Tiruvalam, C. J. Kiely and G. J. Hutchings, *Phys. Chem. Chem. Phys.*, 2008, **10**, 1921.
10 J. Turkevich and G. Kim, *Science*, 1970, **169**, 873.
11 S. J. Mejía-Rosales, C. Fernández-Navarro, E. Pérez-Tijerina, D. A. Blom, L. F. Allard and M. José-Yacamán, *J. Phys. Chem. C*, 2007, **111**, 1256.
12 V. I. Pârvulescu, V. Pârvulescu, U. Endruschat, G. E. Filoti, F. Wagner, C. Kübel and R. Richards, *Chem.–Eur. J.*, 2006, **12**, 2343.
13 D. Wang, A. Villa, F. Porta, L. Prati and D. Su, *J. Phys. Chem. C*, 2008, **112**, 8617.
14 M. R. Knecht, M. G. Weir, A. I. Frenkel and R. M. Crooks, *Chem. Mater.*, 2008, **20**, 1019.
15 S. Mandal, A. B. Mandale and M. Sastry, *J. Mater. Chem.*, 2004, **14**, 4.
16 M. L. Wu, D. H. Chen and T. C. Huang, *Langmuir*, 2001, **17**, 3877.
17 Y. Mizukoshi, K. Okitsu, Y. Maeda, T. A. Yamamoto, R. Oshima and Y. Nagata, *J. Phys. Chem. B*, 1997, **101**, 7033.
18 J. W. Hu, Y. Zhang, J. F. Li, Z. Liu, B. Ren, S. G. Sun, Z. Q. Tian and T. Lian, *Chem. Phys. Lett.*, 2005, **408**, 354.
19 G. Schmid, A. Lehnert, J. O. Malm and J. O. Bovin, *Angew. Chem., Int. Ed. Engl.*, 1991, **30**, 874.
20 R. Ferrando, J. Jellinek and R. L. Johnston, *Chem. Rev.*, 2008, **108**, 845.
21 C. Kan, *J. Phys. D: Appl. Phys.*, 2003, **36**, 1609.
22 R. Harpeness and A. Gedanken, *Langmuir*, 2004, **20**, 3431.
23 D. Ferrer, A. Torres-Castro, X. Gao, S. Sepulveda-Guzman, U. Ortiz-Mendez and M. Jose-Yacaman, *Nano Lett.*, 2007, **7**, 1701.
24 D. Ferrer, D. A. Blom, L. F. Allard, S. Mejia, E. Perez-Tijerina and M. Jose-Yacaman, *J. Mater. Chem.*, 2008, **18**, 2442.
25 N. Toshima, M. Harada, T. Yonezawa, K. Kushihashi and K. Asakura, *J. Phys. Chem.*, 1991, **95**, 7448.
26 M. Harada, K. Asakura and N. Toshima, *J. Phys. Chem.*, 1994, **98**, 2653.
27 N. Toshima, H. Yan, Y. Shiraishi, B. Corain and G. Schmid in *Metal Nanoclusters in Catalysis and Materials Science*, Elsevier, Amsterdam, 2008, p 49.
28 N. Toshima, Y. Yonezawa and K. Kushihashi, *J. Chem. Soc., Faraday Trans.*, 1993, **89**, 2537.
29 N. Toshima and T. Yonezawa, *New J. Chem.*, 1998, **22**, 1179.

30 Y. Mizukoshi, T. Fujimoto, Y. Nagata, R. Oshima and Y. Maeda, *J. Phys. Chem. B*, 2000, **104**, 6028.
31 A. S. Shirinyan and M. Wautelet, *Nanotechnology*, 2004, **15**, 1720.
32 *Phase Diagrams*, 2nd edition Vol. 1, ASM International: Materials Park, OH, 1992.
33 W. J. Tang and G. Henkelman, *J. Chem. Phys.*, 2009, **130**, 194504.
34 J. R. Kitchin, J. K. Norskov, M. A. Barteau and J. G. Chen, *Phys. Rev. Lett.*, 2004, **93**, 156801.
35 J. Zhang, H. M. Jin, C. B. Sullivan, F. Chiang, H. Lim and P. Wu, *Phys. Chem. Chem. Phys.*, 2009, **11**, 1441.
36 L. Guczi, *Catal. Today*, 2005, **101**, 53.
37 A. Groß, *J. Phys.: Condens. Matter*, 2009, **21**, 084205.
38 N. Dimitratos, J. A. Lopez-Sanchez, D. Morgan, A. F. Carley, R. Tiruvalam, C. J. Kiely, D. Bethell and G. J. Hutchings, *Phys. Chem. Chem. Phys.*, 2009, **11**, 5142.
39 M. Haruta, *CATTECH*, 2002, **6**, 102.
40 S. Tsubota, T. Nakamura, K. Tanaka and M. Haruta, *Catal. Lett.*, 1998, **56**, 131.
41 A. Mackay, *Acta Crystallographica*, 1962, **15**, 916.
42 J. Pritchard, L. Kesavan, M. Piccinini, Q. He, R. Tiruvalam, N. Dimitratos, J. A. Lopez-Sanchez, A. F. Carley, J. K. Edwards, C. J. Kiely and G. J. Hutchings, *Langmuir*, 2010, **26**, 16568.
43 S. Meenakshisundaram, E. Nowicka, P. J. Miedziak, G. L. Brett, R. L. Jenkins, N. Dimitratos, S. H. Taylor, D. W. Knight, D. Bethell and G. J. Hutchings, *Faraday Discuss.*, 2010, **145**, 341.
44 M. Comotti, W. C. Li, B. Spliethoff and F. Schüth, *J. Am. Chem. Soc.*, 2006, **128**, 917.

   This journal is © The Royal Society of Chemistry 2011

Nanoporous gold: a new gold catalyst with tunable properties

Arne Wittstock,[*a] Andre Wichmann,[a] Jürgen Biener[b] and Marcus Bäumer[a]

Received 17th February 2011, Accepted 3rd March 2011
DOI: 10.1039/c1fd00022e

Nanoporous gold (np-Au) represents a novel nanostructured bulk material with very interesting perspectives in heterogeneous catalysis. Its monolithic porous structure and the absence of a support or other stabilizing agents opens up unprecedented possibilities to tune structure and surface chemistry in order to adapt the material to specific catalytic applications. We investigated three of these tuning options in more detail: change of the porosity by annealing, increase of activity by the deposition of oxides and change of activity and selectivity by bimetallic effects. As an example for the latter case, the effect of Ag impurities will be discussed. The presence and concentration of Ag can be correlated to the availability of active oxygen. While for the oxidation of CO the activity of the catalyst can be significantly enhanced when increasing the content of Ag, we show for the oxidation of methanol that the selectivity is shifted from partial to total oxidation. In a second set of experiments, two different metal-oxides were deposited on np-Au, praseodymia and titania. In both cases, the surface chemistry changed significantly. The activity of the catalyst for oxidation of CO was increased by up to one order of magnitude after modification. Finally, we used adsorbate controlled coarsening to tune the structure of np-Au. In this way, even gradients in the pore- and ligament size could be induced, taking advantage of mass transport phenomena.

Introduction

In recent years, there has been an increasing interest in monolithic nanoporous materials for various applications, such as for energy storage and conversion materials, sensors and actuator materials and catalysts.[1–4] Among various strategies to generate nanoporosity, the corrosion of alloys turned out to be especially suitable. In this context, nanoporous gold (np-Au), in particular, gained considerable interest. The corrosion of a Au alloy containing at least one less noble constituent, such as Ag or Cu, results in a bulk nanostructured material consisting of almost pure gold.[5,6] The structure of the material is characterized by a three-dimensional porous network of ligaments in the range of typically tens of nanometers (see Figure 1). Starting about ten years ago, an increasing number of publications dealt with the preparation of the material and its promising applications e.g. as catalysts, sensor materials, and high surface area electrodes.[4,5,7,8] One key element for these applications is the open cell morphology of the material that provides a high surface area and makes it penetrable for gases and liquids. The chemical interaction of molecules with the

[a]Institute of Applied and Physical Chemistry, University Bremen, Leobener Str., NW2, 28359 Bremen, Germany. E-mail: awittstock@uni-bremen.de
[b]Nanoscale Synthesis and Characterization Laboratory, Lawrence Livermore National Laboratory, Livermore, California, USA

Fig. 1   Scanning electron micrographs of a cross section of the np-Au material.

surface, *i.e.*, the surface chemistry, can then even lead to an observable macroscopic response. When the material is, for example, exposed to an oxidizing agent, such as ozone, that roughly generates one monolayer of surface bonded oxygen, the material can be reversibly strained on a macroscopic scale by altering the surface stress state.[9]

The high surface area also renders np-Au interesting for catalytic applications. In 2006, Zielasek *et al.* and, in 2007, Ding *et al.* independently discovered that this gold material is a highly active catalyst for the oxidation of CO with molecular oxygen at temperatures as low as −20 °C.[10,11] This finding came as a bit of surprise, as the majority of active gold catalysts is typically comprised of gold nanoparticles in the range of only a few nanometers on a suitable oxide support, such as titania or ceria.[12–15] Since the first reports of the unexpectedly high activity of gold-based catalysts in the 1980's the interaction of gold and the support has been discussed to be a crucial factor for catalytic activity. Consequently, the origin of the catalytic activity of unsupported np-Au has been a matter of debate in the recent literature.

In contrast to supported catalysts, only two factors can play a role. On the one hand, low coordinated gold atoms at steps and kinks - probably constituting a large part of the surface atoms - may be able to activate the molecular oxygen – the crucial step for gold based oxidation catalysts.[16–18] On the other hand, the material still contains traces of a less noble metal such as Ag – being the other constituent of the starting alloy - in the range of ∼ 1 at%.[19] If this second metal is enriched at the surface – in the case of Ag with surface concentration in the range of up to several atomic percent -, it can very well contribute to the catalytic activity.[20] In comparison to Au, molecular oxygen tends to bind more strongly to metals, such as Ag and Cu and thus is more easily activated.[21] It was thus discussed to what extend this ad-metal contributes to the catalytic activity especially with respect to the activation of O_2.[20,22] Recently, UHV model experiments and DFT calculations were also employed to further understand this aspect of the surface chemistry of np-Au.[23]

When dealing with catalysis, one has to not only consider the surface chemistry but also mass transport phenomena, which become particularly pronounced in cases of materials with high porosity. In the case of np-Au, the porous structure is very homogeneous with narrow ligament and pore size distributions throughout the material. This well-defined structure provides a good starting point for quantifying mass transport phenomena in np-Au based on straightforward assumptions.[20] For example, the efficiency of a 200-micron-thick np-Au disk for CO oxidation is only about 10% as the reaction of CO is faster than the mass transport limited supply from the outer gas phase, leading to a fast decrease in the concentration within the material.[20] In order to find an optimum between low mass transport limitation (large pores) and high surface area (small pores), a controlled tuning of the ligament sizes is essential.

By using the intrinsic instability of nanostructures, one can induce coarsening of the ligaments by treatment at elevated temperatures.[24-26] For np-Au, such a coarsening sets in at annealing temperatures above ~200 °C.[27,28] In this way, the ligament and pore sizes can be tailored within several orders of magnitude, starting from ~5 nm up to microns.[29] Interestingly, the morphology of the material in terms of relative density and porosity stays constant, while the size of the ligaments and pores is increased. This behaviour is the result of curvature driven diffusion of surface atoms.[25,29] Accordingly, adsorbates, such as atomic oxygen, which are able to stabilize low coordinated Au surface atoms[30] do have an impact on the diffusion of surface atoms and thus on the coarsening during heat treatment of np-Au. When, for example, np-Au is annealed under an ozone containing atmosphere generating surface bonded oxygen, the size of the ligaments and pores of np-Au can be conserved, at least below the desorption temperature of oxygen on the gold surface (~500 K).[25,30]

In the present study, we describe different strategies to deliberately modify the properties of np-Au aiming at an optimization for specific applications with the emphasis on heterogeneous catalysis. On the one hand, we will show that the surface chemistry of np-Au can be greatly changed by a second component, such as a metal or a metal-oxide. On the other hand, it will be demonstrated how the feature size of np-Au and thus the mass transport properties within the pores can be tuned by adsorbate controlled coarsening.

Taking advantage of the fact that some residual Ag is still contained in the material after preparation, we systematically investigated the dependence of the activity and selectivity of two different oxidation reactions on the amount of residual Ag. We used CO and methanol oxidation as two different probe reactions, showing that a high amount of residual Ag can be beneficial for total oxidation reactions but a disadvantage in case of partial oxidations. We also investigated the effect of two different reducible oxides, praseodymia and titania, on the activity of oxidation reactions, using molecular oxygen. Titania was chosen because of its widespread use for supported Au nanoparticle catalysts and its possible use also in photo-catalysts.[31] Praseodymia on the other hand was chosen because of its rich redoxchemistry and high lattice oxygen mobility.[32] In both cases, we find a considerable increase of activity for CO oxidation of nearly one order of magnitude. We conclude that these changes are linked to the availability of active oxygen on the catalyst surface. Third, we used annealing under environmental control to investigate in more detail the influence of atomic oxygen on the stability of the np-Au structures, using ozone to generate surface bonded atomic oxygen.[25] The results reveal that the stability of the structures is strongly linked to the coverage by atomic oxygen. In combination with mass transport limited availability of ozone inside the pores, we can generate gradients in the porosity along the cross section of the np-Au sample.

Materials and methods

Disks of np-Au with a diameter of 5 mm and a thickness of 200 to 300 μm were prepared by corrosion of Ag(70at%)-Au(30at%) alloys in nitric acid. Two different methods were used: *free corrosion* where no external potential is applied and *galvanic corrosion* where the sample is held on a constant potential during the entire etching procedure. For *free corrosion*, the samples were placed on a glass frit in conc. nitric acid for 48 hours (HNO$_3$, 65 wt%, Fluka Chemical Corp.). For *galvanic corrosion*, the sample was mounted in a typical three electrode setup. A 5 M solution of nitric acid was used as electrolyte (conc. HNO$_3$ diluted with ultra pure water (> 18 MΩ)). The potential was held at 60 mV *versus* a platinum foil used as a reference electrode (Potentiostat, Wenking POS 88, BANK Electronics, counter electrode platinum foil). Samples containing different amounts of residual Ag were prepared by monitoring the current during etching and systematically varying the duration of corrosion, accordingly. For details see also ref. 33.

For doping the np-Au samples with metal oxides two different methods were used. In the case of titania, the samples were coated with a colloidal solution of titania particles. For this purpose, a sample of np-Au was activated first and characterized in terms of catalytic activity for CO oxidation. A suspension of 2.5 g L^{-1} titania particles (Sigma–Aldrich, *nanopowder*, particles < 100 nm) in ethanol (Sigma–Aldrich, p.A.) was prepared. For the sake of homogenisation and to generate a fine dispersion of particles, the suspension was sonicated for 30 minutes. Two droplets of the colloidal suspension were applied on each side of the np-Au disk. After ~1 minute the sample surface was rinsed with ethanol and eventually dried in nitrogen. In case of praseodymia, an impregnation and precipitation method was used. First, the np-Au sample was activated for CO oxidation. Afterwards, the sample was immersed in a solution of ethanol containing 20 g L^{-1} $Pr(NO_3)_3$ (Sigma–Aldrich, 99.9%) for 15 minutes. After drying, the sample was heat treated (calcined) in air, starting at 60 °C and constantly heating up to 500 °C within 2 hours.

The catalytic experiments were performed using a continuous flow reactor especially designed for catalytic experiments with the np-Au disks (for details see also ref. 20,33). The feed gases were He (*Linde AG*, 5.0), O_2 (*Linde AG*, 4.5), and CO (*Linde AG*, 4.7). The total flow was always set to 50 sccm. The amount of methanol was controlled *via* its vapour pressure. First, a gas stream was saturated with methanol at room temperature (around 20 °C). Subsequently, the amount of methanol in the gas phase was adjusted by guiding the stream of saturated gas through a condenser. The stream of gases at the exit of the reactor was analyzed online by IR-Gas-Analyzers (URAS 3G, Hartmann und Braun) and GC-MS (100 µl sample pipe, Fison 8000 series (Column FS-innopeg-1000) coupled with a mass spectrometer TRIO 1000).

Annealing experiments with np-Au were carried out in a tube furnace under either an atmosphere of ultrapure He (5.0, *Linde AG*) or an ozone-oxygen mixture (~ 7 vol% O_3 in 4.5 O_2, *Linde AG*) at a flowrate of 50 sccm (ozone generator type 802 N, ozone analyzer type 964, BMT Messtechnik Berlin). The coarsening of np-Au during annealing was analysed by cross-sectional scanning electron microscopy (XSEM). The average ligament diameter for each sample was then determined from ligament diameter distributions that were obtained by geometrical evaluation (*i.e.*, measuring the diameter of randomly selected ligaments at their center).

Results and discussion

Bimetallic np-Au

The catalytic results for CO and methanol oxidation using bimetallic np-Au with different amounts of residual Ag are depicted in Figure 2. The reactant gases CO and O_2 were supplied and varied while keeping a stoichiometric ratio (CO: O_2; 2:1). As deduced from Figure 2 (a), the production of CO_2 continuously increases with increasing supply of reactants. The fact that the conversion does not linearly depend on the supply of reactants points to mass transport limitation as detailed in ref.20 The characteristics are exemplarily shown for four different samples prepared by *free* as well as *galvanic corrosion* and containing different amounts of residual silver between ~ 0.5 at% and 11 at% (as determined by AAS). All measurements were performed directly after activating[20] the particular sample to exclude any variation of activity due to, for example, possible deactivation over time. Comparing the activities, it is obvious that the total activity for the sample containing 11 at% Ag is about two times higher than the activities of the other samples with Ag concentrations between 0.5 and 1.2 at%. As the concentration of Ag within the bulk and the concentration of Ag on the surface—the latter being decisive for the catalytic

    This journal is © The Royal Society of Chemistry 2011

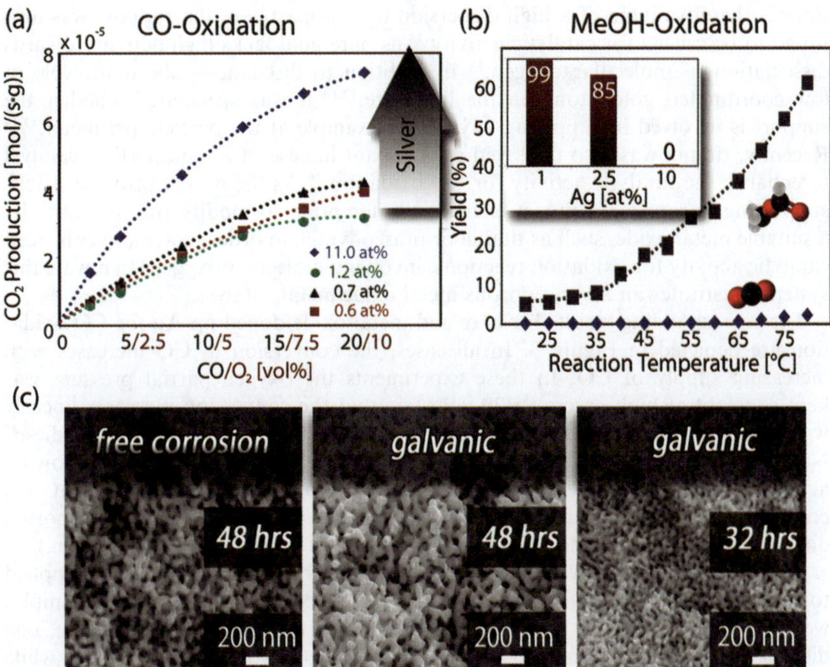

Fig. 2 Catalysis with np-Au: influence of Ag on the activity and selectivity for two different reactions. (a) Conversion of CO to CO_2 using np-Au samples, containing different amounts of Ag (at 40 °C). (b) The oxidation of methanol with molecular oxygen (2 vol% methanol + 1 vol% O_2) in the temperature regime between 20 °C and 80 °C. Only two products are formed, methyl formate (grey squares) and the total oxidation product CO_2 (blue diamonds). In the inset, the selectivity (in percent) for the favoured partial oxidation product methyl formate is shown as a function of the Ag content of np-Au (2 vol% methanol + 1 vol% O_2 at 80 °C). (c) SEM from the middle of the cross section of differently prepared np-Au disks containing different amounts of residual Ag (from left to right 0.5 at%, 2.5 at%, and 10 at%).

behaviour, of course—are related but not equal,[20] one has to be careful when drawing conclusions based on the Ag bulk content. The trend, however, is clear and underlines that the catalytic activity of np-Au for CO oxidation greatly benefits from high residual Ag concentrations and can be enhanced by about 100% when changing the Ag content from ∼1 at% to 11 at%.

The results for the oxidation of methanol in the temperature regime between 20 °C and 80 °C are depicted in Figure 2 (b). Starting at 20 °C, partial oxidation is observed leading to methyl formate as the only product.[33] The unfavourable total oxidation to CO_2 is largely suppressed. When the temperature is raised, the conversion of methanol increases considerably. A different picture emerges when using np-Au samples containing higher amounts of residual Ag. For example, when using a sample containing 2.5 at% residual Ag, the total oxidation becomes more pronounced and the selectivity (fraction of methyl formate) decreases to about 85%. In the case of samples very rich in Ag (∼10 at%), no formation of methyl formate takes place at all. Here, only the total oxidation product CO_2 can be observed. Generally, the activity (sum of all products) tends to decrease with an increase of Ag. Thus, in contrast to CO, in case of methanol increasing concentrations of Ag are unfavourable, leading to more pronounced total oxidation and a decrease of total activity.

Oxide modification of nanoporous gold

Most of the highly active gold catalysts studied in the past were comprised of small Au nanoparticles supported on a porous metal oxide, such as titania or ceria.[15,34]

Besides the advantage of a high dispersion of nanoparticles, the support was also found to be crucial for catalytic activity,[13] as pure gold lacks high activity towards dissociation of molecular oxygen.[35] In addition to discussions about the role of low coordinated gold atoms in the literature,[17,18] it was speculated whether the support is involved in supplying oxygen for example at the particle perimeter.[13,14] Recently, titania was also deployed as a dopant in case of a nanoparticle catalyst to enhance the catalytic activity for CO oxidation.[36] As the np-Au provides a self-supporting porous structure, it is especially tempting to modify this structure by a suitable metal oxide, such as titania or praseodymia, in order to achieve enhanced catalytic activity for oxidation reactions, involving molecular oxygen. In a way, this system constitutes an inverse porous metal oxide/metal catalyst.

The corresponding results for bare and metal-oxide doped np-Au for CO oxidation are depicted in Figure 3. In all cases, the conversion of CO increases with increasing supply of CO. In these experiments the oxygen partial pressure was kept constant at high excess (> 20 vol%) so that the conversion increases linearly with CO indicating a first order reaction as was also observed in previous studies.[20] Notably, the samples modified by either titania or praseodymia deposition show an activity which is enhanced by nearly one order of magnitude. (Note that this comparison is only qualitative and—because of the influence of mass transport— does not directly reflect the differences in the rate of catalytic surface reaction.)

In order to exclude any influence from the preparation (*e.g.* by ethanol applied together with the oxide nanoparticles or the precursors, respectively), samples were also measured after being treated with pure ethanol. For this purpose, one disk of np-Au was broken into 3 pieces. One piece was left as a reference, while another one was treated with a suspension of titania particles and yet another one with pure ethanol. In all cases, an enhanced activity for CO oxidation was *only* observed in the case of titania particles or praseodymia precursor containing suspension. A contribution arising from the treatment of samples with ethanol can thus be excluded.

Measurements with energy dispersive x-ray absorption (EDX) carried out to quantify the amount of material deposited inside the np-Au material revealed no elements apart from Au, meaning that the titania and praseodymia concentrations were below the detection limit. Considering that this method is not surface sensitive and rather probes the entire volume of a 30–50 nm thick ligament, one can conclude that the coverage must be well below 1 monolayer. (Note that Ag could not be detected in these EDX experiments, either).

To gain further insight into the distribution of either titania or praseodymia, np-Au samples were annealed under an inert gas atmosphere. Due to the high

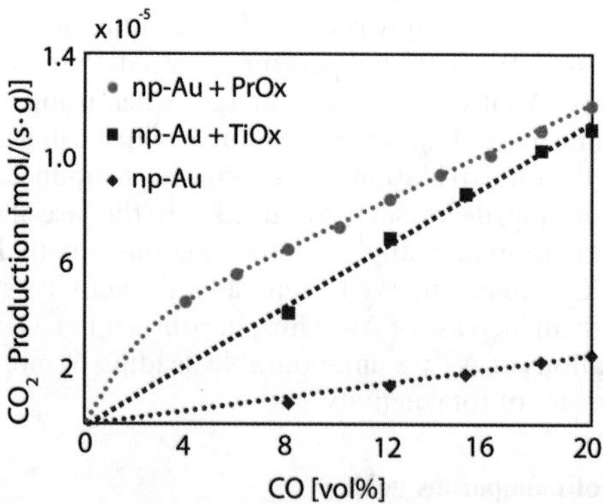

Fig. 3 Activity of pristine np-Au in comparison to PrOx and TiO₂ modified np-Au samples at 60 °C (+ excess of oxygen, *i.e.* > 20 vol% O₂).

    This journal is © The Royal Society of Chemistry 2011

surface-to-bulk ratio, np-Au is structurally unstable and shows coarsening of the porous structure upon thermal treatment. This coarsening is entirely due to the surface self-diffusion of Au atoms on the surface,[37] leading to an increase in the ligament diameter from the nm regime to the μm regime.[29] Previously, it has been observed that adsorbates (see also next section) can hamper this diffusion process and thus stabilize the nanoporous structure.[25] Accordingly, it can be assumed that also oxide deposits on the surface - if localized at step edges *e.g.* which are natural sources of diffusing atoms – should be able to stabilize the structure.

The results for the praseodymia coated sample after annealing to 500 °C are depicted in Figure 4. Clearly, the ligaments close to the outer surface (*e.g.* 10 μm) are indeed stabilized and no coarsening can be observed (average ligament size 40 nm). In the inner sections of the sample, meaning in the middle of a cross section of a 200 μm sample disc, the size distribution was found to be bimodal, indicating that some ligaments are stabilized while others are not (Figure 4 (b)). Based on this finding one can conclude that in case of praseodymia even (sub-) monolayer amounts of deposited material can stabilize the structure during annealing. While this effect is most pronounced close to the outer surface, the deposition techniques applied here (impregnation) results also in small amounts deposited even in deeper sections (100 μm).

In case of titania, no similar stabilization was observed. Already sections close to the outer surface (*e.g.* ~5 μm) showed coarsening of the ligaments. We attribute this to the different deposition techniques (impregnation *vs.* depostion of nanoparticles). In case of the titania, the particles cannot penetrate so deeply, resulting in a significantly larger gradient.

The fact that in *both* cases - TiO_2 and PrO_x modification - an enhanced activity for CO oxidation was detected points towards a contribution of the particular oxide in the catalytic cycle. A scenario similar to supported Au nanoparticles is possible where the oxide contributes to the supply of active oxygen at the perimeter between the oxide and Au.[13] The increased activity for CO oxidation is presumably linked to an enhanced activation of molecular oxygen. (Please note that neither of the oxides are catalytically active for CO oxidation at temperatures below 200 °C.[38,39]) Interestingly, the samples treated with a titania particle suspension show an increased activity for CO oxidation in the same order as praseodymia, revealing that a simple treatment with nanoparticles—leading to a high coverage with oxide within a thin layer—can be as effective as a the more extended coverage obtained with impregnation.

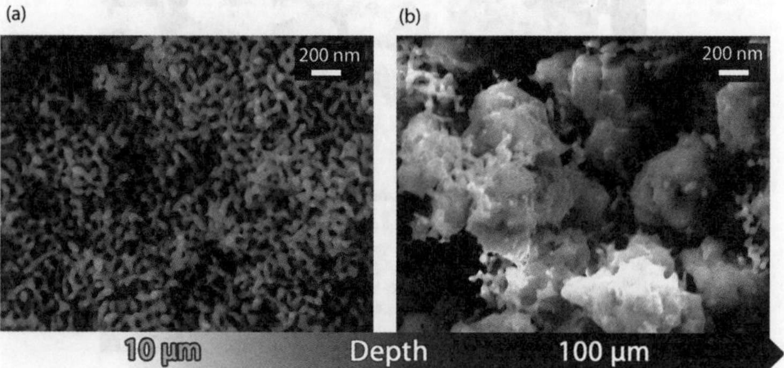

Fig. 4 Cross sectional scanning electron micrographs (XSEM) of a praseodymia coated sample after annealing to 500 °C. (a) Areas close to the outer surface (~10–15 μm) are apparently stabilized, ligament size (~40 nm) and morphology are not affected by the heat treatment. (b) Inner sections of the sample (~100 μm) clearly reveal a bi-modal ligament size distribution. Some ligaments are still ~40 nm in size while other grew to more than 100 nm.

Tailoring structures using adsorbates

Using nanoporous gold in a monolithic form not only offers advantages (such as increased mechanical and thermal stability as well as good thermal conductivity), but also the disadvantage of more severe mass transport limitation reducing the effectiveness of the catalyst.[20] This is due to the pores in the nanometer regime which limit the transport of the reactants into the structure and of the products out of the structure, respectively. This problem becomes even more pronounced, the faster the catalytic reaction proceeds. The increased activity of for example oxide modified np-Au will lead to an even lower efficiency, if diffusion of reactants is not enhanced as well. One possibility to optimize mass transport limitation is of course a tailoring of the pore structure by thermally induced coarsening. In the following we will show results obtained by annealing in inert and oxidative gas atmospheres.

The resulting averaged ligament diameters obtained after thermal treatment are summarized in Figure 5. When annealing at 450 K in a helium atmosphere, the size of the ligament increase from initially 40 nm to roughly 80 nm after 3 hours. When further increasing the temperature to 650 K, the average ligament diameter increases to more than 500 nm.

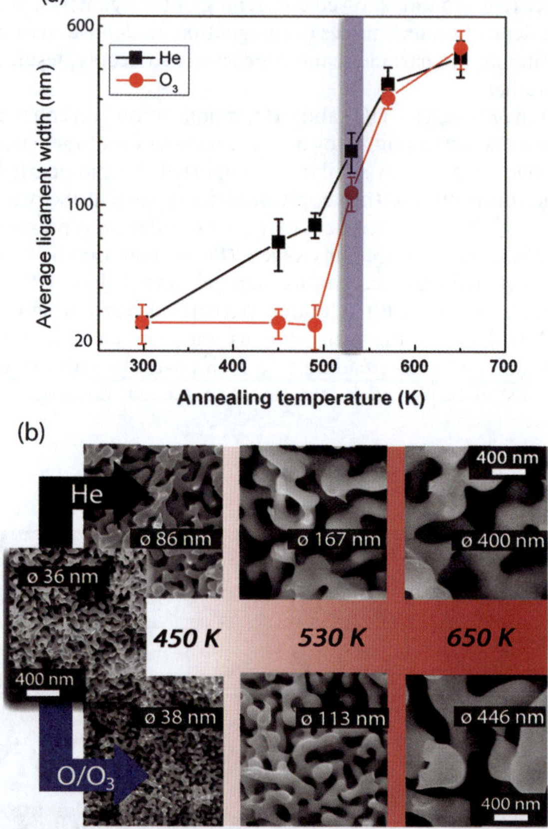

Fig. 5 (a) The development of the average ligament diameter after annealing of np-Au under O_3 containing ($\sim$7 vol%) and inert gas (He) atmospheres for 3 hours (only ligaments close to the outer surface were included in the evaluation). Below the desorption temperature of atomic oxygen on gold ($\sim$530 K) its surface coverage is high enough to prevent coarsening of the nanoporous structure. (b) Corresponding XSEM from the cross section of a np-Au disk after annealing to the particular temperature for 3 hrs. All images are scaled to the same magnification.

This finding is well in line with previous studies on coarsening of np-Au.[25] In case of self-diffusion of the Au atoms on the surface—which is presumably the main pathway for thermally induced coarsening at these temperatures—one would expect an Arrhenius type temperature dependency.[37,40] Indeed, the ligaments size as a function of temperature follows an exponential trend and becomes far more pronounced for higher temperatures.

In case of annealing under an ozone containing atmosphere, the behaviour is greatly changed, though. Below 500 K, where oxygen forms a stable adsorbate layer,[25] the ligament size apparently does not increase. The origin for this adsorbate induced stabilization can be understood in terms of an increased activation barrier for oxygen bonded Au atoms when "jumping" between surface sites, which is the precondition for surface self-diffusion.[37] At temperatures higher than 500 K, however, the residence time of atomic oxygen on Au is drastically reduced from about 1 second at 530 K to only milliseconds at 650 K (assuming first order desorption kinetics).[35,41] Accordingly, the coverage of the surface with atomic oxygen becomes too small to stabilize the structures furthermore. At temperature of 570 K and 650 K the results for the samples exposed to ozone or an inert gas atmosphere are almost indistinguishable, indicating that the coverage of oxygen is negligibly small at these temperatures.

All results discussed so far do not consider mass transport phenomena and are thus only applicable to regions close to the outer surface of the np-Au material. Interestingly, when investigating the cross section of thick np-Au samples (> 200 μm), gradients in the ligament and pore size distribution are obvious (see Figure 6). After annealing at 450 K under ozone containing atmosphere, the structure close to the outer surface is obviously stabilized, as discussed above. On the contrary, the inner sections of the sample exhibit larger structures (ligament diameters > 100 nm) as one would expect for a sample annealed under an inert gas atmosphere. Obviously, the supply of ozone from the outer gas phase is restricted by mass transport through the pores of the material. These characteristics can be described semi-quantitatively, employing a simple kinetic model. The coverage of the surface by oxygen is a function of the impingement rate (ozone molecules hitting the surface per unit of time), the sticking coefficient (probability of a hit leading to O_{ads}), and the desorption rate (oxygen atoms desorbing from the surface per unit of time). The impingement rate of ozone molecules on the surface of a particular ligament is obviously a function of depth. The net impingement rate J_{CF} can be described by the impingement rate on a flat surface J modified by the transmission factor (molecules only hitting the outer surface are repelled) and a factor CF (Clausing factor)[42] which takes into account the aspect ratio of the particular pores:

$$J_{CF} = \Lambda \cdot J \cdot CF = \Lambda \cdot \frac{N_a \cdot P}{\sqrt{2\pi \cdot M \cdot R \cdot T}} \cdot \frac{1}{1 + \dfrac{3\lambda}{4d}},$$

where N_a is the Avogadro constant, P the (partial-) pressure in the ambient gas phase, M is the molecular mass, R the universal gas constant, T the temperature, λ the pore length and d the pore diameter. In this way, the supply of ozone by the outer gas phase can be described as a function of depth. The resulting coverage of the surface with atomic oxygen at 450 K as a function of distance from the outer gas phase is shown in Figure 6 (b) (assuming first order desorption rate of atomic oxygen,[25,35] a sticking coefficient of 0.01,[43] and a transmission factor of 0.7, corresponding to about 70% void volume of the material). The supply of ozone close to the outer surface generates roughly one monolayer of oxygen which is the upper limit observed in experimental studies.[44] However, after about 5 μm distance from the outer sample surface, the O coverage of the surface strongly decreases. At a depth of around 50 μm, the surface coverage accounts for only about one tenth of a monolayer. Since only beyond this depth an increase of the ligament diameter is observed, one can conclude that not only a full oxygen monolayer but also submonolayers

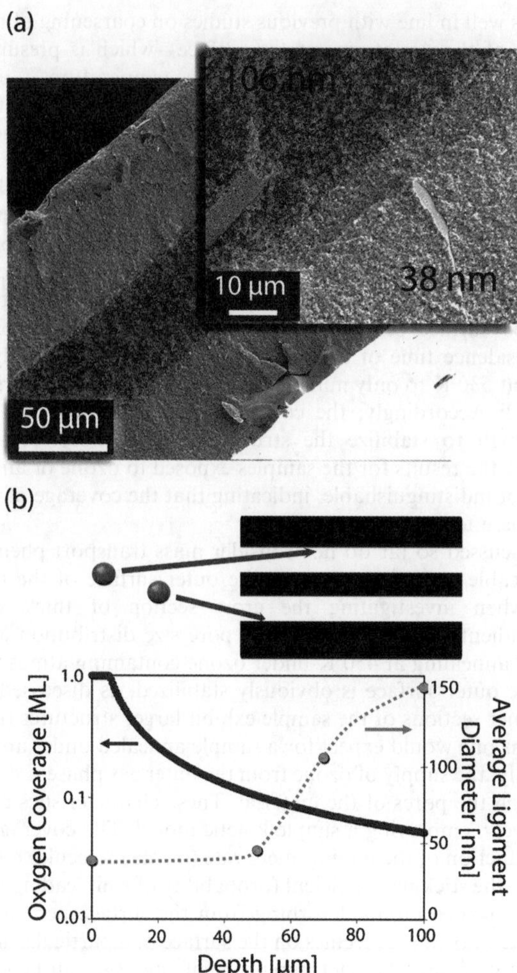

Fig. 6 Structural tuning of np-Au based on surface chemistry. (a) Scanning electron micrographs of a cross section of a np-Au disk after annealing at 450 K for 3 hrs under an ozone containing oxygen atmosphere (~7 vol%). The regions close to the outer surface are obviously stabilized by adsorbed atomic oxygen, while the ligaments in the inner section of the disk show coarsening. (b) Development of the (averaged) ligament diameter as a function of distance (depth) from the outer surface. In the lower section a semi-quantitative description of the oxygen surface coverage at 450 K and the ligament size across a np-Au disk after annealing to 450 K are shown.

(above $\theta \sim 0.1$) lead to a stabilization of the ligaments. This finding is in line with a kinetic model of surface diffusion based dealloying.[2,5] Atoms at or close to step edges are most unstable and thus prone to diffusion. Presumably, the stabilization of step edges and low coordinated atoms by atomic oxygen is crucial for the stabilization of structures during annealing.

Conclusions

Three different methods of controlled modification of the properties of np-Au were employed with the emphasis on catalysis. As np-Au exhibits a well reproducible and self-supporting porous nanostructure without the need of a support, it is particularly suitable for structural and chemical modifications. For example, by variation of dealloying conditions, one can vary the content of Ag remaining in the material after preparation from 0.5 at% to even more than 10 at%. As the Ag content controls the availability of surface oxygen, total oxidation reactions, such as the oxidation of CO, greatly profit from an increased Ag content. This finding is well in line with reports from bimetallic supported Au-Ag particles.[19] Partial oxidation reactions, such as the oxidation of methanol, however, benefit from lower concentrations of Ag. It is thus important to tailor the amount of residual Ag for a specific catalytic application. (Using the dealloying route to prepare different bimetallic catalysts, one is limited to concentrations in the order of 10–15 at%. The reason is that a certain amount of Ag has to be dissolved in order to achieve homogeneous porosity.)

Furthermore, we employed post-dealloying modification by adding metal-oxides to the porous structure of np-Au. In a first attempt, two different metal-oxides were chosen. Titania (TiO_2) is a commonly used support material for gold based catalysts, which itself is very unreactive towards oxidation of CO.[39] Additionally, we chose praseodymia, a less frequently used metal-oxide in catalysis. In the group of rare earth oxides, which includes also for example ceria, it is the one with the highest oxygen mobility, which, however, is only active for CO oxidation at temperatures starting around 300 °C.[38] We observe a strongly enhanced catalytic activity for CO oxidation at temperatures below 100 °C for both oxides, pointing to a strong synergistic effect. The different preparation conditions and the involved mass transport limitations yet prevent a straightforward comparison of activities between both metal-oxide modified systems.

To reduce mass transport limitations without destruction of the monolithic and porous structure a tailored design of the ligament and pore sizes is mandatory. We show that, by using thermally induced coarsening under the influence of adsorbates, one can not only increase the pore diameter but also induce gradients in the pore size distribution along the cross section of a sample. This opens the door for introducing a bimodal and hierarchical pore and ligament size distribution and thus an optimized mass transport.

Acknowledgements

We thank the University Bremen for financial support. Work at LLNL was performed under the auspices of the U.S. DOE by LLNL under Contract DE-AC52-07NA27344. We gratefully acknowledge the experimental support (SEM) of Petra Witte (Prof. Willems, Historical Geology – Palaeontology, Geology department of the University Bremen).

References

1 J. Lee, O. K. Farha, J. Roberts, K. A. Scheidt, S. T. Nguyen and J. T. Hupp, *Chem. Soc. Rev.*, 2009, **38**, 1450–1459.
2 J. Erlebacher and R. Seshadri, *MRS Bull.*, 2009, **34**, 561–568.
3 J. S. King, A. Wittstock, J. Biener, S. O. Kucheyev, Y. M. Wang, T. F. Baumann, S. K. Giri, A. V. Hamza, M. Bäumer and S. F. Bent, *Nano Lett.*, 2008, **8**, 2405–2409.
4 J. Biener, A. Wittstock, T. Baumann, J. Weissmüller, M. Bäumer and A. Hamza, *Materials*, 2009, **2**, 2404–2428.
5 J. Erlebacher, M. J. Aziz, A. Karma, N. Dimitrov and K. Sieradzki, *Nature*, 2001, **410**, 450–453.
6 A. J. Forty, *Nature*, 1979, **282**, 597–598.
7 J. Erlebacher, *J. Electrochem. Soc.*, 2004, **151**, C614–C626.
8 R. C. Newman, S. G. Corcoran, J. Erlebacher, M. J. Aziz and K. Sieradzki, *MRS Bull.*, 1999, **24**, 24–28.
9 J. Biener, A. Wittstock, L. A. Zepeda-Ruiz, M. M. Biener, V. Zielasek, D. Kramer, R. N. Viswanath, J. Weissmuller, M. Bäumer and A. V. Hamza, *Nat. Mater.*, 2009, **8**, 47–51.

10 V. Zielasek, B. Jürgens, C. Schulz, J. Biener, M. M. Biener, A. V. Hamza and M. Bäumer, *Angew. Chem., Int. Ed.*, 2006, **45**, 8241–8244.

11 C. X. Xu, J. X. Su, X. H. Xu, P. P. Liu, H. J. Zhao, F. Tian and Y. Ding, *J. Am. Chem. Soc.*, 2007, **129**, 42–43.

12 M. Haruta, T. Kobayashi, H. Sano and N. Yamada, *Chem. Lett.*, 1987, 405–408.

13 G. C. Bond and D. T. Thompson, *Gold Bull.*, 2000, **33**, 41–51.

14 G. C. Bond and D. T. Thompson, *Catal. Rev. Sci. Eng.*, 1999, **41**, 319–388.

15 A. S. K. Hashmi and G. J. Hutchings, *Angew. Chem., Int. Ed.*, 2006, **45**, 7896–7936.

16 Y. Ding and M. W. Chen, *MRS Bull.*, 2009, **34**, 569–576.

17 H. Falsig, B. Hvolbaek, I. S. Kristensen, T. Jiang, T. Bligaard, C. H. Christensen and J. K. Norskov, *Angew. Chem., Int. Ed.*, 2008, **47**, 4835–4839.

18 J. K. Norskov, T. Bligaard, B. Hvolbaek, F. Abild-Pedersen, I. Chorkendorff and C. H. Christensen, *Chem. Soc. Rev.*, 2008, **37**, 2163–2171.

19 B. Jürgens, C. Kübel, C. Schulz, T. Nowitzki, V. Zielasek, J. Biener, M. M. Biener, A. V. Hamza and M. Bäumer, *Gold Bull.*, 2007, **40**, 142–149.

20 A. Wittstock, B. Neumann, A. Schaefer, K. Dumbuya, C. Kübel, M. M. Biener, V. Zielasek, H.-P. Steinrück, J. M. Gottfried, J. Biener, A. Hamza and M. Bäumer, *J. Phys. Chem. C*, 2009, **113**, 5593–5600.

21 C. B. Wang, G. Deo and I. E. Wachs, *J. Phys. Chem. B*, 1999, **103**, 5645–5656.

22 M. Haruta, *ChemPhysChem*, 2007, **8**, 1911–1913.

23 L. V. Moskaleva, S. Rohe, A. Wittstock, V. Zielasek, T. Kluner, K. M. Neyman and M. Bäumer, *Phys. Chem. Chem. Phys.*, 2011, **13**, 4529–4539.

24 X. B. Ge, X. L. Yan, R. Y. Wang, F. Tian and Y. Ding, *J. Phys. Chem. C*, 2009, **113**, 7379–7384.

25 J. Biener, A. Wittstock, M. M. Biener, T. Nowitzki, A. V. Hamza and M. Bäumer, *Langmuir*, 2010, **26**, 13736–13740.

26 M. Hakamada and M. Mabuchi, *J. Mater. Res.*, 2009, **24**, 301–304.

27 J. Biener, G. W. Nyce, A. M. Hodge, M. M. Biener, A. V. Hamza and S. A. Maier, *Adv. Mater.*, 2008, **20**, 1211–1217.

28 X. Y. Lang, L. Y. Chen, P. F. Guan, T. Fujita and M. W. Chen, *Appl. Phys. Lett.*, 2009, **94**, 213109.

29 A. M. Hodge, J. R. Hayes, J. A. Caro, J. Biener and A. V. Hamza, *Adv. Eng. Mater.*, 2006, **8**, 853–857.

30 J. Biener, M. M. Biener, T. Nowitzki, A. V. Hamza, C. M. Friend, V. Zielasek and M. Bäumer, *ChemPhysChem*, 2006, **7**, 1906–1908.

31 C. C. Jia, H. M. Yin, H. Y. Ma, R. Y. Wang, X. B. Ge, A. Q. Zhou, X. H. Xu and Y. Ding, *J. Phys. Chem. C*, 2009, **113**, 16138–16143.

32 A. Schaefer, A. Sandell, L. E. Walle, V. Zielasek, M. Schowalter, A. Rosenauer and M. Bäumer, *Surf. Sci.*, 2010, **604**, 1287–1293.

33 A. Wittstock, V. Zielasek, J. Biener, C. M. Friend and M. Bäumer, *Science*, 2010, **327**, 319–322.

34 G. J. Hutchings, *Catal. Today*, 2005, **100**, 55–61.

35 J. M. Gottfried, K. J. Schmidt, S. L. M. Schroeder and K. Christmann, *Surf. Sci.*, 2003, **525**, 184–196.

36 R. Guttel, M. Paul and F. Schüth, *Catal. Sci. Technol.*, 2011, **1**, 65–68.

37 G. Antczak and G. Ehrlich, *Surf. Sci. Rep.*, 2007, **62**, 39–61.

38 Y. Borchert, P. Sonström, M. Wilhelm, H. Borchert and M. Bäumer, *J. Phys. Chem. C*, 2008, **112**, 3054–3063.

39 U. Diebold, *Surf. Sci. Rep.*, 2003, **48**, 53–229.

40 A. S. Dalton and E. G. Seebauer, *Surf. Sci.*, 2007, **601**, 728–734.

41 J. M. Gottfried, K. J. Schmidt, S. L. M. Schroeder and K. Christmann, *Surf. Sci.*, 2003, **525**, 197–206.

42 R. G. Gordon, D. Hausmann, E. Kim and J. Shepard, *Chem. Vap. Deposition*, 2003, **9**, 73–78.

43 J. Kirn, E. Samano and B. E. Koel, *Surf. Sci.*, 2006, **600**, 4622–4632.

44 N. Saliba, D. H. Parker and B. E. Koel, *Surf. Sci.*, 1998, **410**, 270–282.

General discussion

Prof. Van Santen opened the discussion of the paper by Professor Haruta: Professor Haruta, I am very impressed by your important contribution. You indicated that for bulk Au, small Ag impurities activate the Au catalyst. Has Ag doping been also investigated for the smaller Au particles that show enhanced reactivity when activated by different supports?

Prof. Haruta answered: As far as I know, there has been no investigation of the catalytic activity change induced by Ag_2O doping into small Au nanoparticles which are supported on different supports. We reported that Au fine powder (particle diameters ranging from 20 nm to 200 nm) contaminated by Ag_2O exhibit markedly enhanced catalytic activity for CO oxidation at 273 K.[1,2] The Au fine powder was prepared by Vaccuum Metallurgical Co. Ltd. in Japan by vacuum evaporation in inert gas. We ordered pure Au powder, however, the sample we received was more active than we had anticipated. Therefore, we analyzed the surface compositions of the sample by XPS and bulk compositions by ICP. The results showed that the sample was a little contaminated by Ag, which was enriched on the surfaces. Finally, it was revealed that the evaporation chamber was not perfectly cleaned.

1. Y. Iizuka, A. Kawamoto, K. Akita, M. Date, S. Tsubota, M. Okumura and M, Haruta, *Catal. Lett.*, 2004, **97**, 203-208.
2. Y. Iizuka, T. Miyamae, T. Miura, M. Okumura, M. Date and M. Haruta, *J. Catal.*, 2009, **262**, 280-286.

Dr Louis asked: I would like to mention the work done by Suzanne Giorgio in Marseille (France) using high resolution environmental electron microscopy.[1,2] She observed reversible changes in gold particle morphology upon oxygen, not only in the case of Au/TiO_2 but also in the case of gold nanoparticles on carbon. In the latter case, gold is not active in CO oxidation.

1. S. Giorgio, S. Sao Joao, S. Nitsche, G. Sitja and C. Henry, *Ultramicroscopy*, 2006, **106**, 503.
2. S. Giorgio, M. Cabié, CR. Henry, *Gold Bull.*, 2008, **41**, 1.

Prof. Haruta Thank you very much for introducing me the preceding work by Professor Giorgio and Professor Henry. It was a pity that due to my computer problems I could not show you the movies of TEM observation which has recently been conducted by Professor Takeda of Osaka University. We used 1.8wt% Au/CeO_2 prepared by deposition-precipitation method followed by calcination at 300 °C, not Au/TiO_2. This is because Au nanoparticles (NPs) have stronger bonding with the support and are very stable. As for Au/amorphous carbon which was not catalytically active, Professor Giorgio and Professor Henry described that the shape of Au NPs got rounded after introducing O_2 and many NPs moved during TEM observation. Accordingly, Au/amorphous carbon is not an appropriate sample to discuss the morphology change by exposure to oxygen.

Dr Louis commented: How do you explain the changes in gold particle morphology when Au/TiO_2 is exposed to oxygen ?

Prof. Haruta answered: The morphology change of Au particles on CeO_2 when exposed to O_2 suggests that O_2 is strongly interacting with the surfaces of Au particles, because no appreciable change in morphology of Au particles was observed under ultrahigh vacuum and under the exposure to N_2. The interaction may be dissociative adsorption or hydroperoxide formation. The fact that Au/TiC which was not active for CO oxidation at temperatures below 373 K did not show the morphology change under O_2 exposure suggests that O_2 may be activated at the perimeter interfaces.

Prof. Campbell remarked: Professor Haruta. Thank you for your excellent introductory talk. I think your beautiful data showing the rate of CO oxidation varies with Au particle size on ceria had a maximum TOF that was still ~10x lower than for Au on titania. Is that correct and if so, do you care to comment? Also, did you normalize these TOFs per surface Au atom or per perimeter Au atom or to what?

Prof. Haruta answered: Thank you very much for clarifying very important points in discussing the catalytic activity. Our TOFs were reaction rates per the total number of metal atoms exposed to the surfaces because it can provide fair comparison among different Au catalysts irrespective of active sites. The number of surface metal atoms were calculated from the actual metal loadings and mean particle diameters assuming a hemi-spherical shape. Cerium oxide is not the best support for room temperature CO oxidation and gives TOF lower than that of TiO_2 by several times, probably due to the stronger interactions with Au NPs.

Prof. Meyerstein asked: I wonder whether the radicals formed as intermediates in the catalytic hydrogenations reacts with the Au^0 NPs. The question is based on the observation that methyl radicals react with Au^0 NPs suspended in aqueous solutions at a rate that approaches the diffusion controlled limit.[1] In these reactions long lived intermediates of the type $(Au^0\text{-NP})\text{-}CH_3$ are formed (T. Zidki, R. Bar-Ziv, H. Cohen and D. Meyerstein poster 14 in this discussion).

1. T. Zidki, H. Cohen and D. Meyerstein, *Phys. Chem. Chem. Phys.*, 2006, **8**, 3552–3556.

Prof. Kiely asked: Your beautiful work shows a clear change in mechanism for the CO oxidation process at around 60 °C, which you interpret as a changeover from periphery sites dominating the action at low temperature, to gold atom sites on the nanoparticle top surface being more important at high temperatures. On which oxide supports have you observed this phenomenon, and does the transition always occur at around 60 °C irrespective of the support identity?

Prof. Haruta responded: We observed a change in reaction mechanism at 60 °C for CO oxidation over two types of Au/TiO_2 catalysts, real powder catalysts and single crystal model catalysts. We have published some data of kinetic investigation of Au/Co_3O_4 and Au/Fe_2O_3 in addition to Au/TiO_2.[1] All these three gold catalysts were prepared by the deposition-precipitation method to exclude raft-like patches as much as possible. We have not yet checked whether the transition of reaction mechanism always occurs at 60 °C or not, however, because the apparent activation energies and reaction orders with respect to CO and O_2 were similar to each other, I assume that transition may take place at around 60 °C irrespective of the type of metal oxide supports.

1. M. Haruta, S. Tsubota, T. Kobayashi, H. Kageyama, M. J. Genet and B. Delmon, *J. Catal.*, 1993, **144**, 175–192.

Prof. Madix said: Professor Haruta, thank you for this very interesting summary of the excellent work on these reactions of gold catalysts. In your paper you show two different rate limiting processes above and below 60 °C, according to your Arrhenius plots. What are the rate limiting steps for these respective processes?

Prof. Haruta answered: For CO oxidation under the partial pressure of CO and O_2 above 0.1 kPa, the rate determining step is the reaction between CO adsorbed

    This journal is © The Royal Society of Chemistry 2011

on Au surfaces and O_2 adsorbed at the perimeter sites (below 60 °C) or at Au surfaces (above 60 °C) . The rate dependency on the concentration was, in rough approximation, zeroth order both to CO and O_2 over Au/Al_2O_3, Au/TiO_2, Au/Co_3O_4, and Au/CeO_2. Zeroth order reaction is surprising because with a decrease in reactant concentration the conversion increases. I really encountered-this phenomena. When partial pressure of CO and O_2 is below 0.1 kPa, the rate dependency is first order with respect to CO and is 0.5th order at temperatures above 60 °C and first order with respect to both CO and O_2 below 60 °C. The rate determining steps under these conditions are being investigated.

Prof. Bowker asked: For the pure gold single crystals you still appear to get the step in activation energy at a particular temperature. It seems to behave similarly to the supported catalyst in this respect, but you consider the low temperature behaviour to be due to a metal–support interaction. How can this be the case if the phenomenon is also there for pure gold?

Prof. Haruta responded: If I understand your comments correctly, you are comparing the kinetic behavior of the two catalyst samples we prepared. One sample is a Au/TiO_2 powder catalyst prepared by deposition precipitation (wet method) and another is not a pure gold single crystal but a model catalyst prepared by depositing Au onto the surfaces of rutile TiO_2 single crystal with a cathodic arc plasma method (dry method). The step in activation energy at around 60 °C, very low (2 kJ mol^{-1}) activation energy at higher temperatures, and relatively high (25-40 kJ mol^{-1}) at lower temperatures are similar to each other, indicating that the two catalyst samples are nearly identical. The behavior of Au/TiO_2 catalysts at temperatures above 60 °C may be the same as that of a pure gold single crystal in a wider range of temperature down to cryogenic temperatures.

Prof. Bowker asked: Regarding CO oxidation on silver, Bob Madix, Mark Barteau and I found that CO oxidation is facile on a Ag(110) crystal.[1] In fact it has a negative activation energy—going faster at low temperature than at room temperature. Perhaps we can use silver nanoparticles for low temperature CO oxidation? Of course it could be that O_2 is too strongly bound, but it should work well with CO in excess.

1. M. Bowker, M.A. Barteau and R.J. Madix, *Surf. Sci.*, 1980, **92**, 564.

Prof. Haruta answered: Thank you for providing valuable knowledge concerning silver. I once worked on silver oxide for the oxidation of CO and H_2 and remembered that it was active just above room temperature. As you have already assumed, in the excess of O_2 silver nanoparticles are readily oxidized into Ag_2O showing decreased catalytic activity in the presence of moisture. It is interesting to know whether oxidized Ag nanoparticles are more or less active than metallic nanoparticles under completely dry conditions. In the excess of CO or in the stream of H_2 (H_2 purification for fuel cells), as you indicated, silver may work efficiently at ambient temperature.

Prof. Friend asked: How do you exclude the possibility that O atoms migrate from the titania support to the Au particle to then be oxidized by CO? I agree that the interface is very important, but that O could migrate to Au from the support. O on Au readily reacts with CO to produce CO_2. Water would potentially facilitate the migration of oxygen *via* OH to Au. OH is unstable with respect to disproportionation to water and O on Au—shown by our group and Prof. Mullins?[1,2] So that this would be a possible mechanism for O migration to Au from the support. Would you care to comment?

1. R. G. Quiller, T. A. Baker, X. Deng, M. E. Colling, B. K. Min and C. M. Friend, *J. Chem. Phys.*, 2008, **129**, 064702.
2. Tae S. Kim, Jinlong Gong, Rotimi A. Ojifinni, J. M. White and C. Buddie Mullins, *J. Am. Chem. Soc.*, 2006, **128**, 6282–6283.

Prof. Haruta replied: I appreciate your comments and references. We don't exclude a possibility that O atoms migrate from TiO_2 support to the Au particles. It happens when temperature is above 60 °C, where reaction between adsorbed CO and adsorbed O atoms can predominantly proceed over the surfaces of Au nanoparticles. Our perimeter mechanism is valid at temperatures below 60 °C, where CO adsorbed on the Au surfaces reacts with negatively charged molecular oxygen at the peripheral sites. The difference in the reaction sites was supported by the linear dependence of reaction rates on the number of surface Au atoms (above 60 °C) and on the number of perimeter Au atoms (below 60 °C).

Prof. Hutchings remarked: In your model you indicate that for a supported gold nanoparticle the oxygen is activated at the gold atoms in contact with the support and the CO is activated initially at the gold atoms in the higher layers. If this is the case then this would suggest that an extended bi layer structure would be a very effective catalyst and the optimal use of gold. These structures need not be large in diameter to ensure maximum effective use of gold. Of course the important factor will be how to make dispersions of these bi layer structures. Could you comment on this?

Prof. Haruta responded: You are smart to assume an extended bi-layer structure as the most active and atom efficient catalyst, because it incorporates direct support contribution to Goodman's bi-layer hypothesis. It is likely that two atom or three atoms layer is optimum, not too thin but not too thick (minimize the number of atoms which do not participate in reaction). Our work on model Au/TiO_2 catalysts showed that three atoms layer had the same electronic state as that of bulk gold and two atoms layer began to deviate a little from bulk gold.[1] I am not sure whether Professor Goodman refrained to discuss or neglected the direct contribution of metal oxide supports. In real powder catalysts raft structured gold patches are surrounded by metal oxide supports and their catalytic activity dramatically changes depending on the kind of supports. Accordingly, it is reasonable to take into account of the support contribution.

1. Y. Maeda, M. Okumura, S. Tsubota, M. Kohyama, M. Haruta, *Appl. Surf. Sci.* 2003, **222**, 409.

Prof. Hutchings opened the discussion of the paper by Prof. Rotello: Is it possible to fine tune the activity of the nanoparticles by alloying gold with other metals. While I appreciate there may be toxicity issues, such an approach may still be useful for *in vitro* studies

Prof. Rotello responded: Alloying particles provides a very nice strategy for functionalising particles with multiple ligand structures. As we demonstrated using FePt particles[1], we can functionalise these particles with very different ligands without phase segregation. Particles featuring controllable mixed monolayers would certainly expand the envelope of surface structures we could generate, potentially enhancing our sensor systems

1. R. Hong, N. O. Fischer, T. Emrick and V. M. Rotello, *Chem. Mater.*, 2005, **17**, 4617.

Prof Dr Bernhardt commented: In your investigations, did you observe an influence of the size of the nano-particles on the selectivity or the sensitivity of the sensing process?

     This journal is © The Royal Society of Chemistry 2011

Prof. Rotello answered: Changing the nanoparticle size has a profound effect on sensor behavior. In our research we focus on 2 nm core diameter gold nanoparticles (7–8 nm overall). These particles are far more resistant to aggregation than larger particles with 4–6 nm diameter cores. This enhanced resistance to aggregation simplifies sensor use and increases sensitivity and selectivity. While not mentioned widely, these smaller particles are at a "sweet spot" as far as biological applications are concerned.

Mr Serapian said: My name is Stefano Serapian and I was a fellow delegate at the Faraday Discussion 152. I have found your talk at the conference very interesting indeed. The aim of my PhD project is to simulate thiol adsorption on pristine gold nanoclusters, therefore the systems I work with are affine to the functionalised biosensing nanoparticles presented by your group.

My question concerns the traditional Brust method you have employed to synthesise your nanoparticles. It has been hotly debated whether, upon adsorption, the thiolic hydrogen is lost or remains bound to the sulfur; in his original paper, Brust does not report hydrogen loss and implies that the hydrogen is retained. The matter is understandably of lesser concern to experimentalists, but it becomes crucial when trying to properly reproduce the adsorption process. Indeed, both theoretical and experimental studies involving methanethiol as the passivating layer have not detected hydrogen loss. However, it appears in several other studies that with longer chains the hydrogen loss is much more facile, as reviewed by Vericat et al.[1], and for example found by Nadler.[2]

I therefore wanted to ask you, out of curiosity, whether during your syntheses you have ever found evidence of hydrogen gas evolution or hydrogen loss, and, if I may, what is your personal view on the matter.

1. Vericat et al., Chem. Soc. Rev., 2010, **39**, 1805.
2. R. Nadler, R. Sánchez-de-Armasa and J. F. Sanz, Comput. Theor. Chem., 2011, in press.

Prof. Rotello replied: In the Brust method sodium borohydride is used as the reducing agent. Addition of borohydride to the gold salt solution generates a considerable among of hydrogen gas through decomposition of the borohydride. As a result, we are unable to observe whether gas evolution occurs upon thiol binding. My personal view is that the structure and reactivity of the thiols suggests that the particle is bound as a thiolate, requiring hydrogen evolution. That being said, my convictions on this matter are flexible.

Prof. Friend commented: Did you study different ligand lengths and different densities of sensing agents, which might affect the fluorescence quenching on one hand or lead to fluorescent resonant energy transfer on the other?

Prof. Rotello responded: Our particles show efficient (85–95%) but not complete quenching of attached fluorophores. This range is ideal for our application, as we have a large dynamic range with some capability for further quenching that can provide additional sensing capability. I think it would be an interesting idea to use slightly longer ligands or smaller particles that would provide less efficient quenching, say 60%, providing a greater range for stronger quenching processes.

Dr Jupille remarked: The study shows that gold nanoparticles that are made cationic by ligands act as biosensors for proteins *via* a pattern recognition process which involves the proteins and pre-adsorbed fluorescent probes. In the buffer solution at a pH of 7.4, the proteins that have an isoelectric point (IEP) higher than 7.4 are negatively charged and should be repelled by the cationic gold. In these conditions, I wonder how these proteins could participate in the negative binding to the fluorescent probes that is pivotal in the sensing process.

Proteins show a large variety of isolelectric points that range from 4.6 to 11.0 (Table 1). Is there a strategy for proteins recognition that could be based on the IEP of other components of the gold polymer/biopolymer complexes such as, for example, the fluorescent molecules?

Professor Rotello responded: Excellent questions. There are two factors that allow binding between cationic particles and cationic proteins. First, protein surfaces are heterogeneous in terms of charge, with *e.g.* anionic patches on overall cationic proteins. Second, many of the ligands have a strong hydrophobic character, allowing interactions between like charged proteins and nanoparticles.

Your second suggestion of varying pH to target isoelectric points would indeed add an additional element to our sensor that could allow differentiation of many more analytes than measurement at a single pH. We are currently exploring this approach now for our protein sensing studies.

Professor Hutchings asked: Is it possible to add a mixture two different colloids that could target different molecules; in such a way perhaps you design a system capable of addressing multiple targets but of course this would be very complicated.

Professor Rotello responded: One of our key goals is multiplexed sensing, *i.e.* having multiple output channels in a single sensor element. Our strategy to accomplish multiplexing is to engineer particles and fluorescent polymers/proteins to be complementary. This would generate a different color of response if the analyze binds to different particles. You are quite correct that generating this sort of sensor is complicated, however we believe sensor systems of this sort are accessible.

Professor Campbell commented: I am surprised that you can detect and quantify proteins in a complex mixture of proteins with a collection of receptors that have such low selectivity differences in their binding constants to different different proteins, especially since their concentrations in physiological mixtures can differ by many orders of magnitude. Have you done some preconcentration in the mixtures you analyze, and if not, can you please explain this? That is, how can a receptor with a binding constant that is only 30 times larger for protein A than protein B not be 'fooled' by a mixture containing 30 times more of protein B?

Professor Rotello replied: For our serum sensing we worked in undiluted and unmodified serum, with no pre-concentration. You are quite correct that sensor responses can become overwhelmed by proteins that are present in much higher amounts. That being said, our noses can detect very small quantities of analytes in complex matrices, *e.g.* the aroma of flowers in a cow pasture. The key to this detection process is data analysis, an area that animals are much more advanced than computer algorithms. As a result, we are actively exploring strategies to remove high concentration proteins such as serum albumin, enhancing our ability to detect less abundant proteins.

Professor Hutchings said: With regards to the ligands used to make the gold colloids, you emphasise the usefulness of the Brust thiol ligands as being particularly effective. Is the effect unique to using sulphur-containing ligands? Can you get effects with other ligands?

Professor Rotello answered: Thiol-based ligands are particularly useful for generating stable nanoparticles. While electrostatic interaction have been used to stabilize gold nanoparticles, these interactions are much less stable in biological fluids. Other aurophilic ligands such as phosphines can be used to stabilize nanoparticles, but they are generally both less effective at stabilizing the particle in biofluids and they are synthetically more challenging.

     This journal is © The Royal Society of Chemistry 2011

Dr Xu commented: This nice approach takes advantage of the interaction between proteins, fluorescent probe molecules, and functionalized Au NPs to differentiate analytes. Experimental training is necessary to determine what fluorescence pattern a particular analyte would generate. Is there any current computational modeling technique sufficiently accurate and fast to allow you to predict such fluorescence patterns and associated statistical margins without training? If not, do you envisage this becoming a reality in the near future? Such an ability could significantly increase the flexibility and applicability of this protein-sensing technique.

Professor Rotello replied: Our current method requires the generation of a training set, a reasonable task for identifying individual components, be they small molecules, proteins, or cells. Analysis of mixtures rapidly becomes more challenging as combinatorial issues arise. Techniques are being developed that apply artificial intelligence strategies such as neural networks to this challenge.[1] We are starting to apply these methods to our sensor work however the results are very preliminary at this point.

1. E. A. Baldwin, J. H. Bai, A. Plotto, and S. E. Dea, *Sensors*, 2011, **11**, 4744–4766.

Professor Hutchings said: In your talk and paper you have tried to be very provocative. What are the factors that will stop your methodology being used in applications? When can we expect it to be applied soon?

Professor Rotello replied: There are two main challenges to real-world implementation of chemical nose sensing. The first issue is data analysis, which requires the proper choice of systems to sense, both in terms of complexity and in the feasibility of artificial olfaction data analysis strategies. The second challenge in implementing "nose/tongue" based strategies is the perceptions of the community. As an example, our sensors can readily distinguish between cell states. We do not, however, know the origin of this differentiation. As scientists are trained to think in terms of cause and effect, it can be difficult for many researchers to accept a non-hypothesis based sensing system.

We and others are working to address both issues. There are already examples of "chemical nose" sensors for volatile chemicals on the market, in particular in the area of security/anti-terrorism. I expect that sensors of this sort will be appearing in the biomedical world in the next 3–5 years addressing a variety of health concerns.

Dr Rodriguez opened the discussion of the paper by Dr O'Mullane: The authors ascribe the peaks at 0.28, 0.55 and 1.08 V to the early oxide or OH formation on "un-coordinated" active sites. How do the authors define "un-coordinated"? Additionally, If the authors read the paper by A. Hamelin[1] , they will find that the peak at 1.08V (anodic) and at 0.78V (cathodic) is ascribed to the oxidation/reduction of the Au(110) sites as is expected on a polycrystalline electrode. However the authors insist that they cannot do a proper comparison with the single crystal. Although they are presenting voltammograms corresponding to a polycrystalline electrode, the surface must present the thermodynamically most stable surface orientation. It is a common practice in the literature, to establish a comparison with single-crystal even for gold thin film deposition, single-crystal electrobeam deposition, gold nano-particles, *etc.*[2,3] The authors must do it as well.

1. A. Hamelin, *J. Electroanal. Chem.*, 1996, **407**, 1–11.
2. J. M. Delgado, J. M. Orts, J. M. Pérez and A. Rodes, *J. Electroanal. Chem.*, 2008, **617**, 130–140.
3. Th. Waldoski, K. Ataka, S. Pronkin and D. Diesing, *Electrochim. Acta*, 2004, **49**, 1233–1247.

Dr O'Mullane answered: The term un-coordinated (in the introduction) should have been written as low-coordinated as detailed later on page 4. It is suggested that the surface must present the thermodynamically most stable surface orientation.

This is not necessarily the case. This is the advantage of creating nanostructured surfaces, as opposed to near defect free single crystal surfaces, in that highly energetic surface orientations can be readily fabricated. See for instance the work of Wang *et al.* who created Pt nanocrystals with electrochemical methods whose surfaces were enclosed with high index facets such as (730) and (520) which have a large density of atomic steps and dangling bonds which are both chemically stable and thermally stable to 800 °C.[1] It is these high energy active sites which were responsible for the electrocatalytic activity observed for the oxidation of small organic molecules. In essence the goal of this work is encompassed by the statement of Somorjai that "rough surfaces do chemistry".[2]

1. Na Tian, Zhi-You Zhou, Shi-Gang Sun, Yong Ding and Zhong Lin Wang, *Science*, 2007, **316**, 732–435.
2. G. A. Somorjai, *Chem. Rev.*, 1996, **96**, 1223.

Dr Rodriguez asked: The authors mention that the electrode is activated at negative potential with hydrogen evolution; however the potential is not mentioned. Could you provide the negative polarization potential?

Dr O'Mullane answered: The activation potential is −1.5 V as stated in the caption of Figure 1.

Professor Bowker asked: You mentioned the activation of the electrode in acid solution. This is a somewhat mysterious process, and can you explain exactly where the hydrogen comes from?

Dr O'Mullane replied: Protons in the solution are reduced to H atoms which combine together to form molecular hydrogen gas. If this reaction occurs within defects on the gold surface, *i.e.* below the surface layers then expulsion of hydrogen gas from the metal lattice will result in severe disruption of the outer layers of the material. See the paper by Burke *et al.* which illustrates by SEM and AFM this type of surface modification after extensive hydrogen evolution at gold electrodes.[1]

1. D. L. Burke, A. P. O'Mullane, V. E. Lodge and M. B. Mooney, *J. Solid State Electrochem.*, 2001, **5**, 319.

Professor Bowker said: During the hydrogen evolution where does the surplus sulphate from the loss of protons go? Is it deposited somewhere, on the Au surface for instance?

Dr O'Mullane answered: During the hydrogen evolution reaction the counter electrode reaction is the oxygen evolution reaction which results in the creation of protons. The sulphate ions are known to specifically adsorb on gold surfaces under potential control. However, *ex situ* XPS measurements of the gold surfaces showed no detectable sulphur present on the surface.

Professor Friend remarked: Does this mean that the oxide has a different stability for different morphologies? Have you tried to use *in situ* tools, *e.g.* EXAFS, to determine the size and bonding of oxides in your material? It should be possible to perform *in situ* measurements using hard X-ray sources. Otherwise, your conclusions seem speculative.

Dr O'Mullane answered: We believe that the oxide formed has different stability in solutions of different pH where the stability of the oxide is greatest at high pH levels as determined by FT-ac voltammetry which shows quasi-reversible processes in the double layer region. It is also most likely to be morphology dependent in that

     This journal is © The Royal Society of Chemistry 2011

hierarchical nanostructures contain many active sites whose activity and ease of oxidation will vary. Unfortunately we have not tried *in situ* measurements which need to be performed as *ex situ* methods such as FTIR, XPS and Raman have proven to be inconclusive.

Professor Hutchings asked: You have indicated that you would like to try to determine the nature of the oxygen species present in your catalytic systems. Is it possible to use EPR spectroscopy and perhaps to combine EPR spectroscopy as an *in situ* technique.

Dr O'Mullane replied: I am unaware if a dynamic *in situ* EPR experiment could be carried out in an electrochemical environment to give the speciation of the gold surface. However, as suggested by Prof. Friend an *in situ* EXAFS experiment will be undertaken.

Professor Rotello asked: Did you use ac voltammetry to further probe the electrochemical behaviour of your materials? If so, were you able to deconvolute at least in part the peaks? And finally, was there any correlationm between the number of peaks and the morphology of the materials?

Dr O'Mullane responded: An FT-ac voltametric study was carried out on all the materials presented. Under alkaline conditions distinct processes could be observed which were resolved from each other. This is demonstrated in Figure 8 which shows the behaviour for honeycomb gold which shows resolved processes at −0.80, −0.20 and 0.20 V. There also seems to be a correlation between morphology and the observed processes which is more evident under acidic conditions. For evaporated gold which has ordered well packed crystallites only one process was observed at 0.80 V (Figure S4). For the anisotropic hedgehog like structures two distinct processes were seen at 0.30 and 1.10V (Figure S3) whereas at honeycomb gold, processes were observed at 0.40, 0.60 and 1.10 V with an indication of a further process at −0.10 V (Figure 7). Therefore the more anisotropic or dendritic like the material the higher number of active site processes that are present.

Professor Rotello asked: In structure–activity correlation, activity is the dependent and structure the independent variable. Have you thought about how you could characterise your structures quantitatively so that you could gain predictive power? Along similar lines, have you generated sufficient intermediate structures to provide interpolation between behaviours?

Dr O'Mullane answered: The ability to gain predictive power in terms of structure–activity relationships is an interesting area of research. Many groups have studied the effect of crystallographic orientation using single crystal electrodes on electrocatalytic activity (R. R. Adzic, S. Trasatti, G. Ertl, N. M. Markovic and M. J. Weaver as some examples) and some have related this to DFT calculations in which effects such as adsorption of reactant and electrolyte anions and bond breaking were investigated. However, relating this information to nanostructured materials is challenging in particular when it has been identified that low co-ordinated atoms such as those that sit at defect sites, kinks, steps and edges dominate the activity of the overall material. Theoretical calculations may offer insights into the optimum surface coverage of atoms that are required at such sites to facilitate electrocatalytic reactions. However, I would suspect that each electrocatalytic reaction would have to be separately studied to determine the optimum surface composition as it can be seen from the presented data that electrocatalytic reactions occur at distinctly different electrode potentials. Regarding the second point generating a series of nanostructures from say spherical to highly anisotropic structures would be extremely useful and would make an interesting study. This type of shape evolution is possible to achieve using electrochemical methods.[1]

1. Yan-Xin Chen, Sheng-Pei Chen, Zhi-You Zhou, Na Tian, Yan-Xia Jiang, Shi-Gang Sun, Yong Ding and Zhong Lin Wang, *J. Am. Chem. Soc.*, 2009, **131**, 10860–10862.

Dr Xu asked: What are the onset potentials for H_2O_2 reduction and oxidation respectively on an Au(111) electrode? It is important to compare them with the results presented in Figure 2 and avoid possible ambiguity that the observed processes are merely catalyzed by the close-packed Au(111) facet and not by special surface sites.

Dr O'Mullane replied: As described by El-Deab *et al.*, the onset potential for oxygen reduction under similar experimental conditions is *ca.* 0.025 V *vs.* Ag/AgCl for a Au(111) single crystal electrode.[1] In our work at porous gold the onset potential is at a significantly more positive potential of 0.22 V (Figure 9b) which indicates that closed packed Au(111) facets are not responsible for the activity observed. In relation to hydrogen peroxide oxidation the onset potential, at porous gold in particular, is *ca.* 0.34 V less positive than at a planar gold electrode (compare Figure 9a and Figure 2d). Given that it is generally accepted that the Au(111) facet is the least electrocatalytically active facet of the low index planes for the majority of reactions it is unlikely that this significant shift in onset potential is due to the creation of more Au(111) facets.

1. M. S. El-Deab, K. Arihara, and T. Ohsaka, *J. Electrochem. Soc.*, 2004, **151**, E213.

Mr Pritchard remarked: Has a tri-metallic system (*i.e.* Au/Pd/Pt) been considered experimentally based on the co-deposition of two metal precursors onto the gold film?

Dr O'Mullane answered: A tri-metallic system has not been tried yet. I believe that a system such as Au/Pd/Pt is certainly achievable. The extent of coverage of Pd and Pt will depend on the thermodynamic driving force of the reaction. It is expected that there may be more Pd decoration as the standard reduction potential for the Pd^{2+}/Pd couple is 0.915 V *vs.* SHE compared to 0.758 V for the $PtCl_4^{2-}$/Pt couple. This would provide an increased driving force of 157 mV for the oxidation of the active sites in the case of Pd.

Mr Pritchard commented: With regard to decorating the gold sites with a secondary metal precursor, electrochemical experiments (0.52V) suggest that not all of the gold active sites are decorated with Pd metal?

Dr O'Mullane replied: That is correct; there are still premonolyer oxidation responses in the double layer region (inset of Figure 12b) after immersion of the sample in the palladium salt solution suggesting that not all the active sites were chemically oxidised.

Dr Rodriguez commented: Regarding the answer of the author, in all the publications that I mentioned (gold nanostructured surfaces) the authors did a proper comparison with single-crystal electrodes and they did the respective assignation of the OH adsorption and oxide formation. Of course any gold surface will have defects, even the single-crystal electrodes present defects, but the surface will have a predominant orientation. The authors mention the work of Wang in Science. I assume that the authors are refereeing this paper trying to justify the presence of low-coordinated atoms on the surface. Well, even the high-index surface of Pt (730) and Pt(520) have periodical domains and the voltammogram of those

   This journal is © The Royal Society of Chemistry 2011

nanoparticles must behave very similar to the single-crystal electrodes with the same orientation (see works of Prof. Juan Feliu for Au and Pt step surfaces and nanoparticles). Tian *et al.* clearly mention and show with HRTEM images that the high-index nanoparticles have periodicity of (111) terraces and (100) steps with terraces length of 3 and 2 atoms.[1] If the authors are suggesting in this paper that their surface, as in the case of the nanoparticles, present a very specific reactive site, they also have to mention which kind of orientation present this site. They also have to demonstrate that they have it. Could the authors provide an STM image and XRD of the surface before and after the polarization treatment?

Moreover, they try to support his answer citing a platinum nanoparticles article. Small and spherical nanoparticles do not present any preferential orientation, the adsorption sites are low coordinated sites. So, gold nanoparticles with these characteristics should present, without no doubt, the sites that they are claiming to have.

1) The features in the cyclic voltammograms of figures 2 and 6 that the authors associate to actives sites are not present in the voltammetric profiles of Au nanoparticles of 4nm.[2] The features in the cyclic voltammograms of figures 2 and 6 that the authors associate to actives sites are not present in the voltammetric profiles of Au nanoparticles of 1 to 20 nm.[3] Additionally and unfortunately a proper comparison in the oxygen reduction region can not be established because the authors are not doing standardization per area.

3) The features in the cyclic voltammograms of figures 2 and 6 that the authors associate to active sites are not present in the voltammetric profiles of Au polycrystalline nanoparticles of 19 nm.[4] I also have to say that the statement "rough surfaces do chemistry" of Prof. Somorjai is cited out of context, because not always the rough surfaces do the chemistry. I can mention several examples of catalytic reaction that takes place on the terraces and defect free surfaces are beneficial: the oxidation of ammonia on Pt(100),[5] the reduction of nitrite[6] or the dimethyl ether oxidation.[7] Additionally the author are trying to ascribe that all the electrochemical reactions reported in the paper takes place at this particular site. The oxygen reduction for example is surface sensitive reaction and Au(100) was shown to be the most active, whereas Au(111) is the least active basal plane. Previous studies with single-crystal electrodes and with nanoparticles (spherical, rods and well-defined nanoparticles) demonstrate that the defects are not required for the reduction of oxygen on gold electrodes and indeed (100) sites are preferred.

1. Na Tian, Zhi-You Zhou, Shi-Gang Sun, Yong Ding and Zhong Lin Wang, *Science,* 2007, **316**, 732.
2. Javier Hernández, José Solla-Gullóna, Enrique Herrero, Antonio Aldaza and Juan M. Feliu, *Electrochimica Acta*, 2006, **52**, 1662–1669.
3. Ave Sarapuu, Margus Nurmik, Hugo Mändar, Arnold Rosental, Timo Laaksonen, Kyösti Kontturi, David J. Schiffrin and Kaido Tammeveski, *J. Electroanal. Chem.*, 2008, **612**, 78–86.
4. María G. Montes de Oca and David J. Fermín, *Electrochimica Acta*, 2010, **55**, 8986–8991.
5. F. J. Vidal-Iglesias, J. Solla-Gullón, P. Rodríguez, E. Herrero, V. Montiel, J. M. Feliu and A. Aldaz, *Electrochem. Comm.*, 2004, **6**, 1080-1084.
6. M. Duca, M. O. Cucarella, P. Rodriguez and M. T. M. Koper, *J. Am. Chem. Soc.*, 2010, **132**, 18042–18044).
7. Yujin Tong, Leilei Lu, Yi Zhang, Yunzhi Gao, Geping Yin, Masatoshi Osawa and Shen Ye, *J. Phys. Chem. C*, 2007, **111**, 18836.

Dr O'Mullane responded: In this work it is not claimed that a single specific reactive site is responsible for the electroactivity observed but rather a distribution of active surface sites whose coverage varies from surface to surface, as demonstrated by the FT-ac responses in acid for electrodeposited gold on ITO, evaporated gold and honeycomb gold. Identifying the nature of these active sites is extremely difficult as conventional XRD does not provide such information. We are trying to pursue HRTEM imaging combined with electron diffraction characterisation to identify these active sites which is an ongoing effort, however making a direct assignment between the polycrystalline surfaces and the active state responses will still be challenging. The features observed in Figures 2a and 6b (obtained in 1 M H_2SO_4) are suggested to not exist on a variety of previously reported gold nanoparticles. Comparison with J. Hernandez *et al.*[1] is difficult as this study was carried out in alkaline conditions. For the other examples we believe that dc voltammetry is not definitive proof that the surface is free of active site behaviour or defect sites, as these responses can be easily masked in the double layer region of the dc voltammograms. For this reason electrochemical activation of electrode surfaces has been pursued to amplify the dc response of active sites which are believed to be present on all surfaces. A major point in this paper is that to gain more insights into the electrochemical behaviour of nano-materials that FT-ac voltammetry is a good complimentary technique which allows Faradaic processes to be observed that are clearly discriminated from capacitive processes. Bond *et al.* have shown that this technique can be used to probe the active site behaviour of gold[2] and copper[3] which show significant Faradaic behaviour in the double layer region of unactivated electrodes which have been correlated with electro-catalytic activity. In general, the role of low index planes and specific adsorption of anions in electrocatalytic reactions is accepted to certainly play a role in determining activity, however, this work strives to throw light on the fact that additional processes (such as oxidation of active sites) are present during electrocatalytic reactions which have a significant influence. It is certainly accepted that not all reactions require a high coverage of defect sites, however I would have to say that in general defective surfaces are very good catalysts.

1. Javier Hernández, José Solla-Gullóna, Enrique Herrero, Antonio Aldaza and Juan M. Feliu, *Electrochimica Acta*, 2006, **52**, 1662–1669.
2. B. Lert-Anantawong, A. P. O'Mullane, W. Surareungchai, M. Somasundrum, L. D. Burke, A. M. Bond, *Langmuir*, 2008, **24**, 2856.
3. M. J. Shiddiky, A. P. O'Mullane, J. Zhang, L. D. Burke, A. M. Bond, *Langmuir*, 2011, **27**, 10302.

Dr Rodriguez asked: The author mentions during the discussion that some authors propose in the literature the adsorption of sulfate, however they do not mention nothing about it in the manuscript. For the knowledge of the authors, the adsorption of sulfate on gold was extensively by thermodynamics studies, electrochemical STM and FTIR. Moreover, it is well-known that the adsorption of sulfate (and other anions) cause the lifting of the reconstruction of the surfaces.[1-5] Additionally the adsorption of anions on the surface play a role in the catalytic activity of gold (i.e the oxygen reduction in HClO4 or H2SO4). In the paper it is not mention neither the effect of the pzc (potential of zero charge) in the catalytic properties of the gold and how the sulfate adsorption and the lifting of the reconstruction affect the catalytic process of the gold.

1. A. Hamelin, *J. Electroanal. Chem.*, 1996, 407, 1–11.
2. A. Hamelin, *J. Electroanal. Chem.*, 1996, 407, 13–21.
3. J. Schneider and D. M. Kolb, *Surf. Sci.*, 1988, **193**, 579–592.
4. D. M. Kolb and J. Schneider, *Electrochimica Acta*, 1986, **31**, 929–936
5. D. M. Kolb, *Prog. Surf. Sci.*, 1996, **51**, 109–173.

Prof Dr Bhargava responded: We acknowledge that the adsorption of electrolyte ions will play a role in certain electrocatalytic reactions by blocking a proportion of surface active sites. It is accepted that electrocatalytic reactions are also likely to be affected by the pzc, however this issue is extremely complicated as pointed out in Faraday Discussion 140—General Discussion of session 1 where it was commented that the onset potentials for the oxygen reduction reaction on Ag(111) and Au(111) surfaces differed significantly from the difference in the pzc of each electrode. Therefore using the pzc as an insight into electrocatalytic activity may be problematic.

Dr Rodriguez said: The authors criticize or question the determination of the area of the electrode by using the charge associated to the gold oxide reduction. In the manuscript the voltammograms are not standardized per area which made impossible to establish a proper comparison with previous results. It is not possible to conclude that the methods reported by the authors create active sites if we can not do a comparison with previous results. Could the author justify this? In the literature, it is possible to find any kind of gold surfaces (films, nanoparticles, etc) and in all the cases the authors determine the area of the electrode by using the reduction of gold oxide charge.

Prof Dr Bhargava replied: The method of calculating the surface area of gold from the charge associated with the oxide reduction process is a valid approach and is not criticised in the paper. The point raised was that when additional cathodic peaks are observed as shoulders on the main oxide reduction process then the value attained using this method may be affected.

Dr Rodriguez remarked: By using FT-alternating current voltametry the authors suggest that at low potential OH adsorption takes place in sulfuric acid 1M. This is thermodynamically imposible, the concentration of protons is too high to form OH on the surface at those potentials (0.28-0.4!).Instead of the OH adsorption in this media, the authors must take into account the adsorption of sulfate and (bi)sulfate which present a higher concentration. Additionally they must establish a quasi-quantitative comparison with results obtained by EC-STM. On the other hand, in alkaline media the adsorption of OH takes place slowly between aprox 0.35 and 1 V *vs.* RHE reaching a coverage of 0.11 (32 C cm^{-2}) before the formation of the gold oxide but this was demonstrated by Lipkowski using thermodynamics analysis and FTIR.[1]

1. J. Lipkowski, *J. Phys. Chem. B*, 1999, **103**, 682–691.

Prof Dr Bhargava replied: It is stated in the paper that the oxidation of active sites is a hydrous oxide type deposit, previously formulated to be $[Au_2(OH)_9]^{3-}$ and not simply as OH adsorption. It can be treated as a hydrated oxide. The adsorption of sulphate may contribute to the dc responses observed but the FT–ac responses clearly demonstrate quasi-reversible electron transfer processes in the double layer region which are unlikely to be due to adsorption of sulphate.

Dr Rodriguez commented: In the introduction the authors must include the papers of Roberts and Kita.[1–3] These authors' report for the very first time the extraordinary catalytic activity of Au towards the oxidation of CO even before the work of Prof. Haruta in gas phase.

1. J. L. Roberts Jr. and D. T. Sawyer, *Electrochim. Acta*, 1965, **10**, 989–1000.
2. J. L. Roberts Jr. and D. T. Sawyer, *J. Electroanal. Chem.*, 1964, **7**, 315–319.
3. H. Kita, H.Nakajima, K. Hayashi, *J. Electroanal. Chem.*, 1985, **190**, 141–156.

Prof Dr Bhargava answered: These references have been noted.

Dr Nijhuis opened the discussion of the paper by Prof. Kiely: This question is related to the catalyst deactivation during the hydrogen peroxide formation and decomposition over your catalysts: Did you analyze the gas phase in your autoclave during or after your catalytic experiments? I am asking this question specifically, since according to my calculations for your hydrogen peroxide decomposition experiments after 2 min most if not all of the hydrogen will be depleted in the autoclave (I estimate 3 mmol of hydrogen being present at the start of the experiment, while 10 mmol of hydrogen peroxide is present). Can the depletion of gas reactants be the cause of the reaction slowing down, rather than a true deactivation occurring ?

Prof. Kiely responded: The reaction system is quite complex and a number of factors are involved. We carry out the reactions in a sealed stirred autoclave with a fixed amount of H_2 at the start of the reaction. As the reaction proceeds, so hydrogen peroxide is produced and the H_2 concentration decreases. The hydrogen peroxide that is generated can then react further by two pathways, namely (a) decomposition and (b) hydrogenation.[1] Clearly, hydrogenation will decrease the concentration of hydrogen further. As the concentration of hydrogen peroxide builds up with time, so the sequential reactions play a progressively more significant role. Hence the rate of formation of hydrogen peroxide is the composite of several reactions as well as the concentration of H_2. In reality it is not readily possible to deconvolute these pathways and obtain a measure of whether true deactivation is occurring. It is for this reason that we have investigated the reaction at very short reaction times, since in this regime the concentration variation in H_2 will be minimal and the effect of sequential reactions is also minimized. These data present a realistic image of the initial activity of these sol-immobilised catalysts and confirm that they are very active for both hydrogen peroxide hydrogenation and decomposition.

1. J. K. Edwards, A.F. Carley, A. A. Herzing, C. J. Kiely and G. J. Hutchings, *Faraday Disc.*, 2008, **138**, 225–229.

Dr Cadete Santos Aires remarked: Could it be that the decrease in intensity is due to a reversal of segregation in the nanoparticles (similar to the one observed for extended surfaces (the surface segregation at equilibrium is modified by the presence of a given environment)?

This would lead the catalyst to a steady state that can however be less active (but stable).

Prof. Kiely answered: You are absolutely correct that the combination of temperature and gaseous environment surrounding the bimetallic particle can dramatically influence the surface composition of the particle. This type of effect could in principle be experimentally investigated by carrying out aberration corrected STEM-HAADF imaging studies in an instrument equipped with a suitable environmental (i.e. *in-situ*) cell or sample holder. Such ETEM instruments are currently in the development stage and it is our firm intention to extend our research in this direction using such an instrument when we are fortunate enough to be able to gain access to one.

We are however convinced that our current 'composition' interpretation of the HAADF intensity variations observed in each of the Au + Pd, Au{Pd} and Pd {Au} nanoparticle variants is correct, since we have double checked the elemental distribution in each case by carrying out STEM-XEDS spectrum imaging experiments. We do recognise however, that these are currently *ex-situ* measurements on samples that have been exposed to air, and that the surface composition of the bimetallic particles may be different under realistic reaction conditions.

Dr Cadete Santos Aires said: Your atomically resolved HAADF images on bimetallic nanoparticles are outstanding. In the case of random alloys do you think you could go a step further in the quantitative analysis and identify variations in the composition of atomic columns (for instance by normalizing the elastic scattering intensity of each type of atoms)?

Prof. Kiely responded: Thank you for your kind words of appreciation for our aberration corrected HAADF images. We are convinced that this is a powerful technique for qualitatively analyzing the elemental distribution in such bimetallic

   This journal is © The Royal Society of Chemistry 2011

systems where there is a sufficient difference in atomic number between the two elemental components. We have also demonstrated for the first time in this paper that measurements of the relative core diameter and shell thickness from such HAADF images of core-shell particles can be used to semi-quantitatively estimate the individual particle composition.

Your question about whether or not we can take this a stage further to estimate the composition variation from atomic column-to-atomic column in a single random alloy particle is an intriguing one. In principle this would be straightforward if the particle were a flat-raft (*i.e.* if each atomic column in the structure has the same projected number of atoms along the viewing direction). We could then use the fact that the integrated HAADF intensity of each column would be related the sum of the $(Z^{1.7})$ scattering powers of the atoms making up that particular column. This will be influenced by the precise configurational ensemble of Au and Pd atoms within the column. However, in reality the shape of the particle is not a simple flat raft—it adopts either a spherical or hemispherical shape on the support. This implies that the projected number of atoms in each column will vary, albeit in a systematic way, from the center to the edge of the particle. This particle shape effect, as well as details of the electron probe shape and convergence, would also have to be taken into account when assessing the column intensity. Analyzing the measuring column intensity to deconvolute out the thickness and composition information is in principle possible, but is a non-trivial challenge which we intend to take up.

Prof. Bond commented: You have performed extensive high-quality characterisation on a number of Pd Au catalysts,most of which fail to meet the desired objectives of activity and selectivity for the chosen test reactions. Do you feel it is possible to exploit this information, to extrapolate it to design the structure of catalysts that would attain acceptable standards of performance?

Prof. Kiely replied: In answering this question it would be worthwhile explaining how we arrived at the logic for making the sets of catalysts by sol-immobilisation that we have investigated in this paper. In our earlier studies we found that supported AuPd catalysts prepared using an impregnation method gave core-shell structures (Au-rich core and Pd-rich shell) on oxide supports[1,2] and homogeneous alloys on carbon supports.[2] However, the particle size range was excessive when catalysts were prepared by impregnation. We suspected that small particles would be the most active based on our work on monometallic gold catalysts for CO oxidation.[3] The use of sol-immobilisation therefore permitted us to test this hypothesis by evaluating catalyst alloy structures with different compositions but with a narrow particle size range. Our study confirms that these structures are indeed very active for the direct synthesis of hydrogen peroxide, but unfortunately they are too active and selectivity is lost. We now consider that larger sized nanoparticles may be the selective active sites and recently we have shown that AuPd catalysts prepared by impregnation on acid pre-treated carbon can give very selective hydrogen peroxide catalysts.[4]

A further benefit derived from our detailed aberration corrected microscopy study of this systematic set of sol-immobilized samples is that we have demonstrated for the first time that fabricating monodisperse core-shell particles with a uniform composition by colloidal methods is much trickier than expected. In order to get a consistent particle-to-particle composition in such catalysts, it is imperative to have a very tight (sub-nm) control over the size of the initial seed (core) particle. Better control of this parameter will lead to better composition control and may allow us to engineer catalysts with specific controlled 'designer' nanostructures. We have also demonstrated that the structure and stability of such sol-immobilised particles is highly sensitive to (i) elemental distribution in the particle, (ii) the heat treatment regime and (iii) choice of support. These factors are therefore prime parameters to consider in the future design of sol-immobilized bimetallic catalysts.

1. D. I Enache, J. K. Edwards, P. Landon, B. Solsona-Espriu, A. F. Carley, A. A. Herzing, M. Watanabe, C. J. Kiely, D. W. Knight and G. J. Hutchings, *Science*, 2006, **311** 362–365.
2. J. K. Edwards, A. F. Carley, A. A. Herzing, C.J. Kiely and G.J. Hutchings, *Faraday Disc.*, 2008, **138**, 225–229.
3. A. A. Herzing, C. J. Kiely, A. F. Carley, P. Landon and G. J. Hutchings, *Science*, 2008, **321**, 1331–1335.
4. J. K. Edwards, B. Solsona, E. Ntainjua N, A. F. Carley, A.A. Herzing, C. J. Kiely and G. J. Hutchings, *Science*, 2009, **323**, 1037–1041.

Prof. Hutchings asked: Following on from the comments of Professor Kiely in response to your question. When we started this work we had a hypothesis that small nanoparticles with specific core structures could be active for H_2O_2 synthesis. However, it is clear that although we could make very active catalysts, they were much more active for H_2O_2 hydrogenation. So we have had to rethink our ideas concerning the optimal particle size and composition for these catalysts.

Dr Xu said: With the various Au, Pd, and Au-Pd catalysts that you have tested, what correlation, if any, have you observed to exist between the O_2 or H_2 consumption rate and the selectivity for H_2O_2?

Prof. Kiely responded: In our studies with the range of catalyst formulations explored, we generally observe that as the catalysts get more active, so the selectivity for H_2O_2 increases. In our recent study[1] we were able to switch-off the H_2O_2 decomposition/hydrogenation reactions and in this case we obtained our most active catalyst.

1. J. K. Edwards, B. Solsona, E. Ntainjua N, A. F. Carley, A.A. Herzing, C. J. Kiely and G. J. Hutchings, *Science*, 2009, **323**, 1037–1041.

Dr Whiston said: What is the thermodynamic driving force fo "dealloying" Au-Pd nanoparticles? Why do the core shell particles behave differently to the initially fully alloyed particles in this respect?

Prof. Kiely replied: It is indeed curious that Pd and Au components in the Au {Pd}/TiO_2 and Pd{Au}/TiO_2 materials readily separate upon calcination at 400 °C, whereas such behavior is not observed in the Au + Pd/TiO_2 variant. This suggests that an initial intimate mixing of the Au and Pd atoms in the nanoparticle (*i.e.* true alloy formation) imparts an additional degree of thermal stability as compared to the situation where one elemental component is simply coated upon the other in the core-shell bimetallic nanoparticles. For the core-shell particles, the Pd component (irrespective of whether it is the shell or the core material) prefers to phase separate form PdO rather than remain associated with the gold. Similarly the Au component prefers to form an extended interface with the TiO_2, rather than retaining it's extensive AuPd interface. The combination of the desire of Pd to oxidize and the tendency of Au to preferentially wet the TiO_2 support seem to be the driving force for phase separation in the core-shell morphology particles upon calcination at 400 °C. Such a phase separation phenomenon is clearly a more difficult to achieve thermodynamically in the random Au + Pd/TiO_2 alloys, where the strong Au-Pd bonding and intimate mixing, clearly hinder the separation of the two elemental components.

Dr Louis asked: Do you know how to explain why after calcination at 400 °C, the structure of the AuPd particles supported on TiO_2 are different from those supported on carbon, *i.e.*, why phase separation is observed on titania and not on carbon?

You wrote in your paper "This unexpected phase separation may possibly be driven by the desire of the Pd component to form PdO at higher calcination temperatures" Did you effectively have evidence for the formation of PdO upon calcination at 400 °C?

Prof. Kiely responded: The response of the AuPd particles immobilized on carbon to a calcination treatment at 400 °C is indeed quite different from the comparable nanoparticles supported on TiO_2. Gross sintering of the AuPd particles occurs in the former case, which is primarily driven by the drastic loss of activated C support material. The agglomeration of many small AuPd alloy particles into much larger (~35nm) composite metal particles as the support surface area collapses, will essentially mask out any potential phase separation effects within the primary AuPd particles.

The Au{Pd}/TiO_2 and Pd{Au}/TiO_2 samples when compared by XPS both before and after calcination at 400 °C, show a large increase in the Pd^{2+} signal and decrease on the Pd^0 signal after the heat treatment. This is interpreted to be due to the formation of some PdO upon heating in air at 400 °C. Furthermore when atomic resolution STEM-HAADF imaging of the fainter tail-like Pd—containing regions of the phase separated particles was attempted, we were never able to image the simple f.c.c arrangement of atoms which would be characteristic of metallic Pd. XPS of the Au + Pd/TiO_2 sample before and after calcination at 400 °C, which do not exhibit the phase separation effect, do not show a large increase in the Pd^{2+} signal after the heat treatment, lending further credence to our interpretation of PdO formation being the driving force for segregation.

Professor Bowker commented: What happens to the PVA which is present as a protective shell on the colloidal particles- does it influence what you see in the microscope? Does it decompose to put C on/in the catalysts when calcined and therefore possibly affect the segregation behaviour in such nanoparticles?

Prof. Kiely answered: In terms of the catalysis, we would expect hydrogen peroxide to decompose the PVA oxidatively. The hydrogen peroxide synthesis reactions are carried out in water methanol solvent mixture and we have recently observed[1] that treatment with water can remove the PVA. We suspect this will be the case in the reactions we describe. So under reaction conditions we expect the PVA ligands to be at least partially removed. Unfortunately we are unable to directly image the ligands on these particles in either bright field (BF) or high angle dark field (HAADF) mode in our probe corrected microscope. There is however a recent report that the ligands can be directly visualized in an image corrected TEM when the particle is supported on a single layer of graphene. This holds out promise that we may in the future be able to acquire direct microscopy evidence of how the protective ligand layer is degrades by the rigours of the calcination process or exposure to the reactant gases. We certainly do not see any gross breakdown of the ligand layer to form graphitic layers or shells around the particles. These would be clearly visible with our current instrumentation. However we cannot rule out the possibility that some carbon atoms arising from scission of the ligand chains are dissolved as interstititial impurities in the Pd lattice structure.[1]

1. J. A. Lopez-Sanchez, N. Dimitratos, C. Hammond, G. L. Brett, L. Kesavan, S. White, P. Miedziak, R. Tiruvalam, R. L. Jenkins, A. F. Carley, D. Knight, C. J. Kiely and G. J. Hutchings, *Nature Chem.*, 2011, **3**, 551–556.

Professor Madix opened the discussion of the paper by C.J. Kiely: Overall, this paper concerns the catalytic behavior of bimetallic alloys, which may have different surface compositions depending on the conditions under which the reaction is conducted. Their catalytic behavior therefore may depend on the reactant concentrations, temperatures, *etc.* I have the following question. Under reaction conditions either gold or palladium would be expected to segregate to the surface depending on the composition of the gas. What is the expected equilibrium structure of the Pd/ Au particles under conditions of equilibrium with the reactant composition? For example, would Pd or Au be expected to be on the external surface?

Prof. Kiely responded: This is an interesting question. We have both oxygen and hydrogen present in the reaction mixture and so both will compete. We expect that Pd is oxidized in the reaction and that eventually some palladium oxide preferentially forms on the surface. In fact we strongly suspect that for the direct production of hydrogen peroxide, the optimal exposed surface (irrespective of whether the particle is Pd_{shell}-Au_{core} or random alloy in character) is a very dilute alloy of Au atoms in a predominantly Pd-rich matrix. Of course, the optimal exposed surface will vary with the reaction conditions, and perhaps also with the substrate, and with glycerol oxidation where there is no hydrogen present in the reaction mixture a random alloy structure is preferred.

Professor Golunski opened the discussion of the paper by Dr Wittstock: By applying a metal oxide overlayer to the nanoporous gold, you are (i) reversing the normal order (in which metal is supported on metal oxide), and (ii) creating a diffusion barrier over the metal surface. And yet, the oxide-modified nanoporous gold shows superior catalytic activity. Can you comment more on the extent and thickness of the overlayer, and suggest where the active sites are located?

Dr Wittstock responded: This very good question touches the very central point of gold catalysis, the interface between gold and (*e.g.*) a metal oxide. In our case we are not depositing a complete overlayer of the metal oxide on the gold substrate. Indeed, in a different publication we used atomic layer deposition which leads to closed sub-nm thick layers of amorphous TiO_2 on the nanoporous gold surface.[1] Those systems are not catalytically active. Only after breaking up the amorphous film at higher temperatures, generating crystalline particles on the gold surface, catalytic activity for CO oxidation was detected. In this contribution, we deposited TiO_2 particles from a liquid suspension into the material. Therefore, we have single TiO_2 particles on the nanoporous gold surface, not a closed overlayer. The catalytic results suggest that by adding metal oxides (such as titania or praseodymia), we can strongly enhance the activation of molecular oxygen. For CO oxidation we detect an altered reaction order for O_2 in the range of 1 as compared to the uncovered surface of 0 (in contrast to the reaction order of CO which does not change). This indicates that the supply of active oxygen in case of oxide modified nanoporous gold is not limited by the surface reaction ("activation") but by the supply from the gas phase. It is likely that the activated oxygen is made available at the interface between the oxide particles and the gold surface reacting with CO chemisorbed on the gold surface. Out work thus suggests that it is primarily the interface between the gold and the metal oxide which is responsible for high catalytic activity.

1. Biener *et al.*, *NanoLett.*, accepted, DOI: 10.1021/nl200993g.

Dr Cadete Santos Aires said: You have shown that in the case of Ag on Au CO oxidation is promoted by Ag up to 11at%. What do you think the trend will be above 11% Ag : will it continue to increase the activity or will it pass through a maximuum of activity (at which composition) before decreasing?

Dr Wittstock replied: This question touches the interplay between the Ag and Au and also a potential synergetic effect (see *e.g.* Liu *et al.*).[1] The surface chemistry of both metals and recent theoretical studies[2] indicate that the Ag on those surfaces is beneficial for the activation of molecular oxygen, yet, it is also involved in the bonding and activation of CO. Experimental work on bimetallic Au-Ag particles showed that at low temperatures (*e.g.* below 80 °C) alloys of Au and Ag are most active for CO oxidation, even more active than either pure metal particles. Depending on the type of support the optimal fraction of Ag was found to be in the order of 10 at% and 40 at%.[3,4] The presence of a support, different experimental procedures and a possible segregation of Ag toward the surface hamper a straight forward forecast of the optimal fraction in Au-Ag alloy catalysts for CO oxidation, though.[5] Due to the limited maximal Ag content residing in a homogeneously dealloyed sample we could not test the development of activity for higher Ag concentrations.

1. Liu *et al.*, *J. Phys. Chem. B*, 2005, **109**, 40–43.
2. Moskaleva *et al.*, *Phys. Chem. Chem. Phys.*, 2011, **13**, 4529–4539.
3. Yen *et al.*, *J. Phys. Chem. C*, 2009, **113**, 17831–17839.
4. Jürgens *et al.*, *Gold Bull.*, 2007, **40**, 142–149.
5. Kim *et al.*, Chem. Phys. Lett. 2005, **403**, 77–82.

Dr Jupille commented: The suggestion that silver strongly affects the chemical activity of the nanoporous gold—although to be demonstrated—is consistent with a segregation of silver at the gold surface. In principle, an extremely small bulk concentration can result in a sizeable surface coverage, provided the bulk-surface equilibrium favors this enrichment. The role of the colloidal titania particles is not such a clear cut. Titania is below the detection limit of EDX which means that the concentration might be on the order of 1%. Particles are said to be less than 100 nm in size. If we take the expanded micrograph shown in figure 1 as a representation of the sample, such concentration and size are expected to result in one or two, may be a few, titania particles spread over the observed surface. Then how to explain that the enhancement in reactivity by a factor 5 of the gold (Figure 3) comes from a vicinity of gold with titania?

Dr Wittstock answered: In our work we demonstrate that the activity as well as the selectivity in case of the methanol oxidation depends on the concentration of the Ag within the npAu material. Notably, also in this case Ag could not be detected by EDX, although XPS indicated a surface enrichment by a factor of 5 up to 10.[1] One has to keep in mind that this technique probes the entire volume of the particular ligaments of the npAu and is not surface sensitive (1 monolayer on the surface accounts to about 1 at% of the entire material). As far as the titania nanoparticles are concerned, we believe that particles with diameters smaller than the pore diameters (which were also contained in the used material as identified by TEM) penetrated the pores. Preliminary Nano-SIMS results show that particles can at least reach depths of about 0.1 µm. Given the bulk sensitivity of EDX, it is not surprising that they could not be detected by EDX either. In a separate publication we demonstrate that, by using atomic layer deposition, nm thick films of titania can be deposited on the npAu structure.[2] After cracking of the film at elevated temperatures leading to titania nanoparticles the material was also very active for CO oxidation. This strongly enhanced activity in titania modified npAu is observed in all cases and thus underlines a common scheme.

1. Wittstock *et al.*, *J. Phys. Chem. C*, 2009, **113**, 5593–5600.
2. Biener *et al.*, *Nano Lett.*, DOI: 10.1021/nl200993g

Professor Bond commented: I wonder whether you can define the term 'nanoporous', as I am unaware of its precise significance. I understand the meaning of the prefixes micro-, meso- and macro- as applied to porosity, but I cannot find in your paper any measurements of pore volume, pore size distribution or BET area which might inform me of the exact nature of the porosity. It is after all within the pores that reaction occurs, and on this point believe you may misjudge the conditions where mass-transport limitation happens. The curves in Figure 2A are in fact linear up to a rate of about 3 in your units, signifying kinetic control in this region.

Dr Wittstock responded: The material we prepared for our studies exhibited pore (and ligament) sizes in the range between 30—50 nm, resulting in a spec. surface area of approximately 5 up to 10 $m^2 g^{-1}$ (see for example Rosner *et al.*, also for an extended morphological characterization of the material in terms of pore and ligament size distribution).[1] According to the BET DIN and IUPAC nomenclature, the correct specification would thus be mesoporous. Note, however, that, by post-preparative thermal treatment of the material, the pore sizes can be considerably increased up to microns (as also shown in our paper) so that a macroporous material is obtained (by templating also a hierarchical pore size distribution, *i.e.* meso- and macropores, can be achieved, see Nyce *et al.*).[2] Nevertheless, the term "nanoporous", has been used throughout the literature denoting porosity and pores in the range of nanometres up to a micron since the first notion of the term "nanoporous" in 2001.[3] As far as the question of mass transport in the material is concerned, we refer to our detailed study published in 2009.[4] Note in this context that a linear curve does not rule out mass transport limitation. Indeed, a reaction of first order also appears as a reaction of first under conditions of mass transport limitation. (For details about the assesment of mass transport and reaction kinetics in catalytic systems see Thomas *et al.*[5]

1. Rosner *et al.*, *Adv. Eng. Mater.*, 2007, **9**, 535–541.
2. Nyce *et al.*, *Chem. Mat.*, 2007, **19**, 344–346.
3. Erlebacher *et al.*, *Nature*, 2001, **410**, 450–453.
4. Wittstock *et al.*, *J. Phys. Chem. C*, 2009, **113**, 5593–5600.
5. J. M. Thomas, W. J. Thomas, Principle and Practice of Heterogeneous Catalysis, VCH Weinheim, New York, Basel, Cambridge, Tokyo, October 1996.

Dr Groszek asked: Can you tell me what is the bulk density of the nanoporous gold that you have prepared?

Dr Wittstock replied: We prepared the npAu material from an alloy containing 70 at% Ag and 30 at% Au. By using the applied corrosion technique this material is known to lead to a npAu material containing 70% void volume and resulting in a relative density of 30% (see for example Hodge *et al.*).[1] Assuming that the density of the gold ligaments is that of bulk gold (19.3 g cm^{-3}) the bulk density of npAu we used accounts to 5.8 g cm^{-3}. This number agrees well with experimental determination of density of our material, too.

1. Hodge *et al.*, *Acta Mater.*, 2007, **55**, 1343–1349

Dr Groszek asked: Do you have any information on the BET surface area of your porous material?

Dr Wittstock answered: BET surface area measurements of the material revealed numbers between about 4 $m^2 g^{-1}$ and 10 $m^2 g^{-1}$. These numbers are in line with electrochemical surface area measurements (see *e.g.* Cattarin *et al.*).[1] Certainly, the most sophisticated technique for determination of specific surface area is based on a 3D transmission electron tomographical study by Rosner *et al.*[2] For a nanoporous gold material used in our studies the specific surface area is calculated to be about 10 $m^2 g^{-1}$.

1. Cattarin *et al.*, *J. Phys. Chem. C*, 2007, **111**, 12643–12649
2. Rosner *et al.*, *Adv. Eng. Mater.*, 2007, **9**, 535–541

Professor Meyerstein said: I wonder whether the Ag^0 remaining in the gold matrix is not situated at sites to which the HNO$_3$ does not reach, *i.e.* at the end of narrow pores or totally surrounded by gold atoms. If so the catalytic effect might be due to the difference in electronegativity between the Au^0 and Ag^0 atoms. Thus the gold matrix will be partially negatively charged which might induce the catalytic properties.

Dr Wittstock answered: This is a very interesting question. Studies from Liu *et al.* focussing on the electrochemistry also indicate that during dealloying some metallic Ag is trapped inside the lattice (similar to the concept of the "parting limit", meaning that the dissolution of Ag is kinetically hampered by coordination of a critical number (*e.g.* 9) of Au atoms) but also some Ag can be present in the form of an oxide (AgO) on the surface being apparently inert for dissolution.[1] This agrees well with our studies. Nevertheless, SEM and TEM measurements in combination with EDX provided no evidence of any significant Ag bulk concentrations ($\sim$> 1 at%) even in deeper layers of the material. As especially emphasised in the paper Ag can be enriched at the surface including the possibilities of subsurface sites. XPS data (in preparation for publication) indicate that oxygen drives Ag to the surface. Theoretical calculations furthermore revealed that the presence of metallic Ag in the gold matrix indeed increases the bonding of molecular oxygen and lowers the activation barrier for its dissociation.[2] Mou *et al.* discussed that it is the possible electron transfer and the proximity of the Au and Ag promoting the electron transfer into the O_2 antibonding states.[3] Our experiments, though, cannot provide evidence whether a charge transfer or a sterical/spacial effect dominates.

1. Z. Lee, K-J. Jeon, A. Dato, R. Erni, T. J. Richardson, M. Frenklach and V. Radmilovic, *Nano Lett*, 2009, **9**, 3365–3369.
2. Moskaleva *et al.*, *Phys. Chem. Chem. Phys.*, 2011, **13**, 4529–4539.
3. Yen *et al.*, *J. Phys. Chem. C*, 2009, **113**, 17831–17839.

Professor Haruta commented: It is very interesting that the modification of Au particles with Pr and Ag oxides appreciably enhances CO oxidation and enables us to tune the product selectivity in methanol oxidation. You have immersed nano-porous Au in ethanol solution of $Pr(NO_3)_3$. In order to obtain markedly large catalytic enhancement by extending the kind of metal oxide promoters, how to make a strong contact with the smooth surfaces of Au is the most important . In this point of view, it seems to be effective to deposit base metal elements as metallic clusters on Au by deposition-reduction method and then oxidize to transform into base metal oxides. Have you ever tried this sort of experiments with Ti, Fe, Co, and Ni?

Dr Wittstock responded: This is an interesting question, also highlighting one of the fascinating aspects of this material system. By the choice of deposition/preparation method, in general, one should be able to generate specific interfaces (*e.g.* metallic *vs.* oxidic). Although we studied different deposition methods for titania as well as praseodymia or iron oxide we cannot definitely answer this question, yet. In general we observe a considerably increased activity in case of doping for example with titania using different deposition routes (*e.g.* depostion of particles *vs.* reaction/deposition of a gaseous precursors during ALD, see Biener *et al.*).[1] In order to achieve a situation similar to the Ag in the material, being reversibly oxidized under reaction conditions, one may try to alloy titanium metal into the material. This however was not yet investigated, one major drawback is that metallic titanium on Au surface is not stable and prefers to segregate into the bulk.[2] Certainly, the design of the interface and the correlation between the chemical and electronic nature of the interface and the catalytic activity is one of the central points in Au catalysis.

1. Biener *et al. Nano Lett.*, DOI: 10.1021/nl200993g.
2. B. Hammer and J. K. Norskov, *Advances in Catalysis*, 2000, 45, 71–129.

Professor Madix commented: You show that the effect of increasing silver concentration on the coupling of methanol is to reduce the selectivity for coupling by increasing combustion. It seems likely that the reason for this is the increased uptake of oxygen at a given pressure, leading to oxidation of the reactive intermediates. Have you considered reducing the oxygen partial pressure to compensate for the increased dissociation probability of dioxygen to see of the selectivity for methyl formate can be recovered at the higher Ag concentrations?

Dr Wittstock replied: Indeed, we observed an increase in combustion with increasing Ag content. This behaviour parallels elevated oxygen pressures at small Ag contents. We thus assume primarily an altered availability of atomic oxygen on the catalyst surface (leading to increased oxidation of methoxy to the surface bonded formate leading to CO_2). As Prof. Madix proposes by reducing the oxygen partial pressure one may compensate this increased activation to some degree (please note that, in general, the reaction order is close to zero for these samples and the dependency on the oxygen partial pressure is weak). Unfortunately, the chosen conditions (1 vol% O_2) already represented the lower limit achievable under our experimental conditions.

Professor Hutchings asked: Gold alloys with many metals, have you tried other alloys? For example AuCu could be a logical starting point as well as AuAg that you have used.

Dr Wittstock responded: Indeed, the combination of npAu with other reactive metals such as Cu, Pd, and Pt is a logical expansion. For example, Kameoka *et al.* prepared npAu/Cu by dealloying an $AuCu_3$ intermetallic.[1] In this work the resulting material contained about 10 at% residual Cu and was found to be very active for CO oxidation as well.

1. S. Kameoka, A.P. Tsai, *Catal. Lett.*, 2008, **121**, 337.

New insight into mechanisms in water-gas-shift reaction on Au/CeO$_2$(111): A density functional theory and kinetic study

Ying Chen,[a] Haifeng Wang,[ab] Robbie Burch,[a] Christopher Hardacre[a] and P. Hu[*a]

Received 11th February 2011, Accepted 24th March 2011
DOI: 10.1039/c1fd00019e

Using density functional theory (DFT) and kinetic analyses, a new carboxyl mechanism for the water-gas-shift reaction (WGSR) on Au/CeO$_2$(111) is proposed. Many elementary steps in the WGSR are studied using an Au cluster supported on CeO$_2$(111). It is found that (i) water can readily dissociate at the interface between Au and CeO$_2$; (ii) CO$_2$ can be produced *via* two steps: adsorbed CO on the Au cluster reacts with active OH on ceria to form the carboxyl (COOH) species and then COOH reacts with OH to release CO$_2$; and (iii) two adsorbed H atoms recombine to form molecular H$_2$ on the Au cluster. Our kinetic analyses show that the turnover frequency of the carboxyl mechanism is consistent with the experimental one while the rates of redox and formate mechanisms are much slower than that of carboxyl mechanism. It is suggested that the carboxyl pathway is likely to be responsible for WGSR on Au/CeO$_2$.

1. Introduction

Recently, there has been renewed interest in the WGSR[1-22] (eq. (1)) because of its potential application in reducing the amount of carbon monoxide from hydrogen fuel cells which contains very low concentrations of CO. Recent studies[3,4,13,23] of WGSR showed that noble metals supported on reducible oxides such as CeO$_2$ are very active catalysts for this reaction:

$$CO + H_2O \rightarrow CO_2 + H_2 \tag{1}$$

There are three main mechanisms proposed in the literature for this reaction: redox mechanism, formate mechanism and carboxyl mechanism. Ceria is well known as an oxygen storage component, taking up of oxygen under oxidizing conditions and releasing oxygen under reducing conditions. Several groups[24,25] investigated CO oxidation on precious-metals (*e.g.* Pt, Pd and Rh) on ceria by temperature programmed desorption (TPD) and they suggested that a bifunctional mechanism operates: CO adsorbed on the metal can react with oxygen from ceria and the O$_2$ reoxidizes reduced ceria. Similarly, by studying WGSR on ceria-supported precious metals, Gorte and co-workers[8,26] proposed the redox mechanism in which CO adsorbed on metal sites reacts with an oxygen atom coming from ceria to product CO$_2$ and subsequently water reoxidizes the reduced ceria to give

[a]School of Chemistry and Chemical Engineering, The Queen's University of Belfast, Belfast, BT9 5AG, UK. E-mail: p.hu@qub.ac.uk
[b]Lab for Advanced Materials, Research Institute of Industrial Catalysis, East China University of Science and Technology, Shanghai, 200237, P. R. China

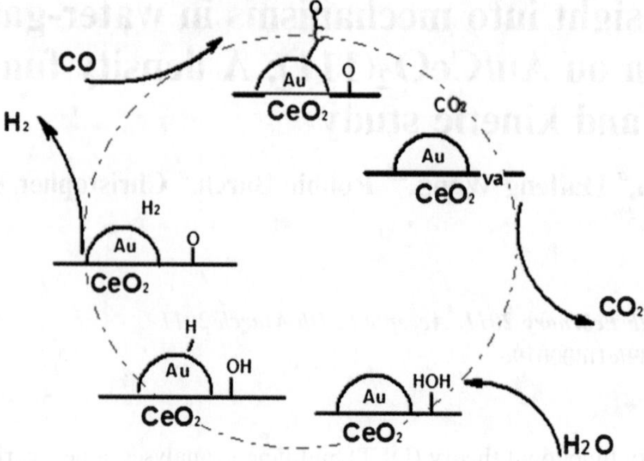

Fig. 1 Catalytic cycle for redox mechanism of WGSR at Au/CeO_2.

hydrogen. This reaction cycle is shown in figure 1. Flytzani-Stephanopoulos and co-workers[3,4] also proposed this mechanism for Au/ceria catalyst.

By means of FT-IR spectroscopy, Shido and Iwasawa[5,6] suggested an alternative formate mechanism. In this mechanism, the main reaction intermediate is a surface formate species produced by the reaction of CO with terminal hydroxyl groups on ceria. Following their studies, many research groups investigated the nature of the main intermediate species. Using infrared measurements and kinetic studies, Davis and co-workers[9–15] favored the formate mechanism to explain WGSR. They suggested that the formate intermediates form on the surface of reduced ceria which is induced by precious metals. Likewise, using CO and water pulse experiments, Sakurai *et al.*[1] also suggested that the formate mechanism is the dominated mechanism in WGSR. On the other hand, based on diffuse reflectance infrared Fourier transform spectroscopy (DRIFTS) experiments, Behm and co-workers[17] proposed a different formate mechanism which can be summarized as follows: (i) CO adsorbed at the active Au sites reacts with OH nearby to form formate; (ii) formates spillover onto the ceria support; and (iii) the surface formates can be reactivated on reverse spillover to the active sites and further decompose to give CO_2 and H_2. Burch and co-workers[27–29] pointed out that the dominant reaction mechanism could be a function of reaction conditions and suggested that at higher temperatures a redox mechanism might be important whereas if a formate mechanism operates this might be dominant at lower temperatures. Subsequently, Meunier and co-workers[30] investigated the proposed formate mechanism by both analyzing the temporal response of formate to a transient isotope switch and by quantifying the rate of formate conversion into nitrogen. They demonstrated that the formate species identified by infrared spectroscopy could not be an important reaction intermediate but was simply a spectator species. In a review of the WGSR over supported gold catalysts, Burch[31] discussed the possibility that since a mechanism involving a formate species was not supported by the experimental evidence, then as an alternative a carboxylate-type species could be important. However, it was not possible to obtain experimental evidence for the formation of such a species, possibly because if it is a real intermediate it reacts so fast that the steady state concentration on the surface is below the limit of detection of the infrared technique.

In a study of the water effect on CO oxidation on Pt, Gong, Hu and Raval presented the first evidence of CO oxidation by OH from DFT calculations.[32] Based on DFT calculations, Liu, Jenkins and King[33] proposed a new mechanism for WGSR, carboxyl mechanism: $H_2O \rightarrow H_{ad} + OH_{ad}$; $CO_{ad} + OH_{ad} \rightarrow COOH_{ad}$; $COOH_{ad} \rightarrow CO_2 + H_{ad}$ and $H_{ad} + H_{ad} \rightarrow H_2$, among which the greatest barrier

   This journal is © The Royal Society of Chemistry 2011

was calculated to be the decomposition of COOH species. They suggested that the active site for the WGSR is on four to six atom Au clusters that is anchored on an O-vacancy of $CeO_2(111)$. Following this mechanism, by calculating the reaction pathway on Cu_{29} and Au_{29}, Rodriguez *et al.*[34] found that the most difficult step was water dissociation and suggested that a reduced $CeO_2(111)$ surface can facilitate dissociation of H_2O. More recently, Gokhale, Dumesic and Mavrikakis[21] used DFT in combination with microkinetic model to investigate WGSR on Cu(111). Their results suggested that the dominant reaction mechanism involves CO oxidation by OH to form carboxyl, followed by the decomposition of COOH through a disproportionation reaction with OH. By comparison of three reaction mechanisms for WGSR on Au(111), Liu *et al.*[35] also showed that the most feasible reaction pathway is $H_2O \xrightarrow{\text{-H}} OH \xrightarrow{\text{+CO}} COOH \xrightarrow{\text{+OH}} CO_2$. In these studies, the reaction mechanisms were investigated on metal surfaces. According to the experimental works,[34] reducible oxides (*e.g.* ceria) play a critical role in WGSR, especially for water dissociation. Therefore, a mechanism study of WGSR occurring at the interface between metal and oxide surface is needed.

In our previous paper,[36] we carefully examined all the possible elementary steps in both redox and formate mechanisms. Based on our DFT calculations, we suggested that it is not energetically easy to break O–H bond in OH group which is a crucial step in both mechanisms. In this work, we examined many possible reaction pathways for WGSR on Au/ceria. It was found that a favored one is as follows: H_2O (g) + $*'' \rightarrow$ $H_2O_{ad''}$; CO(g) + $*' \rightarrow CO_{ad'}$; $H_2O_{ad''}$ + $*' \rightarrow H_{ad'}$ + $OH_{ad''}$; $CO_{ad'}$ + $OH_{ad''} \rightarrow$ $COOH_{ad'}$ + $*''$; $COOH_{ad'}$ + $OH_{ad''} \rightarrow CO_2$ (g) + $*'$ + $H_2O_{ad''}$ and $H_{ad'}$ + $H_{ad'} \rightarrow$ H_2(g) ($*'$ and $*''$ are the free adsorption sites on the metal surface and on the oxides, respectively; ad$'$ and ad$''$ represent adsorption on the metal and on the oxide, respectively). Furthermore, using kinetic model, we obtain numerical result for the turnover frequency (TOF) of WGSR on Au/ceria which is very similar to experimental results.

The paper is organized as follows. Calculation details used in this study and kinetic theory are described in the next section. We will present the results from DFT calculations in sections 3 and the results will be discussed in section 4. Conclusions will be summarized in the last section.

2. Calculation details and theoretical methods

The total energy calculations were performed using the SIESTA code.[37] The standard DFT supercell approach with GGA-PBE[38] functional was implemented, and the Kohn–Sham wave function was expanded with localized basis sets. In the calculations, Troullier–Martins norm-conserving pseudopotentials were used for all the elements. The energy cutoff for the real space grid was 180 Ry. The localization radii of the basis functions were determined from an energy shift of 0.01 eV. Spin polarization was included whenever necessary. Recent computational studies of ceria[36,39,40] using the SIESTA code gave good results for the bulk and surface properties of ceria. The transition states (TSs) were searched using a constrained optimization scheme. The TS was verified when (i) all forces on atoms vanish; and (ii) the total energy is a maximum along the reaction co-ordinate but a minimum with respect to the rest degrees of freedom.[41,42]

$CeO_2(111)$ was modeled by a p(3 × 3) unit cell with 9-layers in which the top 3 layers were relaxed. Under experimental conditions, the ceria surface was partially reduced. Therefore, the ceria surface was modeled with OH coverage of 1/3 ML. An Au_3 cluster was built as follows: one Au on ceria vacancy and the other two on top side of OH (figure 2). Test calculations have shown that a sampling of the surface Brillouin zone using a (2 × 2 × 1) *k*-point mesh results in a converged total energy. Adsorbates and defects were situated on one side of the slab with 10 Å of vacuum separation. Some test results of using plane-wave VASP code with a U term were reported previously.[43–46] In this work, some of the structures and the

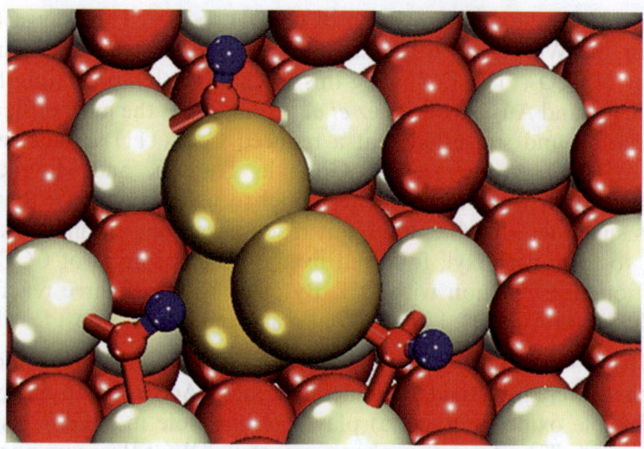

Fig. 2 Side view of the geometrical structures of Au$_3$/CeO$_2$(111) (white atoms are Ce; red atoms O; the yellow atom Au; blue atoms H; this notation is used throughout the paper).

energetic results of the reaction CO$_{ad'}$ + OH$_{ad''}$ → COOH$_{ad'}$ + *'' obtained from SIESTA were further checked by using the plane-wave VASP code[47,48] (see Table 1). In the calculations using the VASP code, the valence electronic states of Ce-5s, 5p, 5d, 4f, and 6s electrons, O-2s, and 2p electrons, and Au-6s, 6p, and 5d electrons were described by a plane-wave basis set with cut-off energy of 500 eV and ion-electron interactions were treated with the projector augmented wave (PAW) method.[49,50] The electron exchange-correlation was described by the GGA-PW91 functional.[51] It was reported in the literature that the U term of 5 eV is a good choice for ceria.[52,53] Therefore, DFT + U approach with a value of U = 5 eV[43,44,52–54] was employed when the VASP code was used. The climbing-image nudged elastic band method was utilized to determine the reaction barrier.[55,56] It can be seen from Table 1 that both codes gave rise to similar results for our systems.

To derive rate expressions, Dumesic's method[57–60] in terms of De Donder relations[61–63] was utilized. According to the De Donder relations, the net rate for elementary step i in term of the forward rate of the step, r_{i+}, and the reversibility z_i can be expressed as

$$r_i = r_{i+} (1 - z_i) \qquad (2)$$

where $z_i = \exp\left(\dfrac{-A_i}{RT}\right)$, A_i is the affinity for step i, equalling to minus the change in the Gibbs free energy G_j with respect to the extent of reaction. Generally, the affinity A_i of step i can be expressed in terms of the standard Gibbs free energies G_j^o and the activity a_j of reactant or product j:

Table 1 Calculated the chemisorptions of CO on Au$_3$ cluster, activation barrier (E$_a$) and enthalpy changes (ΔH) of CO$_{ad'}$ + OH$_{ad''}$ → COOH$_{ad'}$ + *'' on Au$_3$/CeO$_2$(111) using SIESTA and VASP codes

E$_{ad}$(CO)/eV		E$_a$/eV		ΔH/eV	
SIESTA	VASP	SIESTA	VASP	SIESTA	VASP
1.05	1.10	0.64	0.70	0.28	0.30

$$A_i = -\sum_j v_{ij} G_j = -\sum_j v_{ij}\left[G_j^o + RT\ln(a_j)\right] \tag{3}$$

where v_{ij} is the stoichiometric coefficient for reactant or product j of step i. This expression can be written in terms of the equilibrium constant K_i for the step,

$$\exp\left(\frac{-A_i}{RT}\right) = \frac{\prod_j a_j^{v_{ij}}}{K_i} \tag{4}$$

Since the equilibrium constant is determined by the change of the standard state Gibbs free energies:

$$K_i = \exp\left(\frac{-\sum_j v_{ij} G_j^o}{RT}\right) \tag{5}$$

we can rewrite the reversibility z_i as follow:

$$z_i = \frac{\prod_j a_j^{v_{ij}}}{K_i} \tag{6}$$

It should be noted that the value of z_i approaches zero when step i is irreversible and z_i approaches unity when step i becomes quasi-equilibrium.

3. Results

3.1 H$_2$O dissociation

There are three possible reaction channels for water dissociation on Au/ceria shown as follows:

$$H_2O(g) + {}^{*''} + O_{ad''} \rightarrow H_2O_{ad''} + O_{ad''} \rightarrow OH_{ad''} + OH_{ad''} \tag{7}$$

$$H_2O(g) + 2{}^{*'} \rightarrow H_2O_{ad'} + {}^{*'} \rightarrow OH_{ad'} + H_{ad'} \tag{8}$$

$$H_2O(g) + {}^{*''} + {}^{*'} \rightarrow H_2O_{ad''} + {}^{*'} \rightarrow OH_{ad''} + H_{ad'} \tag{9}$$

In reaction (7), water dissociates on ceria. According to our previous work,[40] the O coverage on the ceria surface should be very low under the steady state. Therefore, the reaction rate of reaction (7) will be very small. Liu, Jenkins and King[33] suggested reaction (8) as the main pathway for water dissociation in their mechanism. Although the barrier of dissociation is very low, the chemisorption of water on Au is very weak. If we take the entropy term into account, the reaction rate of this step will be very small. Indeed, Rodríguez et al.[34] showed that a Au$_{29}$ cluster is very inactive for water dissociation. Thus, reaction (9), in which water dissociation takes place at the interface between the Au cluster and the ceria surface with an oxygen-vacancy nearby, is the only favorable channel for water dissociation. We calculated the feasibility of reaction (9). The initial state (IS) of the reaction we located is shown in figure 3(a) in which water adsorbs on an oxygen vacancy that is close to the Au cluster. The adsorption energy is 0.85 eV which is slightly higher than the adsorption energy of water on an oxygen vacancy over the pure ceria surface (0.66 eV). Our results are consistent with the DFT result of Kumar and

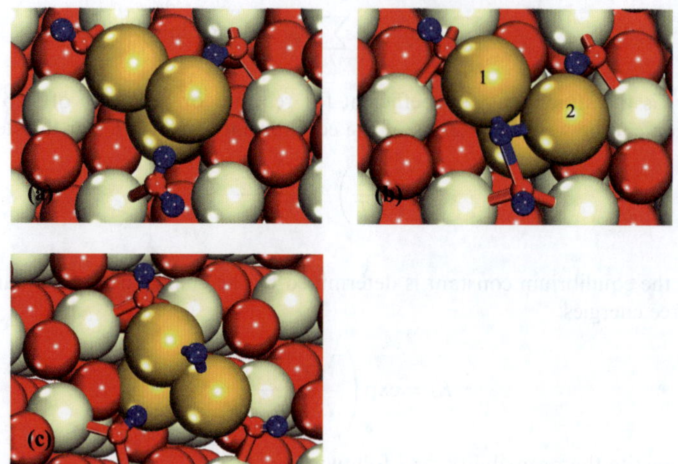

Fig. 3 Illustration of the geometrical structures of elementary steps in $H_2O_{ad''} + *' \rightarrow OH_{ad''} + H_{ad'}$ (a) Side view of IS for H_2O dissociation. (b) Side view of TS of H_2O dissociation. (c) Side view of structures of FS.

Schelling (0.64 eV)[64] and the experimental one which was determined from an analysis of TPD experiments (0.53 eV).[65] The TS we located is shown in figure 3(b) and the distance between the H and the Au_1 is 1.649 Å. We can see that in the final state (FS) (figure 3(c)) H prefers to adsorb at the bridge site of Au_3 which is in agreement with the result obtained by Barrio *et al.*[66] The FS ($OH_{ad} + H_{ad}$) is less stable than the IS (H_2O_{ad}) on $Au_3/CeO_2(111)$ with 0.50 eV energy gap.

3.2 CO oxidation

As discussed in the introduction, CO molecules can adsorb on metal sites to react with oxygen supplied by the metal-ceria interface or neighboring OH groups. The reactions can be written as follows:

$$CO_{ad'} + O_{ad''} \rightarrow CO_2 \text{ (g)} + *'' \tag{10}$$

$$CO_{ad'} + OH_{ad''} \rightarrow COOH_{ad'} + *'' \rightarrow CO_2 \text{ (g)} + H_{ad'} + *'' \tag{11}$$

$$CO_{ad'} + 2OH_{ad''} \rightarrow COOH_{ad'} + OH_{ad''} + *'' \rightarrow CO_2 \text{ (g)} + H_2O_{ad''} + *' + *'' \tag{12}$$

However, in our previous paper,[40] we suggested that it is energetically unfavorable to break O–H bond in OH group to regenerate oxygen due to high activation barrier. Therefore, at the steady state the surface of ceria especially the reaction zone should be covered by high coverages of OH groups and oxygen vacancies. As a result, the most probably reaction pathway could be adsorbed CO reacting with OH at cerium oxide to form COOH. According to the DFT results of Liu, Jenkins, and King,[33] the barrier of the second step in reaction (11) (COOH drops its H to Au) is high. Hence, reaction channel (12) could be the favored pathway to produce CO_2 under the steady state.

The whole pathway of CO oxidation can be described as follows. Firstly, CO adsorbs on the top side of Au_3 cluster with 1.05 eV chemisorption energy (figure 4 (a)). Figure 4(b) shows the TS in which CO tilts towards the OH group nearby

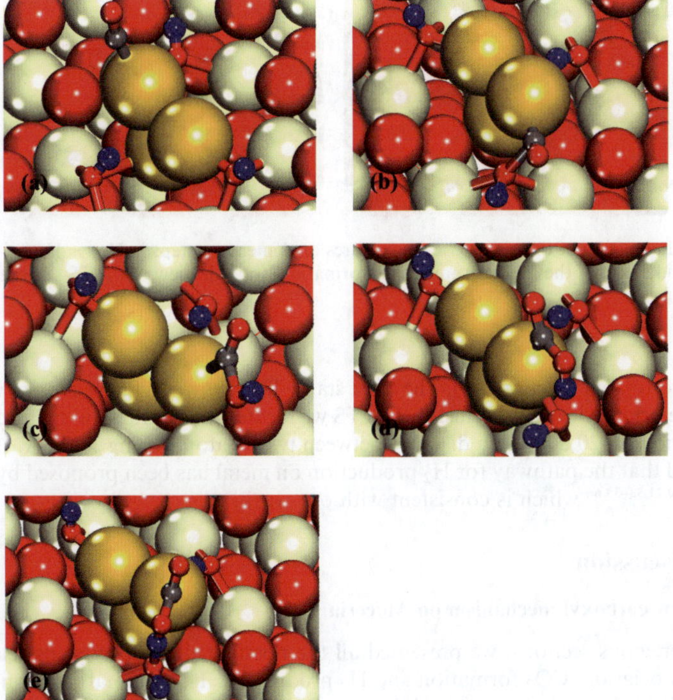

Fig. 4 Illustration of the geometrical structures of elementary steps in $CO_{ad''} + OH_{ad'} \rightarrow COOH_{ad''} + OH_{ad'} \rightarrow CO_2$ (g) + $H_2O_{ad''}$ + *' + *'' (grey atom is carbon). (a) Side view of the IS structures of COOH formation. (b) Side view of the TS structures of COOH formation. (c) Side view of the structures of *cis* COOH. (d) Side view of IS structures of *trans* OCO–H bond breaking. (e) Side view of the TS structures of OCO–H bond breaking.

and the distance between the C in the CO and the O in the OH is 1.890 Å. After the TS, the CO and the OH move together to form *trans*-COOH. It should be noted that there are two COOH configurations in the gas phase,[18] *cis* and *trans* isomers, and the conversion between these two isomers is facile as the energy difference is only 0.03 eV. Then, the *trans*-COOH passes its H to OH nearby to release CO_2. The dissociation barrier is 0.1 eV if the chemisorbed *trans*-COOH is considered to be the IS. The FS is shown in figure 3(a).

3.3 H_2 formation

Three pathways for H_2 formation on Au/ceria are summarized as follows:

$$OH_{ad''} + H_{ad'} \rightarrow H_2(g) + O_{ad''} + *' \tag{13}$$

$$OH_{ad''} + OH_{ad''} \rightarrow H_2(g) + 2O_{ad''} \tag{14}$$

$$H_{ad'} + H_{ad'} \rightarrow H_2(g) + 2*' \tag{15}$$

We already examined reactions (13) and (14) in our previous work[36] and the barriers of these two reactions are more than 1 eV. Therefore, in this work we

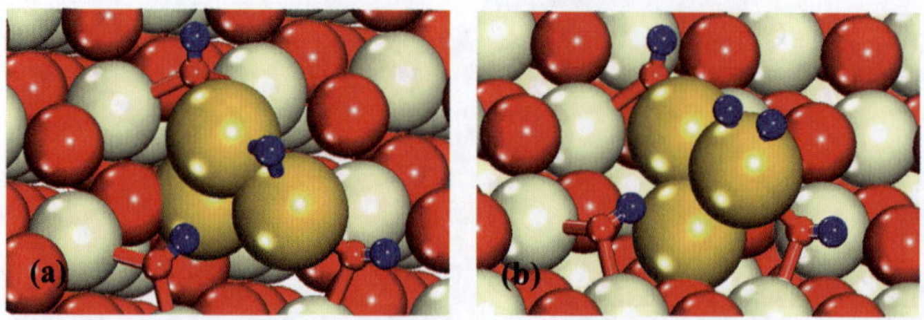

Fig. 5 Illustration of the geometrical structures of elementary steps in $H_{ad'} + H_{ad'} \rightarrow H_2$ (g) + 2*′. (a) Side view of the IS structures of H_2 formation. (b) Side view of the TS structures of H_2 formation.

only investigated reaction (15). The most stable IS is shown in figure 5(a), in which H is on the bridge site of Au_3 cluster. The TS we located is illustrated in figure 5(b) and the barrier is 0.20 eV with a distance between two H atoms being 1.237 Å. It should be noted that the pathway for H_2 production on metal has been proposed by several groups[20,21,33–35,66] which is consistent with our work.

4. Discussion

4.1 New carboxyl mechanism on Au/ceria

In the previous sections, we presented all the possible elementary steps relating to H_2O dissociation, CO_2 formation and H_2 production. The optimal reaction scheme of WGSR is drawn in figure 6 and the structures of those elemental reactions can be seen in the result section. It should be noted that the effect of finite temperature and pressure, which will particularly affect the chemical potential of gas phase molecules, has not been taken into account in figure 6.

It can be seen from figure 6 that there is no particular high barrier in our pathway. It is worth to emphasize some key points in our mechanism: (i) Water can dissociate at the interface between the Au cluster and the ceria with the low barrier (0.64 eV) which is much lower than that on Au(100), as well as ionic and neutral Au_{29} nanoparticles (>1.2 eV). From our previous analysis and DFT results, we can see that neither pure ceria nor active Au nanoparticles are active for water dissociation and water can only dissociate at the interface of Au/ceria. (ii) It significantly reduces the barrier for breaking O–H bond in COOH group when COOH reacts with OH compared with the COO–H bond breaking on Au.

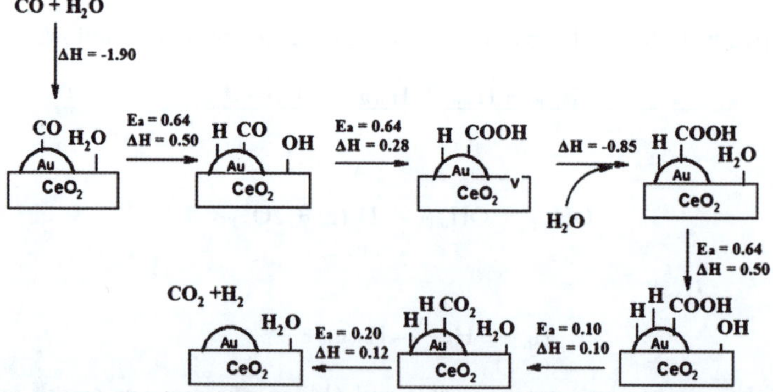

Fig. 6 The optimal reaction scheme for $CO + H_2O \rightarrow CO_2 + H_2$ on $Au_3/CeO_2(111)$.

4.2 Kinetic analyses of the carboxyl mechanism

Based on the DFT results, we propose the following elementary steps for the carboxyl mechanism:

$$CO_{(g)} + *' \leftrightarrow CO_{ad'} \tag{16}$$

$$H_2O_{(g)} + *'' \leftrightarrow H_2O_{ad''} \tag{17}$$

$$H_2O_{ad''} + *' \leftrightarrow OH_{ad''} + H_{ad'} \tag{18}$$

$$CO_{ad'} + OH_{ad''} \leftrightarrow COOH_{ad'} + *'' \tag{19}$$

$$COOH_{ad'} + OH_{ad''} \leftrightarrow CO_{2(g)} + H_2O_{ad''} + *' \tag{20}$$

$$2H_{ad'} \rightarrow H_{2(g)} + 2*' \tag{21}$$

Then, according to equation (6), the reversibilities z_i are given by the following expressions:

$$z_1 = \frac{[CO_{ad'}]}{\frac{p_{CO}}{p^o}[*']K_1} \tag{22}$$

$$z_2 = \frac{[H_2O_{ad''}]}{\frac{p_{H_2O}}{p^o}[*'']K_2} \tag{23}$$

$$z_3 = \frac{[OH_{ad''}][H_{ad'}]}{[H_2O_{ad''}][*']K_3} \tag{24}$$

$$z_4 = \frac{[COOH_{ad'}][*'']}{[CO_{ad'}][OH_{ad''}]K_4} \tag{25}$$

$$z_5 = \frac{\frac{P_{CO_2}}{P^o}[H_2O_{ad''}][*']}{[COOH_{ad'}][OH_{ad''}]K_5} \tag{26}$$

$$z_6 = \frac{\frac{P_{H_2}}{P^o}[*']^2}{[H_{ad'}]^2 K_6} \tag{27}$$

It is reasonable to assume that CO and water adsorb on the catalysts approximately in equilibrium state, *i.e.* $z_1 = z_2 = 1$. At steady state, the net rates of intermediate formation are equal to zero, *i.e.*

$$\frac{d[OH_{ad''}]}{dt} = r_3 - r_4 - r_5$$
$$= k_3[H_2O_{ad''}][*'](1 - z_3) - k_4[CO_{ad'}][OH_{ad''}](1 - z_4)$$

$$-k_5[COOH_{ad'}][OH_{ad''}](1 - z_5) = 0 \tag{28}$$

$$\frac{d[COOH_{ad'}]}{dt} = r_4 - r_5 = k_4[CO_{ad'}][OH_{ad''}](1 - z_4) - k_5[COOH_{ad'}][OH_{ad''}](1 - z_5)$$

$$= 0 \tag{29}$$

$$\frac{d[H_{ad'}]}{dt} = r_3 - 2r_6 = k_3[H_2O_{ad'}][*'](1 - z_3) - 2k_6[H_{ad'}]^2(1 - z_6) = 0 \tag{30}$$

where k_i is the forward reaction rate constant for the elementary step (i) which can be determined by the Arrhenius equation and the transition state theory.

$$k_i = A * e^{-\frac{E_{a(i)}}{RT}} \tag{31}$$

where A is the pre-exponential factor which is assumed to be 10^{13} for all the reactions and $E_{a(i)}$ is the barrier.

Considering the site balance on the Au clusters, we can have:

$$[*'] + [H_{ad'}] + [COOH_{ad'}] + [CO_{ad'}] = 1 \tag{32}$$

On the ceria surface, we assume that the concentration of reactive sites is equal to the concentration of Au. Thus, the site balance is

$$[*''] + [H_2O_{ad''}] + [OH_{ad''}] = 1 \tag{33}$$

In our system, we take the experimental conditions from Fu *et al.*:[67] $T = 450$ K, $0.44Au/Ce(La)O_x$ (DP, NaCN), $P_{CO}/P^o = 11\%$, $P_{H_2O}/P^o = 26\%$, $P_{H_2}/P^o = 26\%$, $P_{CO_2}/P^o = 7\%$. Using our DFT results (Table 2) and equations (22)–(33), we can obtain the numerical results for all the concentrations of the species TOF of WGSR:

$$R_{(COOH+OH)} = k_5[COOH_{ad'}][OH_{ad''}](1 - z_5) \approx k_5[COOH_{ad'}][OH_{ad''}] = 3.5 \text{ (s}^{-1}) \tag{34}$$

Table 2 The reaction barriers ($E_{a(i)}$) and the standard state Gibbs free energies ΔG_i^o of the elementary steps ((4-1)-(4-6)). Here, $\Delta G_i^o = \Delta H_i^o - T\Delta S_i^o$, in which, ΔH_i^o is obtained from our DFT calculations

Activation Energy($E_{a(i)}$)	(eV)	ΔG_i^o	(eV)
$E_{a(3)}$	0.64	ΔG_1^o	−0.11
$E_{a(4)}$	0.64	ΔG_2^o	0.08
$E_{a(5)}$	0.10	ΔG_3^o	0.50
$E_{a(6)}$	0.20	ΔG_4^o	0.28
		ΔG_5^o	−1.11
		ΔG_6^o	−0.54

Under the experimental condition, the gold atom fraction on the surface of $0.44Au/Ce(La)O_x$ (DP, NaCN) is 0.61at.%[67] and the Au cluster dispersion is close to 1. Therefore, we can convert our TOF to 0.02 s^{-1} which is very similar to the experimental one (0.05 s^{-1}) [ref. 59 (figure 4)].

4.3 Kinetic analyses of other mechanisms

Based on the DFT and kinetic results, we also estimated the TOFs of the formate and redox mechanisms. According to our previous work,[36] the key steps for formate and redox mechanisms are as follows:

$$CO_{ad'} + OH_{ad''} \leftrightarrow HCO_{ad'} + O_{ad''} + HCOO_{ad''} + *' \tag{35}$$

$$H_{ad'} + OH_{ad''} \leftrightarrow H_{2(g)} + O_{ad''} + *' \tag{36}$$

Therefore, the TOFs of formate and redox mechanisms can be expressed as follows:

$$R_{(formate)} = k_{formate}[CO_{ad'}][OH_{ad''}](1 - z_{formate}) < k_{formate}[CO_{ad'}][OH_{ad''}] = 4*10^{-12}(s^{-1}) \tag{37}$$

$$R_{(redox)} = k_{redox}[H_{ad'}][OH_{ad''}](1 - z_{redox}) < k_{redox}[H_{ad'}][OH_{ad''}] = 2*10^{-10}(s^{-1}) \tag{38}$$

It is clear that the rates of formate and redox mechanism are much smaller than that of our carboxyl mechanism. By investigating the reactivity of the species formed at the surface of a $Au/Ce(La)O_2$ catalyst in the WGSR, Meunier et al.[30] indicated that formates should be labeled minor reaction intermediates and also the rate of redox route is too low to be observable, which is consistent with our results. From CO and water pulse experiments, Sakurai et al.[1] suggested that CO_2 and H_2 were formed at the same time when H_2O was added, supporting the formate mechanism. Their results may also be understood from our carboxyl mechanism: COOH intermediates, which are generated from adsorbed CO and surface OH groups on pre-reduced ceria, decompose along with the reaction of water dissociation to produce CO_2 and H_2 simultaneously. Recently, Leppelt et al. suggested that CO adsorption on active Au reacts with $OH_{ad''}$ nearby to form formate.[17] Energetically, according to our DFT results, it will be much easier to form COOH than that of formate via the reaction of CO + OH. It should be noted that no experimental evidence of the existence of COOH intermediate was reported in the WGSR. This may be due to the reason that the lifetime of COOH is very short and hence it is very difficult to detect COOH on the surfaces experimentally. This is in fact consistent with our results: According to our kinetic analyses, under steady state conditions, the concentration of COOH is very low and the reaction of $COOH_{ad'} + OH_{ad''} \rightarrow CO_{2(g)} + H_2O_{ad''}$ is nearly barrierless.

5. Conclusion

Based on DFT calculations and detailed kinetic analyses, we propose a carboxyl mechanism for WGSR on small Au-cluster supported by ceria. Regarding the mechanism, we have the following conclusions:

(i) The DFT calculations show that water can readily dissociate at the interface between Au cluster and ceria surface with oxygen-vacancies nearby.

(ii) The DFT results suggest that adsorbed CO can readily react with OH to form COOH and the COOH passes its H to OH nearby with very small barrier.

(iii) The calculated data from DFT also indicate that the H_2 formation occur on Au and H_2 can be produced easily by coupling two weakly adsorbed H on Au.

With respect to the kinetic results, we estimate the turnover frequencies of three mechanisms including the two well established mechanisms in the literature: formate mechanism and redox mechanism. Our kinetic data show that the rate of carboxyl mechanism proposed in this work is very close to experimental value and is much faster than those of formate and redox mechanisms.

References

1 H. Sakurai, T. Akita, S. Tsubota, M. Kiuchi and M. Haruta, *Appl. Catal., A*, 2005, **291**, 179–187.
2 P. Bera and M. S. Hegde, *Catal. Lett.*, 2002, **79**, 75–81.
3 Q. Fu, A. Weber and M. Flytzani-Stephanopoulos, *Catal. Lett.*, 2001, **77**, 87–95.
4 Q. Fu, H. Saltsburg and M. Flytzani-Stephanopoulos, *Science*, 2003, **301**, 935–938.
5 T. Shido and Y. Iwasawa, *J. Catal.*, 1992, **136**, 493–503.
6 T. Shido and Y. Iwasawa, *J. Catal.*, 1993, **141**, 71–81.
7 J. M. Zalc, V. Sokolovskii and D. G. Loffler, *J. Catal.*, 2002, **206**, 169–171.
8 T. Bunluesin, R. J. Gorte and G. W. Graham, *Appl. Catal., B*, 1998, **15**, 107–114.
9 G. Jacobs, L. Williams, U. Graham, G. A. Thomas, D. E. Sparks and B. H. Davis, *Appl. Catal., A*, 2003, **252**, 107–118.
10 G. Jacobs, P. A. Patterson, U. M. Graham, D. E. Sparks and B. H. Davis, *Appl. Catal., A*, 2004, **269**, 63–73.
11 G. Jacobs, S. Khalid, P. M. Patterson, D. E. Sparks and B. H. Davis, *Appl. Catal., A*, 2004, **268**, 255–266.
12 G. Jacobs, E. Chenu, P. M. Patterson, L. Williams, D. Sparks, G. Thomas and B. H. Davis, *Appl. Catal., A*, 2004, **258**, 203–214.
13 E. Chenu, G. Jacobs, A. C. Crawford, R. A. Keogh, P. M. Patterson, D. E. Sparks and B. H. Davis, *Appl. Catal., B*, 2005, **59**, 45–56.
14 G. Jacobs, P. M. Patterson, U. M. Graham, A. C. Crawford, A. Dozier and B. H. Davis, *J. Catal.*, 2005, **235**, 79–91.
15 G. Jacobs, S. Ricote, P. M. Patterson, U. M. Graham, A. Dozier, S. Khalid, E. Rhodus and B. H. Davis, *Appl. Catal., A*, 2005, **292**, 229–243.
16 T. Tabakova, F. B. Boccuzzi, M. Manzoli and D. Andreeva, *Appl. Catal., A*, 2003, **252**, 385–397.
17 R. Leppelt, B. Schumacher, V. Plzak, M. Kinne and R. J. Behm, *J. Catal.*, 2006, **244**, 137–152.
18 D. L. Trimm and Z. I. Onsan, *Catal. Rev. Sci. Eng.*, 2001, **43**, 31–84.
19 G. Bond, *Gold Bull.*, 2009, **42**, 337–342.
20 L. C. Grabow, A. A. Gokhale, S. T. Evans, J. A. Dumesic and M. Mavrikakis, *J. Phys. Chem. C*, 2008, **112**, 4608–4617.
21 A. A. Gokhale, J. A. Dumesic and M. Mavrikakis, *J. Am. Chem. Soc.*, 2008, **130**, 1402–1414.
22 D. Cameron, R. Holliday and D. Thompson, *J. Power Sources*, 2003, **118**, 298–303.
23 Q. Fu, S. Kudriavtseva, H. Saltsburg and M. Flytzani-Stephanopoulos, *Chem. Eng. J.*, 2003, **93**, 41–53.
24 S. Hilaire, X. Wang, T. Luo, R. J. Gorte and J. Wagner, *Appl. Catal., A*, 2001, **215**, 271–278.
25 T. Bunluesin, R. J. Gorte and G. W. Graham, *Appl. Catal., B*, 1997, **14**, 105–115.
26 H. Cordatos, T. Bunluesin, J. Stubenrauch, J. M. Vohs and R. J. Gorte, *J. Phys. Chem.*, 1996, **100**, 785–789.
27 D. Tibiletti, A. Goguet, F. C. Meunier, J. P. Breen and R. Burch, *Chem. Commun.*, 2004, 1636–1637.
28 A. Goguet, F. C. Meunier, D. Tibiletti, J. P. Breen and R. Burch, *J. Phys. Chem. B*, 2004, **108**, 20240–20246.
29 F. C. Meunier, D. Tibiletti, A. Goguet, D. Reid and R. Burch, *Appl. Catal., A*, 2005, **289**, 104–112.
30 F. C. Meunier, D. Reid, A. Goguet, S. Shekhtman, C. Hardacre, R. Burch, W. Deng and M. Flytzani-Stephanopoulos, *J. Catal.*, 2007, **247**, 277–287.
31 R. Burch, *Phys. Chem. Chem. Phys.*, 2006, **8**, 5483–5500.
32 X. Q. Gong, P. Hu and R. Raval, *J. Chem. Phys.*, 2003, **119**, 6324–6334.
33 Z. P. Liu, S. J. Jenkins and D. A. King, *Phys. Rev. Lett.*, 2005, **94**, 4.

34 J. A. Rodriguez, P. Liu, J. Hrbek, J. Evans and M. Perez, *Angew. Chem., Int. Ed.*, 2007, **46**, 1329–1332.

35 X. M. Liu, Z. M. Ni, P. Yao, Q. Xu, J. H. Mao and Q. Q. Wang, *Acta Phys.-Chim. Sin.*, 2010, **26**, 1599–1606.

36 Y. Chen, J. Cheng, P. Hu and H. F. Wang, *Surf. Sci.*, 2008, **602**, 2828–2834.

37 J. M. Soler, E. Artacho, J. D. Gale, A. Garcia, J. Junquera, P. Ordejon and D. Sanchez-Portal, *J. Phys.: Condens. Matter*, 2002, **14**, 2745–2779.

38 J. P. Perdew, J. A. Chevary, S. H. Vosko, K. A. Jackson, M. R. Pederson, D. J. Singh and C. Fiolhais, *Phys. Rev. B: Condens. Matter*, 1992, **46**, 6671–6687.

39 D. Tibiletti, A. Amieiro-Fonseca, R. Burch, Y. Chen, J. M. Fisher, A. Goguet, C. Hardacre, P. Hu and A. Thompsett, *J. Phys. Chem. B*, 2005, **109**, 22553–22559.

40 Y. Chen, P. Hu, M. H. Lee and H. F. Wang, *Surf. Sci.*, 2008, **602**, 1736–1741.

41 A. Alavi, P. J. Hu, T. Deutsch, P. L. Silvestrelli and J. Hutter, *Phys. Rev. Lett.*, 1998, **80**, 3650–3653.

42 C. J. Zhang, P. Hu and M. H. Lee, *Surf. Sci.*, 1999, **432**, 305–315.

43 M. Nolan, S. C. Parker and G. W. Watson, *Phys. Chem. Chem. Phys.*, 2006, **8**, 216–218.

44 V. Shapovalov and H. Metiu, *J. Catal.*, 2007, **245**, 205–214.

45 M. Nolan, S. C. Parker and G. W. Watson, *Surf. Sci.*, 2006, **600**, L175–L178.

46 M. Nolan and G. W. Watson, *J. Phys. Chem. B*, 2006, **110**, 16600–16606.

47 G. Kresse and J. Hafner, *Phys. Rev. B: Condens. Matter*, 1994, **49**, 14251–14269.

48 G. Kresse and J. Furthmuller, *Comput. Mater. Sci.*, 1996, **6**, 15–50.

49 G. Kresse and D. Joubert, *Phys. Rev. B: Condens. Matter Mater. Phys.*, 1999, **59**, 1758–1775.

50 P. E. Blochl, *Phys. Rev. B: Condens. Matter*, 1994, **50**, 17953–17979.

51 J. Perdew, In *Electron Structure of Solides' 91*, ed. P. Ziesche, H. Eschrig, Akademie Verlag, Berlin, 1991.

52 M. Nolan, S. Grigoleit, D. C. Sayle, S. C. Parker and G. W. Watson, *Surf. Sci.*, 2005, **576**, 217–229.

53 M. Nolan, S. C. Parker and G. W. Watson, *Surf. Sci.*, 2005, **595**, 223–232.

54 M. Nolan, S. C. Parker and G. W. Watson, *J. Phys. Chem. B*, 2006, **110**, 2256–2262.

55 G. Henkelman, B. P. Uberuaga and H. Jonsson, *J. Chem. Phys.*, 2000, **113**, 9901–9904.

56 G. Henkelman and H. Jonsson, *J. Chem. Phys.*, 2000, **113**, 9978–9985.

57 J. A. Dumesic, *J. Catal.*, 1999, **185**, 496–505.

58 R. D. Cortright and J. A. Dumesic, in *Advances in Catalysis*, Vol. 46, Academic Press Inc, San Diego, 2001, pp. 161–264.

59 C. T. Campbell, *J. Catal.*, 2001, **204**, 520–524.

60 J. A. Dumesic, *J. Catal.*, 2001, **204**, 525–529.

61 T. De Donder, in *L'Affinité*, Gauthier–Villers, Paris, 1927.

62 M. Boudart, *J. Phys. Chem.*, 1983, **87**, 2786–2789.

63 W. L. Holstein and M. Boudart, *J. Phys. Chem. B*, 1997, **101**, 9991–9994.

64 S. Kumar and P. K. Schelling, *J. Chem. Phys.*, 2006, **125**, 204704.

65 M. A. Henderson, C. L. Perkins, M. H. Engelhard, S. Thevuthasan and C. H. F. Peden, *Surf. Sci.*, 2003, **526**, 1–18.

66 L. Barrio, P. Liu, J. A. Rodriguez, J. M. Campos-Martin and J. L. G. Fierro, *J. Chem. Phys.*, 2006, **125**, 164715.

67 Q. Fu, W. L. Deng, H. Saltsburg and M. Flytzani-Stephanopoulos, *Appl. Catal., B*, 2005, **56**, 57–68.

A periodic DFT study of the activation of O_2 by Au nanoparticles on α-Fe_2O_3

Kara L. Howard and David J. Willock*

Received 25th February 2011, Accepted 31st March 2011
DOI: 10.1039/c1fd00026h

Oxidation chemistry with supported Au nanoparticles as catalysts is an area of intense research. Even so there is still much discussion as to the nature of Au species generated on the complex surfaces of these catalysts and the types of oxygen species that are present. Recent experimental work has highlighted Au bi-layers with dimensions of 0.5 nm supported on iron oxide as a very efficient catalyst system for CO oxidation. This size scale implies clusters containing only 10 Au atoms, making the simulation of the nanoparticles, oxide surface and their interface amenable to periodic density functional theory calculations. We present simulation results which demonstrate that the dissociation of O_2 is energetically favourable at the interface between nanoparticle and oxide, with both surface Fe cations and Au atoms taking part in the adsorption site. Here the barrier to dissociation of O2 is found to be lower than the energy required for molecular desorption which is not the case for isolated Au clusters. This reaction also produces oxidised Au atoms, as confirmed by Bader charge analysis. For isolated clusters we show that such oxidised Au species give rise to empty d-band states, whereas molecular adsorption of O_2 does not.

Introduction

The use of supported Au nanoparticles as oxidation catalysts has continued to excite interest since the first report of room temperature CO oxidation around 20 years ago.[1] It was soon recognized that the size of the Au particles was critical, with average particle sizes below 5nm being required.[2] Indeed, Landman and co-workers used size selected metal particles deposited onto MgO thin films to show that significant CO oxidation activity can be detected for clusters as small as Au_8.[3] More recently Hutchings, Kiely and co-workers[4] employed aberration corrected transmission electron microscopy to compare Au/FeO_x catalysts which showed markedly differing activities in CO oxidation. They concluded that cluster sizes containing around 10 Au atoms as bi-layer particles were present in active catalysts but absent from the inactive material.

The possible structure of Au particles on this scale have been studied using both experimental and theoretical methods by a number of groups. For isolated Au clusters the preferred geometry is a planar two dimensional (2D) array of atoms for suprisingly large cluster sizes. From experimental measurements of ion mobiltiy Furche et al.[5] have estimated the collision cross section at room temperature for Au_n^- anions and have deduced that 2D structures are preferred over 3D for $n \leq 12$. Perdew Wang (PW91) DFT calculations by Xiao et al.[6] are in close agreement giving 2D clusters lower energies than 3D for Au_{13} and below. However, higher level calculations (CCSD(T)) by Gordon and co-workers put the switch over from 2D to

Cardiff Catalysis Institute, School of Chemistry, Cardiff University, Main Building, Park Place, Cardiff, CF10 3AT. E-mail: willockdj@cf.ac.uk

$3D^7$ at the $n = 6$ cluster. For supported clusters interaction with the support material also has an influence on the preferred structure. For TiO_2 as a support Boronat et al.[8] have found that 2D clusters of Au_{13} have lower calculated energy than 3D for stoichiometric surfaces but that the 3D clusters are more strongly stabilised by defects on the reduced surface and become the preferred geometry.

Both experimental and theoretical studies have also considered aspects of the mechanism of oxidation for Au catalysts. Early theoretical work demonstrated that CO oxidation can take place over an isolated Au particle with CO co-adsorbed with molecular oxygen. The O_2 bond only breaking as the new O-(CO) bond is formed.[9] This is consistent with the idea that Au is a difficult metal to oxidise, with DFT suggesting that O_2 dissociation is unfavorable compared to molecular desorption even on high index surfaces of the bulk metal.[10] Oxygen adsorption in both molecular and dissociated form has been studied on 2D clusters up to Au_8 by Yoon et al.[11] For molecular adsorption they show that charge transfer from Au clusters to the O_2 molecule results in an increase of the dioxygen bond length, indicative of a superoxide species, O_2^-. They also noted a remarkable oscillation of the adsorption energy with the number of atoms in the cluster, particularly for anionic clusters, with clusters containing an odd number of electrons giving the more favourable binding energies. For example, the adsorption energies of O_2 to Au_8^- and Au_8 were reported as -84 kJ mol^{-1} and -27 kJ mol^{-1}, respectively (following the sign convention of our definition given in Equation 1). The dissociated state was found to have more favourable adsorption energies, however the calculated dissociation barrier was considerably higher than the adsorption energy, suggesting that O_2 desorption should occur in preference to dissociation. Extending their work to a 1D Au_{20} cluster supported on an MgO overlayer on Mo the same researchers[12] suggest molecular adsorption of O_2 occurs with CO oxidation following a similar pathway to that suggested by Lopez and Nørskov.[9] In contrast, recent work of Illas and co-workers[13] using DFT calculations with the PW91 functional compares the dissociation barrier for O_2 on four 3D Au clusters from Au_{25} to Au_{79}. They suggest that O_2 dissociation occurs in preference to molecular desorption only on the smaller clusters, Au_{25} and Au_{38}, in their work O_2 bond cleavage is accompanied by quite small changes in the cluster geometry.

Some experimental investigations have also assumed that O_2 is activated to form more reactive O adatoms prior to CO oxidation. For example, temporal analysis of products kinetic studies have shown Au/SiO_2 catalysts have an ability to store O atomically.[14] Additional weight has also been leant to this argument by Lambert and co-workers who have considered the partial oxidation of styrene using Au_{55} particles supported on BN. Their results suggest that the oxidising species is atomic O which can only arise from the activation O_2 supplied from the gas phase.[15]

Another point of discussion is the oxidation state of Au in active catalysts. Weiher et al. have used XANES measurements on Au/TiO_2 exposed to O_2, O_2/CO and CO atmospheres to conclude that reduced Au particles form "gold-oxygen complexes" with O_2 and that this results in depletion of the d-band for the Au clusters in the sample.[16] The authors are careful to note that this observation on its own does not suggest O_2 dissociates since it may be the case that the charge transfer to form the superoxide species is sufficient to cause the observed reduction in band occupancy. Kobayashi et al. working with ^{197}Au Mössbauer spectroscopy and samples of $Au/Mg(OH)_2$ have shown a correlation between catalyst activity and the concentration of Au(I).[17] Guzman and Gates have demonstrated a similar dependence of oxidation activity on the concentration of oxidised Au species using XANES spectra for Au/MgO.[18] The combination of XANES and XPS data has also been used to correlate activity with cationic Au in the $Au/\alpha\text{-}Fe_2O_3$ catalysts that form the subject of this modelling study.[19]

In a previous publication[20] we have reported calculations on the energetics of the stable structures involved with O_2 adsorption and dissociation along with CO at the interface between a Au particle and $\alpha\text{-}Fe_2O_3(0001)$, or hematite, in conjunction with

 This journal is © The Royal Society of Chemistry 2011

experimental results from XPS and TAP measurements. In this contribution we present a more detailed set of calculations comparing isolated Au_{10} clusters and Au_{10} supported on slab models of α-Fe_2O_3(0001) demonstrating the importance of the flexibility of small clusters in the activation of O_2. It is also shown that, in addition to favourable thermodynamic factors, the calculated barrier to dissociation of O_2 is sufficiently low to ensure that dissociation is favoured over molecular desorption for Au_{10}/α-Fe_2O_3 but not for similar clusters in isolation.

Methods

All calculations were carried out using the plane wave basis set code VASP,[21] with a plane wave cutoff of 500 eV and the Perdew, Burke and Ernzerhof (PBE) functional.[22] For spin-polarized calculations, the interpolation for the correlation part of the exchange correlation functional was carried out using the method of Vosko, Wilk and Nusair.[23] For the core states of each atom the projector augmented-wave (PAW)[24,25] approach was employed. In our calculations, the core states represented in this way consist of orbitals up to and including the $3p$ orbitals for iron, leaving the $3d$ and $4s$ electrons to be treated explicitly. For oxygen and carbon only the $1s$ orbital is included in the core. The core states for gold consist of orbitals up to and including the $4f$ orbital, so the $5d$ and $6s$ electrons are treated explicitly. The optimization of the atomic coordinates is performed *via* the conjugate gradients technique, structures are accepted once the greatest force on any atom is below 0.05 eV Å^{-1}.

Calculations for α-Fe_2O_3 were performed using the DFT + U approach of Dudarev *et al.*[26] applied to the iron centres only. The DFT + U method introduces an intra-atomic electron-electron interaction with coulomb, U, and exhange, J, terms as on site corrections in order to describe systems with localized d and f electrons. In the Dudarev approach these corrections are formulated so that only U–J need be set as a model parameter. The effect of including these terms in PW91 + U calculations for hematite has been considered by Rohrbach *et al* who showed that U–J = 4 gave the best over all agreement with the experimental lattice parameters, magnetic moments, bulk moduli, band gaps and density of states.[27] This value is used throughout our own work and we note that it is also in line with the choice of U–J = 4.3 made in the recent study of the tensile properties of α-Fe_2O_3 by Liao and Carter.[28,29] To ensure calculations arrived at a given magnetic ordering the initial magnetic moment of the iron atoms was set to ± 5 in the appropriate pattern. The total energy of the hexagonal unit cell of α-Fe_2O_3 was converged for a k-point mesh of $5 \times 5 \times 3$ and this was used for optimising the bulk cell volume and comparing anti-ferromagnetic orderings as described in the results section. The simulation of the (0001) surface was carried out using the slab method with a vacuum gap of 15 Å introduced along the c-direction to expose the surface. For calculations of the (0001) surface energy k-point sampling was reduced to $5 \times 5 \times 1$, *i.e.* only the Γ-point for the direction perpendicular to the surface. To accommodate the Au clusters with the slab method $p(3 \times 3)$ supercells were constructed and for these larger real space cells only the Γ-point was used in all directions. For all slab calculations dipole corrections in the direction perpendicular to the exposed surface were applied.

Atomic charges, were estimated using Bader analysis on the charge density grid provided from VASP using a program developed by Henkelman *et al.*[30] Bader analysis divides the calcualted charge density into atomic basins which are delimited by minima in the electron density in each direction. To identify the atom centres Bader analysis assumes that they are located at charge density maxima. To enable the location of these maxima we add back in the core charge that is replaced by the atomic pseudopotential prior to Bader analysis.

The reported adsorption energies, E_{ads}, are calculated following the equation:

$$E_{ads} = E(SX) - E(S) - E(X) \tag{1}$$

where $E(S)$ and $E(X)$ are the energies of the surface and adsorbate species in their optimised reference states and $E(SX)$ is the optimised energy for the surface with the adsorbate present. For O_2 the reference state is the isolated molecule in the triplet ground state. For isolated or supported Au clusters any constraints discussed in the text are imposed on both the reference state and SX system. In this definition of E_{ads} negative values correspond to favorable adsorption energies.

Results

Bulk Fe_2O_3

α-Fe_2O_3 has the corundum structure with anion and cation layers in the c-direction in the sequence $-O_3$-Fe-Fe-O_3-as shown in Figure 1a. The Fe^{3+} centres have a magnetic moment due to the d^5 configuration in the weak field of the anion centres. Overall the crystal is anti-ferromagnetic and Rohrbach et al.[27] took a setting of $++O_3--O_3++$ (+/− indicating opposite Fe spin states), so that the Fe atoms in between the O_3 layers have the same spin. In figure 2b we confirm that this leads to a lower calculated energy than the alternative anti-ferromagnetic ordering with Fe atoms between layers having opposite spins, $+ -O_3- +O_3+-$. Table 1 summarises the structural data for the different magnetic orderings and shows that the preferred $++O_3--O_3++$ arrangement gives good agreement with the available experimental cell volume, bulk modulus and atomic magnetic moment. The optimised bulk α-Fe_2O_3 cell parameters of $a = b = 5.083$Å, $c = 13.850$Å, are within 1% of the experimental values,[31] $a = b = 5.035$Å, $c = 13.747$Å and agree well with earlier calculations.[27] We also carried out simulations using the PBE functional without the Hubbard model corrections. The effect of the U-parameter on the bulk properties is to introduce a band gap and increase the bulk modulus from 173 GPa (PBE) to 190 GPa (PBE + U) giving closer agreement with the experimental value.[32] In addition the magnetic moment on Fe centres is 4.20 μ_B using PBE + U compared to 3.52 μ_B at the PBE level. Interestingly the calculated bandgap for the alternative

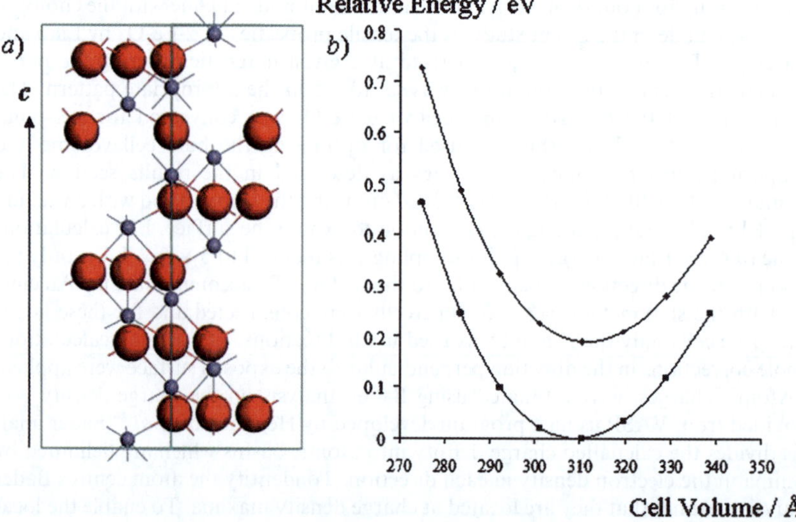

Fig. 1 a) The hexagonal unit cell of α-Fe_2O_3 showing the.Fe–O_3-Fe-Fe-O_3-Fe. layered structure along the c-direction. b) The calculated relative energy per formula unit of Fe_2O_3 as a function of cell volume for alternative anti-ferromagnetic arrangements, lower curve (squares) $++O_3--O_3++$ and upper curve (diamonds) $+-O_3-+O_3+-$. Atom colours: Fe; blue and O; red.

    This journal is © The Royal Society of Chemistry 2011

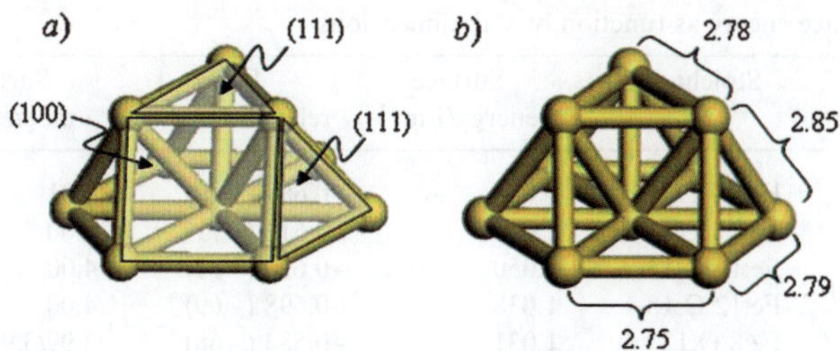

Fig. 2 a) The $Au_{10}(7,3)$ structure used as the starting point for simulations of two layer Au clusters in this work. The cluster is orientated so that the z-direction is perpendicular to the Au layers. b) Structure of optimised cluster with Au–Au bond lengths given in Å, the central Au atom in the base is 2.77 Å from each of the other Au atoms in the base.

$+ -O_3- +O_3+ -$ magnetic ordering is some 0.4 eV below that for the ground state, which itself is in good agreement with the available electrical conductivity[33] and optical measurements.[34]

$Fe_2O_3(0001)$ surface

The (0001) surface of corundum type oxides have low surface energies as a non-polar termination is possible by cutting the oxide between adjacent metal ion layers to give, in this case, an $Fe-O_3-Fe-Fe-O_3$ stacking down the c-direction. Surface science experiments using STM and infrared reflection absorption spectroscopy have shown that ferryl (Fe=O) groups may be formed on the surface which co-exist with this Fe termination.[35] However, under the levels of oxygen partial pressure expected during oxidation catalysis, calculations suggest that the Fe termination is thermodynamically preferred[27] and so our model of the surface consists of a slab with single layer Fe termination on both sides.

The unit cell shown in figure 1a offers six positions for cutting this surface the resulting slab thickness is conveniently described by the number of O_3 layers it contains. To assess the influence of the thickness of the slab on the properties of the surface we calculated the surface energy, the relaxation of the outermost Fe layer and the surface Fe magnetic moments for a series of $p(1 \times 1)$ slabs (Table 2). The surface energy, E_{surf}, for a fully relaxed slab can be obtained using;

$$E_{surf} = \frac{E_{slab} - nE_{bulk}}{2S} \tag{2}$$

where E_{slab} is the calculated energy for the slab representation of the surface, E_{bulk} is that for a reference bulk calculation and n is the ratio of stoichiometries for the slab and bulk calculations. The division by twice the surface area (S) takes account of the two surfaces exposed by creating the slab.

Ruberto et al.[36] and A. Kiejna et al.[37] have suggested ways to use calculations on a series of slabs of the same surface area but differing thicknesses to estimate a value

Table 1 Comparison of calculated material properties for α-Fe_2O_3 with alternative anti-ferromagnetic ordering and comparison with literature values

	Cell volume/Å³	Bulk Modulus/GPa	Fe magnetic moment μ/μ_B	E_{gap}
$+ +O_3- -O_3+ +$	310.6	190	4.15	1.9
$+ -O_3- +O_3+ -$	313.2	198	4.20	1.6
PW91 + U[27]	308.7	192	4.11	2.0
Expt	301.2[31]	231[32]	4.6–4.9	2.0[33]–2.2[34]

Table 2 Surface energy as function of slab dimensions

Layers (surf. vectors)	Stoichiometry	Surface energy/J m^{-2}	Fe–O relaxation/Å (%)	Surface layer magnetisation/μ_B
3 (1 × 1)	Fe:6 O:9	1.035	−0.687 (−80)	4.01
4 (1 × 1)	Fe:8 O:12	1.060	−0.612 (−71)	4.00
5 (1 × 1)	Fe:10 O:15	1.050	−0.620 (−72)	4.00
6 (1 × 1)	Fe:12 O:18	1.038	−0.598 (−69)	4.00
4 (1 × 1)[a]	Fe:8 O:12	1.031	−0.584 (−68)	3.99/3.91[b]
4 (3 × 3)[a]	Fe:72 O:108	1.033	−0.592 (−69)	4.14/3.98[b]

[a] Lower surface held fixed at bulk positions, surface Fe-O$_3$-Fe–Fe layer optimised all other atoms fixed at bulk positions. [b] The second value is that of the fixed surface.

of E_{bulk} without recourse to a separate calculation. A. Kiejna *et al.* start by rearranging (2) to emphasis the expected linear relation between slab energy and the value of n;

$$E_{slab} = 2SE_{surf} + nE_{bulk} \tag{3}$$

For a series of slabs of the same S, n will be proportional to slab thickness and so a linear fit to a graph of E_{slab} *vs* n will yield an estimate of E_{bulk} from the slope and the surface energy from the intercept. The value of E_{bulk} obtained using this approach was adopted for the surface slabs from 3 to 6 layers reported in Table 2. The intercept of the plot corresponded to a surface energy of 1.046 J m^{-2}, the average of the values in Table 2. In our earlier publication[20] the surface energy was calculated using equation 1 with E_{bulk} taken from a reference calculation and we find the surface energies based on this series of slabs are around 0.06 J m^{-2} lower in this case.

For all slabs the greatest relaxation occurs for the surface Fe atoms which move in toward the centre of the slab so that the difference in their z-coordinate and that of the first layer O atoms reduces by around 80% for the 3 layer slab (3l) and 70% for 4–6 layers (Table 2). The surface relaxation is accompanied by a lowering of the magnetic moment for the surface Fe atoms compared to the bulk value reported in Table 1. We note that the degree of relaxation is converged to within 2% for slabs of four O layers or more.

For the calculations on Au clusters supported on α-Fe$_2$O$_3$(0001) we require a larger surface area than these 1 × 1 surfaces and so slabs with a $p(3 × 3)$ surface area (22.4 Å^2 per side) where also constructed. In the interests of computer time we originally used the minimal 3 layer slab geometry and calculations were carried out with all atoms relaxed during optimisation runs. In this work we decided to compare this methodology to a thicker slab (4 layers) with the interface to the bulk oxide represented by constraining atoms in the lower layers of the slab at their optimised bulk positions. This approach is analogous to that adopted in several other studies, for example the recent calculations by Boronat *et al* in their study of Au particles supported on the anatase phase of TiO$_2$.[38] In such a model the lower surface of the slab is left at the bulk termination structure. To obtain the surface energy in this case we first calculate the value for bulk termination, E_{surf}^{term}, from a single point calculation on the unrelaxed slab. The energy of the slab for the constrained optimisation would then be expected to be;

$$E_{slab} = SE_{surf}^{term} + SE_{surf}^{opt} + nE_{bulk} \tag{4}$$

where the first term accounts for the contribution of the fixed lower surface and E_{surf}^{opt} is the surface energy of the optimised upper surface and so,

$$E_{\text{surf}}^{\text{opt}} = \frac{E_{\text{slab}} - nE_{\text{bulk}}}{S} - E_{\text{surf}}^{\text{term}} \tag{5}$$

Table 2 shows that, with optimisation of only the upper $Fe\text{-}O_3\text{-}Fe\text{-}Fe$ layers, the surface energy for the 4 layer (1×1) slab is within 3% of that for the fully relaxed 4 layer slab. In addition the surface relaxation of the outermost Fe atoms and the surface magnetic moment is reproduced well. Using the same methodology with a $p(3 \times 3)$ super cell also gives good agreement in terms of surface energy and relaxation although the magnetic ordering becomes more complex since Fe atoms in the same layer as one another need not have parallel spins.

Au bilayer nanoparticles

In the Introduction section we discussed the transmission electron microscopy data which identified Au_{10} bi-layer clusters as candidates for typical active particles for CO oxidation over Au/Fe_2O_3.[4] For a model of such a cluster we start with the fcc lattice of bulk Au and take an hexagonal seven atom section along with the three Au atoms in the next (111) layer to form the $Au_{10}(7,3)$ shown in figure 2a. When working with this cluster in isolation, the constraint of interactions between the cluster and the support surface are represented by holding the z-co-ordinate of the seven layer atoms frozen. After optimisation (figure 2b) the Au–Au distances within the two planes have all reduced compared to the bulk Au–Au distance at the same level of theory of 2.95 Å. However, the inter-layer Au–Au distances are considerably longer than the intra-layer bonds, showing that the interaction between the layers is weaker than within them. Indeed, a calculation on a 2D equilateral triangle Au_{10} structure yielded a total energy 6.2 kJ mol^{-1} (per atom) lower than $Au_{10}(7,3)$. For some oxygen adsorption calculations we also considered $Au_9(6,3)$ clusters in which an Au atom has been removed from the base layer opposite to the O_2 adsorption site. This allows the effect of the number of atoms in the cluster on the energies of adsorption to be considered without influencing the local geometry of the adsorption site prior to structure optimisation.

For calculations with supported Au clusters the $Au_{10}(7,3)$ cluster was placed onto the relaxed $\alpha\text{-}Fe_2O_3(0001)$ surface. Initial calculations on the adsorption of a single Au atom on the 3 layer slab showed that the calculated adsorption energies for Au over a surface Fe or O ion were actually quite close to one another, -88 kJ mol^{-1} (Fe) and -86 kJ mol^{-1} (O), marginally favouring the Fe site. The $Au_{10}(7,3)$ bi-layer was positioned on the surface with the 7 atom layer placed in contact with the oxide surface so that the central atom was over a surface Fe ion. Several orientations were compared on the 3l slab and it was found that the most favourable adsorption energy for the cluster corresponds to an initial position in which the edges of the (100) faces are aligned with surface O atoms (figure 3a). On optimisation the cluster changes shape significantly with the well defined fcc structure being lost and the Au atoms in contact with the surface moving toward surface Fe atoms. This restructuring of the cluster is also apparent for the 4l slab (figure 3c and d). The relaxation of the surface of the oxide is also affected by the presence of the cluster. For example, in the 4l simulation Fe atoms interacting with Au in the cluster move out of the surface 0.28 Å further than Fe atoms away from the Au cluster so that the inward relaxation noted for the clean surface is less pronounced for these atom. There is also a small electron transfer from the cluster to the surface with a resulting total charge on the cluster of around 0.1 e. Figure 3 also gives the binding energy per Au atom to the surface, with the reference for the Au atoms being the isolated cluster. The four layer cluster has a lower binding energy than the 3l, possibly due to the more restricted response of the surface to the supported Au cluster. Both simulations show a marked distortion of the Au cluster from the initial geometry, but the end point structures are quite different. This is likely to be due to small differences in the initial alignment

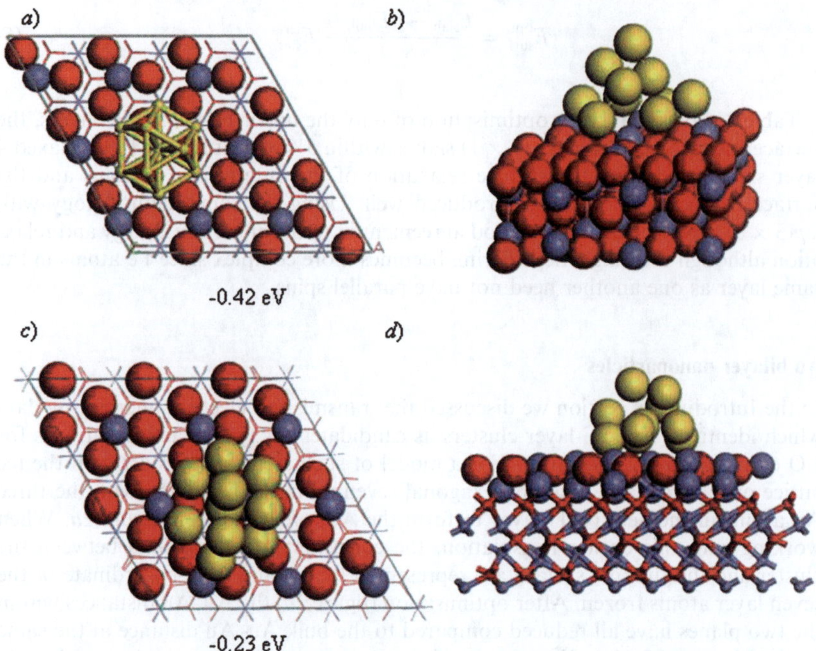

Fig. 3 $Au_{10}(7,3)$ on α-$Fe_2O_3(0001)$ surface a) starting orientation with (100) face edges aligned with surface O atoms, b) The relaxed structure after optimisation from starting point in (a) for 3l slab model. c) Plan and d) side view of $Au_{10}(7,3)$ relaxed structure on α-$Fe_2O_3(0001)$ surface using the 4l slab, atoms free to move during optimisation are shown as spheres in this case. Values shown in (a) and (c) are the relaxed adsorption energy per Au atom. Atom colours: Au; yellow, Fe; blue and O; red.

of the cluster with the surface since it was not possible to use exactly the same starting structure for the two calculations.

Oxygen adsorption

For the isolated $Au_{10}(7,3)$ attempting to adsorb O_2 on the upper three atom face gives a very low adsorption energy of -12 kJ mol^{-1} with the molecule moving away from the cluster on optimisation, further details of adsorption to the three layer section of the cluster will not be reported. Table 3 gives the adsorption energies and some geometric features for O_2 adsorption at the edge of the seven atom base of the cluster. In each case the geometry was initiated with the molecular axis parallel to the edge of the cluster with the two oxygen atoms close to Au atoms in the cluster base. There are two alternative positions for the isolated cluster; on the (100) or (111) edge (figure 2a).

The O_2 bond length in the triplet ground state at the PBE level of theory is 1.232 Å and so it is clear from Table 3 that molecular adsorption at any of the sites discussed below results in an increased O_2 bond length consistent with superoxide formation. For the $Au_{10}(7,3)$ cluster with O_2 on the (111) edge optimisation results in one end of the molecule moving across the triangular edge toward the bridge site with the three atom layer (figure 4a). The triangular face of the cluster has expanded (see figure 2b for starting geometry) with the Au atoms in the base 3.257 Å apart, the bridged Au atoms at 2.962 Å and the third edge measuring 3.099 Å. Table 3 shows that the adsorption energy for this geometry is 50 kJ mol^{-1} less favourable than for the (100) face alternative. In the (100) case the O atoms are each co-ordinated to an Au atom in the base of the cluster (figure 4b), the Au atoms involved have moved

This journal is © The Royal Society of Chemistry 2011

Table 3 Results of O_2 adsorption calculations for supported Au clusters

System	E_{ads}/kJ mol^{-1}	OO/Å	AuO/Å	FeO/Å
Au_{10} O_2(111)	-34	1.405	2.093 2.250/2.324[a]	—
Au_{10} O_2(100)	-84	1.347	2.135 2.137	—
Au_9 O_2(111)	-110	1.328	2.172 2.208	—
Au_9 O_2(100)	-81	1.344	2.118 2.144	—
Au_{10} 2O	-134	3.909	1.956/1.959[b] 2.038/2.046[c]	—
Au_{10}/Fe_2O_3(4l) O_2(100)	-130	1.498	2.104 2.142	2.071 2.129
Au_{10}/Fe_2O_3(3l) O_2(100)	-85	1.495	2.123 2.142	2.053 2.075
Au_{10}/Fe_2O_3(4l) O_2(111)	-108	1.481	2.102 2.167	2.088 2.103
Au_9/Fe_2O_3(4l) O_2(111)	-56	1.373	2.146 2.306	2.022 2.848
Au_{10}/Fe_2O_3(4l) 2O	-405	4.161	1.982/2.078[b] 2.087/2.118[c]	1.982 1.996
Au_{10}/Fe_2O_3(3l) 2O	-279	4.119	2.061/2.068[b] 2.110/2.122[c]	1.937 1.951

[a] First value for Au atom in the base and second for Au in the three atom layer. [b] Au–O distances for the Au atom in between the two O atoms. [c] For the Au–O distances from each O atom to difference Au atoms in the cluster.

apart, from separation of 2.746 Å in the relaxed Au_{10}(7,3) cluster to 3.149 Å with co-ordinated O_2. The optimised structure for the Au_9 (6,3) cluster with O_2 adsorbed on the edge of the (100) face is shown in figure 4d. The geometry local to the O_2 molecule reported in Table 3 is similar to that found for the equivalent site on the Au_{10}(7,3) cluster and the adsorption energies for the two systems differ by only 3 kJ mol^{-1}. In contrast the Au_9(6,3) cluster with O_2 adsorbed on the edge of the (111) face has a binding energy some 76 kJ mol^{-1} lower than its Au_{10}(7,3) counterpart. In this case the structures are also quite different since in the Au_9(6,3) case the base distorts to give the same motif of each O atom binding to a separate Au atom in the base of the cluster that is seen for the (100) face adsorption with the two Au atoms 2.993 Å apart. In this case it appears that removing an atom in the hexagonal base gives enough flexibility to accommodate this binding mode by distortion of the cluster.

On dissociation of O_2 on Au_{10}(7,3), one Au atom moves away from the central atom of the cluster base to form a near linear O–Au–O structure (angle at Au 173°), figure 4e. The two Au–O bonds for this Au atom are the shortest for any of the isolated cluster calculations and each O atom also interacts with a second Au atom in the base of the cluster. The dissociated state also gives a considerably more favourable binding energy than for any of the molecular adsorbed structures on the Au_{10}(7,3) cluster.

Figure 5 gives a comparison of the calculated density of states (DOS) and the partial DOS (PDOS) from the Au $5d$ states for bulk Au and the isolated Au_{10}(7,3) cluster. These are compared with the most favourable molecularly adsorbed O_2 structure ((100) facet, figure 4b) and the dissociated state (figure 4e). Looking to the occupied states from the Fermi level, E_F, in bulk Au (figure 5a)

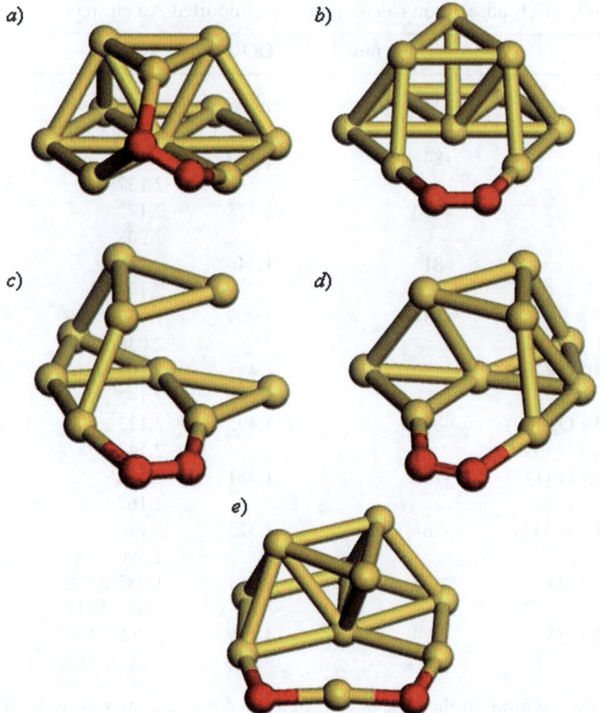

Fig. 4 Optimised structures for Au clusters with adsorbed O_2, a) $Au_{10}(7,3)$ O_2 on (111), b) $Au_{10}(7,3)$ O_2 on (100), c) $Au_9(6,3)$ O_2 on (111), d) $Au_9(6,3)$ O_2 on (100) and e) dissociated O2. Atom colours: Au; yellow and O; red.

the d-levels start at around -1.2 eV with the first peak at -1.91 eV. There are states at E_F, giving the expected metallic character, but the contributions here are mainly from s and p states, consistent with the closed $5d$ shell expected from the atomic electronic configuration. In figure 5b, the $Au_{10}(7,3)$ cluster alone has d-levels starting around -0.9 eV with, initially, a slow rise in the PDOS and the first peak occurring at -1.80 eV. The PDOS for the d-states is also symmetric for the up and down spin components. The DOS near to E_F, is broken up into a series of peaks, indicating molecular orbital like, rather than band like, electronic states. When molecular oxygen is adsorbed at the base of the $Au_{10}(7,3)$ cluster on the (100) facet we find the DOS given in figure 5c. Here the up spin DOS of the occupied states show a new peak at -0.89 eV with a significant d-state contribution that is not present in the down spin DOS. The main peak in d-state region is shifted down in energy compared to $Au_{10}(7,3)$ by 0.20 eV. The asymmetry of the DOS is consistent with the partial transfer of an electron from the cluster to one of the π^* orbitals of the O_2 molecule to form a superoxide like species. Indeed, a Bader analysis of the charge density for this cluster assigns a charge of -0.63 e to the oxygen molecule. We note however that this transfer is made while maintaining a full occupancy of the Au $5d$ states.

The DOS for the dissociated O_2 structure is given in figure 5d. The main peak in the filled states and the up spin is broader than seen for the cluster alone. This peak also occurs at higher energy (-1.67 eV) compared to either the cluster alone or the molecularly adsorbed state. The tail of the main d-band region now extends to the Fermi level and for the levels immediately above E_f, the Au d-states are involved. Focusing on the contribution of the atom forming the near linear structure in figure 4d, confirms that these unoccupied states are associated with this Au atom. A charge

   This journal is © The Royal Society of Chemistry 2011

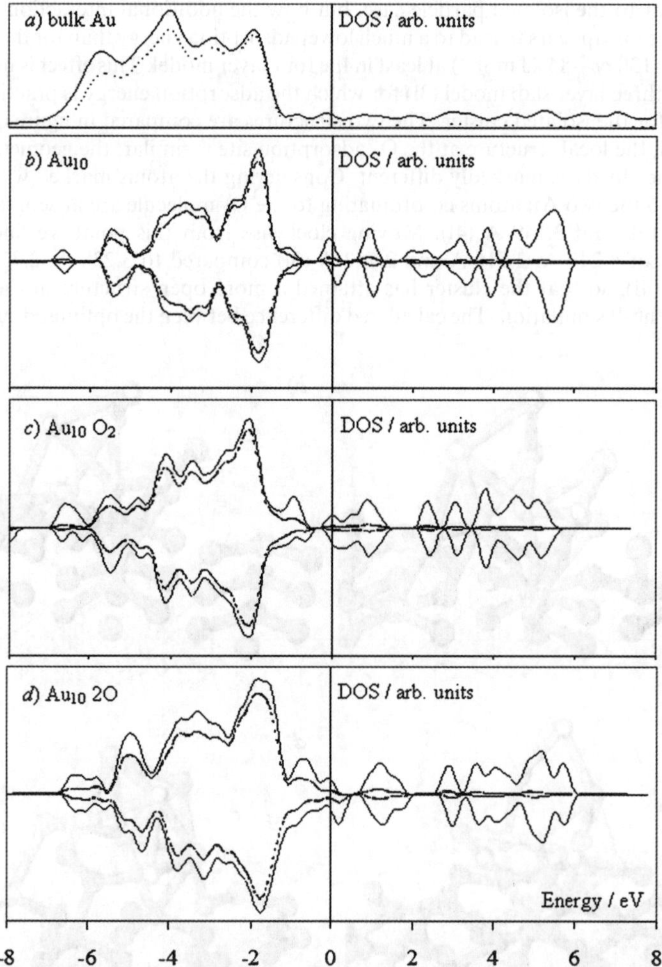

Fig. 5 Calculated Density of states for a) bulk Au, b) the isolated Au$_{10}$(7,3) cluster, c) the isolated Au$_{10}$(7,3) cluster with O$_2$ adsorbed at the base on the (100) facet and d) the isolated Au$_{10}$(7,3) cluster with 2 O atoms adsorbed, one at the base on the (100) facet and one at the base of (111). Solid lines are full system DOS, bold dotted lines the Au d-state PDOS. Energy scales are aligned with that shown for plot d) with the scale zero marking the Fermi level in each case. In b), c) and d) contributions from up and down spins are shown by positive and negative DOS values respectively.

analysis for this Au atom returns 0.69 e, and $-0.73/-0.74$ e for the two O atoms. Using the same methodology, the Bader charges on the gold atoms in AuCl and AuCl$_3$ are 0.33 e and 0.81 e respectively. We conclude that dissociation of O$_2$ results in the oxidation of atoms at the cluster edge which leads to a depletion in occupancy of the Au d-band states.

For Au clusters supported on α-Fe$_2$O$_3$(0001) the adsorbed O$_2$ molecule can also co-ordinate to a surface Fe ion. For both the 3l fully relaxed slab and the 4l with only the surface optimised very similar geometries are found in the immediate vicinity of the O$_2$ molecule. Table 3 gives the structural parameters for these two models when the O$_2$ molecule is adsorbed on the (100) face of the nanoparticle. In both cases the O$_2$ bond is elongated compared to the gas phase O$_2$ molecule and is also longer than found for adsorption in the same position on the isolated Au$_{10}$(7,3) particle. The distances for the co-ordinated Au atoms to the oxygen atoms of the O$_2$ molecule are similar to the isolated particle case, but now the additional interaction with the surface Fe ion appears to lead to a much lower adsorption energy than for the isolated cluster (-130 cf -84 kJ mol^{-1}) at least in the four layer model. This effect is not found with the three layer slab model (3l) for which the adsorption energy is practically the same as for the isolated cluster. The two structures are compared in figure 6a and b. Although the local structure at the O$_2$ adsorption site is similar, the geometry of the rest of the cluster is markedly different. Considering the atoms marked with circles in figure 6 the two Au atoms co-ordinating to the O$_2$ molecule are at separations of 3.042 Å (3l) and 3.149 Å (4l). Moving clockwise from this point we find AuAu distances of 4.247 Å, 2.716 Å and 2.752 Å (3l) compared to 3.238 Å, 2.734 Å and 2.793 Å (4l), so that the cluster has attained a more open structure in the 3l case than for the 4l simulation. The calculated difference between the optimised supported

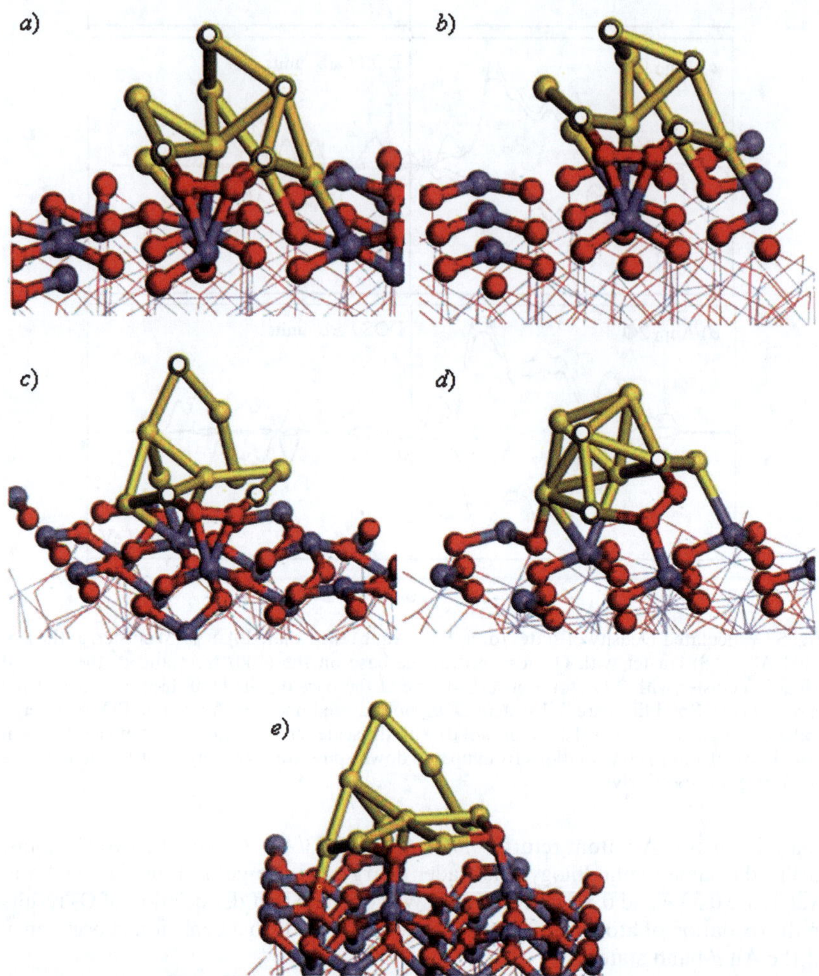

Fig. 6 The optimised structures of O$_2$ adsorbed at the interface of Au nanoparticles and α-Fe$_2$O$_3$(0001). a) Au$_{10}$(7,3) supported on the three O-layer fully relaxed slab with O$_2$ positioned on the edge of a (100) facet, b) as (a) with the four O-layer slab as support with surface only relaxation, c) Au$_{10}$(7,3) supported on the four layer slab with O$_2$ positioned on the edge of a (111) facet and d) Au$_9$(6,3) supported on the four layer slab with O$_2$ positioned on the edge of a (111) facet. e) The dissociated O$_2$ structure for Au$_{10}$(7,3) supported on the four layer slab. For (a) – (d) the Au atoms marked with white circles are those originally in the facet of the nanoparticle to which O$_2$ is co-ordinated. Atom colours: Au; yellow, Fe; blue and O; red.

cluster with and without O_2 adsorbed will include these changes in the cluster structure and so will influence the calculated adsorption energy. The differences in structure emphasizes the flexibility of the Au_{10} cluster on the surface and its ability to change geometry in response to the adsorption of the O_2 molecule, small differences in the initial alignment of the cluster and surface in the two calculations giving quite different outcomes for the cluster geometry.

Figure 6c and d show structures for O_2 adsorption to the edge of the (111) facet of $Au_{10}(7,3)$ and $Au_9(6,3)$ on the 4l slab model, respectively. The $Au_{10}(7,3)$ cluster gives a similar local geometry to that observed for the (100) facet and an adsorption energy 12 kJ mol^{-1} more positive (Table 3). The geometry of the cluster has opened up more than in the (100) calculation. The atoms originally in the (111) facet of the adsorption site are marked with circles in figure 6c and it can be seen that the top layer Au atom is now removed from its former neighbours in the base of the cluster. For the supported $Au_9(6,3)$ the disruption of the cluster is not as great but the local geometry at the adsorbed O has only a single O atom interacting with the surface Fe ion and a considerably less favourable adsorption energy is reported in Table 3.

The dissociated adsorption state for the 4l model of supported $Au_{10}(7,3)$ is shown in figure 6e. As we saw for the isolated cluster, a single Au atom has moved out of the base of the cluster to insert between the O atoms forming a close to linear O–Au–O motif (angle at Au 172°). On the interface site both O atoms also interact with surface Fe ions and the calculated adsorption energy with respect to gas phase O_2 is 275 kJ mol^{-1} more negative than the most favourable molecular adsorption calculated. Once again the calculated structure for the 3l slab local to the adsorption site is similar to that shown in figure 6e, but the adsorption energy is much more negative for the 4l model (Table 3).

To estimate the reaction barrier for the dissociation of O_2 on the isolated $Au_{10}(7,3)$ cluster and for $Au_{10}/Fe_2O_3(3l)$ we use the nudged elastic band method (NEB).[39] In order to obtain the initial reaction pathway a set of structures was generated from the two optimised end points of $Au_{10}(7,3)$ with O_2 on the (100) edge and $Au_{10}(7,3)$ 2O for the isolated cluster and for $Au_{10}/Fe_2O_3(3l)$ $O_2(100)$ and $Au_{10}/Fe_2O_3(3l)$ 2O for the supported particle. In each case, moving between these two structures involves one of the Au atoms in the base of the cluster inserting between the oxygen atoms and simple linear interpolation will lead to a clash between Au and O. To avoid this problem, the atoms of the O_2 molecule and the inserting Au atom were treated as a group.[40] This group was interpolated such that the positions of the two O atoms were changed relative to that of the Au atom by varying the Au–O distances smoothly to generate the initial pathway. All other atoms in the structure were interpolated linearly. A series of 8 images was produced between the start and end structures for use in the NEB calculations. The energies of these structures were used to plot out the reaction barrier and then further points were added near the maximum using constrained optimisation calculations in which the positions of the oxygen atoms from the O_2 molecule were held fixed.

The calculated energies for these NEB calculations as a function of O–O separation are shown in figure 7. The estimated barrier to dissociation for the isolated $Au_{10}(7,3)$ is 141 kJ mol^{-1} and occurs at an O–O separation of 2.468 Å, this point was confirmed as a transition state by a frequency calculation which found only one imaginary mode. For the supported $Au_{10}/Fe_2O_3(3l)$ the barrier is considerably lower at 70 kJ mol^{-1} and the O atoms are further apart at the maximum energy point (2.967 Å). A frequency calculation at this point on the supported cluster pathway gave an imaginary frequency of 129 cm^{-1} for the O_2 dissociation eigenvector. This calculation required a numerical hessian calculation and only the degrees of freedom associated with the atoms of the O_2 molecule and Au cluster were included. Three additional imaginary modes were found with imaginary frequencies below 60 cm^{-1}. These eigenmodes were to do with movements of the atoms included in the hessian relative to those that were excluded and should not influence the estimation of the barrier height significantly.

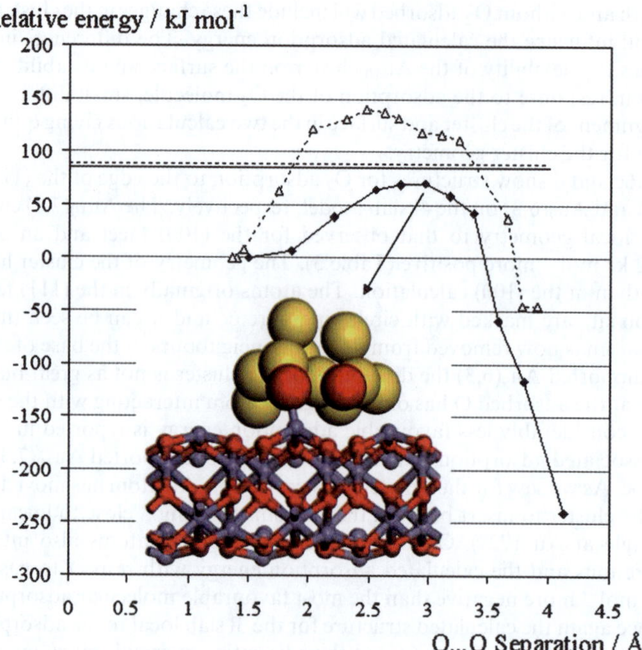

Fig. 7 Calculate energy as a function of O,O separation for O_2 adsorbed to Au_{10}(7,3) in isolation (dotted line and open diamonds) and Au_{10}/Fe_2O_3 (solid line and closed diamonds). The energy is given relative to adsorbed O_2 at the base of the (100) facet in each case and the horizontal lines correspond to the calculated adsorption energy (dotted isolated Au_{10}(7,3), solid Au_{10}/Fe_2O_3) to give an indication of the desorption barrier. The inset structure is that of the highest point on the pathway for the Au_{10}/Fe_2O_3 calculation. Atom colours: Au; yellow, Fe; blue and O; red.

The adsorption energies reported in Table 3 for the molecularly adsorbed states are around 85 kJ mol^{-1} for both starting structures. Accordingly, we can conclude that a barrier height above this value will mean O_2 desorption takes place in preference to dissociation while a barrier below this level would imply O_2 dissociation is likely. The barrier calculations show that dissociation is only favoured for the supported Au_{10} cluster.

For molecular adsorption and dissociation the O atoms interact with surface Fe ions at every stage. In the initial structure (figure 4a) the Bader charges on the superoxide O_2^{-} species are very similar, $-0.55\ e$ and $-0.51\ e$. The inset structure on Figure 7 shows the highest energy structure for the barrier calculation of the supported Au particle. At this point one oxygen atom is still in contact with the Fe ion that co-ordinated the O_2 adsorbate initially, this has a calculated charge of $-0.93\ e$, while the other O atom, that is now remote from Fe centres carries $-0.51\ e$. The Au atom that will insert between the O adsorbate atoms in the dissociated state is still in the main part of the cluster. In the dissociated end point structure both O atoms co-ordinate two Au atoms and a surface Fe ion and have charges of $-0.95\ e$ and $-0.98\ e$. This shows that the ionicity of the O centres has increased as the O_2 bond is broken and one of the Au atoms on the edge of the cluster is oxidised.

Conclusions

In this series of calculations we have attempted to compare O_2 adsorption and dissociation on isolated and supported Au clusters. This allows the influence of the support on the stability of oxygen species to be highlighted.

For the structure of the isolated $Au_{10}(7,3)$ it was found that the Au–Au bond lengths in the relaxed structure indicate that interaction between Au atoms within the two atomic layers is stronger than between layers. This observation is certainly in line with literature data on the relative stability of 2D and 3D clusters of this size.[6] However, the two layer particles have been observed by high resolution transmission electron microscopy[4] for Au on iron oxide catalysts and so were used throughout this study. The binding of molecular O_2 to these clusters favours binding sites on the base of the cluster with molecular adsorption such that each O atom is co-ordinated to a different Au atom. Comparing adsorption on the edge of the (100) facet for $Au_{10}(7,3)$ and $Au_9(6,3)$ clusters shows very little difference in the adsorption energy or geometry so that the observation of more favourable adsorption energies for clusters containing odd number of electrons see for 2D geometries[11] is not upheld here. There is a considerable difference between adsorption on the edge of the (111) facet for $Au_{10}(7,3)$ and $Au_9(6,3)$ with the more negative binding energy found for $Au_9(6,3)$. However, the difference in geometries found in the two calculations makes direct comparison difficult.

Comparison of the density of states for molecular and dissociated O_2 adsorption (figure 5) only shows depletion of the d-band states for the atomically adsorbed states, suggesting that literature XANES data[16] in which such depletion is observed may be due to oxidised Au species generated by O_2 dissociation.

We have confirmed that the DFT + U model with U-$J = 4$ eV gives a reasonable representation of the bulk and surface properties of α-Fe_2O_3, showing that surface cation relaxation on the (0001) surface is quite large but in line with other calculations on corundum surfaces.[41] The calculated surface energy is the same to within 2.5% for all of the slab thicknesses presented, even when limited surface relaxation is allowed. However, the calculated adsorption energies for Au_{10} clusters on a fully relaxed 3l slab and that for the 4l slab with only surface relaxations do differ considerably. In both cases the surface relaxation of Fe ions near to the Au cluster is less than seen for the clean surface. However the geometries of the Au cluster produced differ notably. This is probably largely due to differences in the initial setting up of the cluster on the slab rather than intrinsic differences between the slab models and demonstrates that there will be a range of Au particle configurations possible for nanoparticles of this size.

Molecular adsorption of O_2 at the interface between an $Au_{10}(7,3)$ particle and α-$Fe_2O_3(0001)$ surface gives a similar arrangement of the molecule at the base of the cluster as that seen in the isolated cluster calculations. Additional stabilisation of the superoxide adsorption at the interface is found through interaction with a surface Fe ion. The geometry of the cluster changes considerably on O_2 adsorption and we again see differences between the optimised structures on 3l and 4l slabs with more open clusters having more negative molecular adsorption energies.

Atomically adsorbed oxygen again shows differences between the two slab models, although the local geometry at the adsorption sites are similar to each other and to the isolated cluster case. In all three calculations a single Au atom is oxidised forming a near linear complex with the two atomic oxygen atoms.

Finally, for the isolated cluster and for the 3l model of the supported $Au_{10}(7,3)$ particle, barriers to O_2 dissociation have been calculated. For the isolated cluster the calculated barrier is around 60 kJ mol^{-1} higher than the molecular adsorption energy, making it unlikely that dissociation will occur. The additional stability for the atomically adsorbed state at the Au/α-$Fe_2O_3(0001)$ leads to a much lower barrier which is 10 kJ mol^{-1} lower than the molecular adsorption energy and so dissociation should be expected.

These calculations have demonstrated that the interface region for supported Au catalysts is important to their reactivity. The modelling of these systems is complicated by the flexibility of the nanoclusters which respond with large conformational changes to interactions with the oxide surface and the adsorbed oxygen species. However, this is also important to the particle reactivity since more rigid particles would not undergo the local geometry changes associated with oxidation as easily.

Acknowledgements

Computing facilities for this work were provided by ARCCA at Cardiff University and *via* our membership of the UK's HPC Materials Chemistry Consortium (MCC). The MCC is funded by EPSRC (EP/F067496), allowing this work to make use of the facilities of HECToR, the UK's national HPC service, which is provided by UoE HPCx Ltd at the University of Edinburgh, Cray Inc and NAG Ltd, and funded by the Office of Science and Technology through EPSRC's High End Computing Programme.

References

1 M. Haruta, N. Yamada, T. Kobayashi and S. Iijima, *J. Catal.*, 1989, **115**, 301.
2 N. Lopez, T. V. W. Janssens, B. S. Clausen, Y. Xu, M. Mavrikakis, T. Bligaard and J. K. Nørskov, *J. Catal.*, 2004, **223**, 232.
3 A. Sanchez, S. Abbet, U. Heiz, W.-D. Schneider, H. Häkkinen, R. N. Barnett and U. Landman, *J. Phys. Chem. A*, 1999, **103**, 9573.
4 A. A. Herzing, C. J. Kiely, A. F. Carley, P. Landon and G. J. Hutchings, *Science*, 2008, **321**, 1331.
5 F. Furche, R. Ahlrichs, P. Weis, C. Jacob, S. Gilb, T. Bierweiler and M. Kappes, *J. Chem. Phys.*, 2002, **117**, 6982.
6 L. Xiao, B. Tollberg, X. Hu and L. Wang, *J. Chem. Phys.*, 2006, **124**, 114309.
7 R. M. Olson, S. Varganov, M. S. Gordon, H. Metiua, S. Chretien, P. Piecuch, K. Kowalski, S. A. Kucharski and M. Musial, *J. Am. Chem. Soc.*, 2005, **127**, 1049.
8 M. Boronat, F. Illas and A. Corma, *J. Phys. Chem. A*, 2009, **113**, 3750.
9 N. Lopez and J. K. Nørskov, *J. Am. Chem. Soc.*, 2002, **124**, 11262.
10 J. L. C. Fajín, M. Natàlia, D. S. Cordeiro and J. R. B. Gomes, *J. Phys. Chem. C*, 2007, **111**, 17311.
11 B. Yoon, H. Häkkinen and U. Landman, *J. Phys. Chem. A*, 2003, **107**, 4066.
12 C. Zhang, B. Yoon and U. Landman, *J. Am. Chem. Soc.*, 2007, **129**, 2228.
13 A. Roldán, S. González, J. M. Ricart and F. Illas, *ChemPhysChem*, 2009, **10**, 348.
14 X. L. Zheng, G. M. Veith, E. Redekop, C. S. Lo, G. S. Yablonsky and J. T. Gleaves, *Ind. Eng. Chem. Res.*, 2010, **49**, 10428.
15 M. Turner, V. B. Golovko, O. P. H. Vaughan, P. Abdulkin, A. Berenguer-Murcia, M. S. Tikhov, B. F. G. Johnson and R. M. Lambert, *Nature*, 2008, **454**, 981.
16 N. Weiher, A. M. Beesley, N. Tsapatsaris, L. Delannoy, C. Louis, J. A. van Bokhoven and S. L. M. Schroeder, *J. Am. Chem. Soc.*, 2007, **129**, 2240.
17 Y. Kobayashi, S. Nasu, S. Tsubota and M. Haruta, *Hyperfine Interactions*, 2000, **126**, 95.
18 J. Guzman and B. C. Gates, *J. Am. Chem. Soc.*, 2004, **126**, 2672.
19 G. J. Hutchings, M. S. Hall, A. F. Carley, P. Landon, B. E. Solsona, C. J. Kiely, A. Herzing, M. Makkee, J. A. Moulijn, A. Overweg, J. C. Fierro-Gonzalez, J. Guzman and B. C. Gates, *J. Catal.*, 2006, **242**, 71.
20 A. F. Carley, D. J. Morgan, N. Song, M. W. Roberts, S. H. Taylor, J. K. Bartley, D. J. Willock, K. L. Howard and G. J. Hutchings, *Phys. Chem. Chem. Phys.*, 2011, **13**, 2528.
21 G. Kresse and J. Hafner, *Phys. Rev. B: Condens. Matter*, 1993, **47**, 558; G. Kresse and J. Hafner, *Phys. Rev. B: Condens. Matter*, 1994, **49**, 14251.
22 J. P. Perdew, K. Burke and M. Ernzerhof, *Phys. Rev. Lett.*, 1996, **77**, 3865.
23 S. H. Vosko, L. Wilk and M. Nusair, *Can. J. Phys.*, 1980, **58**, 1200.
24 P. E. Blöchl, *Phys. Rev. B: Condens. Matter*, 1994, **50**, 17953.
25 G. Kresse and J. Joubert, *Phys. Rev. B: Condens. Matter Mater. Phys.*, 1999, **59**, 1758.
26 S. L. Dudarev, G. A. Botton, S. Y. Savrasov, C. J. Humphreys and A. P. Sutton, *Phys. Rev. B: Condens. Matter Mater. Phys.*, 1998, **57**, 1505.
27 A. Rohrbach, J. Hafner and G. Kresse, *Phys. Rev. B: Condens. Matter Mater. Phys.*, 2004, **70**, 125426.
28 P. Liaoa and E. A. Carter, *J. Mater. Chem.*, 2010, **20**, 6703.
29 N. J. Mosey, P. Liao and E. A. Carter, *J. Chem. Phys.*, 2008, **129**, 014103.
30 E. Sanville, S. D. Kenny, R. Smith and G. Henkelman, *J. Comput. Chem.*, 2007, **28**, 899.
31 L. W. Finger and R. M. Hazen, *J. Appl. Phys.*, 1980, **51**, 5362.
32 Y. Sato and S. Akimoto, *J. Appl. Phys.*, 1979, **50**, 5285.
33 S. Mochizuki, *Phys. Status Solidi A*, 1977, **41**, 591.
34 R. K. Quinn, R. D. Nasby and R. J. Baughman, *Mater. Res. Bull.*, 1976, **11**, 1011–1017.
35 C. Lemire, S. Bertarione, A. Zecchina, D. Scarano, A. Chaka, S. Shaikhutdinov and H. J. Freund, *Phys. Rev. Lett.*, 2005, **94**, 166101.

36 C. Ruberto, Y. Yourdshahyan and B. I. Lundqvist, *Phys. Rev. B: Condens. Matter*, 2003, **67**, 195412.
37 A. Kiejna, T. Pabisiak and S. W. Gao, *J. Phys.: Condens. Matter*, 2006, **18**, 4207.
38 M. Boronat, P. Concepción and A. Corma, *J. Phys. Chem. C*, 2009, **113**, 16772.
39 G. Mills, H. Jonsson and G. K. Schenter, *Surf. Sci.*, 1995, **324**, 305.
40 A. Thetford, G. J. Hutchings, S. H. Taylor and D. J. Willock, Proc. R. Soc. A in press.
41 A. Marmier and S. C. Parker, *Phys. Rev. B: Condens. Matter Mater. Phys.*, 2004, **69**, 115409.

Free gold clusters: beyond the static, monostructure description

Elizabeth C. Beret, Luca M. Ghiringhelli and Matthias Scheffler*

Received 25th February 2011, Accepted 24th March 2011
DOI: 10.1039/c1fd00027f

The thermodynamical stability of free, pristine gold clusters at finite temperature, and of cluster+ligands complexes at finite temperature and in the presence of an atmosphere composed of O_2 and CO, is studied employing parallel tempering and *ab initio* atomistic thermodynamics. We focus on Au_{13}, which displays a significant fluxional behavior: Even at low temperature (100 K) this cluster exhibits a multitude of structures that dynamically transform into each other. At finite temperature, the preference of this cluster for three-dimensional *versus* planar structures is found to result from entropic effects. For gold clusters containing one to four gold atoms in an O_2 + CO atmosphere, we apply *ab initio* atomistic thermodynamics. On the basis of these considerations, we single out a likely reaction path for CO oxidation catalyzed by gold clusters.

1 Introduction

The discovery that nanosized gold particles are good catalysts for carbon monoxide oxidation at rather low temperatures,[1,2] as well as for other oxidation reactions,[3–7] has been attracting the interest and excitement of a large number of researchers in the scientific community for more than twenty years.[8–23]

Carbon monoxide oxidation, $CO + \frac{1}{2}O_2 \rightarrow CO_2$, does not happen spontaneously in the gas phase because it breaks the spin conservation rule (O_2 has a triplet ground state, while both CO and CO_2 are singlets). In recent years, several structures at various cluster sizes and stoichiometries have been studied by means of density-functional theory (DFT).[1,10,13,22] Such studies share a common strategy: They aim at finding energetically ordered lists of isomers at the chosen compositions, and the lowest energy structures are often used as starting points for discussing a catalytic process mediated by the cluster. However, as we will stress below, a static, mono-structure description of nanosized particles is likely to miss some important properties and processes. Furthermore, knowing the energetic hierarchy of the cluster+ligands complexes at a given stoichiometry is not sufficient to understand the system at the realistic conditions where catalysis occurs. In heterogeneous catalysis, the catalyst is exposed to an atmosphere of the gas-phase reactants at finite temperature and pressure. The present contribution shows the necessity of accounting for both temperature and pressure effects, even to determine the structure of the catalyst. This is a preliminary but necessary step before taking chemical reaction kinetics into consideration.

To this purpose, we proceed in steps. We first show that the apparently simple question *"what is the structure of the pristine gold cluster of size N?"* requires an answer that involves ensemble averages at the temperature of interest. As an example, we present the case of Au_{13}, which exhibits a significant *fluxional* behavior. The term fluxional reflects that the potential-energy surface (PES) of the cluster has

Fritz-Haber-Institut der Max-Planck-Gesellschaft, Faradayweg 4-6, D-14195 Berlin, Dahlem, Germany. E-mail: scheffler@fhi-berlin.mpg.de; Fax: +49 30 8413 4701; Tel: +49 30 8413 4700

several minima differing by just a few meV, and that these minima are separated by small barriers. When the height of the barriers is comparable with the average kinetic energy of the cluster at a given temperature, the cluster visits many minima in short timescales, say a few ps. At increasing temperatures, such clusters will exhibit first an ensemble of different rigid structures whose population is dictated by their Boltzmann weight, then fluxional behavior, and then the liquid state, where a continuous rearrangement of bonds takes place.

Furthermore, we show how the presence of a single ligand (Au_{13} + CO $\rightarrow$ $Au_{13}CO$) subverts the energetic hierarchy of Au_{13} geometries leading to a very different statistical ensemble.

In a third step, we also consider finite temperature and finite pressure of the reactants in the gas phase. A theoretical approach for the prediction of the thermodynamically stable and metastable catalyst+ligand structures has been developed initially for bulk semiconductors[24,25] and more recently for surface oxides.[26,27] Here we extend this method, called *ab initio* atomistic thermodynamics, to gas-phase clusters in an environment of O_2 and CO. The candidate structures are extensively sampled by means of the basin hopping technique.[28] This is rather efficient for not too big system sizes. In the original method conceived for bulks and surfaces, the temperature dependence was considered only in the partition function of the reactant gas; in other words, the free energy of the catalyst was approximated in terms of its electronic energy. While this is typically a good assumption for bulk materials,[26] the free energy of gas-phase clusters and cluster+ligands complexes depends significantly on temperature.

Ab initio atomistic thermodynamics yields phase diagrams, which identify structures that are probably relevant for catalysis at the realistic temperature and pressure of interest. We further extend the method by considering the possibility of CO_2 desorption. On the basis of this thermodynamic analysis, we can single out likely reaction mechanisms.

In many experimental studies, gold clusters are deposited on an oxide surface, and the nature of the support has a noticeable influence on the catalytic activity (see for instance Ref. 17). Other experiments aim at a "clean room" analysis of the *intrinsic* cluster properties and are conducted for clusters in the gas phase. Due to instrumental constraints, experiments with free clusters usually dealt with ionized species, and whether the cluster bore a positive or a negative electric charge also affected its catalytic properties (see Ref. 18 and references therein). Recently, also experimental studies of neutral clusters became possible.[29] In this contribution we study free, neutral gold clusters by DFT-based thermodynamics and statistical mechanics, in order to isolate their intrinsic catalytic properties from the influence of a support.

2 Calculation methods

For various compositions containing different numbers of Au, CO, and O, we sampled their structures using basin hopping[28] and parallel tempering[30,31] on the basis of DFT total energies and forces. We employed the FHI-aims code[32] using the PBE exchange–correlation functional.[33,34] Scalar–relativistic corrections have been applied.[32,35] Dispersion interactions, which play a noticeable role in describing the structure of gold clusters,[36] as well as their interaction with CO ligands,[37] are included in our calculations by means of a $C_6[n]/r^6$ tail correction to the PBE energy, where the $C_6[n]$ coefficients are derived from the self–consistent electron density n.[38] We will refer to this functional as PBE+vdW.

We have studied $Au_N O_x(CO)_y$ clusters containing between one and four gold atoms, and up to eight O atoms and/or four CO molecules. The structures of these clusters were sampled by means of basin hopping. In their ground electronic states, gas-phase Au_N clusters are doublets (clusters with odd N) and singlets (even N), CO is a singlet, and O, O_2 are triplets. Thus, the cluster+ligands structures can have different spin states, and we have sampled several possibilities ($2S + 1 = 2, 4, 6$

for clusters with odd number of total electrons, *i.e.* odd N, and 1, 3, 5 for clusters with even N). The stabilities of the structures produced in this sampling are obtained as a function of the environmental conditions (temperature and partial pressures of the reactive gases) by means of *ab initio* atomistic thermodynamics, which will be described in the following.

We consider that the Au_N clusters reach thermal equilibrium with the O_2 and CO gases, and $Au_NO_x(CO)_y$ structures are formed. However, CO and O_2 are not in equilibrium with each other. Such a *constrained thermodynamic equilibrium* neglects the non-catalytic $CO + \frac{1}{2}O_2 \rightarrow CO_2$ reaction, which is forbidden by a spin selection rule.

For each set of environmental conditions (T, p_{O_2}, p_{CO}) the most stable $Au_NO_x(CO)_y$ compositions are those with the minimum free energy of formation:

$$\Delta G_f(T,p_{O_2},p_{CO}) = G_{Au_NO_x(CO)_y}(T) - G_{Au_N}(T) - x\mu_O(T,p_{O_2}) - y\mu_{CO}(T,p_{CO}) \qquad (1)$$

$G_{Au_NO_x(CO)_y}$ and G_{Au_N} are the free energies of the cluster+ligands and of the pristine cluster, and μ_O, μ_{CO} are the chemical potentials of oxygen and carbon monoxide. The chemical potential of oxygen is $\mu_O = \frac{1}{2}\mu_{O_2}(T,p_{O_2})$. The free energies of the cluster+ligands and of the pristine cluster are obtained from their partition functions[39,40] including translational, rotational, vibrational, electronic, and configurational degrees of freedom. In the case of surface oxides, the change in the translational, rotational, and vibrational components due to ligand adsorption is typically negligible, and it can be disregarded in eqn (1). This is not the case of metal clusters in the gas phase, for which these contributions can change significantly upon ligand adsorption. The chemical potentials of CO and O_2 are also obtained from their partitions functions. Using eqn (1) we find out which $Au_NO_x(CO)_y$ compositions are the most stable ones at given environmental conditions (T, p_{O_2}, p_{CO}). The results can be plotted as phase diagrams.

For many $Au_NO_x(CO)_y$ compositions considered in the sampling we find that the lowest-energy structure at 0 K is closely followed by several other isomers within energies just a few tens of meV higher. This implies that at finite temperature the $Au_NO_x(CO)_y$ cluster cannot be properly described by only a single structure, but rather consists on a distribution of isomers. The population of a certain conformer of free energy G_i with respect to the structure with the lowest free energy G_0 at a given temperature T is given by the Boltzmann factor

$$B_i(T) = \exp\left(-\frac{G_i - G_0}{k_B T}\right) \qquad (2)$$

For each cluster composition we consider those isomers with a relative population of at least 5% ($B_i > 0.05$).

Parallel tempering was used in conjunction with Born–Oppenheimer molecular dynamics (BO-MD) to study the structures of pristine clusters and cluster+ligands at larger sizes. In this contribution, we discuss results from parallel tempering only for clusters containing thirteen gold atoms. The time step for the BO-MD runs was 10 fs for the pristine Au_{13} cluster and 2 fs for the $Au_{13}CO$ complex. The shorter time step for the cluster + CO system is dictated by the C–O bond stretching frequency, that is much higher than the highest vibrational frequencies in pristine gold clusters. The temperature of the simulations was controlled by means of the Bussi-Donadio-Parrinello thermostat.[41] We have used 16 replicas with temperatures ranging from 100 K to 1000 K, and the swap between configurations was attempted every five MD steps. In comparison to a low frequency of attempt swapping, such a high attempt frequency has been shown to yield a better efficiency of phase-space sampling.[42] Further details about the parallel tempering simulations and their analysis will be published elsewhere. The use of parallel tempering for sampling clusters' phase space is useful for two reasons. By selecting structures that are visited at low temperature and geometrically optimizing them, one gathers a large set of isomers in a fully unbiased manner. Furthermore, the sampling at each temperature is by construction canonical, hence the average value of any generalized coordinate is an estimator of the canonical expectation value of that variable.

We have assessed the accuracy of our PBE+vdW approach by comparing results for certain clusters with those of CCSD(T) calculations. The details of these calculations will be discussed elsewhere.[43] The error in the binding energy of Au_2 at PBE+vdW with respect to the CCSD(T) value is small ($-$ 0.06 eV). There is a significant and well-known error of DFT-GGA when describing the total energy of an O atom or O_2 molecule, and we find a similar error in the total energies of the Au_NO_x clusters. Since we deal with energy differences, $\Delta E_f = E(Au_NO_x) - E(Au_N) - \frac{x}{2}E(O_2)$, both errors are partially compensated. The PBE+vdW binding energies of gold clusters to O/O_2 ligands are about 0.45 $\pm$ 0.1 eV too strong, in comparison to CCSD(T) energies. Since the formation energies of all the clusters containing oxygen are subjected to an overestimation of 0.45 eV, the conclusions derived from comparison of different clusters will not be affected by this error. There is only an uncertainty of $\pm$ 0.1 eV in the PBE+vdW formation energies. As a consequence, the regions in the phase diagrams might actually be wider or thinner than predicted from our calculations, and very thin regions might even disappear.

In a separate contribution[44] we have computed the vibrational spectra of free gold clusters at finite temperature using this same PBE+vdW calculation level. The comparison of these theoretical spectra with experimental ones is more than satisfying, which constitutes a proof that gold clusters are well described at the PBE+vdW level.

3   Fluxional clusters

Identifying the lowest-energy isomer for Au_{13} has been the subject of several publications in the past years.[45–50] The reason of the curiosity raised by this particular size is that the most symmetric structures with 13 particles, the cuboctahedron and the icosahedron, are not at all the most stable isomers: Their total energies are more than 1 eV higher than the most stable structures.[50] Several low symmetry structures, both three-dimensional and planar, have been proposed. At the DFT-GGA level [50] some planar structures have the lowest energies. In particular for charged gold clusters, the relative stability of planar *versus* three-dimensional structures has been heavily debated.[51,52] For these, collision cross-section experiments could be performed. The analysis in Ref. 52 indicates that, albeit DFT-GGA finds as local minima candidate structures that match the experimental measurements, it is poor in describing the relative energetic stability between topologically "very different" structures. In particular, at some of the sizes at which the experiments indicate the presence of a three-dimensional structure, DFT-GGA finds planar structures with lower total energy. On the other hand, possibly more accurate functionals (according to the so-called Jacob's ladder of Perdew) yield a 2D-3D energy hierarchy which is consistent with experiments. For neutral clusters, the discussion in the literature is less rich,[50] since basically no experimental data are available for comparison. Here, we show that the energetic stability of DFT-GGA three-dimensional structures becomes more pronounced by the inclusion of dispersion interactions, and that the free energy, rather than only the total energy, must be taken into consideration for identifying the structures that are present at finite temperature.

3.1   The importance of dispersion interactions

In Fig. 1 we show the geometries of all the mechanically stable isomers that we found with a total energy within about 0.1 eV from the most stable structure. The lowest-energy three-dimensional structures included in the red box (isomers *a–c*) are

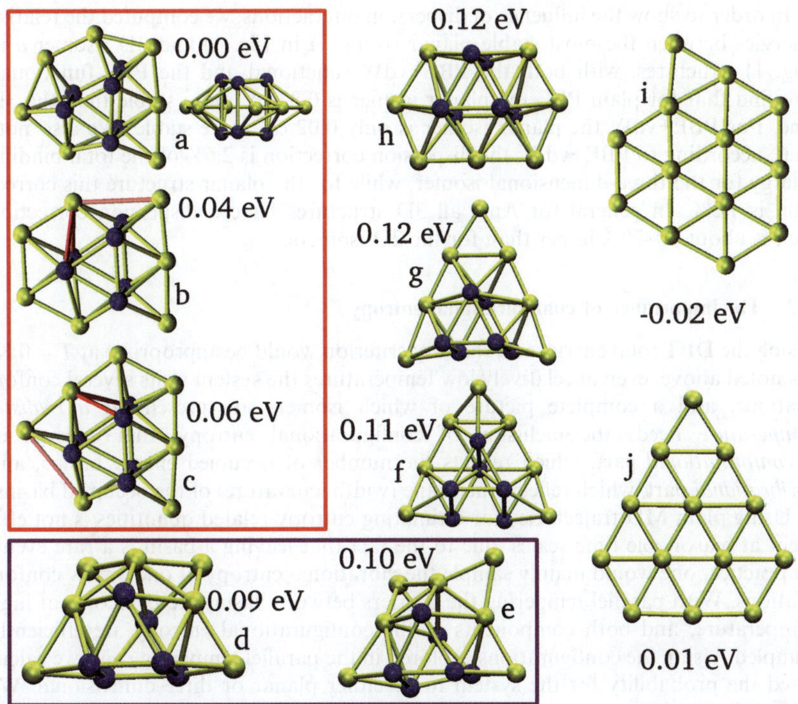

Fig. 1 Low energy isomers for Au$_{13}$. The boxes gather members of a set of isomers with low interconversion energy barrier (see text). The numbers indicate the difference in PBE+vdW energy from the lowest-energy three-dimensional isomer. Highlighted in blue are the gold atoms that constitute the triangular prism.

separated by very low barriers. The fact that barriers are low follows from few hundred ps-long constant-temperature Born–Oppenheimer MD runs. Only barriers that are low when compared to the average kinetic energy of the system can be overcome in a few ps, so that structural change can be observed in a MD run at low temperature. These structures can be described as a six-membered triangular prism and an incomplete ring of 7 atoms, all laying in a plane cutting half-way the prism (in the figure we have highlighted in blue the six atoms constituting the prism). This prism motif for Au$_{13}$ was firstly identified by Gruber *et al.*[50] In the figure we mark in red/pink the bonds of isomers *b* and *c* that need to break/form in order to transform the structure into the lowest-energy one (*i.e.* isomer *a* in Fig. 1). In a MD trajectory at temperature as low as 100 K, these three structures do change into each other. In the purple box there are two different structures (isomers *d* and *e* in Fig. 1) yet retaining the prism motif and again separated by a very low barrier, so that they also transform into each other at 100 K. The barriers separating the structure in the red box and those in the purple box are slightly higher; one needs a MD trajectory at 150 K in order to visit all those five structures (isomers *a–e* in Fig. 1). To our knowledge, structures like isomers *d* and *e* were not found in previous investigations. In a 500 ps long run at 200 K we also identify other 3D structures (isomers *f–h* in Fig. 1), which are again based on the prism motif. The parallel tempering sampling finds also two planar structures (isomers *i* and *j* in Fig. 1) and their PBE+vdW energies are very close to that of the most stable three-dimensional isomer. As it is the case for all planar gold clusters at all sizes, they appear as fragments of a hexagonal planar lattice.

In order to show the influence of dispersion interactions, we computed the relative energies between the most stable planar (isomer i in Fig. 1) and 3D (isomer a in Fig. 1) structures, with both the PBE+vdW functional and the PBE functional. We find that for plain PBE the planar isomer is 0.33 eV more stable than the 3D one. For PBE+vdW the planar isomer is only 0.02 eV more stable. We also note that, according to PBE+vdW, the dispersion correction is 2.6% of the total binding energy for the three-dimensional isomer, while for the planar structure this correction is 1.5%. In general for Au_{13} all 3D structures have a dispersion correction that is about 60–70% larger than for the 2D isomers.

3.2 The importance of configurational entropy

Using the DFT total energy as a stability criterion would be appropriate at $T = 0$ K. As noted above, even at relatively low temperatures the system visits several conformations, and a complete picture of which isomers are present at *a realistic temperature* needs the inclusion of configurational entropy, that consists of a *conformational* part, which reflects the number of occupied energy basins, and a *vibrational* part, which reflects the shape (width, curvature) of the occupied basins.

Using plain MD trajectories for estimating entropy-related quantities is not efficient at reasonable time scales, due to the fact that leaving a basin is a rare event. In practice, one would mainly sample the vibrational entropy of one or few conformations. With parallel tempering the barriers between basins are overcome at high temperature, and both components of the configurational entropy are efficiently sampled. Using the configurations explored in the parallel tempering runs, we calculated the probability for the system to be either planar or three-dimensional. We found that at 100 K the probability for a cluster to be planar is about 10%, while at 300 K the probability drops to 3%.

In summary, Au_{13}, as an example of a rather *fluxional* cluster, has at low temperature (*i.e.* around 100 K) the structure of a triangular prism with a surrounding planar ring of atoms which continuously rearranges (*a–c* in Fig. 1). At higher temperatures (*i.e.* around 150 K) also structures with a slightly distorted prism are sampled (*d* and *e* in Fig. 1). At even higher temperatures (more than 200 K) other prism+ring structures appear (*f–h* in Fig. 1). Finally, at 300 K the cluster is observed to melt into a liquid droplet: The atoms are still covalently bound, but the lifetime of each particular bond is of the order of only one ps. Even as a liquid droplet, Au_{13} does not have a spherical shape. The instantaneous configuration is oblate (one inertia moment is much smaller than the other two) thus reminding the common feature of all the most stable isomers at $T = 0$ K. Although some of the planar isomers (*i* and *j* in Fig. 1) have a total energy comparable to that of the lowest-energy three-dimensional structures, they are rarely seen due to entropic reasons. In Fig. 2 we explain this finding in a sketch of the configurational phase space of Au_{13}, where all the $13 \times 3 - 6 = 33$ degrees of freedom are collapsed into one coordinate. The planar clusters are in a small region of the phase space, separated by relatively high barriers (the planar clusters melt rapidly in a MD run at 800 K) from the region of three-dimensional clusters. The basins populated by three-dimensional clusters are shallow, due to the presence of low barriers separating the different minima.

Our conclusions differ from the argument put forward by Koskinen *et al.*[53] for anionic gold clusters. These authors suggested that some regions of the phase space (namely, planar clusters) might be difficult to reach when quenching high-temperature clusters: Liquid clusters are three-dimensional and a rapid quenching might lead the system to thermal energies that are too low to overcome the barrier(s) to go to the planar structure. In contrast, our parallel tempering studies imply that an equilibrium argument can be used, that is, the population of planar structures is low due to the topology of the free-energy landscape. Analyzing the free-energy landscape of a gold cluster was already attempted for Au_{12} by Vargas *et al.*[23] by means of metadynamics.[54] They found nearly equal probability for planar and three-dimensional

     This journal is © The Royal Society of Chemistry 2011

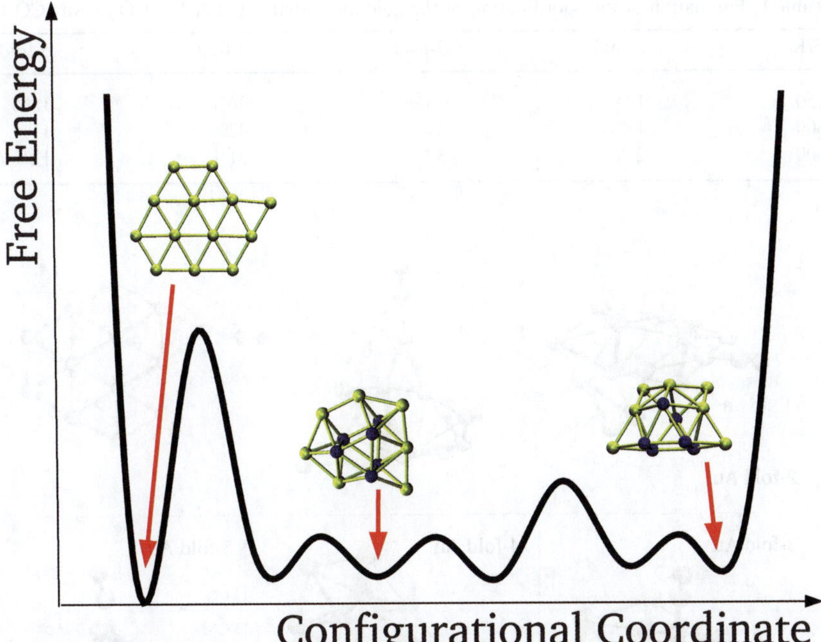

Fig. 2 Schematic view of the free-energy landscape as emerging from our parallel tempering calculations. The planar clusters are confined to a small and stiff basin (left), which is separated by a relatively high energy barrier from the basins containing the most likely three-dimensional structures. The latter structures appear at several shallow minima, separated by barriers that are easily overcome with a modest thermal activation.

structures, whereas the lowest-energy planar structure is 0.2 eV more stable than the lowest-energy three-dimensional one. Here with Au_{13} we show a neat example where structures with nearly the same total energy have very different statistical weight.

The fluxional character of Au_{13} is rather extreme within the cluster sizes that we investigated (Au_3–Au_{20}) but not at all unique. Clusters as small as Au_3 and Au_4[44] have two low-energy stable isomers, which transform into each other at room temperature. Au_7[44] presents one stable isomer that nonetheless continuously transforms into an isomer with identical shape and shuffled atoms, and so on.

4   Ligand adsorption on gold clusters

4.1   The influence of ligands on cluster fluxionality

The fluxional behavior of clusters like Au_{13} is not restricted to the pristine cluster. Also the configurational phase space of the cluster with ligands is rather rich. As an example, we discuss the configurations of $Au_{13}CO$ complexes, as analyzed by means of parallel tempering. The results show that CO does never dissociate and only one-fold coordinated bonding between the carbon atom and one gold atom occurs in the ensemble. In order to summarize the results, we chose as a descriptor the coordination (actually, the number of Au neighbors) of the gold atom that is bonded to the carbon atom. Table 1 lists the probabilities of finding coordination from two- to five-fold at different temperatures, and in Fig. 3 we show some of these structures. At lower temperature, three- and four-fold coordinated isomers dominate, while at higher temperatures also two-fold coordinated clusters become important, whereas higher-fold coordination becomes

Table 1 Population of the coordination of the gold atom that is bonded to CO in $Au_{13}CO$

T/K	2-fold	3-fold	4-fold	5-fold
150	13%	33%	36%	18%
300	16%	31%	42%	11%
600	25%	32%	31%	12%

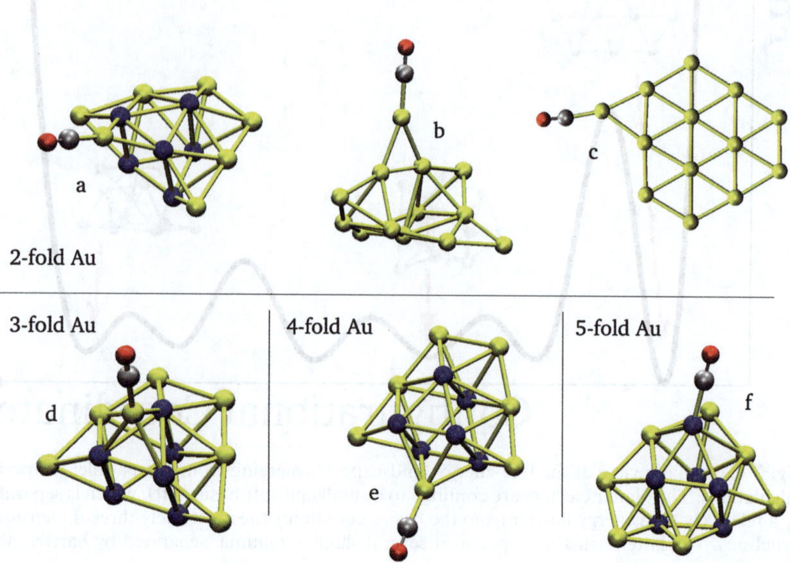

Fig. 3 Low-energy isomers for $Au_{13}CO$ grouped according to the coordination of the gold atom that is bonded to the carbon atom. Highlighted are the gold atoms that constitute the triangular prism.

less likely. For planar $Au_{13}CO$ structures like isomer *c* in Fig. 3, it holds an argument similar to pristine clusters: Despite some of the planar structures having a total energy comparable to that of the most stable 3D structures, they are largely unfavored on entropic grounds. In Table 2 we show the energy differences among the clusters displayed in Fig. 3. In the third column of this table we report the energy difference between the Au_{13} remainder after removal of CO and the lowest energy Au_{13} (*i.e.* isomer *a* in Fig. 1).

In practice, CO is removed from each isomer appearing in Fig. 3 and the remaining Au_{13} is geometrically optimized. In all cases, the relaxed Au_{13} structure maintains the same topology as in $Au_{13}CO$. By comparing the numbers in columns number 2 and 3 of Table 2, it is clear that a conspicuous reordering of the energy hierarchy amongst structures takes place upon adsorption of one single CO. For instance, isomer *d* without CO is not one of the most stable Au_{13} isomers, but its stronger bonding with CO makes it the lowest-energy structure.

The different kinds of bonding topology are related to different bonding strengths between CO and the clusters. In this respect, the role of each isomer in a possible catalytic reaction (*e.g.* after adsorption of a oxygen molecule) is certainly different. Only by considering the population of different structures at the temperature of the reaction, one can think to model the process in an accurate (and predictive) way.

$Au_{13}CO$ clusters are fluxional, but they are overall more rigid than pristine Au_{13} clusters. In other words, they start showing fluxional behavior at higher temperatures than Au_{13} clusters. In particular, in the presence of certain bonding motifs,

     This journal is © The Royal Society of Chemistry 2011

Table 2 The first column refers to the $Au_{13}CO$ isomers depicted in Fig. 3. The second column is the difference in total energy of $Au_{13}CO$ isomers from the lowest-energy one (namely isomer d in Fig. 3). The third column contains the energy difference between the Au_{13} remainder after removal of CO (geometrically optimized, but still at the $Au_{13}CO$ geometry) and the lowest-energy Au_{13}, $i.e.$ isomer a in Fig. 1

Isomer	ΔE [eV]	ΔE^* [eV]
a	0.08	0.16
b	0.11	0.27
c	0.01	−0.02
d	0.00	0.19
e	0.03	0.06
f	0.07	0.00

like those of isomers d and e in Fig. 3, the structure is particularly rigid (in 20 ps long MD run at $T = 300$ K these structures remain unchanged, while the corresponding pristine cluster isomers change into different structures in few ps). Clusters with other bonding motifs, like that of isomer f, are more fluxional, even though the temperature at which the fluxional behavior appears is significantly higher than for the corresponding Au_{13} structure.

4.2 Stoichiometries at 300 K

When gold clusters are in contact with an atmosphere composed of CO and O_2 gases, these ligands may adsorb onto the cluster, and the cluster may restructure completely. The preferred compositions $Au_NO_x(CO)_y$ adopted by the clusters in different experimental conditions are obtained using eqn (1), and the results are summarized in phase diagrams as the one shown in Fig. 4a. In the following we will focus our discussion on the case of Au_2, but we will also briefly comment on trends that we find for Au, Au_3, and Au_4. A detailed analysis of the results obtained for all cluster sizes will be published elsewhere.[43]

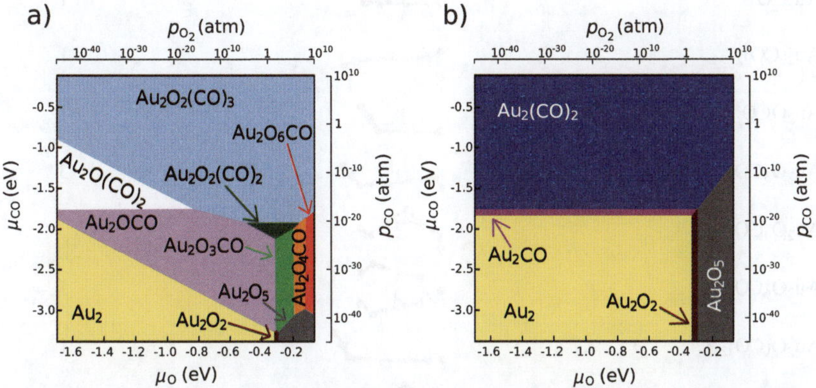

Fig. 4 Phase diagrams obtained for $Au_2O_x(CO)_y$ clusters at 300 K. Each region shows the preferred composition under the partial pressures of O_2 and CO given by the p_{O_2}, p_{CO} bars. The μ scales do not correspond to absolute chemical potentials, but rather to the change in free energy at (T, p) with respect to the absolute zero temperature and standard pressure: $\Delta\mu(T, p) = \mu(T, p) - \mu(T, p°)$, with $\mu(T, p°) = E^{DFT} + E^{ZPE}$. The phase diagram has been built considering a) all isomers found in the sampling, b) only those isomers from which CO_2 dissociation does not take place (see main text).

Fig. 4a shows the phase diagram obtained for Au_2 at 300 K and under a wide range of partial pressures of CO and O_2. The phase diagram reports, for each (p_{O_2}, p_{CO}) pair, the composition with the lowest free energy. This is done with a color code, such that an area with a given color identifies univocally a composition, which is indicated on (or near) the area. The approach behind Fig. 4a is also called *constrained thermodynamics* as it does not allow O_2 and CO to react in the gas phase, nor does it allow CO_2 to desorb (at least in Fig. 4a). The molecular structures corresponding to the different regions of Fig. 4a are shown in Table 3. In the region of low p_{O_2} and low p_{CO} the most stable structure is the pristine cluster. Adsorption of O_2 and CO takes place when the partial pressures of the reactive gases increase.

Contrary to intuition, the first stable composition that is found by increasing p_{CO} at very low p_{O_2} (*i.e.* following upwards the left edge of the phase diagram) is not Au_2CO, but rather Au_2OCO (violet field in Fig. 4a). This apparent oddity is due to the presence of the CO_2 moiety in Au_2OCO (see Table 3) whose formation energy stabilizes the cluster more than the adsorption energy of a single CO. Consistently, at very low p_{CO} (bottom edge), Au_2OCO becomes stable overcoming species containing only oxygen. Although Au_2OCO is the most stable composition even at very low partial pressures of one of the two reactant gases, its observation in experiment may be hindered: At pressures lower than those found in intergalactic space, the collision probability of either O (near the left edge of the phase diagram) or CO (near the bottom edge) with the Au_2 cluster is vanishingly small, and reaching thermodynamical equilibrium may require timescales longer than the age of the universe. However, according to thermodynamics the co-adsorption of O and CO is preferred to adsorption of CO alone, and Au_2CO does not appear in the phase diagram.

Table 3 Preferred isomers and spin multiplicities for all the compositions appearing in Fig. 4a

Composition	Isomer	$(2S + 1)$
Au_2		1
Au_2O_2		3
Au_2O_5		5
Au_2CO		1
$Au_2(CO)_2$		1
Au_2OCO		1
Au_2O_3CO		3
Au_2O_4CO		3
Au_2O_6CO	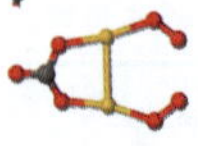	5
$Au_2O(CO)_2$		1
$Au_2O_2(CO)_2$		1
$Au_2O_2(CO)_3$		1
	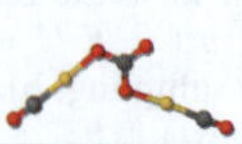	1

The phase diagram in Fig. 4a shows the existence of Au_2O_2 and Au_2O_5 clusters, while Au_2O, Au_2O_3, and Au_2O_4 are absent. Thus, a gradual increase in p_{O_2} does not translate into successive adsorption of O atoms. The reason is that the thermodynamical stability of the Au_2O_x structures does not increase linearly with the number of O atoms in the cluster. Instead, some stoichiometries are noticeably more stable than others, and only the most stable ones at a given p_{O_2}, p_{CO} appear in the phase diagram. A similar behavior is found for Au, Au_3, and Au_4 clusters. The presence in Fig. 4a of Au_2O_5 indicates that dissociative adsorption of O_2 is possible for Au_2 clusters at room temperature. However, it might be a slow process, due to the dissociation barrier and spin selection rules (Au_2O_5 is a quintuplet, see Table 3). Dissociative adsorption of O_2 appears in the phase diagrams of Au_3 and Au_4, but not in the phase diagram of the single, neutral gold atom, for which molecular adsorption of O_2 is energetically more favorable than dissociative.

We also note here that the region corresponding to Au_2O_2 in Fig. 4a is defined within a range of μ_O of about 0.04 eV, below the accuracy in our calculations (our computational level has an uncertainty of about $\pm$ 0.1 eV, see Section 2) so that we cannot ensure its appearance or absence in the phase diagram.

In the p_{O_2}, p_{CO} region between 0.01 and 100 atm the preferred structure $Au_2O_2(CO)_3$ results from co-adsorption of both CO and O_2. The presence of co-adsorbed species in the mentioned pressure region is a feature that is also found in the phase diagrams of Au, Au_3, and Au_4. Interestingly, in $Au_2O_2(CO)_3$ the Au–Au bond is broken (see Table 3). We find that Au_3 and Au_4 clusters also undergo noticeable distortions due to ligand adsorption, which stresses the need of a sampling method that allows the system to go through major structural changes when searching for candidate reaction intermediates. This finding is in agreement with the work of Zhai et al.,[21] who arrived to a similar conclusion for one particular size of anionic gold cluster, Au_6^-, with up to 6 CO molecules adsorbed onto it.

4.3 Implications for catalysis

Many of the stoichiometries in Fig. 4a correspond to structures on which the reaction gases have formed CO_2 and CO_3 arrangements (see Table 3). All these clusters can be considered intermediates from which CO_2, the reaction product, may desorb into the atmosphere. In this section we consider the possibility of CO_2 desorption from these structures. Again, we focus our discussion on the p_{O_2}, p_{CO} region in the phase diagram between 0.01 and 100 atm, as this is the most relevant pressure region for catalysis at room temperature. In the case of Au_2 the preferred structure in this region is $Au_2O_2(CO)_3$. It is worth noting here that the stability of $Au_2O_2(CO)_3$ extends for orders of magnitude outside the pressure region of interest. Therefore, our predictions will not be affected by the mentioned DFT-GGA error in the formation energies (see Section 2).

If CO_2 is formed on the cluster and released into the atmosphere, eventually there will be a certain partial pressure of CO_2 in equilibrium with the cluster: $R–CO_2 \rightleftharpoons R + CO_2$. Using the desorption energy and the partition function of CO_2, we can compute $p_{CO_2}^{eq}$, the partial pressure at which CO_2 is in thermodynamical equilibrium with the isomer $R–CO_2$. If in the experimental setup the pressure of CO_2 is below this $p_{CO_2}^{eq}$, CO_2 dissociation will be thermodynamically favorable. In the case of $Au_2O_2(CO)_3$ we can write:

$$Au_2O_2(CO)_3 \rightleftharpoons Au_2O(CO)_2 + CO_2: p_{CO_2}^{eq} = 4.6 \times 10^9 \text{ atm.}$$

Taking as a reference the value of p_{CO_2} estimated in the Earth's atmosphere, $p_{CO_2} \cong 10^{-3}$ atm, it follows that CO_2 will desorb from $Au_2O_2(CO)_3$. Let us consider the remaining cluster $Au_2O(CO)_2$:

$$Au_2O(CO)_2 \rightleftharpoons Au_2CO + CO_2\text{: } p_{CO_2}{}^{eq} = 8.9 \times 10^{11} \text{ atm},$$

from which CO_2 will also readily desorb at $p_{CO_2} \cong 10^{-3}$ atm.

For all structures in Fig. 4a that contain CO_2, we calculate the equilibrium partial pressure of CO_2 and remove from the phase diagram all the isomers from which CO_2 desorbs. In the particular case of Au_2, all clusters containing the CO_2 moiety are not stable with respect to CO_2 dissociation at the reference p_{CO_2}. With this criterion, we construct a second phase diagram, shown in Fig. 4b, in which all the isomers are stable with respect to both adsorption of O_2 and CO, and desorption of CO_2 (at the reference pressure of 10^{-3} atm). In this second phase diagram, Au_2CO, *i.e.* the remaining cluster after dissociation of two CO_2 molecules as described above, is not the most stable structure in the pressure region of interest, and it must adsorb one CO molecule in order to reach the most stable composition, $Au_2(CO)_2$.

The latter stoichiometry is to be regarded as the thermodynamically stable catalyst at the mentioned pressure conditions. Indeed, this is the most stable composition which does not contain CO_2 moieties. Further adsorption of CO and O_2 molecules will eventually lead to the $Au_2O_2(CO)_3$ structure (light blue in Fig. 4a) from which CO_2 would desorb, and so forth.

We can identify a *cycle* that runs between the two extreme compositions, $Au_2(CO)_2$ and $Au_2O_2(CO)_3$, by successive adsorptions of O_2 and CO molecules and desorption of CO_2 molecules. Taking into account that the total spin momentum must be conserved in each reaction step, the cycle can be written as in Fig. 5, where we indicate the energy difference (in eV) between products and reactants for each elementary step. Those species in a triplet spin state are marked with the symbol †, while the rest of species are in a singlet state. There are two elementary step not energetically favorable, with energy differences of + 0.70 and + 0.10 eV. Such energy differences can typically be overcome in the course of a catalytic process at 300 K. The rest of reaction steps in the cycle are energetically favorable. As a consequence, the proposed mechanism is driven by an overall gain of energy. Although there is not a unique possibility for a reaction path connecting the $Au_2(CO)_2$ and $Au_2O_2(CO)_3$ structures, we present the above mechanism as an example of likely reaction path according to thermodynamic considerations. Our predictions will be assessed with a complementary kinetic study of energy barriers and the consideration of alternative reaction paths.

Comparing the two phase diagrams in Fig. 4, we can distinguish two classes of $Au_2O_x(CO)_y$ clusters: On one hand, the *intermediates*, shown in Fig. 4a. The formation of these structures is very favored thermodynamically from adsorption of O/O_2 and CO. They are decomposed by means of CO_2 desorption, and again reformed by O_2 and CO adsorption, *i.e.* for this group of intermediates the reaction would follow a Langmuir-Hirshfeld mechanism. On the other hand, the clusters in Fig. 4b constitute the equilibrium structures of the *catalyst*. In the p_{O_2}, p_{CO} region between 0.01 and 100 atm, the catalyst is $Au_2(CO)_2$. This class of clusters may participate in

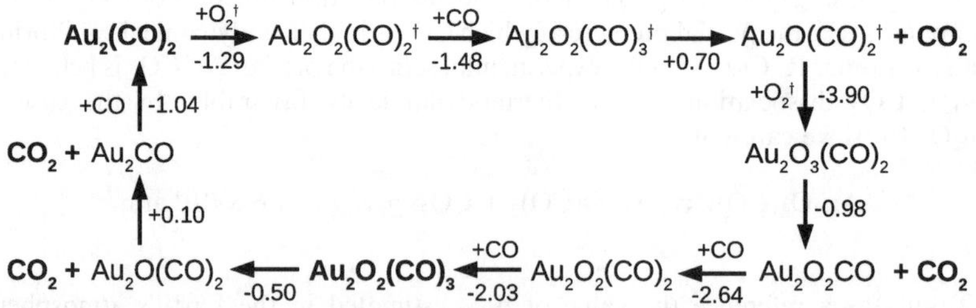

Fig. 5 Reaction cycle connecting the catalyst ($Au_2(CO)_2$) and the main reaction intermediate ($Au_2O_2(CO)_3$) at 300 K.

the reaction if either additional ligands adsorb on their surface to form one of the *intermediate* structures, or if the already adsorbed ligands react with another ligand from the gas phase (Eley-Rideal mechanism).

At higher temperatures, both the intermediate and the catalyst can be different from the structures identified at room temperature (see discussion of section 3). Different reaction pathways might then dominate; the study of the influence of temperature on the identification of the relevant species for the reaction will be part of a future contribution. In addition, it must be stressed that the temperature of a small cluster in the gas phase undergoes very strong fluctuations: When a ligand is adsorbed, the energy released is rapidly distributed among the vibrational degrees of freedom (on a time scale of picoseconds) and the cluster heats up. In contrast, thermal equilibration with the buffer gas requires many collisions and a timescale of several nanoseconds. Therefore, clusters in the gas phase cannot be regarded as a system at a constant temperature.

As we have illustrated in this section, our phase diagrams do not only provide information on the relative stabilities of different cluster compositions, but they can also suggest those reaction paths that may be propitious at the environmental conditions of interest. However, these predictions are based on the assumption that during the experiment the system is able to reach thermodynamical equilibrium. This might not be the case if the catalytic reaction is too fast. Furthermore, the system might need to surmount large energy barriers in order to reach the thermodynamically preferred reaction intermediates, so that those structures might be kinetically hindered during the catalytic reaction. Our thermodynamical predictions need to be complemented with an extensive kinetic study considering reaction paths, barriers, and the conservation of spin in the processes of ligand adsorption, desorption, and reaction. This will be the subject of a future contribution from our group.

5 Conclusions

We have studied free pristine gold clusters and cluster+ligands complexes at finite temperature and in contact with an atmosphere composed of O_2 and CO. The preferred stoichiometries and structures are determined as a function of the environmental temperature and partial pressures of the reacting gases. Pristine as well as complexed gold clusters show a fluxional character when their finite temperature behavior is explicitly considered. Even at low temperatures (*e.g.* 100 K) gold clusters such as Au_{13} and complexes such as $Au_{13}CO$ appear as a statistical ensemble of structures, dynamically transforming into one another, rather than as a dominant structure. On the basis of the analysis of Au_{13}, we find a statistical argument for explaining why two-dimensional structures that have similar bonding energies to three-dimensional structures are less likely to be seen in a canonical sampling.

Our statistical argument is not limited to Au_{13}; the necessary condition to invoke it is having a fluxional system in which classes of topologically different structures have members with almost the same energy.

The adsorption energy of CO on Au_N is larger than that of O_2. However, if both O_2 and CO are in the surrounding gas phase the co-adsorption of both species is energetically favored. Adsorption of O_2 (without CO) happens only at high O_2 partial pressures, and is preferred in a dissociative manner. Contrary to what might be the general chemical intuition, a progressive increase in p_{O_2} does not lead to the appearance of clusters containing a linearly increasing number of O atoms. Instead, certain structures are favored, due to their marked energetic stability.

When both O_2 and CO are adsorbed, these ligands often react on the cluster surface to form adsorbed CO_2 and CO_3 species. Such structures are considered reaction intermediates in a mechanism of the Langmuir-Hinshelwood type. A second type of metastable structures is identified as those clusters from which CO_2 desorption is not possible or not thermodynamically favorable. This last class of structures represents the equilibrium structures of the catalyst, and can either undergo ligand adsorption to form reaction intermediates, or react with ligands in the gas phase *via* an Eley-Rideal mechanism.

In addition, at 300 K and in the p_{O_2}, p_{CO} pressure region between 0.01 and 100 atm, we can propose a reaction path which cycles between the equilibrium catalyst and the reaction intermediate, on the basis of thermodynamical arguments. A thorough study of different reaction pathways, completing our thermodynamical information with kinetic studies, will be the subject of a future contribution.

Acknowledgements

E.C.B. thanks the Alexander von Humboldt foundation for a research fellowship. This work is supported by the Deutsche Forschungsgemeinschaft through the Cluster of Excellence UNICAT hosted by the Technical University Berlin.

References

1 H. Huber, D. McIntosh and G. A. Ozin, *Inorg. Chem.*, 1977, **16**, 975–979.
2 M. Haruta, T. Kobayashi, H. Sano and N. Yamada, *Chem. Lett.*, 1987, **2**, 405–408.
3 G. C. Bond and D. Thompson, *Catal. Rev. Sci. Eng.*, 1999, **41**, 319.
4 S. Carrettin, P. McMorn, P. Johnston, K. Griffin and G. J. Hutchings, *Chem. Commun.*, 2002, 696.
5 A. Abad, P. Concepcion, A. Corma and H. Garcia, *Angew. Chem., Int. Ed.*, 2005, **44**, 4066.
6 M. D. Hughes, Y. J. Xu, P. Jenkins, P. McMorn, P. Landon, D. I. Enache, A. F. Carley, G. A. Attard, G. J. Hutchings, F. King, E. H. Stitt, P. Johnston, K. Griffin and C. J. Kiely, *Nature*, 2005, **437**, 1132.
7 M. Turner, V. B. Golovko, O. P. H. Vaughan, P. Abdulkin, A. Berenguer-Murcia1, M. S. Tikhov, B. F. G. Johnson and R. M. Lambert, *Nature*, 2008, **454**, 981.
8 L. Lian, P. A. Hackett and D. M. Rayner, *J. Chem. Phys.*, 1993, **99**, 2583–2590.
9 T. H. Lee and K. M. Ervin, *J. Phys. Chem.*, 1994, **98**, 10023–10031.
10 H. Häkkinen and U. Landman, *J. Am. Chem. Soc.*, 2001, **123**, 9704–9705.
11 W. T. Wallace and R. L. Whetten, *J. Am. Chem. Soc.*, 2002, **124**, 7499–7505.
12 G. Mills, M. S. Gordon and H. Metiu, *Chem. Phys. Lett.*, 2002, **359**, 493–499.
13 L. D. Socaciu, J. Hagen, T. M. Bernhardt, L. Wöste, U. Heiz, H. Häkkinen and U. Landman, *J. Am. Chem. Soc.*, 2003, **125**, 10437–10445.
14 H. Häkkinen, S. Abbet, A. Sanchez, U. Heiz and U. Landman, *Angew. Chem., Int. Ed.*, 2003, **42**, 1297–1300.
15 B. Yoon, H. Häkkinen and U. Landman, *J. Phys. Chem. A*, 2003, **107**, 4066–4071.
16 R. Meyer, C. Lemire, S. K. Shaikhutdinov and H. J. Freund, *Gold Bull.*, 2004, **37**, 72–124.
17 S. Arrii, F. Morfin, A. J. Renouprez and J. L. Rousset, *J. Am. Chem. Soc.*, 2004, **126**, 1199–1205.
18 T. M. Bernhardt, *Int. J. Mass Spectrom.*, 2005, **243**, 1–29.
19 M. L. Kimble, N. A. Moore, G. E. Johnson, A. W. Castleman, Jr., C. Bürgel, R. Mitric and V. Bonačič-Koutecký, *J. Chem. Phys.*, 2006, **125**, 204311–14.
20 S. Chrétien, S. K. Buratto and H. Metiu, *Curr. Opin. Solid State Mater. Sci.*, 2007, **11**, 62–75.
21 H. J. Zhai, L. L. Pan, B. Dai, B. Kiran, J. Li and L. S. Wang, *J. Phys. Chem. C*, 2008, **112**, 11920–11928.
22 F. Wang, D. Zhang, X. Xu and Y. Ding, *J. Phys. Chem. C*, 2009, **113**, 18032–18039.
23 A. Vargas, G. Santarossa, M. Iannuzzi and A. Baiker, *Phys. Rev. B: Condens. Matter Mater. Phys.*, 2009, **80**, 195421.
24 C. M. Weinart and M. Scheffler, *Defects in Semiconductors, Mat. Sci. Forum 10–12*, 1986, 25–30.
25 M. Scheffler and J. Dabrowski, *Philos. Mag. A*, 1988, **58**, 107.
26 K. Reuter and M. Scheffler, *Phys. Rev. B: Condens. Matter*, 2001, **65**, 035406.
27 K. Reuter and M. Scheffler, *Phys. Rev. B: Condens. Matter*, 2003, **68**, 045407.
28 D. J. Wales and J. Doye, *J. Phys. Chem. A*, 1997, **101**, 5111–5116.
29 P. Gruene, D. M. Rayner, B. Redlich, A. F. G. van Der Meer, J. T. Lyon, G. Meijer and A. Fielicke, *Science*, 2008, **321**, 674.
30 E. Marinari and G. Parisi, *Europhys. Lett.*, 1992, **19**, 451.
31 Y. Sugita and Y. Okamoto, *Chem. Phys. Lett.*, 1999, **314**, 141.
32 V. Blum, R. Gehrke, F. Hanke, P. Havu, V. Havu, X. Ren, K. Reuter and M. Scheffler, *Comput. Phys. Commun.*, 2009, **180**, 2175–2196.

33 J. P. Perdew, K. Burke and M. Ernzerhof, *Phys. Rev. Lett.*, 1996, **77**, 3865–3868.
34 J. P. Perdew, K. Burke and M. Ernzerhof, *Phys. Rev. Lett.*, 1997, **78**, 1396–1396.
35 E. van Lenthe, E. J. Baerends and J. G. Snijders, *J. Chem. Phys.*, 1993, **99**, 4597–4610.
36 R. M. Olson, S. Varganov, M. S. Gordon, H. Metiu, S. Chrétien, P. Piecuch, K. Kowalski, S. A. Kucharski and M. Musial, *J. Am. Chem. Soc.*, 2005, **127**, 1049–1052.
37 F. Mendizabal, *Organometallics*, 2001, **20**, 261–265.
38 A. Tkatchenko and M. Scheffler, *Phys. Rev. Lett.*, 2009, **102**, 073005.
39 R. Fowler and E. A. Guggenheim, *Statistical thermodynamics*, Cambridge University Press, Cambridge, 1949.
40 J. Rogal and K. Reuter, *Experiment, Modeling and Simulation of Gas-Surface Interactions for Reactive Flows in Hypersonic Flights*, Educational notes RTO-EN-AVT-142, Neuilli-sur-Seine, France, 2007, pp. 2-1–18.
41 G. Bussi, D. Donadio and M. Parrinello, *J. Chem. Phys.*, 2007, **126**, 014101.
42 D. J. Sindhikara, D. J. Emerson and A. Roitberg, *J. Chem. Theory Comput.*, 2010, **6**, 2804.
43 E. C. Beret, L. M. Ghiringhelli and M. Scheffler, in preparation.
44 L. M. Ghiringhelli, P. Gruene, J. T. Lyon, G. Meijer, A. Fielicke and M. Scheffler, in preparation.
45 D. Häberlen, S. C. Chung, M. Stener and N. Rösch, *J. Chem. Phys.*, 1997, **106**, 5189.
46 J. Oviedo and R. E. Palmer, *J. Chem. Phys.*, 2002, **117**, 9548.
47 M. Chang and M. Y. Chou, *Phys. Rev. Lett.*, 2004, **93**, 133401.
48 M. Fernandez, J. M. Soler, I. L. Garzon and L. C. Balbas, *Phys. Rev. B: Condens. Matter Mater. Phys.*, 2004, **70**, 165403.
49 L. Xiao, B. Tollberg, X. K. Hu and L. C. Wang, *J. Chem. Phys.*, 2006, **124**, 114309.
50 M. Gruber, G. Heimel, L. Romaner, J. Brédas and E. Zojer, *Phys. Rev. B: Condens. Matter Mater. Phys.*, 2008, **77**, 165411.
51 S. Gilb, P. Weis, F. Furche, R. Ahlrichs and M. M. Kappes, *J. Chem. Phys.*, 2002, **116**, 4094.
52 M. P. Johansson, A. Lechtken, D. Schooss, M. M. Kappes and F. Furche, *Phys. Rev. A: At., Mol., Opt. Phys.*, 2008, **77**, 053202.
53 P. Koskinen, H. Häkkinen, B. Huber, B. von Issendorff and M. Moseler, *Phys. Rev. Lett.*, 2007, **98**, 015701.
54 A. Laio and M. Parrinello, *Proc. Natl. Acad. Sci. U. S. A.*, 2002, **99**, 12562.

Aurophilic attractions between a closed-shell molecule and a gold cluster

Pekka Pyykkö,[*ab] Xiao-Gen Xiong[b] and Jun Li[b]

Received 11th February 2011, Accepted 10th March 2011

DOI: 10.1039/c1fd00018g

The attractions between a closed-shell gold cluster and a closed-shell Au(I) molecule are theoretically studied, and related to monomer properties. The results suggest that the Au(I) mainly interacts with the nearest gold atoms and that the interaction is roughly proportional to the number of nearest neighbours. Different functionals are compared. The SCS-MP2 results are close to the CCSD (T) ones for the systems studied. The question of ionic contributions to the stability of 'staple' structures is raised.

1 Introduction

The 'aurophilic'[1] or 'metallophilic'[2] attraction is a relatively recently identified cohesive contribution that appears to occur between two closed-shell metal atoms inside a molecule, or between such atoms in different molecules. The system can contain dimers, oligomers, infinite chains, or infinite, two-dimensional sheets, such as the Au(I) planes in solid AuCN. The gold compounds usually have Au(I) although Au(III) may occur as well.

The earlier literature in this area was summarised in our reviews,[3–6] and in the introduction of our recent paper on the dependence of aurophilicity in $[ClAuL]_2$ systems on the neutral ligand, L.[7] Some key steps were the identification of the attraction as an electron correlation effect,[8,9] the London-like R^{-6} behaviour at large distances[10,11] and the detailed analysis of the dispersion-type plus other bonding contributions using localised orbitals.[12,13] In the virtual one-electron excitations, the initially occupied orbital i is typically an Au 5d, but the final orbital f may have considerable ligand character.[8,12] In that sense the idea of a monoatomic 'metallophilicity' at each monomer is not accurate.

The metallophilic interaction strength oscillates along the series MP2, MP3, MP4, CCSD, CCSD(T).[13–15] This had an unexpected consequence: while the relativistic effects strengthened the interaction at MP2 level,[14] as verified by O'Grady and Kaltsoyannis,[16] they found that at the higher correlated levels QCISD, CCSD and CCSD (T) the interaction weakened, when passing from Ag to Au, which is almost tantamount to adding relativistic effects to non-relativistic gold, silver being approximately 'non-relativistic gold'.[17] The trends in perpendicular $[XAuL]_2$ model systems[2,7] suggest that softer 'halogens' X will have stronger aurophilic interactions. As a final link to metallophilicity, we quote the so-called 'no-pair' bonding in systems like a predicted, tetrahedral $S = 2$ high-spin energy minimum of Cu_4. In that case there is no electron-pair bonding, all valence electrons having the same spin, but electron correlation from the $3d^{10}$ shell explains most of the bonding which is calculated to be 19 kJ mol^{-1} per Cu–Cu pair.[18]

[a]Department of Chemistry, University of Helsinki, POB 55 (A. I. Virtasen aukio 1), 00014 Helsinki, Finland. E-mail: Pekka.Pyykko@helsinki.fi; Fax: +358 9 19150169; Tel: +358 9 19150171

[b]Department of Chemistry, Tsinghua University, 100 084 Beijing, China

What seems to be missing, are theoretical studies on metallophilic attractions for systems that contain, on one hand, a basically closed-shell gold cluster, in an average Au oxidation state far below +I and, on the other hand, a closed-shell molecular group, containing Au(I), such as the bridging surface groups (called *staples*, for a sketch, see Fig. 8 below), of the type [-SR-Au-SR-]$^-$ in recent cluster work. The need for such studies appeared, when it was found that many thiolate-protected gold clusters consisted of a compact inner Au_n cluster and outer groups of that type, or of the longer [-SR-Au-SR-Au-SR-]$^{n-}$ type. The first example was the [$Au_{102}(SR)_{44}$], experimentally studied by Jadzinsky *et al.*[19] Theoretical modelling of that experimental structure was undertaken by Walter *et al.*[20] using density functional theory (DFT), and gave a closely similar result. Another example is [$Au_{25}(SR)_{18}$]$^-$, where both experiments[21] and DFT[22] suggest a formal inner Au_{13}^{5+} inner icosahedron and an outer layer containing [(SR-Au-SR-Au-SR)]$^{6-}$ units.

While DFT describes well the covalent Au–S bonds of the system and also the Coulomb attractions between the cationic core and the anions on the surface—an aspect not explicitly discussed yet—regular exchange–correlation (XC) functionals, commonly used today, are not able to reliably describe the predominantly dispersion-type interactions. Therefore even the simplest model studies on the aurophilic interactions between a formal Au(I) and a gold cluster are of interest. A typical question to ask is, will the molecular Au(I) center interact with only the nearest gold atoms of the cluster, or with the entire cluster, and what is the order of magnitude of the total interaction. Further possible bonding contributions are the different induction terms.

We address here the simplest possible model cases, modelling the inner part by Au_2, Au_6 or Au_8 and the outer part with AuH or $AuCl_2^-$. The wave-function theoretical (WFT) methods applied range from MP2 to CCSD(T). A number of density functional theories are tested, as well. The calculations were performed using the codes Molpro2010[23] and NWChem6.0.[24] The latest Au pseudopotential of Figgen *et al.*[25] and the related, correlation-consistent basis sets of Peterson and Puzzarini[26] were used. For comparison, an MP2-level study of the basis-set limits for two other aurophilic models exists,[27] including an extrapolation to the infinite-basis limit.

2 Results

2.1 The gold hydride models

One of the simplest models for intermolecular aurophilicity is the end-on Au–Au··· Au–H. An even simpler system is the monomolecular C_{2v} HAu_2^+ of Berger.[28] Recall also the mass-spectroscopic observation of linear AuAuH$^-$ and AuHAu$^-$ by the group of L. S. Wang,[29] the matrix-spectroscopic observations of further, neutral gold hydride species, such as $(H_2)AuH$ and $(H_2)AuH_3$ by the group of Andrews[30–33] and the theoretical study on AuH_3.[34]

2.1.1 Calibration results on monomers. Some calibration results for the monomers are given in Table 1.

2.1.2 Linear Au_2···AuH, 1. The binding-energy curves, calculated with various WFT methods, are shown in Fig. 1. The structures for the dimer models studied are shown in Fig. 2

The end-on Au_2 results are shown in Table 2.

We note that CCSD gives for **1** the weakest interaction, while CCSD(T) and SCS-MP2[35] are comparable, thus justifying SCS-MP2 as a 'poor man's CCSD (T)'. MP2 gives the strongest interaction. We chose SCS-MP2 as a tool for studying larger systems.

The side-on planar (Au_2)···AuH collapses to the single molecule **2**, consisting of an Au_3 triangle, with a H bound to a corner, and was omitted.

Table 1 Calculated bond lengths, R_e, (in picometres, pm) for the monomers at triple-zeta level

Case	Method	R_e
AuH	MP2	149.1
	SCS-MP2	149.1
	CCSD	152.5
	CCSD(T)	152.3
	Exp	152.4
Au_2	MP2	242.9
	SCS-MP2	242.9
	CCSD	249.1
	CCSD(T)	248.4
	Exp	247.2

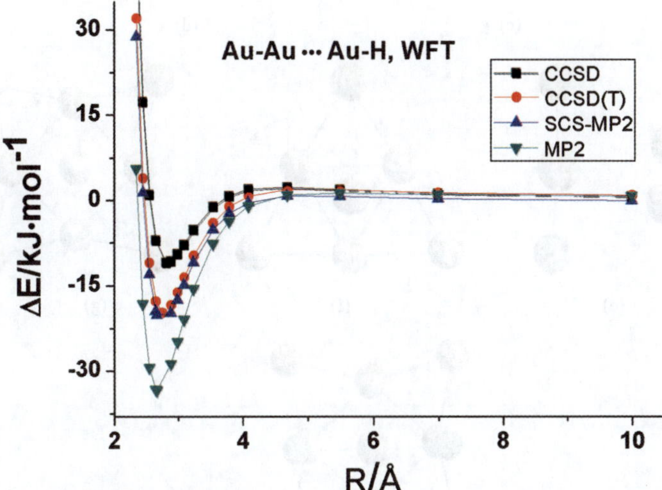

Fig. 1 The various wave-function-theory (WFT) curves for Au–Au⋯Au–H, **1**. The distance R is the intermolecular, Au_2⋯Au_3 one.

2.2 Comparison with DFT

The interaction energies for **1** with various functionals are given in Table 2 and Fig. 3, in order of strength. We conclude that, for the particular model system **1**, B3LYP and revPBE underbind, while the last five functionals, notably PBE mostly used by Häkkinen's group (see *e.g.* ref. 20, 22, 36) overbind, in this case by about 10 kJ mol⁻¹ or 50%. Wang and Schwarz[37] find for a number of aurophilic systems that B3LYP underbinds while the S and SV LDF overbind, in line with the present results.

2.3 The electron-localization functions

The 'electron-localization functions' (ELF) for species **1** and **2** are shown in Fig. 4. They also correspond to closed-shell and partially covalent interactions between the Au_2 and AuH moieties, respectively.

2.4 The large-R limit

It is often instructive to relate the potential-energy curve $V(R)$ to the various terms $V = C_n/R^n$ at the large-R limit.[7,10,11] We do it here for the colinear Au–Au⋯Au–H

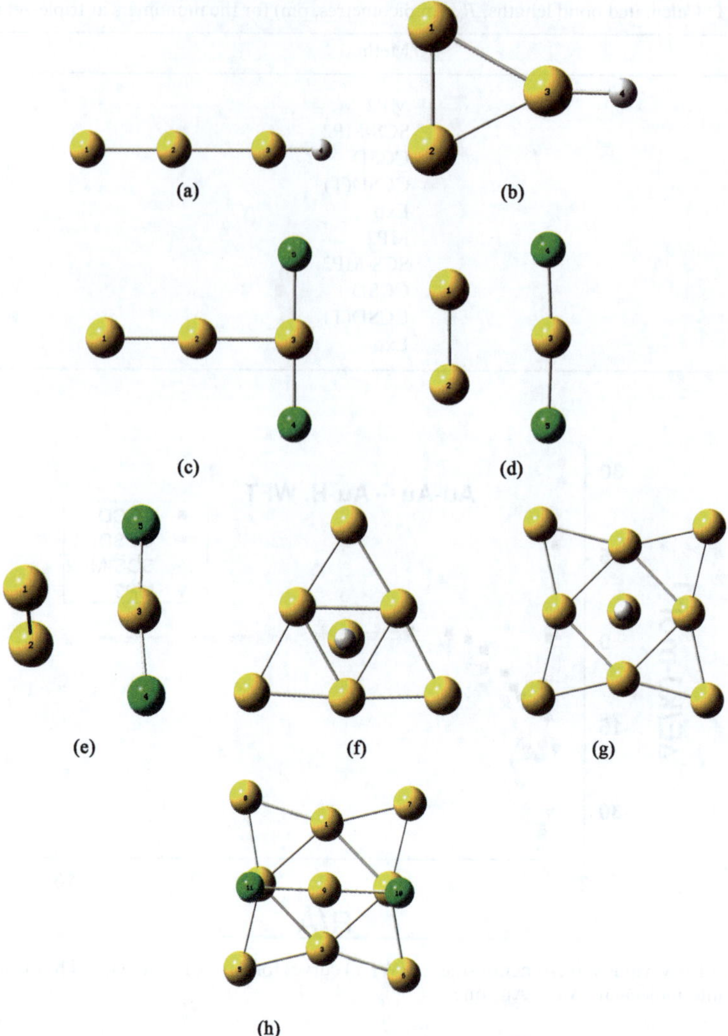

Fig. 2 The structures for the dimer models **1** (a), **2** (b), **3a** (c), **3b** (d), **3c** (e), **4** (f), **5** (g) and **6** (f).

(=A···B) case. Denoting the dipole and quadrupole moments by μ and Θ, respectively, the formulae of Stone[38] and Muñiz *et al.*[7] yield (here in a.u.)

$$C_4 = 3\Theta^A \mu^B, \tag{1}$$

$$C_5 = 6\Theta^A \Theta^B, \tag{2}$$

$$C_6 = -\frac{1}{2}\frac{I_A I_B}{I_A + I_B}\left[\alpha_\perp^A \alpha_\perp^B + 2\alpha_\parallel^A \alpha_\parallel^B\right]. \tag{3}$$

Denoting the distance between the Au_2 midpoint and the Au_3 atom as R, we can write the dominant, repulsive, long-distance interaction as

Table 2 Aurophilic interaction distances, R_e, (in pm) and energies, $D(R_e)$, (in kJ mol^{-1}) for end-on Au$_2\cdots$AuH **1**

Case	Method	R_e	$D(R_e)$
1	MP2	265.7	34.8
	SCS-MP2	278.7	20.6
	CCSD	280.8	10.7
	CCSD(T)	276.7	20.5
	B3LYP	284.3	14.3
	revPBE	277.2	15.7
	BLYP	280.0	20.4
	TPSSh	268.5	25.9
	SV-BP	270.5	29.5
	PBE	269.2	31.4
	TPSS	267.3	32.2
	PW91	268.5	33.7
	M06	274.9	34.3
	X α	264.5	57.7
	SVWN	259.1	68.1

$$V_4(R) + V_5(R) = C_4/(R + x_B)^4 + C_5/R^5 \tag{4}$$

while for the short-distance attraction we can experiment with a single Au$_2$ polarisability, α, at the molecular midpoint, or half of it on each Au atom:

$$V_6^{\text{Single}}(R) = -C_6/R^6, \tag{5}$$

$$V_6^{\text{Split}}(R) = -\frac{1}{2}C_6/(R - x_A)^6 - \frac{1}{2}C_6/(R + x_A)^6. \tag{6}$$

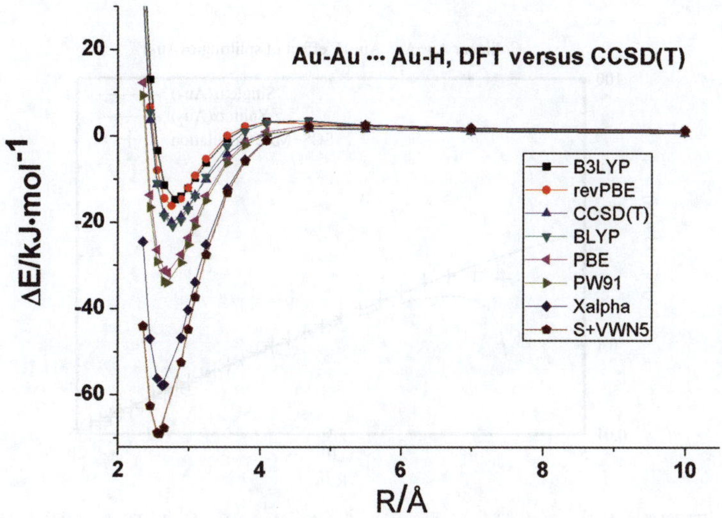

Fig. 3 The various DFT curves *versus* the CCSD(T) one for Au–Au$\cdots$Au–H, **1**.

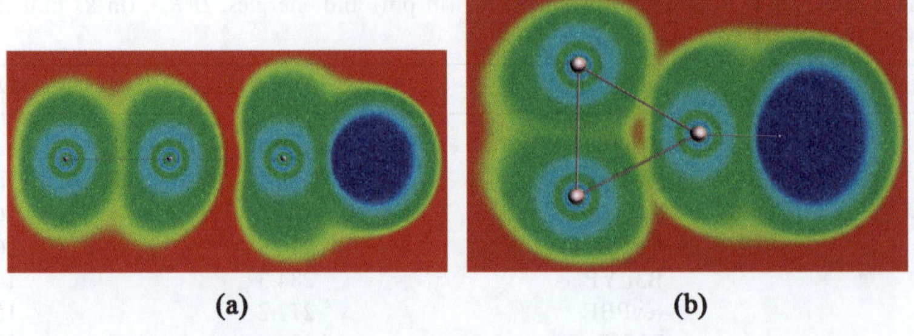

(a) **(b)**

Fig. 4 The ELFs for species **1** (a) and **2** (b).

Table 3 MP2-level monomer properties, used for fitting the aurophilic attraction potential, V (R) in Fig. 5, 6. 'PW' = present work. If not otherwise said, atomic units are used

Molecule	R_e/pm	μ	Θ	$\alpha_{\parallel}$	$\alpha_{\perp}$	α_{av}	Source
Au	242.6	—	6.272	49.38	112.28	70.35	PW
				65.7	114.2	81.9	[a]
AuH	151.	0.405	2.809	36.06	41.65	37.92	PW

[a] Ref. 39, Dirac-Fock level.

$$V(R) = V_4(R) + V_5(R) + V_6(R) \tag{7}$$

The required monomer properties are given in Table 3. The tacit assumptions are that the AuH dipole moment and the Au_2 quadrupole moment reside at the molecular midpoints, while the AuH quadrupole moment and polarisability reside at gold atom 'Au$_3$'. The obtained, total interactions are compared in Fig. 5 and 6. The conclusion becomes that the 'split' polarisabilities perform quite well for **1**, at the MP2 level used. Moreover, note that this 'local London' estimate is not far from the MP2 $V(R_e)$, the work done against the intermolecular Pauli repulsion being small. One can almost continue the large-R extrapolation to R_e. To the contrary,

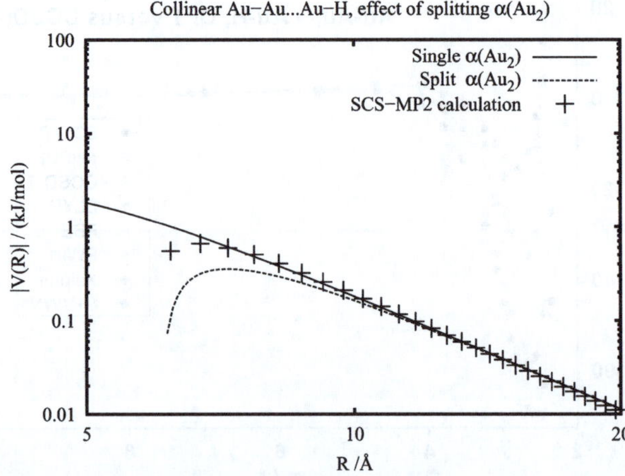

Fig. 5 The long-distance repulsive interaction for Au–Au···Au–H, **1**. The points come from SCS-MP2 calculations. The curves correspond to (eqn (5,6).

     This journal is © The Royal Society of Chemistry 2011

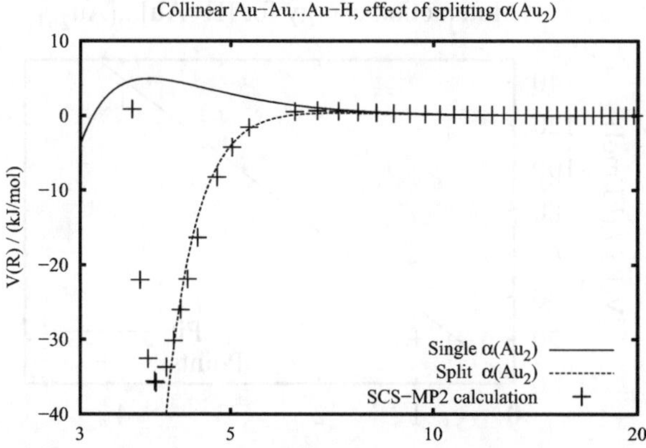

Fig. 6 The short-distance interaction for Au–Au$\cdots$Au–H, **1**. The points come from SCS-MP2 calculations. The curves correspond to (eqn (5,6).

if one tries to do the same with the single polarisability of the entire Au_2 molecule, the interaction remains much too weak, as seen from Fig. 5 and 6.

It would be premature to characterise this success of the 'split model' as an example on local, virtual surface plasmons, the atom Au_1 of the gold cluster representing the bulk atoms and Au_2 representing the surface atoms of the gold cluster, nearest to the approaching Au(I) atom.

2.5 Larger models

We also considered the case $Au_2\cdots AuCl_2^-$ in three geometries, all C_{2v}, see Fig. 2. The results are given in Table 4. The T-shaped planar structure **3a** has an SCS-MP2 interaction energy of -122 kJ mol^{-1}. The planar **3b** and the perpendicular **3c** have energies in the aurophilic range. Moreover, these energies between one Au(I) and two Au(0) of **3** are larger than those between one Au(I) and one nearest-neighbour Au(0) of the previous example **1**.

The Au(I) of the approaching AuH unit has three nearest Au neighbours in Au_6 in **4** system and four ones in Au_8 in system **5**. The increase of the interaction energy with the number of nearest neighbours is shown in Fig. 7.

A further aurophilic model where one Au(I) interacts with an Au_n unit ($n > 2$), we let the $AuCl_2^-$ interact with a quasiplanar D_{4h} Au_8 in **6**. The structure is given in Fig. 2 and the interaction described in Table 4.

3   A bond-energy-plus-Coulomb model for the staple-covered clusters

It is interesting to ask whether differential Coulomb effects could contribute to the transition from a homogeneous gold cluster, covered by simple thiolates,

Table 4 Aurophilic interaction distances, R_e, (in pm) and energies, $D(R_e)$, (in kJ mol^{-1}) for larger models at SCS-MP2 level. Here n is the number of nearest-neighbour Au atoms to the approaching Au atom and R_e is taken as the distance between them

Case	n	R_e	$D(R_e)$
3a	1	259.7	122
3b	2	299.8	54
3c	2	297.1	30
4	3	281.6	122
5	4	280.9	132
6	4	350.3	76

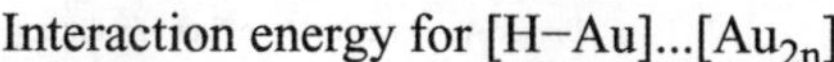

Fig. 7 The increase of the [H–Au]$\cdots$[Au$_{2n}$] interaction energy with the number of nearest neighbours, n for species **1**, **4** and **5**. The points come from SCS-MP2 calculations.

[Au$_N$(SR)$_{2M}$] to an [Au]@[SRAuSR] 'core+shell' structure. Note that the former S atoms are two-coordinated and the latter three-coordinated. This argument for the L = S–Au–S 'staples' can be generalized for the L = S–Au–S–Au–S staples.

To demonstrate the principle, let $M = 1$ and consider the bond energies (all taken as positive) in Fig. 8. They yield the thermochemical diagram in Fig. 9, from which we can read out the energy change

$$\Delta E = 2E_1 - 2E_2 + E_{Au} + IP - EA - E_C - 2E_3 - E_4. \tag{8}$$

Here E_1, E_2, E_3 are the energies of the Au–S bonds, explained in Fig. 8, IP is the ionisation potential of the Au cluster and EA the electron affinity of Au(RS)$_2$. E_4 is the aurophilic attraction energy between the ligand Au atom and the remaining cluster and E_C the Coulomb attraction between the staple ligand and the Au cluster.

Some 'educated guesses' for these bond energies can be obtained from the compilation of Luo.[40] A typical Au–S bonding energy is quoted as 253 kJ mol^{-1} but we can let the terms $2E_1$ and $2E_2$ cancel out. For the third, Au–S, bond to two-coordinated sulfur, Luo quotes values of the order of 50 kJ mol^{-1}, taken as estimate for E_3. Our own current estimate for E_4 in system '**4**' is 76 kJ mol^{-1}. A typical Au–Au bond

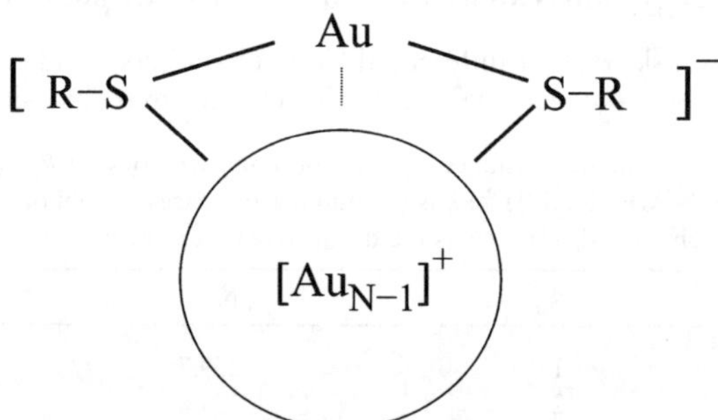

Fig. 8 A schematic structure for the simplest 'staple' structure on a gold cluster. We denote the bond energies by E_2 for the S–Au(staple), E_3 for S–Au(cluster) and E_4 for the aurophilic interaction Au(staple)$\cdots$Au(cluster). E_C is the Coulomb attraction between the staple and the cluster. The Au–S bond energy of the original [(Au$_N$)(SR)$_2$] cluster is denoted by E_1.

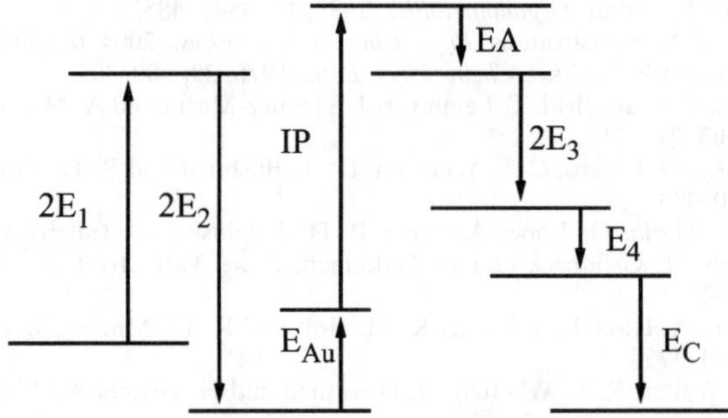

Fig. 9 A thermochemical diagram for the system in Fig. 8.

energy E_{Au} is 225 kJ mol^{-1}. The adiabatic IA and EA for Au are reported as 602 and 368 kJ mol^{-1}, respectively.[41] Then eqn (8) estimates that ΔE turns negative for an additional Coulomb attraction, E_C, of only 225 +602 $-$368 $-$100 $-$76 = 283 kJ mol^{-1}. This would correspond to two opposite unit charges at a distance of 491 pm. This very rough order-of-magnitude estimate would permit a role for Coulomb interactions between the staples and the core, as one of the driving forces.

4 Conclusions

This paper presents more questions than definitive answers. We are not aware of earlier studies on the aurophilic attraction between a closed-shell Au(I) molecular moiety and an Au_N cluster at WFT level. Even the simplest models considered here may provide valuable calibration points for existing DFT studies. The idea of local virtual excitations in the cluster may be useful. Moreover, the possible role of Coulomb attractions between the staples and the remaining cluster may be worthy of future consideration.

Acknowledgements

PP belongs to the Finnish Centre of Excellence in Computational Molecular Science and thanks Tsinghua University for their hospitality. XGX and JL are supported by NKBRSF (2011CB932400) and NSFC (20933003) of China. The calculations were performed by using supercomputers at the Computer Network Information Center, Chinese Academy of Sciences and the Shanghai Supercomputing Center.

References

1 F. Scherbaum, A. Grohmann, B. Huber, C. Krüger and H. Schmidbaur, *Angew. Chem.*, 1988, **100**, 1602–1604.
2 P. Pyykkö, J. Li and N. Runeberg, *Chem. Phys. Lett.*, 1994, **218**, 133–138.
3 P. Pyykkö, *Chem. Rev.*, 1997, **97**, 597–636.
4 P. Pyykkö, *Angew. Chem., Int. Ed.*, 2004, **43**, 4412–4456.
5 P. Pyykkö, *Inorg. Chim. Acta*, 2005, **358**, 4113–4130.
6 P. Pyykkö, *Chem. Soc. Rev.*, 2008, **37**, 1967–1997.
7 J. Muñiz, C. Wang and P. Pyykkö, *Chem.–Eur. J.*, 2011, **17**, 368–377.
8 P. Pyykkö and Y.-F. Zhao, *Angew. Chem.*, 1991, **103**, 622–623.
9 P. Pyykkö and Y.-F. Zhao, *Angew. Chem., Int. Ed. Engl.*, 1991, **30**, 604–605.
10 J. Li and P. Pyykkö, *Chem. Phys. Lett.*, 1992, **197**, 586–590.
11 P. Pyykkö and F. Mendizabal, *Chem.–Eur. J.*, 1997, **3**, 1458–1465.
12 N. Runeberg, M. Schütz and H.-J. Werner, *J. Chem. Phys.*, 1999, **110**, 7210–7215.
13 L. Magnko, M. Schweizer, G. Rauhut, M. Schütz, H. Stoll and H.-J. Werner, *Phys. Chem. Chem. Phys.*, 2002, **4**, 1006–1013.
14 P. Pyykkö, N. Runeberg and F. Mendizabal, *Chem.–Eur. J.*, 1997, **3**, 1451–1457.

15 P. Pyykkö and T. Tamm, *Organometallics*, 1998, **17**, 4842–4852.
16 E. O'Grady and N. Kaltsoyannis, *Phys. Chem. Chem. Phys.*, 2004, **6**, 680–687.
17 J. P. Desclaux and P. Pyykkö, *Chem. Phys. Lett.*, 1976, **39**, 300–303.
18 M. Verdicchio, S. Evangelisti, T. Leininger, J. Sánchez-Marín and A. Monari, *Chem. Phys. Lett.*, 2011, **503**, 215–219.
19 P. D. Jadzinsky, G. Calero, C. J. Ackerson, D. A. Bushneit and R. D. Kornberg, *Science*, 2007, **318**, 430–433.
20 M. Walter, J. Akola, O. Lopez-Acevedo, P. D. Jadzinsky, G. Calero, C. J. Ackerson, R. L. Whetten, H. Grönbeck and H. Häkkinen, *Proc. Natl. Acad. Sci. U. S. A.*, 2008, **105**, 9157–9162.
21 M. W. Heaven, A. Dass, P. S. White, K. M. Holt and R. W. Murray, *J. Am. Chem. Soc.*, 2008, **130**, 3754–3755.
22 J. Akola, M. Walter, R. L. Whetten, H. Häkkinen and H. Grönbeck, *J. Am. Chem. Soc.*, 2008, **130**, 3756–3757.
23 H.-J. Werner, P. J. Knowles, F. R. Manby, M. Schütz, P. Celani, G. Knizia, T. Korona, R. Lindh, A. Mitrushenkov, G. Rauhut, T. B. Adler, R. D. Amos, A. Bernhardsson, A. Berning, D. L. Cooper, M. J. O. Deegan, A. J. Dobbyn, F. Eckert, E. Goll, C. Hampel, A. Hesselmann, G. Hetzer, T. Hrenar, G. Jansen, C. Köppl, Y. Liu, A. W. Lloyd, R. A. Mata, A. J. May, S. J. McNicholas, W. Meyer, M. E. Mura, A. Nicklass, P. Palmieri, K. Pflüger, R. Pitzer, M. Reiher, T. Shiozaki, H. Stoll, A. J. Stone, R. Tarroni, T. Thorsteinsson, M. Wang and A. Wolf, *MOLPRO, version 2010.1, a package of ab initio programs*, 2010, see http://www.molpro.net/info/authors.
24 M. Valiev, E. J. Bylaska, N. Govind, K. Kowalski, T. Straatsma, H. van Dam, D. Wang, J. Nieplocha, E. Apra, T. Windus and W. de Jong, *Comput. Phys. Commun.*, 2010, **181**, 1477–1489.
25 D. Figgen, G. Rauhut, M. Dolg and H. Stoll, *Chem. Phys.*, 2005, **311**, 227–244.
26 K. A. Peterson and C. Puzzarini, *Theor. Chem. Acc.*, 2005, **114**, 283–296.
27 P. Pyykkö and P. Zaleski-Ejgierd, *J. Chem. Phys.*, 2008, **128**, 124309.
28 R. J. F. Berger, *Z. Naturf.*, 2009, **64b**, 388–394.
29 H.-J. Zhai, B. Kiran and L.-S. Wang, *J. Chem. Phys.*, 2004, **121**, 8231–8236.
30 X.-F. Wang and L. Andrews, *J. Am. Chem. Soc.*, 2001, **123**, 12899–12900.
31 X.-F. Wang and L. Andrews, *J. Phys. Chem. A*, 2002, **106**, 3744–3748.
32 X.-F. Wang and L. Andrews, *Angew. Chem., Int. Ed.*, 2003, **42**, 5201–5206.
33 L. Andrews, X.-F. Wang, L. Manceron and K. Balasubramanian, *J. Phys. Chem. A*, 2004, **108**, 2936–2940.
34 N. B. Balabanov and J. E. Boggs, *J. Phys. Chem. A*, 2001, **105**, 5906–5910.
35 S. Grimme, *J. Chem. Phys.*, 2003, **118**, 9095–9102.
36 O. Lopez-Acevedo, J. Akola, R. L. Whetten, H. Grönbeck and H. Häkkinen, *J. Phys. Chem. C*, 2009, **113**, 5035–5038.
37 S.-G. Wang and W. H. E. Schwarz, *J. Am. Chem. Soc.*, 2004, **126**, 1266–1276.
38 A. J. Stone, *The Theory of Intermolecular Forces*, Clarendon Press, Oxford, 1996.
39 T. Saue and H. J. Aa. Jensen, *J. Chem. Phys.*, 2003, **118**, 522–536.
40 Y.-R. Luo, *Chemical Bond Energies*, CRC Press, Boca Raton, FL, 2007.
41 H.-Q. Wang, X.-Y. Kuang and H.-F. Li, *Phys. Chem. Chem. Phys.*, 2010, **12**, 5156–5165.

Theoretical insights into the superior activity of gold catalysts and reactions of organogold intermediates with electrophiles

A. Stephen K. Hashmi,* Markus Pernpointner and Max M. Hansmann

Received 2nd March 2011, Accepted 5th April 2011
DOI: 10.1039/c1fd00029b

Three fundamental steps of homogeneous gold catalysis, the activiation of substrates by coordination to gold, the protodeauration after the nucleophilic addition and the transmetalation to palladium in palladium-catalysed C-C bond forming reactions using organogold intermediates have been studied and discussed in detail.

Introduction

Although in the field of homogeneous gold catalysis the development of new synthetic methodologies[1,2] and their application[3] have been growing exponentially in the last ten years, the number of mechanistic studies[4] is still limited. Apart from labelling experiments, in situ spectroscopy, the trapping of intermediates and kinetic studies, the field of computational chemistry can help to gain *fundamental* insights into the factors determining the reactivity of homogeneous gold catalysts.

With regard to the substrates in homogeneous gold catalysis, the use of alkenes is still rare, the more reactive allenes are slightly more popular, but alkynes still dominate the field.[3]

The most representative catalytic cycle in homogeneous gold catalysis composes of the following steps (Scheme 1).

In the absence of the catalyst, the substrate **1** would typically not react with the nucleophile NuH, activation by the gold complex [Au] as in **2** changes this situation. Now the gold catalyst rapidly reacts with NuH. This delivers a vinylgold species **3**. From the nucleophile, which is now positively charged as it is bound to the vinyl group, a proton has to be removed (**4**) for the liberation of the active gold catalyst. The catalytic cycle closed by the substitution of the gold at the double bond by a protodeauration reaction.

Results and discussion

a) The nucleophilic addition

The first intermediate **2** is most reactive in the case of activation by gold complexes. The kinetic activity of the gold catalysts is by far superior to that of other p-acidic Lewis acids like platinum, here the rate difference can easily reach a factor of 1000. It is simple to measure this rate difference by kinetic experiments, but the reason for this rate difference cannot be determined by experiments, only theory can explain that different activity. Here a calculation of the isoelectronic species Au(III)

Universität Heidelberg, Organisch-Chemisches Institut, Im Neuenheimer Feld 270, 69120, Heidelberg, Germany. E-mail: hashmi@hashmi.de; Fax: +49-6221-544205; Tel: +49-6221-548413

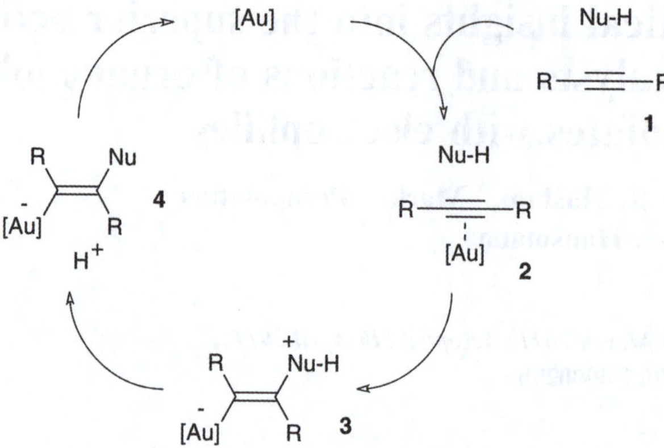

Scheme 1 Representative catalytic cycle of a gold-catalysed nucleophilic addition to an alkyne.

(d^8 configuration, **5**) and Pt(II) (d^8 configuration, **6**) coordinated to propyne was conducted.[5] The use of propyne as the alkyne also addresses the question of the regio-selectivity. A fully relativistic ab initio calculation of both the gold(III) and the platinum(II) alkyne-catalyst complexes was conducted at the Dirac–Hartree–Fock self-consistent field (DHF-SCF) as well as at density functional theory (DFT/B3LYP) and Green's function (GF) level in order to properly account for the large relativistic effects of gold and platinum. The neutral $AuCl_3$ and the neutral $PtCl_2(H_2O)$, two species present in solution, were used. For the alkyne/catalyst complexes both the perpendicular and in-plane[6] conformations were studied as these possess very similar ground state energies and may easily transform into each other (Scheme 2). Strongly varying orbital populations together with sizable energetic and structural differences of the frontier orbitals are found and can be considered as a major source of the differing catalytic activity.

These frontier orbitals mainly comprise vanishing LUMO densities at the carbon centers in the platinum complex together with increased LUMO energies, making

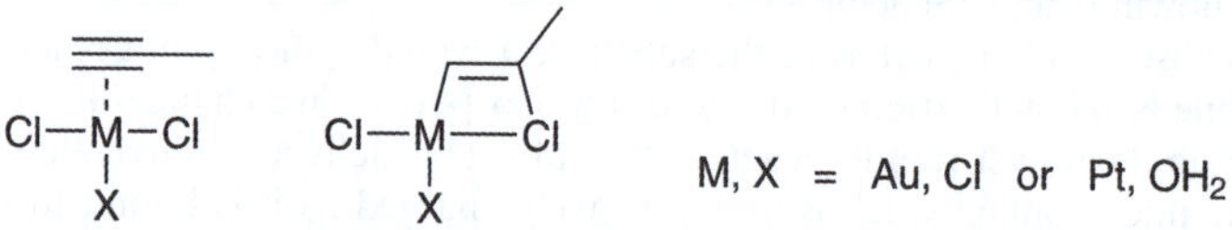

Scheme 2 Similar energies are found for the perpendicular and the in-plane conformation in the case of both gold and platinum.

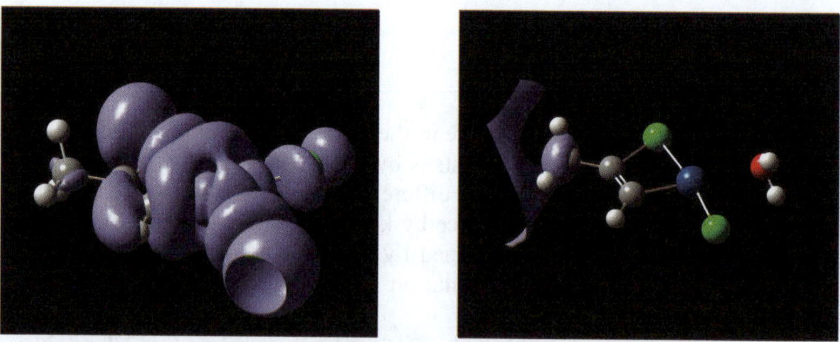

Fig. 1 Lumo orbital densities for the in-plane coordination of propyne to gold (left) and platinum (right).

a nucleophilic attack harder than in the corresponding gold compound (Figure 1). As Green's function calculations show, DFT/B3LYP seems to overestimate correlation contributions leading to an unphysical energetic lowering of many unoccupied orbitals (Table 1).

The region-selectivity of the nucleophilic attack, following the Markovnikow rule, can be easily explained by the Mulliken charges (Table 2).

Looking at more details of the calculations, it became clear that spin-orbit coupling has a considerable influence on the orbital populations in the Pt-containing complexes but leaves the outer valence compositions in the corresponding gold species nearly unaltered.

The orbital energies of both complexes are hardly modified when including SO coupling which is understandable due to their closed-shell character. Very differing metal d populations in the outer valence space lead to large structural alterations in

Table 1 Orbital energies at various levels of theory in eV of (a) the in-plane Au/propyne complex and (b) the in-plane Pt/propyne complex (IOTC = infinite-order two-component,[7] DKINF = infinite-order Douglas–Kroll Hamiltonian,[8] ADC = algebraic diagrammatic construction[9])

Orbital	IOTC/HF	DKINF/HF	IOTC/DFT	DKINF/DFT	IOTC/ADC(3)
(a) $AuCl_3\cdots$propyne					
LUMO + 3	1.445	1.442	−0.604	−0.607	—
LUMO + 2	1.443	1.438	−1.305	−1.308	—
LUMO + 1	0.884	0.880	−1.457	−1.453	—
LUMO	0.107	0.055	−3.613	−3.640	—
HOMO	−10.720	−10.743	−7.236	−7.277	−9.608
HOMO-1	−10.957	−10.920	−7.417	−7.361	−9.907
GAP	10.827	10.798	3.623	3.637	—
(b) $PtCl_2(H_2O)\cdots$propyne					
LUMO + 3	1.990	1.988	−0.304	−0.308	—
LUMO + 2	1.947	1.954	−0.383	−0.406	—
LUMO + 1	1.863	1.860	−0.543	−0.509	—
LUMO	1.264	1.262	−1.107	−1.119	—
HOMO	−9.252	−9.336	−5.980	−6.182	−8.072
HOMO-1	−9.705	−9.932	−6.077	−6.242	−8.244
GAP	10.516	10.598	4.872	5.063	—

Table 2 Mulliken charges determined at the DFT/B3LYP and HF level using relativistic ECPs[a]

method	C1	C2	C3
Au⊥ B3LYP	+0.001	+0.0522	−0.273
Au⊥ HF	−0.049	+0.0529	−0.274
$Au_{in\text{-}plane}$ B3LYP	−0.122	+0.0317	−0.263
$Au_{in\text{-}plane}$ HF	−0.155	+0.0319	−0.244
Pt⊥ B3LYP	−0.175	+0.0489	−0.274
Pt⊥ HF	−0.221	+0.0393	−0.259
$Pt_{in\text{-}plane}$ B3LYP	−0.226	+0.0201	−0.275
$Pt_{in\text{-}plane}$ HF	−0.315	+0.0387	−0.253
C_3H_4 HF	−0.199	−0.0747	−0.2752

[a] Au (Pt) represents for the gold (platinum)/propyne complex and ⊥ for the perpendicular orientation. The charges of free propyne are listed in the last line. The metal-induced strong activation of the C2 atom for a nucleophilic attack becomes obvious.

the LUMOs which can be considered as one main factor for the strongly different catalytic efficiency. These methods show that no alkynyl carbon LUMO density was found for the platinum complexes making an overlap with a frontier orbital of an attacking nucleophile unfavorable. In contrast to that both gold complexes possess a high LUMO density together with considerably lower LUMO energies accelerates a nucleophilic attack both from a structural and from an energetic point of view. The orientation of the propyne moiety (perpendicular or in-plane) is irrelevant. The calculations also prove that electron correlation treated at the relativistic DFT level has a major impact on orbital energies, populations, and densities. The interpretation of the DFT results especially with respect to virtual orbital energies is not straightforward, and relativistic propagator calculations reveal a considerable overestimation of correlation effects by DFT. Therefore the obtained DFT LUMO densities cannot serve as an unambiguous basis for the structural analysis.

b)   The protodeauration

The next step after the nucleophilic addition is the migration of a proton from the nucleophile to the position of the gold for the protodeauration (Scheme 1, intermediates **3** and **4**). When trying to simulate this process by computational methods, unexpected high and unrealistic barriers of activation were observed. These high barriers were not in agreement with the kinetic data of the catalysis reactions.

Homogeneous gold catalysis is known to work under "open flask conditions", neither oxygen (air) nor water need to be excluded. Calculations with small water clusters, composing of only six water molecules (one of these as nucleophile, Figure 2 for a gold(I) complex), gave significantly lower barriers of activation for the proton transfer.[10] These clusters serve as efficient proton shuttles. A further important outcome of these calculation was the confirmation of the *anti*-addition of the nucleophile and the gold catalyst to the triple bond. Previously a *syn*-addition had been suggested on the basis of gas phase calculations but the experimental data had always indicated an *anti*-oxyauration process.[11] If the counter ion of the cationic

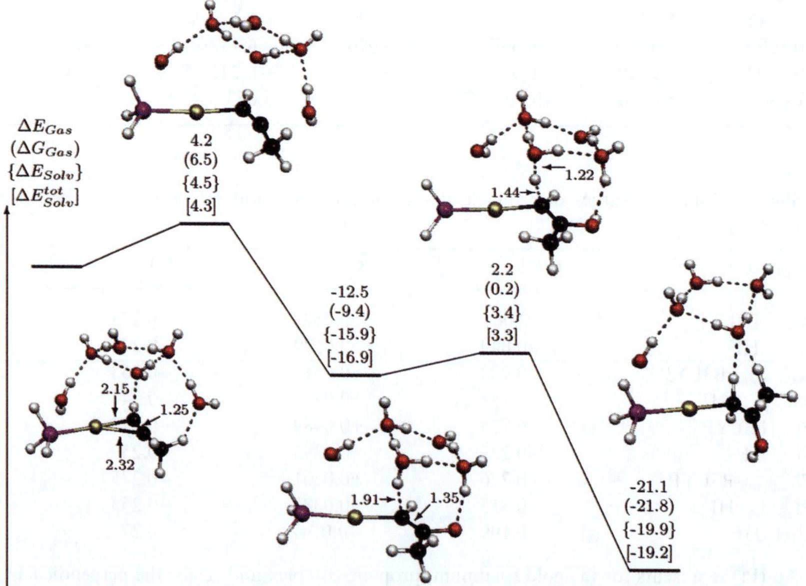

Fig. 2 Water clusters as efficient proton shuttles in gold catalysis: water is not only tolerated but needed.

gold(I) complex can accommodate protons, for example trifluoromethansulfonate, these can participate and lead to a similar lowering of the barriers of activation, too.

c) Palladium electrophiles

A significant extension of the synthetic potential of gold catalysis would be the combination with palladium-catalysed coupling reactions. Both of the individual catalytic cycles, the gold-catalysed nucleophilic addition and the palladium-catalysed cross coupling are known to work well. The crucial question is, whether an efficient transmetalation from the organogold(I) intermediates to the palladium(II) intermediates is possible. Here the palladium(II) intermediate of a cross-coupling reaction would replace the proton from the protodeauration reaction. This was first verified experimentally and proceeded very well in reactions that applied aryl iodides as coupling partners for stoichiometric amount of organogold compounds, less successfully with bromides or chlorides.[12] Overall, the transmetalation from the more electronegative (more noble) gold(I) (in **6**) to the less electronegative (less noble) palladium(II) (in **5**) seemed problematic. We conducted a series of calculations focussing on this critical transmetalation step (the calculations showed that the overall reactions were exothermic). Figure 3.

The starting material and product structures with varying ligands X were energetically optimized at the DFT/B3LYP and MP2/RECP level of theory together with the cc-pVDZ, cc-pVTZ and LANL2DZ basis sets. The tightness of structure optimization was also checked but has only little influence on the results.

Table 3 shows that the isolated transmetalation step is slightly endergonic at all levels of theory, with comparable results for the individual halides at the DFT level. Increasing the tightness of structure optimization, namely the threshold of the energy and norm of gradient vector, had only little influence on the thermodynamics but uses considerably more computational resources. The transition from a VDZ to

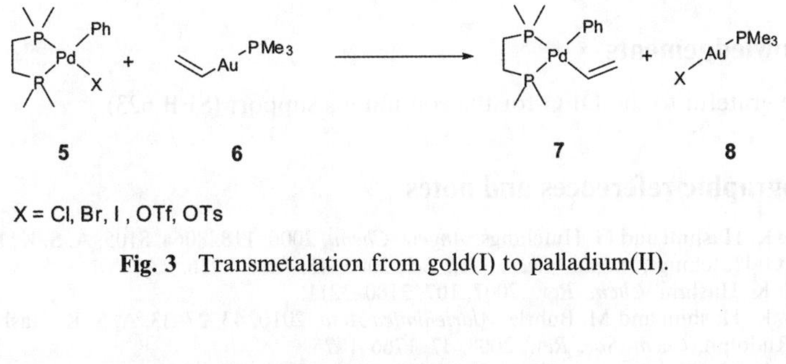

Fig. 3 Transmetalation from gold(I) to palladium(II).

Table 3 Orbital energies at various levels of theory in eV of (a) the in-plane Au/propyne complex and (b) the in-plane Pt/propyne complex (IOTC = infinite-order two-component,[7] DKINF = infinite-order Douglas-Kroll Hamiltonian,[8] ADC = algebraic diagrammatic construction[9])

Ligand	E/kcal mol$^{-1\ a}$	E/kcal mol$^{-1\ b}$	E/kcal mol$^{-1\ c}$	E/kcal mol$^{-1\ d}$
a (Cl)	2,316	2.291	2.172	0.414
b (Br)	2,129	2.664	2.093	1.971
c (I)	1,857	1.856	1.738	4.246
d (OTf)	14,174	—	—	—
e (OTs)	13,991	—	—	—

^{a} DFT/B3LYP, cc-pVDZ, normal convergence. ^{b} DFT/B3LYP, cc-pVDZ, tight convergence. ^{c} DFT/B3LYP, cc-pVTZ, normal convergence. ^{d} MP2/RECP, cc-pVDZ, normal convergence.

a VTZ basis did not influence the DFT results significantly, justifying the use of the double zeta basis for all further calculations, which is important, since a triple zeta basis can not be employed for the large two-center complexes due to numerical bottlenecks. Interestingly, a methodological change from DFT to MP2 leads to an inverted predictions of the reactivity trend with respect to the halides but retains the overall endergonic character and is an indication that the transmetalation is the rate-determining step. The oxygen-bridged ligands are significantly raised in energy, which indicates a smaller tendency for the gold-oxygen bond formation as observed for the overall thermodynamics. No accurate experimental bond dissociation energies for the $PMe_3Au–X$ systems are available which would be of great value in order to gain further insight into the transmetalation step. In order to extend the systematic basis set test, we also tested the smaller LANL2DZ Pd basis set and an aug-cc-pVDZ Pd basis. Both variations led to very similar results in the transmetalation step from which we deduced that an augmentation of the double zeta basis is not relevant and would lead to serious computational bottlenecks.

At the current stage of investigation, no concerted one-step transmetalation mechanism was localized. Furthermore, a dissociative mechanism under cleavage of the Pd–X bond was found to be too high in energy and thus can be ruled out. Instead, a dissociative two-step procedure which occurs under opening of the chelate ring in case of dmpe yielded well-defined transition states and energetic accessibility.

Conclusion

Computational chemistry in the last years has provided important insights into reaction mechanisms of gold-catalysed reactions, namely the activation of triple bonds, the protodeauration and the transmetalation from gold to palladium. These insights could not have been obtained by experimental methods. This stresses the important role which theory will play in the future for the detailed understanding of elementary steps in gold catalysis.

Acknowledgements

We are grateful to the DFG for the continuous support (SFB 623).

Bibliographic references and notes

1 A. S. K. Hashmi and G. Hutchings, *Angew. Chem.*, 2006, **118**, 8064–8105; A. S. K. Hashmi and G. Hutchings, *Angew. Chem., Int. Ed.*, 2006, **45**, 7896–7936.
2 A. S. K. Hashmi, *Chem. Rev.*, 2007, **107**, 3180–3211.
3 A. S. K. Hashmi and M. Bührle, *Aldrichimica Acta*, 2010, **43**, 27–33; A. S. K. Hashmi and M. Rudolph, *Chem. Soc. Rev.*, 2008, **37**, 1766–1775.
4 A. S. K. Hashmi, *Angew. Chem.*, 2010, **122**, 5360–5369; A. S. K. Hashmi, *Angew. Chem., Int. Ed.*, 2010, **49**, 5232–5241.
5 M. Pernpointner and A. S. K. Hashmi, *J. Chem. Theory Comput.*, 2009, **5**, 2717–2725.
6 The in-plane conformation resembles the gold-catalysed addition of chloride to alkynes: G. J. Hutchings, *J. Catal.*, 1985, **96**, 292.
7 M. Ilias and T. Saue, *J. Chem. Phys.*, 2007, **126**, 064102.
8 M. Barysz and A. J. Sadlej, *J. Chem. Phys.*, 2002, **116**, 2696.
9 J. Schirmer, L. S. Cederbaum and O. Walter, *Phys. Rev. A: At., Mol., Opt. Phys.*, 1983, **28**, 1237.
10 C. M. Crauter, A. S. K. Hashmi and M. Pernpointner, *ChemCatChem*, 2010, **2**, 1226–1230.
11 See for example: A. S. K. Hashmi, J. P. Weyrauch, W. Frey and J. W. Bats, *Org. Lett.*, 2004, **6**, 4391–4394.
12 A. S. K. Hashmi, C. Lothschütz, R. Döpp, M. Rudolph, T. D. Ramamurthi and F. Rominger, *Angew. Chem.*, 2009, **121**, 8392–8395; A. S. K. Hashmi, C. Lothschütz, R. Döpp, M. Rudolph, T. D. Ramamurthi and F. Rominger, *Angew. Chem., Int. Ed.*, 2009, **48**, 8243–8246; A. S. K. Hashmi, R. Döpp, C. Lothschütz, M. Rudolph, D. Riedel and F. Rominger, *Adv. Synth. Catal.*, 2010, **352**, 1307–1314.

A computational investigation of H_2 adsorption and dissociation on Au nanoparticles supported on TiO_2 surface

Andrey Lyalin†* and Tetsuya Taketsugu

Received 9th February 2011, Accepted 23rd February 2011
DOI: 10.1039/c1fd00013f

The specific role played by small gold nanoparticles supported on the rutile $TiO_2(110)$ surface in the processes of adsorption and dissociation of H_2 is discussed. It is demonstrated that the molecular and dissociative adsorption of H_2 on Au_n clusters containing $n = 1, 2, 8$ and 20 atoms depends on cluster size, geometry structure, cluster flexibility and the interaction with the support material. Rutile $TiO_2(110)$ support energetically promotes H_2 dissociation on gold clusters. It is demonstrated that the active sites towards H_2 dissociation are located at corners and edges on the surface of the gold nanoparticle in the vicinity of the support. The low coordinated oxygen atoms on the $TiO_2(110)$ surface play a crucial role for H_2 dissociation. Therefore the catalytic activity of a gold nanoparticle supported on the rutile $TiO_2(110)$ surface is proportional to the length of the perimeter interface between the nanoparticle and the support.

1   Introduction

Since the pioneering work of Haruta on the oxidation of carbon monoxide by small gold nanoparticles supported by metal oxides,[1] an extensive interest has been devoted to understanding the catalytic properties of gold. Such an interest is stipulated by the fact that gold nanoparticles are active even at room temperatures that makes them unique catalysts for many industrial applications.[2,3]

The most explored type of catalytic reactions with gold nanoparticles are reactions of oxidation and epoxidation, including the oxidation of carbon monoxide at mild temperatures, alcohol oxidation, the direct synthesis of hydrogen peroxide and alkene epoxidation; see, *e.g.*, ref. 1,4–10 and references therein. In spite of intensive theoretical and experimental studies the origin of catalytic activity of gold remains highly debated.

It is commonly accepted that several factors can influence the catalytic activity of gold. The most important factor is the size effect. It has been shown that the unique properties of gold in oxidation reactions emerge when the size of catalytic particles decreases down to 1–5 nm or even less; while larger sized particles and the bulk form of gold are catalytically inactive.[1,4,5,7] It has been shown that small gold clusters consisting of a few atoms can also possess extraordinarily high catalytic activity.[11,12] On the one hand, the size effects in gold nanocatalysis are determined by quantum effects, resulting from the spatial confinement of the valence electrons in the cluster;[13] on the other hand, in such clusters a dominant fraction of atoms are

Division of Chemistry, Graduate School of Science, Hokkaido University, Sapporo, 060-0810, Japan. E-mail: lyalin@mail.sci.hokudai.ac.jp; Fax: +81-(0)11-706-4921; Tel: +81-(0)11-706-3535

† On leave from: V. A. Fock Institute of Physics, St Petersburg State University, 198504 St Petersburg, Petrodvorez, Russia.

under-coordinated (in comparison with the bulk), hence they exhibit an enhanced chemical reactivity.[13–15]

Interaction with the support material is another important factor that considerably influences the chemical reactivity of the gold nanoparticles.[4,16–18] Most experimental studies on the catalytic properties of gold have been performed for gold nanoparticles supported on the surfaces of metal oxides; see, *e.g.*, ref. 4,19 and references therein. It has been demonstrated that the catalytic activity of gold nanoparticles depends on the presence of defects in the support material (*e.g.* F-center defects), charge transfers from the support, the chemical state of the surface, doping, moisture, special additives, and the presence of adsorbates on the cluster surface, including the reactant molecule itself; see, *e.g.*, ref. 4,10,16,19–24 and references therein. However, a work by Turner *et al.* presents strong experimental evidence that small gold entities (~1.4 nm) derived from Au_{55} clusters and deposited on an inert support can also be efficient and robust catalysts for oxidation reactions.[5] Hence, the interaction with the support is important but not the determining factor in catalytic activity of gold; therefore even free clusters can be effective catalysts.

Heterogeneously catalyzed hydrogenation is another type of reaction where gold nanoparticles have shown their great potential as catalysts. It has been demonstrated experimentally that gold nanoparticles supported on metal oxides such as SiO_2, Al_2O_3, TiO_2, ZnO, ZrO_2, and Fe_2O_3 are effective catalysts for selective hydrogenation of several classes of organic molecules, including α,β-unsaturated aldehydes, unsaturated ketones, and unsaturated hydrocarbons.[25–34] Moreover, supported gold nanoparticles are very selective for the direct formation of hydrogen peroxide from H_2/O_2 mixtures.[8] Mechanisms of such catalytic reactions are largely not understood. Experimental results demonstrate that molecular hydrogen does not bind to the clean gold extended surface, and molecular or dissociative hydrogen adsorption occurs on the supported gold nanoparticles.[32,35] It has been suggested that hydrogen reacts and dissociates on low coordinated gold atoms in the corner or at edge positions on the surface of gold nanoparticles.[30,35] It has been also shown that the shape of the gold particles plays a role on their catalytic activity.[36] However, in a recent work by Haruta *et al.* it has been found that the catalytic activity of the gold nanoparticles supported on the rutile $TiO_2(110)$ surface depends on the number of gold atoms located at the perimeter interface of the supported gold nanoparticles, irrespective of the cluster size.[37] It was also supposed that active sites for H_2 dissociation may be formed by a combination of gold atoms and oxygen atoms from TiO_2 at nanoparticle–surface interface.[37]

Surprisingly, very little attention has been paid to theoretical investigations of the hydrogenation reactions on gold nanoparticles. Interaction of molecular hydrogen with gold atoms and small gold clusters has been theoretically studied in ref. 38–45. It has been demonstrated, that the flexibility of the cluster structure plays a key role in the bonding and dissociation of H_2 on Au_{14} and Au_{29}.[43] Furthermore, authors in ref. 43 have concluded that dissociation of H_2 is only possible when a cooperation between four active low coordinated Au atoms on the cluster surface is allowed. However, in ref. 46 the compelling evidence has been presented that H_2 adsorbs and dissociates with small activation barriers on low coordinated Au atoms situated in corner positions of different Au_{25} nanoparticles without further conditions. Thus, there is no need for cooperation between several active Au atoms to dissociate H_2 and dissociation can occur spontaneously on a single low coordinated Au atom, when coordination number is four.[46]

There are relatively few theoretical studies on the catalytic hydrogenation reactions on gold clusters. These include works on formation of hydrogen peroxide from H_2 and O_2 over small gold clusters;[47,48] acrolein hydrogenation on Au_{20} nanoparticle;[49] and combined experimental and theoretical investigation of the unusual catalytic properties of gold nanoparticles in the selective hydrogenation of 1,3-butadiene toward *cis*-2-butene.[50]

     This journal is © The Royal Society of Chemistry 2011

Despite of the clear experimental evidence indicating a strong influence of the support materials on the processes of adsorption and dissociation of molecular hydrogen and related hydrogenation processes on the supported gold nanoparticles,[32,37] there are no theoretical studies concerning the role of the support with an exception of recent works by Boronat *et al.*[51] and Florez *et al.*[52] In ref. 51, the authors performed an elegant theoretical study on elucidation of active sites for H_2 adsorption and activation on Au_{13} cluster supported on the anatase TiO_2 (001) surface. It has been shown that Au atoms that are active for H_2 dissociation must have a net charge close to zero, be located at corner or edge low coordinated positions, and not be directly bonded to the support.[51] The Au_{13} isomers consisting of two layers of gold atoms have the largest number of potentially active sites where H_2 can be adsorbed and activated. These active atoms are located on the top layer of the Au_{13} cluster that does not have contact with the support.[51] It has been demonstrated that the presence of O vacancy defects on the anatase TiO_2 (001) surface can stabilize the most active two-layer Au_{13} particles.[51] In ref. 52 it was shown that small two-dimensional Au nanoparticles supported on TiC(001) can dissociate H_2 in a more efficient way than when supported on oxides. The active sites for H_2 dissociation on Au/TiC(001) are located at the particle edge and in direct contact with the underlying substrate.[52]

In the present paper we report the results of a systematic theoretical study of the adsorption and dissociation of H_2 on free and $TiO_2(110)$ supported Au_n clusters containing $n = 1, 2, 8$ and 20 atoms. In our work we want to clarify how hydrogen molecule adsorbs and dissociates on the gold nanoparticles. What factors are the most important for H_2 dissociation on the surface of gold? Does the dimensionality of the cluster play a role in H_2 adsorption and dissociation? What is the role of the support? Is it possible to promote dissociation of H_2? Whether or not the charge transfer between the adsorbed H_2 and the gold cluster plays any role in H_2 catalytic activation and dissociation? Why the catalytic activity of gold nanoparticles supported on the rutile $TiO_2(110)$ surface depends on the number of gold atoms located at the perimeter interface, irrespective to the particle size in the case of H_2 dissociation, but it depends on cluster size in the case of O_2 dissociation? Finally, what are the important directions for further theoretical investigation of the considered processes? Answering these questions is the central aim of the present work.

2 Computational details

The calculations are carried out using density-functional theory (DFT) with the gradient-corrected exchange–correlation functional of Perdew, Burke and Ernzerhof (PBE).[53] The atom-centered, strictly confined, numerical basis functions[54,55] are used to treat the valence electrons of H, Au, Ti and O atoms with the reference configurations $1s^1 2p^0 3d^0 4f^0$ for H, $6s^1 6p^0 5d^{10} 5f^0$ for Au, $4s^2 4p^0 3d^2 4f^0$ for Ti, and $2s^2 2p^4 3d^0 4f^0$ for O. Double-ζ plus polarization function (DZP) basis sets are used for Au, Ti and O, and triple-ζ plus polarization function (TZP) for H. The remaining core electrons are represented by the Troullier–Martins norm-conserving pseudopotentials[56] in the Kleinman-Bylander factorised form.[57] Relativistic effects are taken into account for Au.

Calculations have been carried out with the use of the SIESTA code.[58–60] Periodic boundary conditions are used for all systems, including free molecules and clusters. In the latter case the size of a supercell was chosen to be large enough to make intermolecular interactions negligible.

Within the SIESTA approach, the basis functions and the electron density are projected onto a uniform real-space grid. The mesh size of the grid is controlled by an energy cutoff, which defines the wavelength of the shortest plane wave that can be represented on the grid. In the present work the energy cutoff of 200 Ry is chosen to guarantee convergence of the total energies and forces. The self-consistency of the density matrix is achieved with a tolerance of 10^{-4}. For geometry optimization the conjugate-gradient approach was used with a threshold of 0.02 eV Å^{-1}.

The range of the basis pseudoatomic orbitals (PAO) is limited by an energy shift parameter which defines the confinement of the basis functions. It has been shown that systematic convergence to physical quantities can be obtained by varying the energy shift.[54,55] A common energy shift of 50 meV is applied for Au, Ti and O on the basis of careful comparison of the obtained theoretical results with the available reference data for the free Au$_2$ dimer, small gold clusters, bulk Au and TiO$_2$. Thus the calculated values of the dissociation energy and bond length in Au$_2$ (2.29 eV, 2.572 Å) are in an excellent agreement with those of earlier experimental studies, (2.31 eV, 2.472 Å).[61] The optimized lattice constant of the face-centered cubic (fcc) structure of bulk gold is 4.208 Å, which is 3% larger than the experimental value of 4.079 Å.[62] This is a general feature of PBE functional to overestimate the lattice constants for various types of solids.[63] In addition we tested the ability of our approach to reproduce the optimized structures and isomer sequences of small free gold clusters consisting of up to 10 atoms. The obtained structures are in a good agreement with those reported in our recent work ref. 24 and previous theoretical studies; see, *e.g.*, ref. 64–68.

Basis set for hydrogen was optimized with the use of the Nelder-Mead simplex method[69] according to the procedure described in ref. 55. The dissociation energy, D_e, and bond length in H$_2$ (4.55 eV, 0.750 Å) are in a good agreement with experimental data (4.74 eV, 0.741 Å).[61] We also tested the gold–hydrogen interaction optimizing AuH dimer. The calculated dissociation energy and bond length in AuH (3.09 eV, 1.547 Å) are in a good agreement with experimental data (3.36 eV, 1.524 Å).[61]

The rutile structure belongs to the $P4_2/mnm$ tetragonal space group. The primitive unit cell contains two TiO$_2$ units. The unit cell parameters determined by neutron diffraction are found to be $a = b = 4.587$ Å, $c = 2.954$ Å at 15 K.[70,71] In the present work the rutile lattice was optimized using the Monkhorst–Pack[72] $6 \times 6 \times 9$ k-point mesh for Brillouin zone sampling. The calculated lattice parameters $a = 4.594$ Å and $c = 2.959$ Å are in excellent agreement with the experimental values.

The optimized lattice of the bulk rutile was used to construct slabs for the TiO$_2$(110) surface. The six-layer slab containing four units of TiO$_2$ represents the element of the (110) rutile face with the surface area of $\sqrt{2}a \times c$. In the present work we study adsorption of Au$_n$ clusters containing $n = 1, 2, 8$ and 20 atoms on the rutile (110) surface. In the case of Au and Au$_2$ the TiO$_2$(110) surface is modeled by the p(2 × 5) slab (40 units of TiO$_2$ per slab), while for Au$_8$ and Au$_{20}$ we use p(3 × 6) (72 units of TiO$_2$) and p(4 × 9) (144 units of TiO$_2$) slabs, respectively. The periodically replicated slabs are separated by the vacuum region of 25 Å in the (110) direction. It has been shown that the surface energy, the energy gap and interlayer distances in thin films of rutile oscillate with decreasing amplitude as the number of layers increases.[73–75] The choice of the six-layer slab provides good convergence of the obtained results at moderate computational cost. Similar slab model has been used in ref. 76 to study an O$_2$ supply pathways in CO oxidation on Au/TiO$_2$(110).

Fig. 1 presents the side view of the TiO$_2$(110) p(2 × 5) slab. The TiO$_2$(110) surface contains atoms with different coordination. The fivefold coordinated Ti atoms are undercoordinated relative to the bulk sixfold coordinated Ti. The O atoms raised above the plane of Ti atoms are twofold coordinated. These atoms are called bridge-site O. The oxygen atoms lying in the plane of Ti atoms are threefold coordinated. In all calculations the bottom two layers in the slabs are fixed, and all other atoms are fully relaxed. Relaxation of the TiO$_2$(110) surface results in a surface rumpling, as it is shown in Fig. 1. The undercoordinated Ti and O atoms are moving inward by 0.10 Å, decreasing the distance with subsurface atoms. The sixfold coordinated Ti atoms and threefold coordinated O atoms are moving outward by 0.1 Å and 0.2 Å, respectively. The similar puckered structure of the rutile (110) surface has been reported in ref. 73,75.

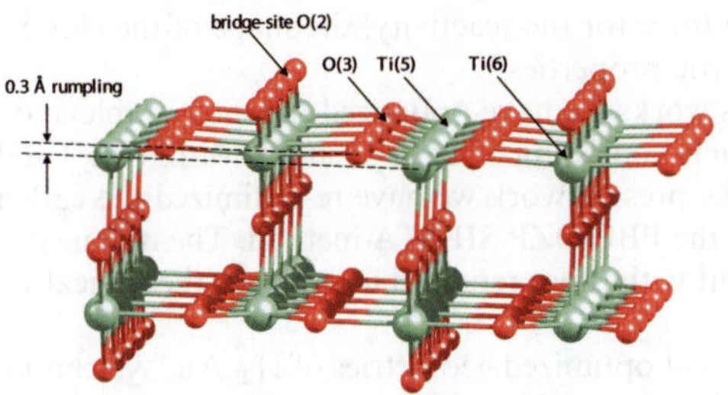

Fig. 1 Side view of the $TiO_2(110)$ p(2×5) slab. The bridge-site twofold coordinated O(2) atom, threefold coordinated O(3) atom, five- and sixfold coordinated Ti(5) and Ti(6) atoms are marked, respectively. Relaxation of the $TiO_2(110)$ surface results in a surface rumpling.

3 Results

3.1 Hydrogen adsorption on free gold clusters

The molecular and dissociative adsorption of H_2 on the free gold clusters of different sizes, chains of gold atoms, and extended gold surfaces have been the subject of a number of theoretical studies.[40,41,43,46,77–79] It was demonstrated that H_2 effectively adsorbs and dissociates at the low coordinated Au atoms regardless if they are in gold clusters or at extended line defects, such as monoatomic rows on Au surface.[43,46] However, in the case of small Au clusters, coordination alone cannot explain all the features of the reactivity of gold clusters for H_2 dissociation.[79] Several major factors are determining the catalytic properties of free clusters: the structure associated with individual atoms (*i.e.* coordination), the binding structure, and the border effects.[79] It was demonstrated that the planar gold structures are not reactive, when the H_2 approaches the cluster from the top. However, dissociation of H_2 can be promoted when planar clusters are folded or the H_2 approaches clusters in their plane.[79] The flexibility can play an important role in H_2 dissociation, because of the large atomic displacements in the cluster.[43] Finally, the size, the structure and the charge state of gold clusters influence on the molecular and dissociative adsorption of H_2.[78]

In the present work we consider adsorption of H_2 on gold clusters Au_n consisting of $n = 1, 2, 8$, and 20 atoms in order to elucidate the size and the geometry effects on hydrogen adsorption. The structural properties of free gold clusters have been intensively studied, see, *e.g.*, ref. 19,68,80 for a review. DFT calculations demonstrate that Au_n clusters favor planar, two-dimensional (2D) structures up to cluster size of 11–13 atoms, while Au_n of larger sizes are three-dimensional (3D).[67,81,82] The high stability of planar structures for small gold clusters has been explained by relativistic effects. The relativistic effects make the $6s$ and $5d$ atomic orbitals closer in energy, resulting in a sd hybridization, which in turn stabilizes planar configurations.[83] The exact value of the critical size for 2D $\rightarrow$ 3D transition in neutral Au_n is not yet clarified. *Ab initio* MP2 and CCSD(T) calculations performed in ref. 84 demonstrate that Au_8 possesses a 3D structure, while the similar CCSD(T) calculations performed with the larger basis set favors a 2D structure for Au_8.[85] Recent CCSD(T) calculations accounting for the basis-set superposition error (BSSE) corrections are in favor of a 3D structure for Au_{10}[86] in accord with DFT predictions. Competition of 2D and 3D structures of the small gold clusters is a very interesting feature that can play an important role in cluster reactivity. The 2D structures contain a large number of low coordinated atoms, hence they might possess higher catalytic activity in comparison with the 3D clusters. Moreover, it has been shown that if the shape of neutral gold clusters is modified, it is possible to change their electron donor–acceptor capacities—the 2D gold clusters are better electron acceptors than 3D ones.[87] Since the electron transfer between the cluster and the adsorbate is an important driving force for the reactivity, the shape of the cluster can directly affect the cluster's catalytic properties.[87]

In our previous works we have optimized the most stable and isomer geometries of small gold clusters consisting of up to 10 atoms within DFT B3PW91/LANL2DZ approach.[9,24] In the present work we have re-optimized the earlier obtained geometries of Au_n using the PBE/DZP SIESTA method. The obtained structures are in an excellent agreement with those reported in previous theoretical studies; see, *e.g.*, ref. 24,64–68.

In Fig. 2 we present optimized geometries of H_2–Au_n systems in the case of molecular and dissociative adsorption of H_2 on free Au_n clusters. In order to obtain the most stable configuration of H_2 adsorbed on Au_n, we have created a large number of starting geometries by adding H_2 molecule in different positions (up to 30) on the surface of the considered clusters. The starting structures have been optimized further without any geometry constraints. We have successfully used a similar approach to find the optimized geometries of O_2 and C_2H_4 molecules adsorbed on small neutral, anionic and cationic gold clusters.[9,10,24,88] In addition to the most stable 2D structure of Au_8 and 3D structure of Au_{20} we consider the closest in energy 3D isomer of Au_8 and 2D isomer of Au_{20} in order to understand how dimensionality and structural properties affect the cluster reactivity. The calculated relative energies for the higher isomeric Au_8 (3D) and Au_{20} (2D) structures with respect to the corresponding lowest energy structures are 0.18 eV and 1.95 eV, respectively. The binding energies for molecular, E_b^{mol}, and dissociative, E_b^{dis}, adsorption of H_2 on free Au_n as well as the H–H bond length r_{H-H}^{mol} in H_2 adsorbed molecularly are summarized in Table 1.

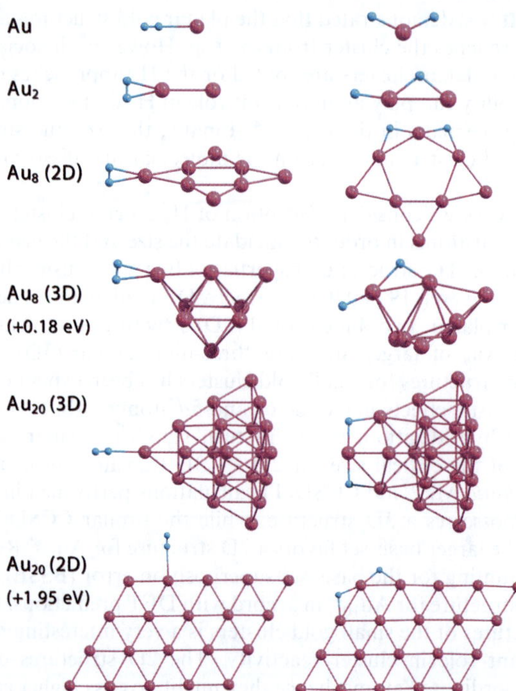

Fig. 2 Optimized geometries of H_2–Au_n clusters in the case of molecular (left) and dissociative (right) adsorption of H_2. The numbers in the parentheses correspond to the relative energy for the higher isomeric free Au_8 (3D) and Au_{20} (2D) clusters with respect to the corresponding Au_8 (2D) and Au_{20} (3D) lowest energy structures.

Table 1 Binding energies for molecular, E_b^{mol}, and dissociative, E_b^{dis}, adsorption of H_2 on free Au_n clusters and the H–H bond length, $r_{\text{H-H}}^{\text{mol}}$, in the case of molecular adsorption of H_2

Cluster	$r_{\text{H-H}}^{\text{mol}}$ (Å)	E_b^{mol} (eV)	E_b^{dis} (eV)
Au	0.780	0.13	0.04
Au_2	0.845	0.59	0.59
Au_8 (2D)	0.806	0.26	0.68
Au_8 (3D)	0.788	0.13	0.52
Au_{20} (3D)	0.768	0.09	0.14
Au_{20} (2D)	0.762	0.09	0.72

The binding energy of H_2 adsorbed on a free Au_n cluster is defined as

$$E_b^{\text{mol,dis}} = E_{\text{tot}}(Au_n) + E_{\text{tot}}(H_2) - E_{tot}^{\text{mol,dis}}(H_2 - Au_n), \tag{1}$$

where $E_{tot}^{\text{mol,dis}}(H_2-Au_n)$ denotes the total energy of the compound system, H_2-Au_n, while $E_{\text{tot}}(Au_n)$ and $E_{\text{tot}}(H_2)$ are the total energies of the non-interacting fragments Au_n and H_2, respectively.

Results of our calculations demonstrate, that H_2 binds weakly to Au as a molecule, with $E_b^{\text{mol}} = 0.13$ eV. Dissociation of H_2 on Au is not favorable energetically. The H–H bond length in H_2 adsorbed on Au is slightly enlarged in comparison with the free H_2. Adsorption of H_2 on Au_2 is relatively strong, with the binding energy equal to 0.59 eV both for the molecular and the dissociative adsorption. Table 1 demonstrates that the H–H bond length increases to 0.845 Å in H_2-Au_2, indicating that H_2 is highly activated.

With the further increase in cluster size the binding energy calculated for molecular adsorption of H_2 decreases to 0.26 eV (0.13 eV) for Au_8 (2D) (Au_8 (3D)) and 0.09 eV both for 3D and 2D isomers of Au_{20}. On the other hand, the dissociative adsorption of H_2 becomes energetically favorable both for Au_8 and Au_{20}. The binding energy of H_2 on free gold clusters strongly depends on the cluster's isomer state. Thus, dissociative adsorption of H_2 on 2D structures of Au_8 and Au_{20} are energetically favorable compared with the corresponding 3D structures. Therefore, dissociation of H_2 on small gold clusters can be promoted if we can find a way to enhance the stability of 2D structures for $n \geq 13$. That should be possible for supported clusters, where an attractive interaction with the surface stabilizes oblate configurations.[89] Thus, by varying the interaction with the support one can change the cluster shape, and hence it can be possible to tune its reactivity. Other factors that influence H_2 dissociation are coordination of Au atoms interacting with hydrogen and flexibility of cluster structure. It is seen from Fig. 2 that H_2 dissociates at the low coordinated corner Au atom with formation of the slightly bent H–Au–H bond. It is worth noting, that the Au–Au bond length in Au_2 does not change noticeably upon molecular adsorption of H_2; but increases up to 2.753 Å for dissociative adsorption. In the case of Au_8 and Au_{20} dissociation of H_2 on the cluster surface is accompanied by the structural (at least local) transformations. Therefore we can confirm the conclusion made in ref. 43 that the structural flexibility of the gold clusters might be an important factor for H_2 dissociation.

3.2 Hydrogen adsorption on Au_n/TiO_2(110)

How does the interaction with the support influence the reactivity of gold clusters? Can a supported nanoparticle exhibit novel properties that are not found for free particles? Recent experiments performed by Haruta *et al.* clearly demonstrate that the perimeter interface of the gold nanoparticles supported on the rutile TiO_2(110)

surface plays a crucial role in the process of hydrogen dissociation.[37] It was suggested that the active sites for H_2 dissociation might be formed by a combination of gold atoms and oxygen atoms from TiO_2 at the nanoparticle–surface interface.[37]

In the present work we perform a systematic theoretical study of the structure and energetics of H_2 adsorbed on Au_1, Au_2, Au_8 (2D), Au_8 (3D), Au_{20} (3D) and Au_{20} (2D) clusters supported on the rutile $TiO_2(110)$ surface. The analysis of the energetics of H_2 adsorption on the supported gold clusters provides deep insights into the nature of bonding and dissociation of the H_2 molecule and reveals the details of the reaction mechanism.

As discussed above the dissociation of H_2 on free gold clusters is governed by the interplay of structural factors, coordination of the adsorption center and cluster flexibility. The complexity of the changes in structural and electronic features of gold nanoparticles becomes more diverse when one deposits nanoparticles on the support.

In order to obtain the most stable geometries of Au_n supported on the rutile $TiO_2(110)$ surface we have created a large number of starting configurations by adding Au_n isomers with different orientation in respect to the surface. These structures have been fully optimized on the rutile surface, with accounting for the relaxation of all the gold atoms as well as the top four layers of the six layer slab of the rutile. The bottom two layers in the slab were fixed. Thus, we have taken into account deformations of the cluster structure due to the interaction with the support as well as structural relaxations on the rutile $TiO_2(110)$ surface due to its interaction with the supported cluster.

Fig. 3 presents the most stable geometries of Au_n clusters ($n = 1, 2, 8,$ and 20) optimized on the rutile $TiO_2(110)$ surface. We found that gold tends to maximize interaction with the O(3) and Ti(5) atoms on $TiO_2(110)$ surface. Thus, Au atom tends to occupy a position above the row of O(3) surface atoms, simultaneously bridging two O(3) atoms in the direction parallel to the O(3) row, as well as Ti(5) and O(2) atoms, in the direction perpendicular to the O(3) row. In the case of Au_2, both of Au atoms occupy positions above of the rows of O(3) atoms, symmetrically in respect to the row of Ti(5) atoms. The Au–Au bond length of 2.57 Å in Au_2 matches the distance between the two rows of O(3) atoms (2.53 Å). The geometry structures of two- and three dimensional isomers of Au_8 on the $TiO_2(110)$ surface are slightly deformed if

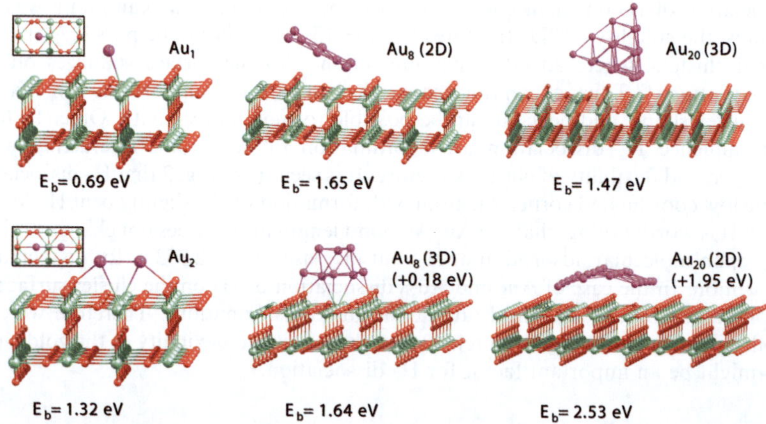

Fig. 3 The most stable geometries of Au_n clusters ($n = 1, 2, 8,$ and 20) optimized on the rutile $TiO_2(110)$ surface. Top views of Au and Au_2 on $TiO_2(110)$ are shown in the inserts. The binding energies of Au_n on $TiO_2(110)$ are marked below the corresponding structures. Numbers in parentheses correspond to the relative energy for the higher isomeric free Au_8 (3D) and Au_{20} (2D) clusters with respect to the corresponding free Au_8 (2D) and Au_{20} (3D) lowest energy structures.

     This journal is © The Royal Society of Chemistry 2011

compared with the free gold clusters. The Au$_8$ (2D) cluster maximizes the interaction between the cluster edge and the row of O(3) atoms on the TiO$_2$(110) surface. The angle between the cluster's plane and the rutile surface is 26°. The supported Au$_8$ (3D) isomer prefers to maximize its interaction with the rutile TiO$_2$(110) surface, orienting the 5-atom face parallel to the support, as shown in Fig. 3. The binding energies of Au$_8$ (2D) and Au$_8$ (3D) isomers on TiO$_2$(110) are 1.65 eV and 1.64 eV, respectively. A tetrahedral Au$_{20}$ (3D) cluster binds to the surface, maximizing interaction between the cluster edge and rows of O(3) and Ti(5) surface atoms; while Au$_{20}$ (2D) isomer binds by two its edges to two parallel rows of Ti(5) surface atoms, slightly bending its planar structure, to avoid interaction with the row of O(2) bridge atoms in the middle. The planar Au$_{20}$ structure binds to the surface much more strongly than the 3D structure, because it has a wider contact area. Although Au$_{20}$ (3D) structure is still energetically favorable when adsorbed on the TiO$_2$(110) surface in comparison with the 2D structure, the energy difference between two configurations decreases from 1.95 eV for free clusters to 0.89 eV for supported clusters, respectively. If we could further increase the interaction with the support, we would make 2D structure energetically favorable. Such an effect has been reported for Au$_{20}$ clusters deposited on thin MgO(100) films supported on Mo(100) substrate.[90] By changing the thickness of the metal-oxide film, one can tune the interaction of the cluster with the support, and thus increase its wetting propensity and induce a dimensionality crossover from 3D cluster structures on MgO(100) to the energetically favored 2D geometries on the metal-supported films.[90]

Fig. 4 presents the most stable geometries calculated for the molecular and dissociative adsorption of H$_2$ on the pure rutile TiO$_2$(110) surface. Our calculations demonstrate that the hydrogen molecule adsorbs on the pure rutile TiO$_2$(110) surface on top of the Ti(5) atom with the binding energy of 0.14 eV. The H–H bond length of 0.78 Å is slightly increased in comparison with the free H$_2$. The dissociative state of H$_2$ on TiO$_2$(110) corresponds to the situation when both H atoms form the OH group with the low coordinated O(2) bridge atoms on the rutile surface. The binding energy of the dissociated configuration of H$_2$ is 1.50 eV; however, the distance between two rows of O(2) atoms on TiO$_2$(110) is 6.50 Å, which is too large to promote dissociation of the H$_2$ adsorbed on top of Ti(5) atom. In this case H$_2$ would dissociate in the vicinity of the adsorption center as a first step, followed by adsorption of H atoms on O(2). This process would require to overcome a dissociation barrier much higher than the energy of the molecular adsorption. In this situation H$_2$ would likely desorb from the surface and fly away rather than dissociate. Therefore, to promote H$_2$ dissociation on TiO$_2$(110), it is necessary to have adsorption centers on the rutile surface in the vicinity of low coordinated O(2) atoms. Supported gold clusters can serve as a source of such centers.

Fig. 5 presents optimized geometries calculated for molecular and dissociative adsorption of H$_2$ on Au/TiO$_2$. It is seen from Table 2 that H$_2$ adsorbs molecularly on Au/TiO$_2$ with a binding energy of 1.15 eV, which is considerably larger than the corresponding energy calculated for H$_2$ adsorbed on free Au atom. The adsorbed H$_2$ is highly activated, with the H–H bond length $r_{\text{H-H}}^{\text{mol}} = 0.906$ Å. As a result of H$_2$ adsorption, the supported Au atom shifts slightly towards the row of the low

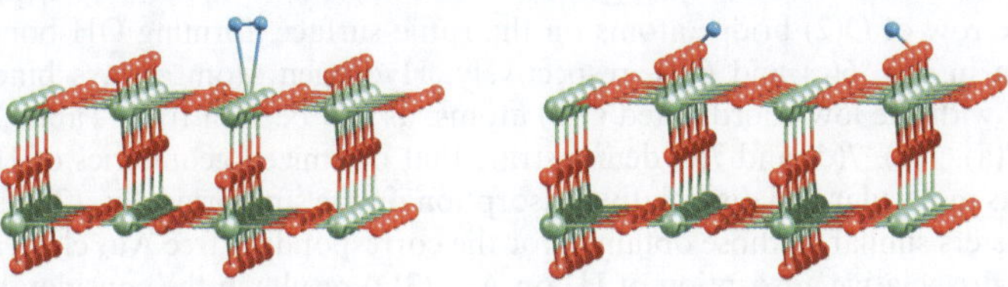

Fig. 4 Optimized geometries for molecular (left) and dissociative (right) adsorption of H$_2$ on the pure rutile TiO$_2$(110) surface.

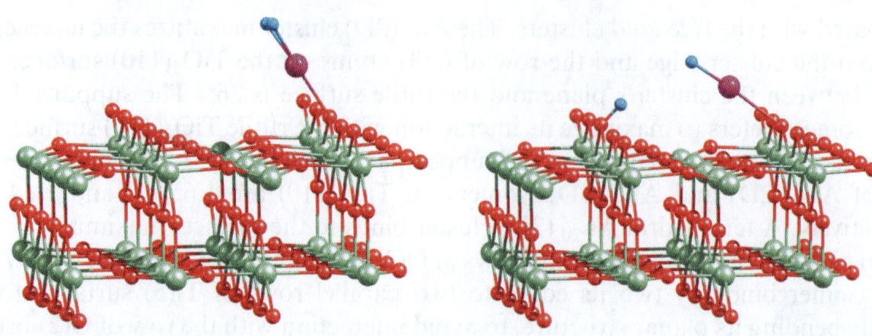

Fig. 5 Optimized geometries for molecular (left) and dissociative (right) adsorption of H_2 on Au/TiO$_2$.

Table 2 Binding energy calculated for molecular, E_b^{mol}, and dissociative, E_b^{dis}, adsorption of H_2 on Au$_n$/TiO$_2$, and the H–H bond length, r_{H-H}^{mol}, in the case of molecular adsorption of H_2

Cluster	r_{H-H}^{mol}(Å)	E_b^{mol}(eV)	E_b^{dis}(eV)
Au	0.906	1.15	2.06
Au$_2$	0.821	0.37	0.73[b] 0.90[c] 1.55[d]
Au$_8$ (2D)	0.792	0.24	0.75[b] 1.55[c]
Au$_8$ (3D)	0.753	0.09	0.55[e] 0.95[f]
Au$_{20}$ (3D)	0.795[a] 0.806[a']	0.17[a] 0.13[a']	0.20[b] 0.83[c]
Au$_{20}$ (2D)	0.798[a] 0.805[a']	0.11[a] 0.11[a']	0.55[e] 1.42[f]

coordinated O(2) bridge atoms. In the previous section we saw that H_2 dissociation is not favorable on the free Au atom. However, in the case of the supported Au atom, dissociation of H_2 can occur with formation of the OH group with the O(2) atom located either in the nearest to Au row of O(2) or in the next row of O(2), as is shown in Fig. 5. In the latter case the calculated binding energy $E_b^{dis} = 2.06$ eV is higher, not only if compared with the binding energy for molecular adsorption of H_2 on Au/TiO$_2$(110), but also if compared with the binding energy of the dissociated state of H_2 on the pure TiO$_2$(110) surface. Thus, we can conclude, from the energetic point of view, that the TiO$_2$(110) support considerably promotes H_2 dissociation on Au atom.

The optimized geometries calculated for molecular and dissociative adsorption of H_2 on Au$_2$/TiO$_2$ are shown in Fig. 6. Fig. 6(a) and 6(b) demonstrate that H_2 binds to the supported Au$_2$ in a similar way as for the free Au$_2$. However the binding energy calculated for the molecular adsorption of H_2 on Au$_2$/TiO$_2$ is lower than the corresponding energy obtained for free Au$_2$, while the dissociated configuration of H_2 on Au$_2$/TiO$_2$ is more stable if compared with dissociative adsorption of H_2 on the free Au$_2$. Therefore, the interaction of Au$_2$ with the support results in the energetic promotion of H_2 dissociation on the supported Au$_2$. We found, however, that the geometry configuration shown in Fig. 6(b) is not the most stable one for H_2 dissociation. We found that the hydrogen atom can migrate to the row of O(3) atoms, or to the row of O(2) bridge atoms on the rutile surface, forming OH bonds, as it is shown in Fig. 6(c) and 6(d), respectively. Hydrogen atom always binds more strongly with the low coordinated O(2) atoms, as can be seen from Table 2.

Fig. 7(a), 7(b), 7(d) and 7(e) demonstrate that optimized geometries of H_2 in the case of its molecular and dissociative adsorption on the supported Au$_8$ (2D) and Au$_8$ (3D) clusters similar to those obtained for the corresponding free Au$_8$ clusters. Note that the dissociative adsorption of H_2 on Au$_8$ (3D) results in the considerable rearrangement of the gold structure as shown in Fig. 7(e). Tables 1 and 2 demonstrate that the binding energies calculated for molecular adsorption of H_2 on free and

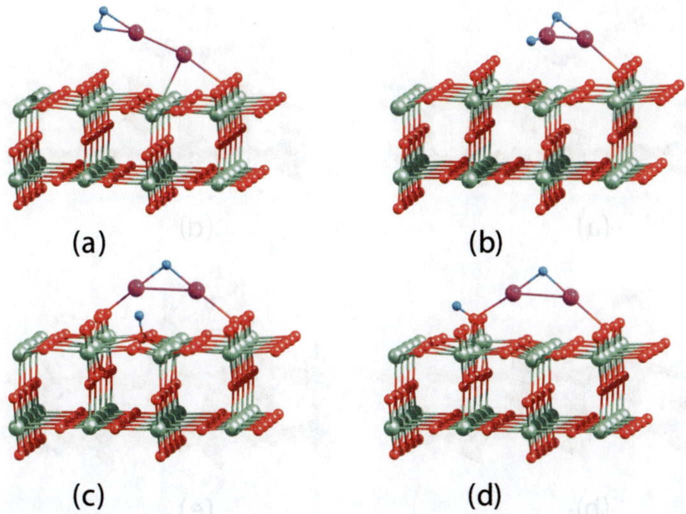

Fig. 6 (a) Optimized geometry in the case of molecular adsorption of H_2 on Au_2/TiO_2. Optimized geometries in the case of dissociative adsorption of H_2 on Au_2/TiO_2: (b) dissociation of H_2 on supported Au_2; (c) dissociation of H_2 with formation of the OH group on O(3) surface atom; (d) dissociation of H_2 with formation of the OH group on O(2) bridge atom.

supported Au_8 (2D) and Au_8 (3D) are quite similar. However, the dissociative configuration of H_2 on the supported Au_8 (2D) and Au_8 (3D) clusters is slightly promoted energetically. Migration of H on the low coordinated O(2) atom as it is shown in Fig. 7(c) and 7(f), results in formation of the OH bond and considerable increase in binding energies calculated for H_2 adsorbed dissociatively. It is seen from Fig. 7(c) that the planar structure of Au_8 undergoes drastic rearrangements upon H_2 dissociation. The dissociative adsorption of H_2 on Au_8 is energetically favorable if compared with the adsorption on the 3D isomer of Au_8.

The Au_{20} (3D) and Au_{20} (2D) clusters supported on TiO_2(110) are the largest systems studied in the present work. In the previous section it was shown that the hydrogen molecule binds weakly to the free Au_{20} (3D) and Au_{20} (2D) clusters. We have found that, in the case of the supported Au_{20}, the hydrogen molecule adsorbs on top of the Ti(5) atoms on the rutile surface in the vicinity of Au_{20}, rather than directly on the Au_{20} clusters, as it is shown in Fig. 8(a), 8(d) and 8(d'). The geometry structure when H_2 adsorbs at the vertex of Au_{20}(3D), bridging Au vertex atom in the cluster and the low coordinated O(2) atom on the rutile surface is also stable (Fig. 8(a')). Further H_2 dissociation can occur either at the vertex of Au_{20} (3D) or at the acute-angled corner of Au_{20} (2D), as is shown in Fig. 8(b) and Fig. 8(e), respectively. H_2 dissociation at the acute-angled corner of Au_{20} (2D) is accompanied by severe local structural rearrangement of Au atoms at the corner of Au_{20} (2D) and migration of one H atom to the O(2) atom on the surface. Note that the dissociation of H_2 at the obtuse-angled corner of the supported Au_{20} (2D) cluster is not energetically favorable, in comparison to the case of free Au_{20} (2D). The most stable configurations of the dissociated H_2 are those where one of the H atoms binds to the edges of Au_{20} (3D) or Au_{20} (2D) that are oriented perpendicularly to the rows of O(2) surface atoms, while another H forms the OH bond with the O(2) atom on the rutile surface, as is shown in Fig. 8(c) and 8(f). We found several geometrical configurations of this type with binding energies of 1.19–1.42 eV. Dissociation at the edge of the supported 2D isomer of Au_{20} is always energetically favorable.

Thus, combination (interplay) of several factors such as geometry structure, cluster dimensionality, presence of the low coordinated oxygen atoms in the vicinity of the cluster-surface interface, *etc.* are important for H_2 dissociation.

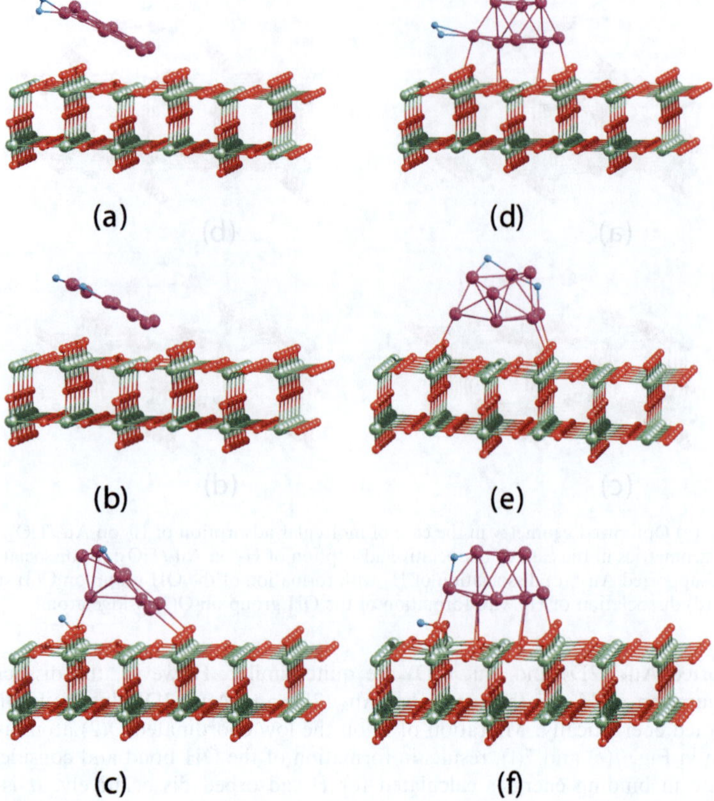

Fig. 7 Optimized geometries in the case of (a) molecular adsorption of H_2 on $Au_8(2D)/TiO_2$; dissociative adsorption of H_2 on $Au_2(2D)/TiO_2$; (b) dissociation of H_2 on supported $Au_2(2D)$; (c) dissociation of H_2 with formation of the OH group on O(2) bridge atom. Optimized geometries in the case of (d) molecular adsorption of H_2 on $Au_8(3D)/TiO_2$; dissociative adsorption of H_2 on $Au_2(3D)/TiO_2$; (e) dissociation of H_2 on supported $Au_2(3D)$; (f) dissociation of H_2 with formation of the OH group on O(2) bridge atom.

It is well known that, in the case of catalytic oxidation reactions by molecular oxygen on free and supported gold nanoparticles, the charge transfer from the gold to the antibonding orbital of O_2 is responsible for the catalytic activation and dissociation of O_2. However, in the case of H_2 dissociation the analysis of the Bader charges[91,92] demonstrate that there is no considerable charge transfer between the adsorbed hydrogen and the gold clusters (free or supported). We found the largest charge transfer occurs for H_2 adsorbed molecularly on Au/TiO_2 (Fig. 5). In this case the calculated Bader charge localized on H_2 is +0.11e, where e is the elementary charge. Although such a charge transfer is relatively small, it might be responsible for the strong enlargement of the H–H bond length in H_2–Au/TiO_2 up to 0.906 Å. Nevertheless, we have not found the direct correlation of the hydrogen dissociation with the charge transfer between hydrogen and gold atoms. On the other hand, formation of the OH group with the bridge O(2) atom on the rutile surface is accompanied by the large charge transfer from H atom to O(2), resulting in a Bader net charge of H in OH equal to +0.75e. Thus, we can conclude that the catalytic activity of gold nanoparticles for O_2 dissociation would depend on the electronic structure and the size of the nanoparticles; however in the case of H_2 dissociation it will be proportional to the number of gold atoms located in the vicinity of the low coordinated O(2) atoms at the nanoparticle – surface interface.

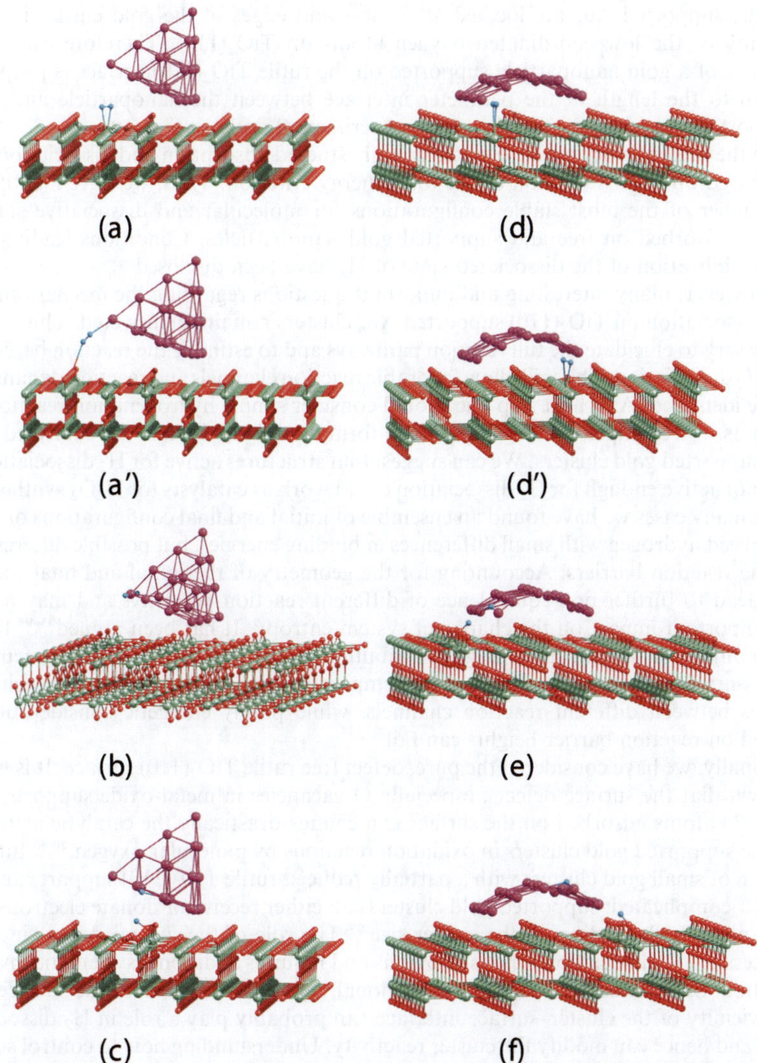

Fig. 8 Optimized geometries for H_2 adsorbed on Au_{20}(3D)/TiO_2 (left) and Au_{20}(2D)/TiO_2 (right). (a) and (a′) Molecular adsorption of H_2 in the vicinity of Au_{20}(3D)/TiO_2; (b) dissociative adsorption of H_2 at the vertex of Au_{20}(3D) with formation of the OH group; (c) dissociative adsorption of H_2 at edge of Au_{20}(3D) with formation of the OH group; (d) and (d′) molecular adsorption of H_2 in the vicinity of Au_{20}(2D)/TiO_2; (e) dissociative adsorption of H_2 at the corner of Au_{20}(2D) with formation of the OH group; (f) dissociative adsorption of H_2 at edge of Au_{20}(2D) with formation of the OH group.

4 Conclusions

The present theoretical study demonstrates that adsorption of H_2 on Au_n ($n = 1, 2, 8,$ 20) strongly depends on cluster size, geometry structure, flexibility and interaction with the support material. Strong interaction with the support can stabilize 2D structures of gold clusters on the surface. This is an important finding, because the energy release due to H_2 dissociation is the largest for the 2D isomers of gold clusters. Rutile TiO_2(110) support energetically promotes H_2 dissociation on gold clusters. The formation of the OH group near the supported gold cluster is an important condition for H_2 dissociation. We have shown that the active sites towards H_2 dissociation on the supported Au_n are located at corners and edges of the gold cluster in the vicinity of the low coordinated oxygen atoms on $TiO_2(110)$. Therefore catalytic activity of a gold nanoparticle supported on the rutile $TiO_2(110)$ surface is proportional to the length of the perimeter interface between the nanoparticle and the support, in accordance with the recent experimental findings.[37]

In the present work we have systematically studied adsorption and dissociation of H_2 based on the assumption of the total energy criterion. Thus, we have identified a number of the most stable configurations for molecular and dissociative states of H_2 adsorbed on free and supported gold nanoparticles. Conditions leading to the stabilization of the dissociated state of H_2 have been disclosed.

However, many interesting and important questions regarding the mechanism of H_2 dissociation on $TiO_2(110)$ supported Au_n clusters remain unanswered. Thus, it is necessary to elucidate the full reaction pathways and to estimate the reaction barriers for H_2 dissociation. This will allow favorable reaction channels and reaction dynamics to be identified. As a next step one should consider simple hydrogenation reactions, such as, for example, hydrogen peroxide formation from H_2 and O_2 catalyzed by the supported gold clusters. We can suggest that structures active for H_2 dissociation, but not active enough for O_2 dissociation could work as catalysts for H_2O_2 synthesis.

In many cases we have found an ensemble of initial and final configurations of the adsorbed hydrogen with small differences in binding energies, but possible difference in the reaction barriers. Accounting for the geometry of the initial and final states can lead to further non-equivalence of different reaction pathways and may have an important impact on the change of system entropy. It has been argued[93–95] that accounting for an entropy change contribution to the free energy of the system is necessary for correct description of the temperature dependencies of the branching ratios between different reaction channels, while purely energetic considerations based on reaction barrier heights can fail.

Finally, we have considered the pure, defect free rutile $TiO_2(110)$ surface. It is well known that the surface defects, especially O vacancies in metal-oxide supports, or extra O atoms adsorbed on the surface can change drastically the catalytic activity of the supported gold clusters in oxidation reactions by molecular oxygen.[20,96] Interaction of small gold clusters with a partially reduced rutile $TiO_2(110)$ support can be rather complicated; supported gold clusters can either receive or donate electrons to the substrate depending on the cluster size.[97] The role of the surface defects in the process of H_2 dissociation is not so obvious and requires additional systematic investigation. Such defects, in particular additional O atoms adsorbed on the surface in the vicinity of the cluster–surface interface can probably play a role in H_2 dissociation and hence can modify the cluster reactivity. Understanding how to control such chemical reactions on a cluster surface is a vital task for nanocatalysis.

We hope that the present discussion will stimulate further experimental and theoretical investigations of the processes considered.

Acknowledgements

This work was supported by the Global COE Program (Project No. B01: Catalysis as the Basis for Innovation in Materials Science) from the Ministry of Education, Culture, Sports, Science and Technology, Japan; the Grant-in-Aid for the Project on Strategic Utilization of Elements and the JSPS Grant-in-Aid for Scientific Research C. The computations were performed using the Research Center for Computational Science, Okazaki, Japan.

References

1 M. Haruta, T. Kobayashi, H. Sano and N. Yamada, *Chem. Lett.*, 1987, **16**, 405–408.
2 M. Haruta, *Chem. Rec.*, 2003, **3**, 75–87.
3 D. T. Thompson, *Nanotaday*, 2007, **2**, 40–43.

4 M. Haruta, *Catal. Today*, 1997, **36**, 153–166.
5 M. Turner, V. B. Golovko, O. P. H. Vaughan, P. Abdulkin, A. Berenguer-Murcia, M. S. Tikhov, B. F. G. Johnson and R. M. Lambert, *Nature*, 2008, **454**, 981–984.
6 M. D. Hughes, Y.-J. Xu, P. Jenkins, P. McMorn, P. Landon, D. I. Enache, A. F. Carley, G. A. Attard, G. J. Hutchings, F. King, E. H. Stitt, P. Johnston, K. Griffin and C. J. Kiely, *Nature*, 2005, **437**, 1132–1135.
7 H. Tsunoyama, H. Sakurai, Y. Negishi and T. Tsukuda, *J. Am. Chem. Soc.*, 2005, **127**, 9374–9375.
8 P. Landon, P. J. Collier, A. J. Papworth, C. J. Kiely and G. Hutchings, *Chem. Commun.*, 2002, 2058–2059.
9 A. Lyalin and T. Taketsugu, *J. Phys. Chem. C*, 2009, **113**, 12930–12934.
10 A. Lyalin and T. Taketsugu, *J. Phys. Chem. Lett.*, 2010, **1**, 1752–1757.
11 A. A. Herzing, C. J. Kiely, A. F. Carley, P. Landon and G. J. Hutchings, *Science*, 2008, **321**, 1331–1335.
12 M. J. Rodríguez-Vázquez, M. C. Blanco, R. Lourido, C. Vázquez-Vázquez, E. Pastor, G. A. Planes, J. Rivas and M. A. López-Quintela, *Langmuir*, 2008, **24**, 12690–12694.
13 *Nanocatalysis*, ed. U. Heiz and U. Landman, Springer, Berlin, Heidelberg, New York, 2007.
14 B. Hvolbæk, T. V. W. Janssens, B. S. Clausen, H. Falsig, C. H. Christensen and J. K. Nørskov, *Nanotoday*, 2007, **2**, 14–18.
15 S. M. Lang, T. M. Bernhardt, R. N. Barnett, B. Yoon and U. Landman, *J. Am. Chem. Soc.*, 2009, **131**, 8939–8951.
16 U. Landman, B. Yoon, C. Zhang, U. Heiz and M. Arenz, *Top. Catal.*, 2007, **44**, 145–158.
17 C. Harding, V. Habibpour, S. Kunz, A. N.-S. Farnbacher, U. Heiz, B. Yoon and U. Landman, *J. Am. Chem. Soc.*, 2009, **131**, 538–548.
18 J. A. Rodríguez, L. Feria, T. Jirsak, Y. Takahashi, K. Nakamura and F. Illas, *J. Am. Chem. Soc.*, 2010, **132**, 3177–3186.
19 R. Coquet, K. L. Howard and D. J. Willock, *Chem. Soc. Rev.*, 2008, **37**, 2046–2076.
20 A. Sanchez, S. Abbet, U. Heiz, W.-D. Schneider, H. Hakkinen, R. N. Barnett and U. Landman, *J. Phys. Chem. A*, 1999, **103**, 9573–9578.
21 M. Daté, M. Okumura, S. Tsubota and M. Haruta, *Angew. Chem., Int. Ed.*, 2004, **43**, 2129–2132.
22 D. Matthey, J. G. Wang, S. Wendt, J. Matthiesen, R. Schaub, E. Lægsgaard, B. Hammer and F. Besenbacher, *Science*, 2007, **315**, 1692–1696.
23 A. M. Joshi, W. N. Delgass and K. T. Thomson, *J. Phys. Chem. B*, 2006, **110**, 23373–23387.
24 A. Lyalin and T. Taketsugu, *J. Phys. Chem. C*, 2010, **114**, 2484–2493.
25 J. Jia, K. Haraki, J. N. Kondo, K. Domen and K. Tamaru, *J. Phys. Chem. B*, 2000, **104**, 11153–11156.
26 T. V. Choudhary, C. Sivadinarayana, A. K. Datye, D. Kumar and D. W. Goodman, *Catal. Lett.*, 2003, **86**, 1.
27 J. E. Bailie and G. J. Hutchings, *Chem. Commun.*, 1999, 2151–2152.
28 S. Schimpf, M. Martin Lucas, C. Mohra, U. Rodemerck, A. Brückner, J. Radnik, H. Hofmeister and P. Claus, *Catal. Today*, 2002, **72**, 63–78.
29 M. Okumura, T. Akita and M. Haruta, *Catal. Today*, 2002, **74**, 265–269.
30 C. Mohr, H. Hofmeister, J. Radnik and P. Claus, *J. Am. Chem. Soc.*, 2003, **125**, 1905–1911.
31 R. Zanella, C. Louis, S. Giorgio and R. Touroude, *J. Catal.*, 2004, **223**, 328–339.
32 P. Claus, *Appl. Catal., A*, 2005, **291**, 222–229.
33 Y. Zhu, H. Qian, B. A. Drake and R. Jin, *Angew. Chem., Int. Ed.*, 2010, **49**, 1295–1298.
34 Y. Zhu, Z. Wu, C. Gayathri, H. Qian, R. R. Gil and R. Jin, *J. Catal.*, 2010, **271**, 155–160.
35 E. Bus, J. T. Miller and J. A. van Bokhoven, *J. Phys. Chem. B*, 2005, **109**, 14581–14587.
36 C. Mohr, H. Hofmeister and P. Claus, *J. Catal.*, 2003, **213**, 86–94.
37 T. Fujitani, I. Nakamura, T. Akita, M. Okumura and M. Haruta, *Angew. Chem., Int. Ed.*, 2009, **48**, 9515–9518.
38 L. Andrews, X. Xuefeng Wang, L. Manceron and K. Balasubramanian, *J. Phys. Chem. A*, 2004, **108**, 2936–2940.
39 L. Andrews, *Chem. Soc. Rev.*, 2004, **33**, 123–132.
40 A. Zanchet, O. Roncero, S. Omar, M. Paniagua and A. Aguado, *J. Chem. Phys.*, 2010, **132**, 034301-1–10.
41 S. A. Varganov, R. M. Olson, M. S. Gordon, G. Mills and H. Metiu, *J. Chem. Phys.*, 2004, **120**, 5169–5175.
42 M. Okumura, Y. Kitagawa, M. Haruta and K. Yamaguchi, *Appl. Catal., A*, 2005, **291**, 37–44.
43 L. Barrio, P. Liu, J. A. Rodríguez, J. M. Campos-Martín and J. L. G. Fierro, *J. Chem. Phys.*, 2006, **125**, 164715-1–5.
44 H. W. Ghebriel and A. Kshirsagar, *J. Chem. Phys.*, 2007, **126**, 244705-1–9.
45 A. M. Joshi, W. N. Delgass and K. T. Thomson, *Top. Catal.*, 2007, **44**, 27–39.

46 A. Corma, M. Boronat, S. González and F. Illas, *Chem. Commun.*, 2007, 3371–3373.
47 D. H. Wells, Jr, W. N. Delgass and K. T. Thomson, *J. Catal.*, 2004, **225**, 69–77.
48 K. A. Kacprzak, J. Akola and H. Häkkinen, *Phys. Chem. Chem. Phys.*, 2009, **11**, 6359–6364.
49 Z. Li, Z.-X. Chen, X. He and G.-J. Kang, *J. Chem. Phys.*, 2010, **132**, 184702-1–5.
50 X.-F. Yang, A.-Q. Wang, Y.-L. Wang, T. Zhang and J. Li, *J. Phys. Chem. C*, 2010, **114**, 3131–3139.
51 M. Boronat, F. Illas and A. Corma, *J. Phys. Chem. A*, 2009, **113**, 3750–3757.
52 E. Florez, T. Gomez, P. Liu, J. A. Rodríguez and F. Illas, *Chem. Cat. Chem.*, 2010, **2**, 1219–1222.
53 J. P. Perdew, K. Burke and M. Ernzerhof, *Phys. Rev. Lett.*, 1996, **77**, 3865–3868.
54 E. Artacho, D. Sánchez-Portal, P. Ordejón, A. García and J. M. Soler, *Phys. Status Solidi B*, 1999, **215**, 809–817.
55 J. Junquera, O. Paz, D. Sánchez-Portal and E. Artacho, *Phys. Rev. B: Condens. Matter*, 2001, **64**, 235111-1–9.
56 N. Troullier and J. L. Martins, *Phys. Rev. B: Condens. Matter*, 1991, **43**, 1993–2006.
57 L. Kleinman and D. M. Bylander, *Phys. Rev. Lett.*, 1982, **48**, 1425–1428.
58 D. Sánchez-Portal, P. Ordejón, E. Artacho and J. M. Soler, *Int. J. Quantum Chem.*, 1997, **65**, 453–461.
59 J. M. Soler, E. Artacho, J. D. Gale, A. García, J. Junquera, P. Ordejón and D. Sánchez-Portal, *J. Phys.: Condens. Matter*, 2002, **14**, 2745–2779.
60 D. Sánchez-Portal, P. Ordejón and E. Canadell, *Structure and Bonding*, 2004, **113**, 103–170.
61 K. P. Huber and G. Herzberg, *Molecular Spectra and Molecular Structure Constants of Diatomic Molecules*, Van Nostrand Reinhold, New York, 1979.
62 B. Morosin, *J. Appl. Crystallogr.*, 1968, **1**, 123–124.
63 P. Haas, F. Tran and P. Blaha, *Phys. Rev. B*, 2009, **79**, 085104-1–10.
64 X. Ding, Z. Li, J. Yang, J. G. Hou and Q. Zhu, *J. Chem. Phys.*, 2004, **120**, 9594–9600.
65 E. Fernández, J. M. Soler, I. L. Garzón and L. C. Balbás, *Phys. Rev. B: Condens. Matter Mater. Phys.*, 2004, **70**, 165403-1–14.
66 A. W. Walker, *J. Chem. Phys.*, 2005, **122**, 094310-1–12.
67 L. Xiao, B. Tollberg, X. Hu and L. Wang, *J. Chem. Phys.*, 2006, **124**, 114309-1–10.
68 H. Häkkinen, *Chem. Soc. Rev.*, 2008, **37**, 1847–1859.
69 J. A. Nelder and R. Mead, *The Computer Journal*, 1965, **7**, 308–313.
70 J. K. Burdett, T. Hughbanks, G. J. Miller, J. W. Richardson, Jr and J. V. Smith, *J. Am. Chem. Soc.*, 1987, **109**, 3639–3646.
71 J. Muscat, V. Swamy and N. M. Harrison, *Phys. Rev. B*, 2002, **65**, 224112-1–15.
72 H. J. Monkhorst and J. D. Pack, *Phys. Rev. B: Solid State*, 1976, **13**, 5188.
73 H. Perron, C. Domain, J. Roques, R. Drot, E. Simoni and H. Catalette, *Theor. Chem. Acc.*, 2007, **117**, 565–574.
74 T. Bredow, L. Giordano, F. Cinquini and G. Pacchioni, *Phys. Rev. B*, 2004, **70**, 035419-1–6.
75 M. Ramamoorthy, D. Vanderbilt and K.-S.R.D., *Phys. Rev. B: Condens. Matter*, 1994, **49**, 16721–16727.
76 L. M. Liu, B. McAllister, H. Q. Ye and P. Hu, *J. Am. Chem. Soc.*, 2006, **128**, 4017–4022.
77 Y. Wang and X. G. Gong, *J. Chem. Phys.*, 2006, **125**, 124703-1–12.
78 G.-J. Kang, Z.-X. Chen, Z. Li and X. He, *J. Chem. Phys.*, 2009, **130**, 034701-1–6.
79 A. Zanchet, A. Dorta-Urra, O. Roncero, F. Flores, C. Tablero, M. Paniagua and A. Aguado, *Phys. Chem. Chem. Phys.*, 2009, **11**, 10122–10131.
80 P. Pyykkö, *Chem. Soc. Rev.*, 2008, **37**, 1967–1997.
81 H. Häkkinen, B. Yoon, U. Landman, X. Li, H. Zhai and L. Wang, *J. Phys. Chem. A*, 2003, **107**, 6168–6175.
82 B. Assadollahzadeh and P. Schwerdtfeger, *J. Chem. Phys.*, 2009, **131**, 064306-1–11.
83 H. Häkkinen, M. Moseler and U. Landman, *Phys. Rev. Lett.*, 2002, **89**, 033401-1–4.
84 R. M. Olson, S. Varganov, M. S. Gordon, H. Metiu, S. Chretien, P. Piecuch, K. Kowalski, S. A. Kucharski and M. Musial, *J. Am. Chem. Soc.*, 2005, **127**, 1049–1052.
85 Y.-K. Han, *J. Chem. Phys.*, 2006, **124**, 024316-1–3.
86 Y. C. Choi, W. Y. Kim, H. M. Lee and K. S. Kim, *J. Chem. Theory Comput.*, 2009, **5**, 1216–1223.
87 A. Martínez, *J. Phys. Chem. C*, 2010, **114**, 21240–21246.
88 A. Lyalin and T. Taketsugu, *AIP Conf. Proc.*, 2009, **1197**, 65–75.
89 V. V. Semenikhina, A. Lyalin, A. V. Solov'yov and W. Greiner, *J. Exp. Theor. Phys.*, 2008, **106**, 678–689.
90 D. Ricci, A. Bongiorno, G. Pacchioni and U. Landman, *Phys. Rev. Lett.*, 2006, **97**, 036106-1–4.
91 R. Bader, *Atoms in Molecules: A Quantum Theory*, Oxford University Press, New York, 1990.

92 G. Henkelman, A. Arnaldsson and H. Jònsson, *Comput. Mater. Sci.*, 2006, **36**, 354–360.

93 C. Bréchignac, P. Cahuzac, M. de Frutos, N. Kébaïli, A. Sarfati and V. Akulin, *Phys. Rev. Lett.*, 1996, **77**, 251–254.

94 C. Bréchignac, P. Cahuzac, N. Kébaïli and J. Leygnier, *Phys. Rev. Lett.*, 1998, **81**, 4612–4615.

95 O. I. Obolensky, A. G. Lyalin, A. V. Solov'yov and W. Greiner, *Phys. Rev. B: Condens. Matter Mater. Phys.*, 2005, **72**, 085433-1–11.

96 B. Yoon, H. Häkkinen, U. Landman, A. S. Wörz, J.-M. Antonietti, S. Abbet, K. Judai and U. Heiz, *Science*, 2005, **307**, 403–407.

97 S. Chrétien and H. Metiu, *J. Chem. Phys.*, 2007, **126**, 104701-1–7.

General discussion

Professor Van Santen opened the discussion of the paper by Professor Hu: The model calculations of the CO shift reaction are presented for one gold particle size. The essential elementary steps are proposed to occur at the interphase of metal particle and support. Do you have information how reaction energy barriers will change when Au particle size is varied? Would you expect large changes in adsorption energies? The micro kinetics results do also depend sensitively on values used for the activation entropies of the adsorption of reagents and desorption of products. Also the values used for the sticking coefficients of adsorption are relevant. Which data did you use?

Professor Hu answered: We investigated a key step (CO + OH → COOH) on several different Au clusters and found that the activation barrier is the Au-cluster-dependent. The CO adsorption energy is also related to the Au cluster size, but mainly depending on the Au coordination number. However, detailed studies on the Au cluster size for elementary-step barriers in general are beyond the scope of this paper. Regarding the barriers of adsorption and desorption, we tried several approaches including using experimental sticking coefficients. It was found that our mechanism was much more favoured irrespective of which approach was utilised.

Professor Golunski said: Do your models cast any more light on the mechanism by which the catalyst deactivates, particularly in the presence of high concentrations of H_2O and CO_2?

Professor Hu responded: In this work, we focused on the WGS mechanism. We have not systematically investigated the deactivation in the system.

Professor Bowker commented: I am surprised by some aspects of the reaction profile you propose. For instance, there is a very large summed heat of adsorption of CO and H_2O. What is the value for CO adsorption on gold?

Professor Hu responded: The chemisorption energy of CO is about 1 eV in total energy. But in the term of free energy, the CO chemisorption is small.

Professor Bond said: I do not find your mechanism in Figure 6 in the paper totally satisfying; basically, it is too long. The more steps you write in a mechanism, the slower it will be. Let me however start with a comment on your rejection on the basis of DFT calculations of the decomposition of carboxyl as Au–COOH → Au–H + CO_2 (1) in favour of one involving a second water molecule. *viz.* Au–COOH + Au–OH → CO_2 + Au + Au–H + H_2O (2) You are not comparing like with like, because in the second process the carboxyl has to await the arrival and decomposition of a water molecule; all this takes time, so the rates of these reactions are not simply given by the energetics of the ultimate steps. Incidentally you do not allow for the possibility that arises early in the mechanism for the process Au–COOH + Au–H → CO_2 + H_2 + 2Au (3) In the mechanism I have published,[1] I opted for process 1, but it is easily modified to include process 3 instead.

1 G. C. Bond, *Gold Bulletin*, 2009, **42**, 337.

Professor Hu replied: It is incorrect that we did not compare like with like. When we differentiated one pathway from others, we did compare their rates, in which not only the barriers but also concentrations were considered. I agree that if a mechanism is too long, it will be slow and possibly it is incorrect. However, I do not think that our mechanism is too long compared to other mechanisms in the literature. As we discussed in our paper, the barrier of reaction 2 is nearly zero and is much smaller than step 1. Our kinetic analyses showed that under the steady state, the rate of step 2 is much faster than step 1. Step 3 was considered and it was found that it is slower than step 2.

Dr Xu asked: In the proposed mechanism the oxygen vacancy in ceria at the Au–ceria interface plays a very important role in activating water (see reaction #9 in the paper), something that does not readily occur on Au itself. Does this imply that those Au clusters that do not have oxygen vacancies at the interface will therefore be inactive for the WGS reaction? Are all Au clusters expected to have oxygen vacancies at the Au–ceria interface?

Professor Hu responded: If there is no oxygen vacancy at the interface of a Au cluster, it is inactive. According to our calculations, not all the Au clusters have oxygen vacancies at their interfaces.

Professor Meyerstein commented: In the first step in the reaction scheme you propose an O–H bond is broken. What does this imply to the strength of the Au–H bond formed?

Professor Hu answered: According to our calculations, the Au–H bond is weak and the reaction energy mainly comes from OH adsorption on the O vacancy.

Dr Ghiringhelli asked: In the kinetic analysis of the mechanisms, you correctly introduce the pressure of the gas phase species and the environmental temperature. Don't you think that pressure and temperature are also fundamental to understand which species are actually likely to be present on the surface, on thermodynamics grounds? In other words, if pressure and temperature are such that certain species that you consider are actually instable, is it consistent that pressure and temperature enter the estimates of the transition rates among species but not the selection of the species?

Professor Hu responded: The following is my answer, if I understand your question correctly: In this work, we considered all the possible intermediates on the surface under the reaction conditions. Indeed, the pressure and temperature are important for the reaction mechanisms. Using the experimental condition as a reference, we checked the temperature and pressure over a reasonable range on the mechanisms and found that our mechanism is always more favourable than other mechanisms.

Professor Hutchings asked: As an experimentalist I am always interested in testing proposed new mechanisms in some way. In the present case this could be done by co-feeding possible intermediate structures. For example one could co-feed HCOOH as this might permit access to the COOH intermediate in your carboxyl reaction mechanism. Is this an approach worth pursuing?

Professor Hu replied: Maybe. It is worth mentioning that some important intermediates may have very short lives during the reaction, while many unimportant intermediates may have very long lives in catalytic systems (essentially they are spectators). This fact may give rise to difficulties to study reaction mechanisms experimentally.

Professor Hutchings asked: You mentioned in your response to this discussion that trying to determine the presence of short lived intermediates would be difficult

   This journal is © The Royal Society of Chemistry 2011

and I totally agree. I would advocate an experimental approach in which the reacting system is perturbed by using pulse experiments. There should be experimental approaches to test the hypothesis of the reaction mechanism you are proposing.

Professor Hu replied: Possibly. According to our calculations, COOH which is an important species in our mechanism should decompose very quickly and its life should be very short in the reaction system.

Professor Friend asked: You considered a limited number of models, not allowing, *e.g.* for changes in catalyst structure. On the other hand, thermodynamic structures were considered. So, in other words, the models are constrained to specific cases. How do you generalize further than your model to be sure you have the correct model?

Professor Hu answered: I should say that this is indeed a general issue in the field of modelling. For the system we studied, we did some calculations using other models, from which we are convinced that our model is correct. But the detailed results are to be published.

Professor Van Santen opened the discussion of the paper by Dr Willock: The promotion of O_2 dissociation at the Au/Fe_2O_3 interphase appears to be related to its stronger interaction than on Au. How does dissociative adsorption to the Fe_2O_3 surface compare? Does the answer imply that the promotional effect is mainly steric? O_2 dissociation implies the activation of a molecule in the triplet state. What happens to spin conservation? Is one of the reasons for the beneficial role of Fe^{3+} the possibility of low energy spin-flip in the oxidation reaction?

Dr Willock replied: You are correct that the interface region offers stronger interactions with the dissociation products than does the isolated cluster. This appears to be because the O species produced are anionic in nature and so can interact with the Fe^{3+} centres electrostatically. The distance from Fe^{3+} to the anionic O atoms that are produced is actually quite similar to the lattice spacing for Fe–O in the oxide. We did not test the dissociation of O_2 on the Fe_2O_3 surface itself and I agree that would be a useful calculation to add to these results. As for the spin state of this system. The O_2 in the gas phase is, of course, a triplet ground state. However, molecular adsorption to the Au cluster (either in isolation or supported) results in donation of an electron from the cluster to the molecule to give a doublet. We think that the spin conservation during the actual cleavage of the bond occurs through a similar interaction (*i.e.* with the Au cluster).

We have not considered the dissociative adsorption of O_2 on the bare Fe_2O_3 surface and I agree that this would be a useful check on the roles played by the oxide surface. The O_2 spin state is a triplet in the gas phase but adsorption to the Au cluster, either in isolation or supported, leads to electron transfer from the metal to the molecule. The molecular adsorbed state is then a O_2^- species in a doublet spin state. In the calculations we use a spin unrestricted approach and the VASP code then also optimises the difference between up and down spin densities. For either the isolated cluster or the supported case the required spin change occurs smoothly and so we think that the Au cluster mediates the required spin changes.

Professor Campbell said: Thank you for the interesting talk. You calculated that there is a lower barrier for the dissociation of O_2 than for its desorption on the Au cluster. If this were the case, then if one dosed O_2 to such a Au cluster even in ultra-high vacuum, most of the transiently adsorbed O_2 molecules would dissociate instead of desorbing, and adsorbed O would build up quickly on the Au cluster surface (*i.e.*, after dosing only a few Langmuirs of O_2). This seems inconsistent with the inability of many investigators to see O buildup on such Au particle surfaces in ultrahigh vacuum surface science studies with even much larger doses, at least for Au on other oxides. Do you know if this has been tried for Au on Fe_2O_3 model (or real) catalysts? Is there something very special about Fe_2O_3 in this respect?

Dr Willock answered: I am not sure that the process described in our paper would lead to a build up of surface O species. The dissociation we observe is specific to the periphery of the cluster where the O atoms interact with both the Au atoms and surface Fe cations. These sites will be limited in the supported catalyst and may become saturated quite quickly on exposure to oxygen. Charge analysis suggests that these O atoms are quite ionic and that the Au that is in between the two O atoms is oxidised. This is in contrast to the sort of surface O atoms that have been studied by Prof. C. M. Friend and others. In those cases the oxygen atoms sit in the three fold hollow sites of the metal surface and I would expect are not as strongly bound as those we see here. The calculations we have made show that the periphery sites of the Au clusters are quite easily oxidised. We are currently looking to see if the oxygen atoms in these sites can be used in oxidation reactions or if the edge of the cluster becomes oxidised and then the normal catalyst system contains hybrid Au oxide/Au metal particles. We are also looking at other oxide supports to try and answer your final point.

Professor Fortunelli asked: When you evaluate the energy barrier for O_2 dissociation on a supported cluster, you should consider the global minimum structure of the cluster on the surface. If you consider any other structure, then you should obviously add the energy difference between this structure and the global minimum to your energy barrier to get the real energy barrier. In our experience, we indeed found that you can decrease the energy barrier of some processes (including O_2 dissociation on Au clusters) by taking into account not the lowest-energy structure but a higher energy isomer.

Dr Willock answered: The shape of the cluster we use in this work is based on the experimental observation that clusters of this size supported on iron oxide form bilayer structures.[1] This information from atomic resolution STEM indicates that the DFT preferred 2D structure is not the one present on the catalyst, even under the vacuum conditions used in the STEM. The $Au_{10}(7,3)$ cluster used as the start point in all of our calculations is only 6.2 kJ mol^{-1} (per Au atom) higher in energy than the 2D structure at the theory level we use. We would argue that the bilayer is observed either as a result of an entropic factor or a surface/cluster interaction that favours the bilayer. This structure is not created from the 2D form as a result of the reaction process; it is just the shape of the clusters observed. This means that the correct reference state for the Au cluster in our calculations is the geometry optimised bi-layer.

We stress that in the calculations for the isolated cluster and for the supported cluster the same initial $Au_{10}(7,3)$ form is used. This is relaxed so that the interaction with the oxide can be taken into account prior to the adsorption/reaction of O_2. The free cluster is also optimised with only a constraint on the z-co-ordinate of the atoms in the 7-layer. This means that the Au clusters have a low energy form that is consistent with the experimental observation.

During the calculation of adsorption and reaction of O_2, the Au atom positions are again optimised at every stage and the results show that the ability of the Au cluster to distort in response to the presence of the O atoms is an important contribution to the calculated barriers.

1 A. A. Herzing, C. J. Kiely, A. F. Carley, P. Landon and G. J. Hutchings, *Science*, 2008, **321**, 1331.

Professor Hutchings asked: In the discussion of the small gold clusters it is apparent that there are many different isomers that could be present. In a real

   This journal is © The Royal Society of Chemistry 2011

catalyst we know that a very broad range of structures are present but only a few can be expected to be active. How fluxional are the small clusters once they are supported, are they still able to access many of the different structures that are available?

Dr Willock answered: We observed changes in the structure of the Au_{10} particle on adsorption to the oxide surface (figure 3 of the paper) and in response to the adsorption and dissociation of O_2 (figure 6 of the paper). In fact dissociation is accompanied by a Au atom moving out of the cluster base to insert between the two O atoms. This indicates to us that the flexibility of the cluster plays a role in its reactivity.

Dr Xu asked: Does the Hubbard U parameter affect the adsorption and activation of oxygen on Fe_2O_3-supported Au clusters?

The barrier for O_2 dissociation on an isolated Au_{10} cluster (1.5 eV) appears too large compared to what has been reported for Au surfaces in the literature. Could you please double-check this result?

Dr Willock answered: In response to your first point. The Hubbard U parameter was only applied for the Fe centres in our calculations. This was done to ensure the band gap of the bulk material and other properties of the oxide were well reproduced. Working without a U parameter gives very different results for even the Fe_2O_3 lattice and so we did not consider calculations without U on the Fe atoms.

The barrier we obtain on the isolated cluster is 1.46 eV relative to the adsorbed O_2 molecule. This is for the equivalent process that we observe for the supported Au cluster, *i.e.* the dissociation is accompanied by oxidation of Au at the periphery of the cluster. We do state in the introduction that lower barriers on isolated clusters have been observed.[1] However in these calculations the cluster geometry is not optimised and the O species produced are more like adatoms than anions.

1 A. Roldan, S. Gonzalez, J. M. Ricart and F. Illas, *Chem. Phys. Chem.*, 2009, **10**, 348.

Professor Campbell remarked: Is there a barrier for the molecular adsorption of O_2 on this cluster?

Dr Willock answered: We did not calculate the molecular adsorption process itself but would expect any barrier to be small compared to the molecular adsorption energy that we use to estimate the desorption barrier.

Dr Jupille said: In this study of the activation of oxygen by gold supported on the (0001) surface of hematite, the Fe termination is preferred (FeO_3Fe_2-, down the c direction). Generally speaking, the termination of $Fe_2O_3(0001)$ is predicted to depend on the chemical potential of oxygen.[1,2] Numerical simulations performed by density functional theory (DFT) using the generalized gradient approximation (GGA) predict a O3 termination at high chemical potential, either on a bulk-like geometry $(O_3Fe_2O_3Fe_2-)$[1] or on a specific geometry $(O_3Fe_3O_3Fe-)$.[2] At lower chemical potential, a ferryl termination $(OFeO_3-)$ appears.[1,2] The above iron termination (FeO_3Fe_2-) is found at very low chemical potential.[1,2] Calculations performed by adding an on-site Coulomb repulsion U (GGA+U method)[2] leads to very different results. The Fe_2O_3- termination is seen to compete with FeO_3- at low chemical potential and $OFeO_3-$ is observed in the limit of the high chemical potential. The O3 termination no longer appears.[2] An experimental analysis performed by Surface X-Ray Diffraction on a natural hematite crystal[3] shows a rather good agreement with the GGA calculation. The Fe_2O_3- termination is observed at high chemical potential, followed by the ferryl termination as the chemical potential decreases. A $O_2Fe_2O_3-$ termination is characterized at lower chemical potential (instead of a Fe-terminated surface).[3] Therefore, in conditions required for catalytic oxidation,

DFT–GGA and diffraction experiments both favor either O_3 or ferryl termination for the hematite surface.

1 W. Bergermayer, H. Schweiger, and E. Wimmer, *Phys. Rev. B*, 2004, **69**, 195409.
2 A. Rohrbach, J. Hafner and G. Kresse, *Phys. Rev. B*, 2004, **70**, 125426. (Reference 27)
3 A. Barbier, A. Stierle, N. Kasper, M.-J. Guittet and J. Jupille, *Phys. Rev. B*, 2007, **75**, 233406.

Dr Willock replied: In our work we took the GGA+U approach you mention from the reference of Rohrbach *et al.* This gives a much better description of the bulk structure and band gap than the GGA method alone. Our simulations were part of a combined TAP, XPS and electron microscopy experimental study of O_2 and CO adsorption on Au supported on iron oxide which was reported in an earlier publication.[1] We were aware of the discussion of the preferred termination of the iron oxide surface and took the FeO_3Fe_2-termination because in the experimental studies the catalyst is only briefly exposed to the reactant mixture and is held under vacuum otherwise. We do comment that the dissociation of the oxygen we observe may be relevant only at early times in a reaction before the catalyst has reached equilibrium. The important point is that these calculations do infer that the catalyst will have oxidised Au atoms at the periphery of the particles even under steady state conditions.

1 A. F. Carley, D. J. Morgan, N. Song, M. W. Roberts, S. H. Taylor, J. K. Bartley, D. J. Willock, K. L. Howard and G. J. Hutchings, *Phys. Chem. Chem. Phys.*, 2011, **13**, 2538, ref. 20 in our paper.

Professor Campbell asked: We have shown that O adatoms on both bulk Au and tiny Au nanoparticles on $TiO_2(110)$ react very rapidly with CO gas, at rates per CO collision with the surface that are >5 orders of magnitude faster than the TOF for CO oxidation.[1] So if O_2 really dissociated so rapidly on Au nanoparticles, it would have a much higher CO oxidation rate than is observed. Doesn't this imply that O_2 does not really dissociate so easily on the Au, and must be as slow as or slower than the rate of CO oxidation? Note too that O adatoms do not desorb as O_2 until much higher temperatures (as also shown in the following paper), so desorption is not a kinetically competitive process.

1 V. Bondzie, S. C. Parker and C. T. Campbell, *Catalysis Letters*, 1999, **63**, 143–151.

Dr Willock responded: From the TPD data in the paper you reference you have estimated the adsorption energy for Oa to be 124.7 kJ mol^{-1} for the smallest particles that were made on a $Au/TiO_2(110)$ material. In our calculations the estimated adsorption energy for the O species created when the periphery of the particles is oxidised is some -250 kJ mol^{-1}. Even given that we are using different oxides this is a large difference and suggests that we are not studying the same surface Oa atoms you observe. In the desorption experiments you would not expect oxygen that is this tightly bound to desorb and so it would not be seen. We are still working on calculations to estimate if CO titration would allow its recovery.

Professor Van Santen opened the discussion of the paper by Dr Beret†: Computed DFT energy differences of gas phase clusters are less than 10 kJ mol^{-1}. This is less than the accuracy of DFT calculations. How can they then be reliably used also in calculations of the relative free energies?

Dr Ghiringhelli responded: Stating that the accuracy of DFT calculations is less than 10 kJ mol^{-1} is a bit simplistic. It depends on the functional and on the system. DFT functionals are grouped in levels (the so-called "Jacob's ladder") from the least

---

† Dr Beret's paper was presented by Dr Ghiringhelli, Fritz-Haber-Institut der Max-Planck-Gesellschaft, Faradayweg 4-6, D-14195 Berlin, Dahlem, Germany.

     This journal is © The Royal Society of Chemistry 2011

accurate (LDA) to the most accurate (exact exchange plus RPA for the correlations). At the top of the ladder the accuracy is better than 10 kJ mol^{-1}. One has to be reminded that with the exact exchange and correlation functional, DFT is an exact theory for the electronic ground state. For our paper, we have used a GGA functional, PBE, *i.e.* belonging to just the second rank of the ladder, plus vdW corrections. In general the accuracy of PBE can be summarized by the cited 10 kJ mol^{-1} figure, but the vdW tail correction greatly improves its performance. In particular, with our settings we produce finite temperature vibrational spectra that compare very well with experimental far-IR multiple photo dissociation spectra (paper to appear soon). This suggests a good description not only of the minima of the PES, but also of the surroundings, where the system spends the most of its time. Inaccuracies of the barriers between basins, even hypothetically large ones, would be of negligible importance for these equilibrium properties.

What it is most important for our paper is that our conclusion can well stand a 10 kJ mol^{-1} inaccuracy. Indeed, the fact that the lowest energy 2D structures, nearly degenerated in energy with the lowest energy 3D isomers, are free-energetically less favored, comes from the shape of the free-energy landscape. This conclusion comes from the fact that 3D structures can easily transform from one to the other, while the 2D ones are much stiffer. In the poster that we presented at the workshop, we also show that the 3D structures that we find have a finite temperature vibrational spectrum in good agreement with the experiment, and the 2D structures that we exclude have spectra non compatible with the experimental measure.

Professor Friend commented: I have a question about thermodynamics *vs.* kinetics? You studied thermodynamic structures, but did not probe TS and possible transients. A classic problem in catalysis is that minority and short-lived species might be the most active. Catalysis uses kinetics to determine selectivities, so how do you justify not including kinetics? Will you do kinetic Monte Carlo or molecular dynamics simulations?

Dr Ghiringhelli replied: Kinetics does not overcome thermodynamics. After all, a catalyst can only change the reaction barriers, not the relative (free-)energies of the (*meta*-)stable structures. Our thermodynamic analysis tells us where the system would like to go, given enough time and if product species would not leave the clusters (remember that our structures include also clusters with reaction products adsorbed on them). Thus, there is a thermodynamic driving force towards the stoichiometries we identify as the most stable ones. Whatever happens, the system aims to our calculated "catalysts" and "reaction intermediates" (following their definition as given in the paper).

It is true that we might miss, in our reaction cycle based on thermodynamics, some unstable short lived species that might play an important role in the reaction cycle. For calculating the reaction barriers (using accelerated MD techniques) we plan to include all possible sequences of adsorption (and desorption) of ligands, yet using as beacons the thermodynamically stable structures. After rates for the elementary steps are known, kMC, as a numerical solver of the master equation related to the catalytic cycle, will tell us which steps are important.

Prof Dr Bernhardt said: In the CO oxidation mechanism that you present in your paper you emphasize the importance of the gold carbonyl in the catalytic reaction cycle. In particular, it appears that the CO adsorption precedes the adsorption of molecular oxygen on the gold clusters. We performed gas phase reaction studies with negatively charged gold clusters, in which the adsorption of molecular oxygen clearly was shown to be the first step and even more, the oxygen pre-adsorption enhanced the CO co-adsorption (see Fig. 1[1,2]). Could you please comment on the

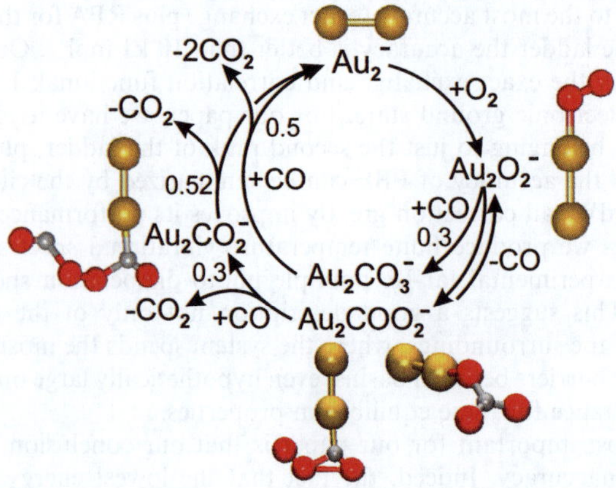

Fig. 1 Schematic representation of the gas-phase catalytic cycle for the oxidation of carbon monoxide by gold dimer anions, based on the reaction mechanism determined by kinetic measurements in conjunction with first principles simulations. The numbers denote calculated energy barriers (in eV). Also displayed are geometrical structures of reactants and intermediate products according to the calculations (large yellow spheres: Au; small grey spheres: C; red spheres: O).[1]

sequence of carbon monoxide and oxygen adsorption, perhaps based on the relative binding energies of these molecules to the gold clusters?

1 L. D. Socaciu, J. Hagen, T. M. Bernhardt, L. Wöste, U. Heiz, H. Häkkinen, U. Landman, *J. Am. Chem. Soc.*, 2003, **125**, 10437.
2 J. Hagen, L. D. Socaciu, M. Elijazyfer, U. Heiz, T. M. Bernhardt, L. Wöste, *Phys. Chem. Chem. Phys.*, 2002, **4**, 1707.

Dr Ghiringhelli responded: As indicated in the text, our proposed reaction cycle has been constructed by always putting the step that was energetically more favored. That cycle must be intended as illustrative of our thermodynamic understanding of the system. In particular, we want to emphasize the so-called (by us) "reaction intermediates" and "catalysts", as milestone that the reaction cycle should contain. Finding the actual preferred ordering of successive adsorptions will be clear after a kinetic analysis which is in progress. For the moment, we know that typically CO adsorption is free-energetically preferred *versus* oxygen adsorption. The fact that gold carbonyl species are not detected in your experiment may not mean that these species are not favored. It can be that they are consumed very fast during the cycle, while oxygen containing species are less reactive.

We are currently investigating also the anionic dimer in O_2 and CO atmosphere. We are currently still completing the analysis, but I can already say that we clearly find as "catalyst" (in our language) Au_2O_2CO-, which is one of the species that you observe in the chamber. It is remarkable how the thermodynamic analysis points directly to this structure. I cannot comment on the ordering of adsorption, O *vs.* CO, on the anionic cluster at the moment.

Professor Fortunelli commented: I understand you used a global optimization approach in your structural search. Did you use any constraint in this search to avoid producing the obvious global minimum. *i.e.*, CO oxidation to CO_2?

Dr Ghiringhelli answered: We have not used any constraint during the basin hopping search. After the scan was completed, we excluded, from the free-energy analysis, only those configurations in which CO_2 was formed and detached from the clusters (or interacting only *via* weak vdW interactions). All other minima

This journal is © The Royal Society of Chemistry 2011

were used for producing the phase diagrams. The phase diagram at high partial pressure of CO_2 shows indeed structures that contain CO_2 (and/or CO_3), but it is absolutely not trivial how many CO_2 moieties the most stable structures contain, and what is co-adsorbed on those clusters. In the phase diagram at low CO_2 pressure most of CO_2 containing structures are not present, but, again it is thermodynamics that tells us which structures are present, not a screening of "obvious" structures.

Ms Mukhamedzyanova remarked: You have said that Au_{13} cluster is fluxional and has a number of isomers, is this cluster stable or not? What is the energy gap of Au_{13}? What do you think, is such fluxionality good or bad in catalysis?

Dr Ghiringhelli replied: Au_{13} is (*meta*-)stable in the sense that it would not spontaneously dissociate into fragments (also upon ligand adsorption). It is fluxional, meaning that its structure undergoes some reconfiguration also in a relatively short time time-scale (few ps) at room temperature. The fact that many isomers are nearly degenerated in energy, means also that an ensemble of such clusters will show, instantaneously, a set of structures, none of them being by far the dominant one.

The calculated HOMO–LUMO gap for the structures we show is between 0.12 and 0.19 eV, where the 2D structures have the smallest gaps.

We believe that the fluxionality of this, and other gold clusters can be quite important for catalysis: The fact that a cluster (also with ligands) can rearrange its shape along short times, might ease the chemical reactions between ligands, for example, by bringing them closer than expected from a static single-structure approach.

Professor Bowker remarked: How relevant are clusters of this size for high activity CO oxidation? Is it possible that the real catalyst is not fluxional in structure?

Dr Ghiringhelli answered: We do not know yet whether these cluster sizes have high activity for CO oxidation. We will know this only after we will have calculated the reaction rates related to the reaction cycle we identified on the basis of *ab initio* atomistic thermodynamics. The choice of these small sizes (in the paper we report results only for Au_2, but we have results also for Au, Au_3, and Au_4, and we are collecting data for Au_7) is motivated by the possibility of having a complete screening of the possible structures. With increasing cluster sizes, it would be technically very difficult, if not impossible, to thoroughly scan the configurational phase space. One thing we have learned with these small sizes is that many of the most likely structures are quite unpredictable, due to the reach chemistry of gold clusters (and indeed many structures were missed by former more chemical-intuition based searches).

We do think that the "real catalyst" is fluxional, also considering that every molecular adsorption brings a sizable amount of energy that most likely goes into the vibrational degrees of freedom as a sort of local increase of temperature, which is likely to lead to reconfiguration of the cluster structure.

Professor Campbell said: It seems to me that heat of adsorption you calculate may be too high in some cases. It would be nice if you and the other theoreticians in this proceedings would include in their papers benchmark calculations of systems with known energies (in this case the bulk cohesive energy of solid Au, the heats of formation of bulk Au oxide and hydroxide, the heat of adsorption of CO on Au(111), *etc.*, to prove the accuracy of their methods.

Dr Ghiringhelli responded: I do not think that bulk and surface properties would be a good benchmark for assessing the accuracy of calculations regarding cluster-sized gold. The electronic properties for periodic systems are too different. The most notable difference is the finiteness of the HOMO–LUMO gap for clusters *versus* the metallic behavior of the bulk.

In our cluster case, we mainly compared our calculations to higher level calculations, namely at the CCSD(T) level, *i.e.* the so-called "gold standard of quantum chemistry". We found remarkable good agreement for pristine clusters' energetics and for the adsorption of CO. The adsorption of O_x (different stoichiometries) was found in general overestimated by our PBE+vdW functional. This does not affect qualitatively our phase diagrams (in particular the areas in which our predicted stable structures are the most stable is larger than the uncertainties in the calculations), while, there can be, quantitatively, a shift of the coexistence lines. These comparisons are quickly discussed in our paper, and will be more extensively described in a subsequent publication.

Professor Bowker continued the discussion of the paper by Professor Hu: 1eV adsorption heat for CO seems unreasonably high to me, and is not that dissimilar to Pd nanoparticles. I suspect that this is not a realistic binding energy. The point about Au for CO oxidation is that CO binding is low, so that it does not get self-poisoned in the reaction, as Pt, for instance, does.

Professor Hu responded: I should point out that the CO chemisorption on flat Au surfaces, say Au(111), is very weak. On the other hand, CO adsorption on low coordination sites of Au is much stronger than those on the flat surfaces, which is consistent with the work in the literature. For example, Liu *et al.* reported that the chemisorption energy of CO on four Au cluster is 1.22 eV.[1] Our previous work also showed that on Au surface defects (*e.g.* steps) the chemisorption energies of CO are similar to what we reported here.[2,3]

1 Z. P. Liu, S. J. Jenkins and D. A. King, *Phys. Rev. Lett.*, 2005, **94**, 4.
2 Z. P. Liu, P. Hu and A. Alavi, *J. Am. Chem. Soc.*, 2002, **124**, 14770.
3 H. F. Wang, X. Q. Gong, Y. L. Guo, Y. Guo, G. Z. Lu and P. Hu, *J. Phys. Chem.*, 2009, **113**, 6124.

Professor Bowker asked: Have you tried varying the initial heat of adsorption in the calculations (preferably to lower values) to see if this gives a change in the preferred mechanism? Don't forget that changing that initial value then changes the thermodynamics between the adsorbed states and final products and so can potentially change the rate limiting step.

Professor Hu answered: Yes, it showed that our mechanism is still much more preferred.

Professor Poliakoff said: You appeared to say that your proposed mechanism was very complicated and therefore could not be tested experimentally. Did I understand you correctly? If I did, it seems to be very worrying and a dead-end in the discussion of this reaction because much of the value of theoretical explanations lies in their ability to make predictions which can be tested experimentally.

Professor Hu responded: I did not say that our mechanism was very complicated. In fact, I believe that our mechanism is very simple. What I said was (in replying to Professor Hutchings) that in many systems in heterogeneous catalysis, some key intermediates are very short-lived. It may be very difficult to detect them using the current experimental techniques. It does not mean that it is not possible to test all the mechanisms proposed from theoretical work.

Professor Poliakoff asked: You propose that formic acid is an intermediate. Graham Hutchings' point seems to have been that you should try to predict what would happen if formic acid were fed into the system.

Professor Hu responded: We did not propose that formic acid is an intermediate.

Professor Poliakoff opened a general discussion of the papers by Professor Hu, Dr Willock and Dr Ghiringhelli: Theoretical analyses will only be of limited value unless they make one or more predictions that can be tested experimentally.

Dr Willock responded: I quite agree that an important goal of theory is to set up a model that can be tested. In our case the structure of the cluster was chosen for the best agreement with high resolution TEM data. So the idea of the work is to take the sort of particles that are seen experimentally and ask what should happen to these when they interact with oxygen? The model predicts that Au atoms in the periphery base of the clusters can be oxidised because the resulting species are stabilised through interaction with the surface. It is well known that catalysts formed on this sort of support show a mix of oxidic and metallic Au species in XPS or XANES experiments. These observations could be explained either with separate areas of metallic and oxidised Au or with particles that contain both. Our results are consistent with the latter interpretation. We are currently working with our electron microscopy colleagues to see if there is evidence for mixed oxidation states in the Au particles from atom–atom spacing information. The model also allows us to explore other questions, such as: Are the O atoms that are incorporated into the cluster available to oxidise adsorbates or are they just spectators in the catalysis? Can other oxygen containing species such as water become involved in oxidising the Au clusters? These calculations will clearly lead to predictive statements that could be checked with suitable isotope labelling experiments.

Professor Hu answered: In principle, I agree with this point.

Professor Hutchings continued the discussion of the paper by Dr Beret: Just to follow on from the comments of Professor Poliakoff concerning mechanistic validation of theoretical models it is not necessary to try to observe short lived intermediates as this will be exceptionally difficult; rather you need to perturb the system in some way. Theory should be able to predict what should be the result of such a perturbation. I would suggest that theory and experiment should be more closely linked and work together.

Dr Ghiringhelli responded: I fully agree.

Professor Fortunelli continued a general discussion of the papers by Professor Hu, Dr Willock and Dr Beret: One can ask what are the perspectives of performing realistic theoretical simulations of catalytic processes, considering that thermodynamic phase diagrams after all are not so useful for phenomena such as catalysis which are often driven by kinetics rather than thermodynamics, and that guessed (biased) reaction paths are not guaranteed to be the real ones.

I believe that these perspectives are good. Actually, we have already recently implemented at the first-principles (DFT) level and applied to catalytic oxidation of CO or propene by (Ag–Au) trimer supported on MgO(100) a novel approach which we named "reactive global optimization" (RGO) and which we believe satisfies all your requests (described in Poster P11 and in a paper in preparation for Nanoscale). In this approach the system starts from a generic structure (say—an adsorption configuration) and is then allowed to reach neighboring local minima *via* a kinetic Monte Carlo algorithm governed by the energy barriers to reach such minima (connected to the temperature of the system). Stoichiometry moves are also considered as a function of the pressure of reacting gases in a fully realistic way. For example, in the case of propene oxidation by Ag3 at room temperature and ambient conditions, the approach correctly predicted the products of partial oxidation of propene (propylene oxide and acrolein) in the appropriate temperature range, and NOT the thermodynamic product which is CO2 and H2O. The limitations at present are: (1) the approach is computationally very demanding: if working at the first-principles level, even exploiting the fact that the approach lends itself to efficient use on massively parallel supercomputers, the number of atoms fully included in the global search cannot exceed—say—20; (2) the approach has so far been applied to a model oxide substrate. The latter limitation will soon be overcome in future work. The former one is intrinsic.

Dr Willock opened the discussion of the paper by Professor Pyykkö: In considering the dispersion interaction between Au(I) complexes, how do you differentiate between the terms involving the Au centres and those from the ligand...ligand interactions?

Professor Pyykkö responded: To begin with, one must use Wave-Function Theory (WFT), not DFT. From the simplest to the most sophisticated practical theories, the WFT results for the metallophilic attraction oscillate along MP2, MP3, MP4, CCSD and CCSD(T). The MP2 exaggerates the interaction but is physically sound. A large-basis CCSD(T) has a chance of being close to the final result. The reference point is then the Hartree–Fock model which has no electron correlation effects, in particular not those of dispersion type. Now this only tells the total intermolecular interaction energy between the interacting (usually frozen) closed-shell molecules.

Your question can be answered by using WFT based on localised orbitals, first introduced by Runeberg *et al.*[1] As known, the simplest dispersion term between molecules A and B involves virtual electronic excitations from the ground state of A to an excited state A', and similarly B-to-B'. Now the most relevant occupied orbitals in the ground state for Au(I) are typically fairly pure 5d, as might be expected for Au$^+$. The most relevant, originally empty orbitals in the excited states A' and B' are not pure Au(6p) orbitals, but may contain substantial ligand character (if those linearly dependent basis functions can be separated). In that sense, your question is highly relevant. As said, the largest term mostly is dispersion and decays at large R as R^{-6}. Another important contribution is the virtual charge transfer or 'ionic' contribution of type A-to-A' plus B-to-A' (or inverse).[1] It decays exponentially at large R. For monomers with large multipole moments, the multipole–multipole or induction terms can also be relevant.[2] In addition to the mentioned metal–metal and metal–ligand interactions, nothing prevents in principle the ligand–ligand interactions, although the distances would be longer for these perpendicular model dimers. The R^{-6} behaviour at large R speaks for classical dispersion.

1 N. Runeberg, M. Schuetz and H. J. Werner, *J. Chem. Phys.*, 1999, **110**, 7210–7215. See also later papers quoting this one.
2 J. Muniz, C. Wang and P. Pyykkö, *Chem. Eur. J.*, 2011, **17**, 368–377.

Dr Allison remarked: This is a very interesting paper which addresses a particularly important topic that is seldom fully investigated, namely the comparison of the performance of wavefunction theory (WFT) methods *versus* density functional theory (DFT) methods for transition metals. For example, you show DFT methods that significantly overestimate and underestimate the interaction energy in the Au2... AuH system. You also mention that the WFT results may be useful to calibrate DFT methods. Given the results reported in your paper, what insights and advice can you offer to those doing calculations on transition metal systems using DFT?

Professor Pyykkö replied: Our material is too limited to be able to give a good answer to Dr Allison's question.

Dr Willock asked: Your results show that PBE over binds for the interaction between Au(I) systems. This suggests that the approach of adding the missing dispersions into DFT calculations through the various DFT+D methods would actually

     This journal is © The Royal Society of Chemistry 2011

require a positive term in this case. In contrast the adsorption of molecules to Au would be expected to require a negative contribution from dispersion. Do you envisage a problem with using DFT+D methods for adsorbates on Au?

Professor Pyykkö answered: To give a good answer, one should explicitly test the DFT+D methods for the problem at hand.

Professor Van Santen said: The figure 7 of your paper is quite interesting. Will a similar correlation be found for the neutral system or when particle size changes?

Professor Pyykkö responded: This Figure 7 is all that we currently have. Future will show.

Dr Ghiringhelli said: Here I wish to make just a comment on reliability of DFT in particular regarding the PBE functional.

In his paper, Professor Pyykkö has presented a case in which PBE overbinds, with respect to the reference (CCSD(T)). Typically PBE would underbind weakly bound system. In our work[1] we have adopted the Tkatchenko-Scheffler (TS, Ref. 38 in our paper) correction to the PBE functional for describing the van der Waals interactions, which proved to be of great accuracy for a large set of weakly bound systems. In particular for the systems of our interest in the paper, this correction is necessary for the understanding of the experimental IR spectrum of Au_13; only when the TS scheme is adopted, the planar and 3D clusters becomes energetically nearly degenerated and an entropic argument can be used to rule the planar out of the population at finite temperature. If the energetics of pure PBE are used the planar clusters are too stable. In conclusion functionals like PBE, when a correct account of long range correlation is adopted, are quite accurate. Exception, like the case described in paper 10 are always possible and this is why one should always check case by case the reliability of the adopted method (see also discussion of question 104).

1 E. C. Beret, L. M. Ghiringhelli and M. Scheffler, *Faraday Discuss.*, 2011, DOI: 10.1039/ C1FD00027F.

Dr Groszek communicated: Binding energies of Au–H and Au–Au reported by Prof Pyykko are considerably lower than the heats of interaction between Au and hydrogen molecules determined experimentally by flow through microcalorimetry by Groszek *et al.*[1] The integral heat of hydrogen displacing nitrogen from the surface of pure Au powder was found to be 406 kJ mol^{-1}. The differential heats reached 549 kJ mol^{-1} at 91% saturation with hydrogen. These high heats were obtained after exposure of 660 mg of the Au sample to 135 micromoles of argon at 125 C, which activated the gold surface.

1 A. J. Groszek *et al.*, *Appl. Surf. Sci.*, 2010, **256**, 5498–5502.

Professor Pyykkö communicated in reply: This question must be left unanswered. Dr Groszek refers to a particular set of experiments on Au powder, while we have done calculations on small molecules.

Professor Campbell remarked: You showed that PBE overestimates Au–Au dispersion interactions by 50%. I find this interesting, since for benzene and naphthalene on Pt(111), we have shown that similar DFT methods (by others) underestimate our measured heats of adsorption by ~80 and 130 kJ mol^{-1}, respectively (see Ref. 1), and many theoreticians have told me was probably due to missing dispersion interactions in DFT. Do you care to comment?

1 J. M. Gottfried, E. K. Vestergaard, P. Bera, and C. T. Campbell, *J. Phys. Chem. B*, 2006, **110**, 17539–17545.

Professor Pyykkö replied: My test case involved a closed-shell Au(I) molecule on one side and a closed-shell gold cluster on the other side. Thus it is safest to limit the conclusion of exaggerated dispersion interactions to the aurophilic case.

In your example you are having a closed-shell aromatic molecule on one side and the two-dimensional surface of metallic gold on the other side.

Dr Willock commented: The interaction of the Au(I) complexes will largely be a balance between the dispersion attraction and the coulombic repulsion. You note that PBE over estimates the attraction by around 50%. Could this be due to under-estimation of the repulsion rather than over-estimation of the attraction?

Professor Pyykkö replied: The monomer charges, or even the effective atomic charges (by any definition) are rather stable quantities, as function of the functional. Therefore the underestimation of coulombic repulsion is unlikely.

Here it is perhaps amusing to mention the apparent violation of Coulomb's law in certain tetrameric chain systems.[1] To begin with, we can disproportionate two L–Au–X monomers into the counterions [L–Au–L]$^+$ and [X–Au–X]$^-$. Here L is a neutral ligand, such as a phosphine, and X for instance Cl. Call these + and −, respectively. Now those ion-pair dimers (+−) can experimentally form chain tetramers in the orders [−++−] and [+−−+], in addition to [+−+−] or a cyclic tetramer. The suggested explanation[1] was that the charge of the gold atoms in both the + and − complex ions was small, about +0.5 at NBO level, and the same in both ions.

1 P. Pyykkö, W. Schneider, A. Bauer, A. Bayler and H. Schmidbaur, *Chem. Comm.*, 1997, 1111–1112.

Ms Mukhamedzyanova commented: I see, that in the table 2 in your paper you have given the magnitudes of interatomic distances in pm, and this value has three signs after the point. Also the dissociation energies are given with one sign after the point in kJ mol^{-1}. What is the accuracy of the results obtained?

Professor Pyykkö replied: The number of digits given is only meaningful within each model. The variation shows the variation between the models.

Ms Mukhamedzyanova said: In the paper it is written "the increase of the interaction energy with the number of nearest neighbours". Do you have any suggestions about the reason of this phenomenon?

Professor Pyykkö answered: Only a hunch. It could come from short-range, pair-wise interactions, as said at the end of chapter 2.4., but this is still unproven.

Dr Ghiringhelli opened the discussion of the paper by Professor Hashmi: In you paper you point out the necessity of the presence of water for lowering the reaction barrier, due to the fact that the water clusters shuttle the proton transfer. Since at finite temperature the water clusters are quite mobile, I was wondering whether the conformational entropy related to the cluster rearrangement wouldn't change the barrier scenario. Here it is not a question of having possibly missed structures that might have lower energy than those that you have found for your proposed pathways, I am saying that the structural rearrangement of the cluster might create an ensemble of pathways whose overall configurational free-energy might lower even more the barriers you find. Of course this calls for a quite involved finite temperature sampling, for which the methodology is not yet really settled. Mine is just a bottom line curiosity whenever "fluffy" structures are involved in catalysis.

Professor Hashmi responded: Our point was to show that even a small water cluster can lower the barrier for proton transfer to realistic values. We completely

    This journal is © The Royal Society of Chemistry 2011

agree that probably lower barriers can be found when either taking into account conformational entropy and structural rearrangements to the cluster or investigating larger water clusters.

At the end the crucial message is that gold catalysis not only tolerating water, but sometimes needs water for proton transfers. We take your statement as an inspiration for future work.

Professor Van Santen asked: Is there a general answer to the question whether Au is preferred over Ag or Cu in the transmetallation reaction?

Professor Hashmi responded: We have not investigated that interesting question yet. From the literature we know that for example in the Sonogashira coupling a direct transmetalation from copper to palladium is highly efficient. And in stoichiometric organometallic chemistry there are numerous examples for the transfer of organic groups from silver to other transition metals. With these two metals an efficient transmetalation is not dependent on the use of iodides, which seems to be the clearly preferred halide (experimentally and by computational chemistry). This observation might suggest that copper and silver are more efficient, we will have a closer look at this interesting and important question in the future.

Professor Meyerstein opened the discussion of the paper by Dr Lyalin: The last overhead you have shown implies equal energies of H_2 and two H atoms adsorbed to TiO_2. Does this imply that the $H–TiO_2$ bond strength is $ca.\,200$ kJ mol^{-1}? Then if H atoms are adsorbed preferentially on Au–NPs adsorbed on TiO_2 surfaces this would mean that the $Au^0–H$ bond strength is larger than 200 kJ mol^{-1}. This is difficult to envisage as H_2 is easily released from Au electrodes.

Alternatively one could argue that the binding of Au^0 NPs to the TiO_2 decreases considerably the $H–TiO_2$ bond strength and therefore the $Au^0–H$ bond strength does not have to be so strong.

Dr Lyalin answered: According to our calculations H_2 adsorbs on the pure rutile $TiO_2(110)$ surface with the binding energy $E_b^{mol} = 0.14$ eV; while the binding energy of the dissociated configuration of H_2 is $E_b^{dis} = 1.50$ eV. In both cases the binding energy is calculated in respect to the total energies of the non-interacting H_2 molecule and TiO_2 slab, $i.e.$, $E_b^{mol} = -(E(H_2 - TiO_2) - E(H_2) - E(TiO_2))$, and $E_b^{dis} = -(E(2H - TiO_2) - E(H_2) - E(TiO_2))$. Indeed, according to our calculations H atom binds strongly on the supported Au atom or small cluster, than on the pure TiO_2 surface. There is no contradiction with experimental findings that H_2 is easily released from Au electrodes. In the present work we study small gold clusters, but not extended Au surfaces. This is the regime where each atom counts, physical and chemical properties of clusters are extremely size sensitive and cannot be deduced from those known for larger sizes and the bulk. Thus, binding of H_2 to the extended Au surface and small clusters can be very different.

Professor Bond asked: I would like to know whether in your opinion H atoms formed by dissociation of molecules at low coordination sites are capable of moving onto plane surfaces, leading to high coverage. If they are confined to corners and edges, then presumably hydrogenations can only occur at these places. It is well known from some quite old work[1] that H atoms can diffuse readily through Au, and I believe their solubility may be quite high. So in addition to exploring surface diffusion, knowledge of the energetics of H atom dissolution would be very helpful.

1 R. S. Yolles, B. J. Wood and H. Wise, *J. Catal.*, 1971, **21**, 66.

Dr Lyalin answered: In the present work we have investigated hydrogen adsorption and dissociation on the relatively small gold clusters consisting up to 20 atoms.

In that case it is difficult to say whether or not H atom can migrate onto the gold plane, because cluster faces are small and all consist of low coordinated atoms. On the other hand we have noticed that dissociation of H_2 on the cluster surface results in a local structural rearrangements of the cluster. That might be related with the tendency of the H atom to "solve" or "dive" in the cluster volume. It is important to note, that the presence of an impurity in the metal cluster can result in alteration of its thermodynamic properties (see, *e.g.*,).[1] Such an effect should be taken into account when considering surface diffusion and dissolution of H atom in the gold cluster. Definitely, more detailed theoretical study of this interesting process is needed.

1 A. Lyalin, A. Hussien, A. Solovyov, and W. Greiner, *Phys. Rev. B*, 2009, **79**, 165403.

Dr Groszek commented: Recently published heats of adsorption of hydrogen on Au/TiO_2 catalysts are considerably higher than the energies of binding between gold surfaces and hydrogen atoms.[1,2] The heat of displacement of nitrogen by molecular hydrogen from the gold catalysts has been determined in a flow-through microcalorimeter ranged from 90 kJ mol^{-1} to 450 kJ mol^{-1} at 50% surface coverage. Even greater heat evolutions were observed when the catalysts containing adsorbed hydrogen atoms were exposed to nanomole pulses of molecular oxygen which produced repeated heat evolutions from 1500 kJ mol^{-1} to 1800 kJ mol^{-1} for a 1% Au/TiO_2 catalyst. For a 7% Au/TiO_2 catalyst the heats were of the order of 4000 kJ mol^{-1} of oxygen. To my knowledge there is no credible theory that could explain these abnormally high heat evolutions.

1 H. H. Kung *et al. J. Phys. Chem. B*, 2005, **109**, 10319–10326.
2 A.J. Groszek *et al. J. Appl. Surf. Sci.*, 2011, 3192–3195.

Dr Lyalin responded: This is a good example, showing that there are plenty of opened questions requiring systematic explanation. Detailed mechanisms responsible for the high catalytic activity of gold clusters have not yet been elucidated. In spite of the fact that catalytic activity of gold nanoparticles has been discovered more than 20 years ago, gold is still full of surprises!

Dr Louis asked: If you come to the conclusion that there is preferential dissociation of H_2 at the $Au-TiO_2$ interface, do you mean that the nature of the support has an influence in reactions of hydrogenation? We prepared gold on various supports (alumina, titania, ceria and zirconia) using the same method of deposition–precipitation with urea. The gold loading was the same (1 wt%) and the metal particle size as well, and we did not observe differences of reaction in selective hydrogenation of butadiene.[1] Could you comment?

1 A. Hugon, L. Delannoy and C. Louis, *Gold Bull.*, 2008, **41**, 127–138.

Dr Lyalin responded: In the present work we suggest that the adsorbed hydrogen dissociates at the perimeter interface of the supported gold cluster in the vicinity of the bridge-site twofold coordinated O atom. On the other hand the threefold coordinated oxygen atom on the TiO_2 (110) surface can also play a similar role (see, *e.g.*, Fig. 6(c) in the paper) for H_2 dissociation. Therefore if the support contains the low-coordinated O atoms on the surface, the H_2 dissociation reaction will occur. In all cases you have mentioned, there are such atoms on the surface, and hence the hydrogenation reaction is observed. I would not expect very large difference in the reaction rates, but some small dependence of the reaction rate on the type of the support should take place. I can suggest that for the oxygen-free supports, such as for example h-BN, the reaction rate can be very different, or reaction will not occur at all.

   This journal is © The Royal Society of Chemistry 2011

Dr Willock asked: It has been mentioned that your calculations imply quite a strong Au–H bond is formed after H_2 dissociation. It seems that the adhesion energy of Au to the stoichiometric (110) surface of rutile TiO_2 is usually found to be quite low, but increases if Au is located at a defect site. For example, PW91 give around 0.6 eV on the perfect surface but this rises to between 2 and 4.5 eV when an oxygen defect is used as the Au location.[1,2] Have you calculated the adhesion energy of the Au atoms to the surface and do you think that the weak surface interaction will give rise to stronger Au–H binding than would be found for Au located at a surface defect?

1 Y. Wang and G. S. Hwang, *Surf. Sci.*, 2003, **542**, 72.
2 A. Vijay, G. Mills and H. Metiu, *J. Chem. Phys.*, 2003, **118**, 6536.

Dr Lyalin answered: We found that the adhesion energy of an Au atom on the defect-free rutile TiO_2 (110) surface is 0.69 eV, which is close to the results of other theoretical works you have mentioned. Although we have considered only the defect-free TiO_2 surface it is quite clear that presence of oxygen defects can result in considerable increase in the adhesion energy of gold clusters. Defects can trap the metal nanoparticle and enhance charge transfer between the support and the nanoparticle. Thus, for example, it is well know that Au_8 clusters supported on the MgO(100) surface rich of F-center defects show high catalytic activity, when Au_8 deposited on defect-poor MgO(100) surface are inert.[1] One can expect that similar effect occurs for gold clusters supported on TiO_2. The important question here is the charge state of the supported clusters. Our calculations demonstrate that gold clusters remain almost neutral upon adsorption on the defect-free rutile surface. However, if the supported cluster trapped by the surface defect, the supported cluster can be charged. The charge state of the cluster can influence the hydrogen adsorption and make Au–H bond stronger. On the other hand, our calculations demonstrate, that even in the case of a week interaction of Au with the defect-free surface, Au–H binding is considerably promoted. Therefore, it is an interesting question whether or not the surface defects will further promote the Au–H binding.

1 B. Yoon *et al.*, *Science*, 2005, **307**, 403.

Professor Haruta asked: Your DFT calculations support the results of Dr Fujitani's experimental work that the rate of H_2–D_2 exchange reaction was in proportion to the periphery length around the Au particles. You assume that the active sites for H_2 dissociation are corners and edges of Au clusters in the vicinity of low-cordinated oxygen atoms on TiO_2(110). In this perimeter mechanism, is hydrogen dissociation heterolytic to form hydride (H^-) on the Au surfaces and H^+ on the TiO_2 surfaces or homolytic to form H radicals on the Au surfaces?

Dr Lyalin replied: Yes, indeed, we found that H_2 preferably dissociates at corners and edges of Au clusters in the vicinity of the bridge-site twofold coordinated O atoms on the TiO_2(110) surface. Our preliminary results show that after H_2 dissociation the H atom migrating to the bridge-site oxygen atom on the TiO_2(110) surface is charged positively. On the other hand, the H atom on the Au cluster surface is charged negatively, with the Bader net charge of -0.1 e or less. More detailed investigation of this interesting process is needed, in order to elucidate mechanisms of such charge distribution. In particular, it would be very interesting to study whether or not the interaction with the support can promote or quench the hydride formation. Hydrides on the cluster surface might play a key role in understanding of the high catalytic activity of gold clusters in a variety of hydrogenation reactions. We hope that the present study will stimulate further investigation of this interesting process.

Dr Lang asked: I am referring to your calculations on free, unsupported clusters as shown in Figure 2 of your paper. Can you please comment on the activation barriers that have to be overcome when going from the molecularly adsorbed hydrogen to dissociated hydrogen. Is it possible to observe the dissociation of molecular hydrogen in the experiment or are the barriers to high to overcome under typical experimental conditions?

Dr Lyalin answered: In the present work we have not calculated dissociation barriers for H_2 adsorbed on free, unsupported gold clusters. However, our preliminary results show that the typical height of such barriers is about 0.5–1.0 eV, depending on the geometry configuration, cluster size, adsorption center on the cluster surface, *etc.* The results of *ab initio* CCSD(T) calculations of the dissociation barriers of H_2 adsorbed on the neutral Au_2 and Au_3 clusters have been reported by S.A. Varganov *et al.*[1] According to this work, the barriers for H_2 dissociation on small gold clusters with respect to Au_nH_2 complexes are 1.10 eV and 0.59 eV for $n = 2$ and 3.

1 S. A. Varganov, R. M. Olson, M. S. Gordon, G Mills and Horia Metiu, *J. Chem. Phys.*, 2004, **120**, 5169.

Professor Bernhardt said: To complement your interesting theoretical investigation, we would like to note that we recently published a combined experimental and theoretical study on the interaction of molecular hydrogen with small size-selected gold clusters.[1] Previous experiments already demonstrated that the adsorption of molecular hydrogen and molecular oxygen on small gold clusters exhibit a strong cluster charge state dependence: Whereas gold cluster cations react with hydrogen, but not with oxygen under thermal reaction conditions, gold cluster

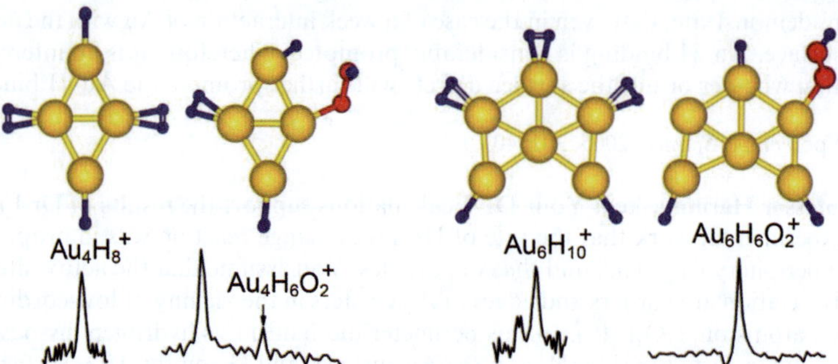

Fig. 2 Products of the reaction of hydrogen alone and a mixture of hydrogen and oxygen with Au_4^+ (left) and Au_6^+ (right), respectively. The upper row displays the calculated structures, the lower row the mass signals. Please note that the oxygen molecules alone do not react with the gold cluster cations (large yellow spheres: Au; small blue spheres: H; red spheres: O).[1]

anions do not react with hydrogen, yet they do react with oxygen. Most interestingly, we found, however, that the pre-adsorption of hydrogen on positively charged gold clusters can promote the co-adsorption and even the activation of molecular oxygen on these clusters (see Fig. 2[1]). From your study, do you expect a promotional effect of hydrogen on the co-adsorption of oxygen at supported clusters and is the dissociation of molecular hydrogen required for the expected peroxide formation on gold clusters?

1 S. M. Lang, T. M. Bernhardt, R. N. Barnett, B. Yoon, U. Landman, *J. Am. Chem. Soc.*, 2009, **131**, 8939.

   This journal is © The Royal Society of Chemistry 2011

Dr Lyalin responded: We found that there is no large charge transfer between the hydrogen molecule and the gold clusters. Nevertheless, when hydrogen adsorbs molecularly on the gold clusters, H_2 gains some small positive charge. Therefore, in the case of multiple adsorption of H_2 on the gold cluster, the cluster become negatively charged, which in turn can promote O_2 adsorption and activation. Similar effects are known not only for H_2 and O_2 co-adsorption, but also for co-adsorption of O_2 and CO or hydrocarbons. Thus, for example, in our recent works we have demonstrated that there is a cooperative effect in O_2 and C_2H_4 adsorption;[1] moreover co-adsorption of C_2H_4 on the neutral gold clusters can promote O_2 dissociation, which is energetically forbidden process on the neutral gold clusters without C_2H_4.[2] On the other hand the dissociative adsorption of H_2 on the free fold clusters results in a small charge transfer to the H atoms ($\sim -0.05e$), and hence the gold cluster possesses small positive charge. That can suppress adsorption of O_2 on the gold cluster directly; however O_2 can possibly interact with the negatively charged H atom. One can speculate that such an effect can possibly lead to H_2O_2 formation; however more theoretical studies are needed to clarify this suggestion. The charge transfer processes become more complicated for the supported gold clusters. For example, in the case of the gold dimer adsorbed on the defect-free rutile surface we found that there is a very small positive charge ($+0.05e$) localized on Au_2. The supported gold dimer becomes neutral as a result of the molecular adsorption of H_2. However, in the case of dissociative adsorption of hydrogen on the supported gold dimer, Au_2 gains the large positive charge, that is equal to $+0.63e$ for the configuration shown in Fig. 6d (in the paper). It is interesting to note, that the H atom in the surface OH group possesses large positive charge of $+0.75e$; however the remaining H atom bridging the two gold atoms (see Fig. 6d) gains the negative charge of $-0.1e$.

1 A. Lyalin and T. Taketsugu, *J. Phys. Chem. C*, 2009, **113**, 12930.
2 A. Lyalin and T. Taketsugu, *J. Phys. Chem. Lett.*, 2010, **1**, 1752.

Ms Mukhamedzyanova remarked: It's really difficult and requires significant computational effort to simulate solid state. We have to use sufficiently large systems for better representation. For this purpose it's important to choose an appropriate model for your simulation. Can you explain the choice of 6-layer slab of titania in your investigation? In plenty of papers this problem is treated and researchers look for an appropriate model to avoid different oscillations and also to represent solid state as good as possible.

Dr Lyalin answered: Indeed, the choice of an appropriate model to simulate various catalytic reactions on the supported nanoparticles is a very important task. Therefore one should always think about the main physical process that you want to describe and neglect some other unessential details. It is always a compromise between an accuracy of calculations and computational feasibility. The simulation of the rutile support for gold clusters is a vivid example. It is well known, that some quantities, such as the surface energy, the energy gap, *etc.* in thin films of rutile oscillate with decreasing amplitude as the number of layers increases. According to Ref. 74 the band gap in the rutile TiO_2 (110) film is largest for a 2 layer of TiO_2 film, 2.50 eV, and smallest for a 3 layer of TiO_2 film, 1.45 eV. Both values are far from the bulk gap which is estimated in Ref. 74 to be 1.93 eV. Does it mean that we have to use a very large slab for the rutile in order to avoid oscillations and perform calculations with high accuracy? The experimental value for the bulk band gap is 3.06 eV.[1] Such a large difference between the theory and experiment is typical for DFT studies of solids and surfaces. As we see this difference is much larger than the oscillations of the band gap as a function of number of layers! Therefore, in order to reproduce an accurate value of the rutile band gap one should first of all concentrate on improving the accuracy of the current DFT models, rather than on increasing the size of TiO_2

slab. Our current work is not focused on calculation of the rutile band gap. It is aimed at investigation of the interaction between gold clusters and the support and a role of cluster-surface interface in H_2 dissociation. Therefore it is important to check how the number of layers influences on adsorption of gold, hydrogen and other molecules on the rutile surface. In Ref. 76 it was demonstrated that the six-layer slab provides good convergence of the obtained results at moderate computational cost when considering an O_2 supply pathways in CO oxidation on $Au/TiO_2(110)$. Therefore we selected the six-layer slab model for all our calculations.

1 P. A. Cox, *Transition Metal Oxides? An Introduction to Their Electronic Structure*, Clarendon, Oxford, 1992.

Ms Mukhamedzyanova commented: Have you varied the number of layers in the slab of titania in your investigation to choose an appropriate model?

Dr Lyalin replied: No, we refer to the previous studies (see, *e.g.*, Ref. 73–76) on this subject. Once again I would like to stress that it is important to study how the number of layers influences the catalytic properties of the supported gold clusters and adsorption processes on the rutile surface in general, rather than how it influences the energy gap of TiO_2.

Ms Mukhamedzyanova asked: Have you estimated a band gap of 6-layer slab of titania?

Dr Lyalin responded: Yes, the calculated value of the energy gap of six-layer slab is 2.23 eV, which is in a good agreement with previous theoretical studies.

Professor Campbell asked: We noted in our paper in this same proceedings that TPD desorption rate measurements give an activation energy of $\sim$52 kJ mol^{-1} for the process:

$$2H(adsorbed) \rightarrow H_2(gas)$$

on Au(110). I am curious how this compares to your calculations of this barrier for Au(110) and for Au clusters on TiO_2.

Dr Lyalin replied: In the present paper we have studied energetics of adsorption and dissociation of H_2 on free and $TiO_2(110)$ supported gold clusters. The full reaction pathways and the reaction barriers for H_2 dissociation on the supported gold clusters will be presented in a separate publication. However, our preliminary results demonstrate, that dissociation barrier of H_2 adsorbed on the rutile surface is $\sim$4.5 eV which is comparable with dissociation energy of a free hydrogen molecule. As the binding energy of H_2 adsorbed on the defect-free rutile surface is only 0.14 eV; hence, H_2 would likely desorb from the surface and fly away rather than dissociate. However, H_2 dissociation energy drastically decreases for gold particles supported on TiO_2. Thus, in the case of H_2 adsorption on Au atom supported on the rutile surface the calculated dissociation barrier of H_2 is equal $\sim$0.6 eV. Barrier for the reverse process $2H(adsorbed) \rightarrow H_2(gas)$ should be higher, because atomic hydrogen binds quite strongly to the low coordinated oxygen atom on the TiO_2 surface.

Dr Willock remarked: You show in your paper that for the supported Au atoms and clusters on TiO_2 there is favourable binding energies for dissociation with one H atom on Au and one forming a surface OH group. Have you tested the electronic structure of these geometries to assess if the H–H bond cleavage is homolytic or heterolytic?

     This journal is © The Royal Society of Chemistry 2011

Dr Lyalin replied: This is a very interesting question. We found, that in general there is no considerable charge transfer between gold and hydrogen. Thus, the largest charge transfer occurs for H_2 adsorbed molecularly on Au/TiO_2. In this case the calculated Bader charge localized on H_2 is +0.11e, where e is the elementary charge. In most cases of molecular adsorption of hydrogen, the adsorbed molecule possesses some small positive charge. However, after dissociation, the hydrogen atoms can get an excess of negative charge. Depending on the geometry configuration, either both of the hydrogen atoms possess some small negative charge, or one H is charged negatively, while another H is charged positively. Thus, for example the H–H bond cleavage on the unsupported Au_2 can be considered as heterolytic—one H possesses negative charge of −0.11e, while another H possesses very small positive charge of +0.01e. In this case the gold dimer becomes positively charged upon H_2 dissociation with the net charge of +0.1e. Of course, in the final stage of H_2 dissociation on the supported Au clusters, when the OH surface group is formed, the hydrogen in OH gains the large positive charge of +0.75e. Another H on Au cluster remains negatively charged. Formation of the hydrogen anions on the gold surface can influence probability of further hydrogenation reactions catalyzed by the supported gold clusters.

Professor Van Santen asked: Can you deduce from your calculations whether at the Au-support interphase H_2 dissociation is homolytic or heterolytic? Presumably analysis of charge density changes on Ti^{4+} will answer this question.

Dr Lyalin responded: We have not found any considerable change in the local charge density of Ti^{4+} atoms caused by the H_2 dissociation. As for the first part of your question, see my answer on the previous question by Dr Willock.

Professor Pyykkö opened a general discussion of Professor Pyykkö's, Professor Hashmi's and Dr Lyalin's papers: Referring to the possibility of heterolytic splitting of H_2, I should like to quote the case of such splitting by boron-nitrogen-based 'hydrogen tweezers'.[1] There the H^- and H^+ make the borate- and ammonia-like halves of the molecule into tetrahydroborate- and ammonium-like moieties, with charges close to −1 and +1, respectively. Then the added Coulomb attraction of −1/R at an R of 3.3 Å is over 400 kJ mol^{-1}, comparable with the H_2 dissociation. As said in that paper, "Coulomb pays for Heitler–London".

Now the difference in the gold case is that the charges of the two effective counterions may be much smaller, and therefore the heterolytic splitting mechanism may be less likely.

1 V. Sumerin, F. Schultz, M. Atsumi, C. Wang, M. Nieger, M. Leskelä, T. Repo, P. Pyykkö and B. Rieger, *J. Am. Chem. Soc.*, 2008, **130**, 14117–14119.

Prof Dr Bhargava responded: Agreed.

Professor Madix commented: The last comment is interesting, because it suggests that gold exposed to air will be covered with oxygen. Has anyone in this group actually taken a gold single crystal or foil from the atmosphere and examined its surface composition? If so, what is seen?

Dr Cadete Santos Aires remarked: It is certainly possible to find oxygen (together with other pollutants such as carbon and others) on pure Au when you analyze a surface coming from air as we did with our samples. However this experiment does not allow you to precisely determine if the presence of oxygen is part of the pollutants (carbonates, ...) or if it is due to dissociation on the pollutants or on the Au surface. In the latter case, if dissociation occurs (which does not seem to be the case from other studies namely cited by Professor Friend) it is orders of magnitude lower that on supported catalysts. I believe the experiment proposed by Professor Campbell can certainly give us better insight on this point to show that oxygen does not dissociate spontaneously on Au pure structured extended surfaces.

Prof Dr Bhargava responded: Agreed.

Professor Friend said: A rough metallic Au surface can, in fact, dissociate O_2. A roughened Au(111) surface dissociates O_2 with a probability of 10^{-3} at ~400 K.[1] The rate may be slower than for small particles on oxide supports; however, it is finite. On the other hand, it is difficult to accumulate O on the surface because the oxygen is so reactive. For example, even at extremely low pressures (UHV), as pointed out by Professor Campbell, the O does not accumulate because it reacts with, *e.g.* CO.

1 X. Deng, B. K. Min, A. Guloy, and C. M. Friend, *J. Am. Chem. Soc.*, 2005, **127**(25), 9267–9270.

Professor Bowker asked Professor Friend: Can you make surface carbonate on gold if there is atomic oxygen on the surface, as you can on Ag?

Professor Friend answered: We have not observed surface carbonate on Au however, we have not made an extensive investigation. CO is readily oxidized to gaseous CO_2 even at low temperature and all O is removed from the surface. There have been reports by others that carbonate can form,[1] for example, indicating that it could be formed transiently under some conditions.

1 Mullins, *et. al.*, *J. Phys. Chem. C*, 2008, **112**(45), 17631–17634.

Professor Hutchings commented: Following on from the comment of Professor Madix, we know that gold nanoparticles do not need to be supported to be active. Work from the Milan group for glucose oxidation showed that the initial rate for gluconic acid formation was very similar for the unsupported and supported gold nanoparticles prepared by a sol-immobilisation method. Supporting the nanoparticles effectively stabilised them and permitted the catalysts to be observed over a longer period. One has to consider is it something other than the gold surface atoms required to activate oxygen; the colloidal nanoparticles have protecting ligands present and maybe these play a role in some way.

Dr Lyalin responded: I think we should have better understanding (experimental and theoretical) what factors can influence the catalytic activity of gold nanoparticles. Support effects, ligand effects, solvent effects can play a crucial role in catalytic activity of gold. In our recent work, presented here, at Faraday Discussion at the poster session we have studied the catalytic activity of small gold clusters supported on the hexagonal boron nitride (h-BN) surface. Such a surface has been believed to be an "inert" support that does not change catalytic properties of gold clusters at all. However, we have demonstrated that such a support can considerably modify properties of gold.

Professor Madix remarked: Angelici, *et al.* have observed complex partial oxidation chemistry in organic solvents on unsupported metallic gold particles. The manner in which dioxygen is activated by these surfaces is not understood. However, it is known that carbonate, sulfite and nitrate act as oxygen transfer agents for partial oxidation on silver single crystals.[1] Depending on the "wash" used to clean metallic gold initially, such species might be present.

1 L. Zhou and R. J. Madix, *Langmuir*, 2010, **26**, 16282–16286).

   This journal is © The Royal Society of Chemistry 2011

Professor Fortunelli said: Concerning the effect of the support on the ability of Au clusters to dissociate O_2, we presented in our poster P11 evidence that Au_3 cannot (or can hardly) dissociate O_2 in the gas phase, in the sense that dissociation is energetically unfavorable, whereas it becomes energetically favorable after adsorption even on a simple (supposedly inert) ionic oxide such as regular MgO(100). This is mainly due to electrostatic effects: the electrostatic field from the surface stabilizes the dissociated configuration because this has a significant dipole moment. In general, the energetics of the supported structures are different from that found in the gas phase, precisely because of these strong electrostatic effects. Furthermore, the energy barrier for O_2 dissociation, which is substantial (about 0.9 eV) when O_2 is only adsorbed on Au_3 supported on MgO(100), becomes much smaller when CO or another ligand such as propene is co-adsorbed on the supported cluster. This is due to the donating properties of CO or propene which significantly facilitate O_2 breaking.

Concerning the presence of carbonated species suggested by Professor Maddix, what we observed in our global optimization search of propene and O_2 absorption on $(Ag-Au)_3$ clusters supported on MgO(100) is that a structure with a formate adsorbed in a bidentate fashion on these clusters was very stable, nearly as stable as that one in which CO_2 was produced. Whether these species can act as a reservoir of oxygen, I do not know for $(Ag-Au)_3$ clusters, but I doubt it because they are energetically very stable and probably unreactive. We do have experience on Ni clusters in which these structures are certainly unreactive.

Professor Madix answered: Surface bound formate is very stable on extended gold surfaces, decomposing to CO_2 and hydrogen at 360 K on the Au(110) single crystal face.[1] Its chemical behavior is not expected to change with different crystallographic orientations. It has never been observed to react with other surface intermediates or to be a reservoir of oxygen.

1 D. A. Outka and R. J. Madix, *Surf. Sci.*, 1987, **179**, 361.

Professor Hutchings addressed Professor Madix and Dr Xu: Professor Friend indicates that in the case of alcohol oxidation with gold nanoparticles the early results required the presence of base and Dr Xu has indicated that under such conditions recent research for Bob Davis suggests that the oxygen addition is initially involving water. However, oxidation of alcohols with gold nanoparticles does not require the presence of base and we now have very active catalysts of solvent-free or just aqueous conditions. Hence the question of oxygen activation on gold surfaces has to be considered more deeply.

Professor Madix responded: As I noted earlier, substrates may be activated by several different sources of oxygen, the details remaining to be understood. It is certainly possible that adsorbed OH, in addition to adsorbed O, OOH, carbonate, sulfite, etc may activate alcohols to initiate the reaction sequence. Thus, depending on the details of the reaction environment, different modes of activation may occur.

Dr Xu responded: I would like to add a follow-up comment: My summary of the work by Profs. Davis and Neurock[1] was less than stellar during the meeting. The mechanism that they proposed involves OH, in both aqueous ($OH^-_{(a)}$) and adsorbed (OH^*, on Au) forms, functioning both to dehydrogenate the organic substrate (alcohol in their study) and to oxidize it *via* nucleophilic addition to the α C. They reported that "the product distribution depends on pH, and little or no activity is seen over Au catalysts without added base". O_2 does not directly participate in the reaction, but instead is reduced by H_2O to H_2O_2 which regenerates OH.

1 B. N. Zope, D. D. Hibbitts, M. Neurock and R. J. David, *Science*, 2010, **330**, 74–78.

Insights into catalysis by gold nanoparticles and their support effects through surface science studies of model catalysts

Charles T. Campbell,[*ab] James C. Sharp,[a] Y. X. Yao,[a] Eric M. Karp[b] and Trent L. Silbaugh[b]

Received 4th March 2011, Accepted 6th May 2011
DOI: 10.1039/c1fd00033k

One important aid in understanding catalysis by gold nanoparticles would be to understand the strength with which they bond to different support materials and the strength with which they bond adsorbed intermediates, and how these strengths depend on nanoparticle size. We present here new measurements of adsorption energies by single crystal adsorption calorimetry, and new analyses of other recent measurements by this technique in our lab, which imply that: (1) small nanoparticles of metals like Au bind much more strongly to supports like titania and iron oxide which are generally observed to be effective in making Au nanoparticles active in catalysis than to supports like MgO which are considered less effective, (2) the thermodynamic stability of adsorbed intermediates for catalytic reactions can either increase strongly or decrease strongly with decreasing metal nanoparticle size below 8 nm, depending on the system, and (3) the reaction to insert O_2 into the Au–H bond of adsorbed H on the Au(111) surface to make Au–OOH ($O_{2,g} + H_{ad} \rightarrow OOH_{ad}$) is exothermic by $\sim$80 kJ mol^{-1}. This adsorbed hydroperoxy species is thought to be a key intermediate in selective oxidation reactions over Au nanoparticle catalysts, but its production by this reaction may also provide a route for O_2 activation in less demanding reactions (like CO oxidation) as well. Its stability would be even higher on Au nanoparticles below 3 nm in diameter, but even there it is too unstable to be formed by combining adsorbed OH with an O adatom ($OH_{ad} + O_{ad} \rightarrow OOH_{ad}$), which is estimated to be endothermic by 175 kJ mol^{-1}. The implications of the stability of metal nanoparticles *versus* particle size on different supports and of the stability and potential reactions of OOH_{ad} in Au catalysis will be discussed.

I. Introduction

Supported gold and gold alloy nanoparticles have been found to be highly active and selective in a variety of important catalytic reactions, particularly when the particles are only a few nm in size.[1–17] The catalytic activity of gold has been ascribed to a variety of mechanisms, including reactions occurring at neutral gold atoms on the gold nanoparticles which differ from atoms on bulk gold, either due to the fact that these gold atoms have far fewer nearest neighbor atoms (*i.e.*, a high degree of coordinative unsaturation),[18–21] quantum size effects which alter the electronic band structure of gold nanoparticles,[3] electronic modification of the charge on the gold atoms by interactions with the underlying oxide,[22–24] or the combined effect

---

[a]Department of Chemistry, University of Washington, Seattle, WA, 98195-1700, USA. E-mail: campbell@chem.washington.edu
[b]Department of Chemical Engineering, University of Washington, Seattle, WA, 98195-1700, USA

---

of sites on the oxide (vacancies or other defects), possibly modified by the presence of nearby gold, working together with sites on the gold nanoparticles.[25-27] For many reactions, it appears that reducible oxides like Fe_2O_3, TiO_2 and CeO_2 serve as the best support materials.[5-9] Water vapor at low concentration is known to have beneficial effects on the catalytic activity of gold nanoparticles,[6,26,28-30] although it inhibits the elementary reaction of CO with oxygen adatoms to make CO_2 on gold.[30] It has been postulated that water may help activate O_2.[29,30] Calculations using density functional theory on gold clusters by Barrio et al.[31] predicted that O_2 can insert into the bond between Au and an adsorbed hydrogen atom to make the adsorbed hydroperoxo intermediate ($-OOH_{ad}$). Assuming that water may dissociate to produce H_{ad} plus OH_{ad}, this suggests the interesting possibility that water may activate O_2 for oxidation reactions on gold and gold alloy nanoparticles *via* this hydroperoxo intermediate. This would also be consistent with the observations that H_2 enhances the rate of CO oxidation and direct propene epoxidation over Au catalysts.[32,33] Indeed, a related mechanism whereby HOOH is produced at gold sites from $H_2 + O_2$ has been proposed and even included in kinetic simulations of net epoxidation rates.[32,34,35] Vibrational spectroscopic evidence for adsorbed hydroperoxo species has been observed,[36] although this does not distinguish whether it is bound to the gold or the oxide surface.

Here we present new calorimetric measurements that compare the thermodynamic stability of silver nanoparticles on reducible oxides with that on non-reducible oxides, specifically addressing the strength of binding of Ag nanoparticles to a very well-defined reduced iron oxide surface, $Fe_3O_4(111)$. The results are analyzed in terms of the effects of such reduced supports on Ag nanoparticles' resistance to sintering and the effect of the oxide on Ag nanoparticles' chemisorption reactivity. Given the similarities between Ag and Au, the results give significant insights into the role of the support in these two respects during catalysis by Au nanoparticles. We also present new analysis of previous calorimetric measurements which provide a good estimate of the heat of formation of the adsorbed hydroperoxo intermediate and the adsorbed hydroxyl species on Au surfaces. The results support the hypothesis that water may act as a catalyst to activate O_2 for oxidation reactions on gold and gold alloy nanoparticles *via* this hydroperoxo intermediate, which we estimate should be exothermically formed from the reaction of O_2 gas with H adatoms, and which we estimate should very exothermically decompose to give $O_{ad} + OH_{ad}$.

II. Experimental

The details of the calorimetric measurement of Ag gas atom adsorption energies onto oxide supports has been presented in detail elsewhere.[37] The calorimeter was housed in an ultrahigh vacuum chamber, with a base pressure of $\sim2 \times 10^{-10}$ mbar (rising to $\sim1 \times 10^{-9}$ mbar, which was mainly H_2, during Ag deposition). It was equipped with low energy electron diffraction (LEED), Auger electron spectroscopy (AES), electron energy loss spectroscopy (EELS), low-energy ion scattering spectroscopy (ISS), a quadrupole mass spectrometer (QMS), and a quartz crystal microbalance (QCM). AES was carried out using 1.6 keV electrons from a PHI LEED system (15–120), and a Leybold-Heraeus EA11 hemispherical energy analyzer. ISS was carried out using He^+ ions with 1 keV primary energy from a de-focused ion gun (Leybold-Heraeus IQE 12/38). The calorimeter consists of a pulsed metal atom beam and a pyroelectric polymer (polyvinylidenefluoride, PVDF) ribbon pressed against the back of the single crystal for heat detection. The 4-mm diameter, chopped Ag atom beam is produced from a high-temperature effusion cell. The Ag was purchased from Alfa Aesar with 99.999% purity. The metal atom beam is chopped to provide 100 ms pulses containing ~0.02 ML of Ag atoms every two seconds. One monolayer (ML) is defined throughout as 1.31×10^{15} atoms cm^{-2}, which is the number of oxygen atoms exposed to the vacuum per unit area for $Fe_3O_4(111)$. The absolute beam flux is measured by using a calibrated QCM. The sticking probability is measured by a modified King-Wells method, using a line-of-sight quadrupole mass spectrometer (QMS) to measure the fraction of metal atoms which strike the surface but do not adsorb. The mass spectrometer signal was calibrated by measuring the integrated desorption signal from a known amount of Ag reflected from a hot Ta foil (1100 K to 1200 K), located at the same position as the sample and corrected for average velocity.

The typical operating temperature of the effusion cell was ~1380 K, which generated some thermal radiation that impinged on the sample and also was detected by the calorimeter. The heat signal was corrected for this and calibrated using light pulses from a He–Ne laser as described elsewhere.[37,38] The amount of light adsorbed per pulse from this He–Ne laser for pulses of fixed energy was measured and found to decrease nearly linearly with Ag coverage, from its starting value on the Ag-free $Fe_3O_4(111)$ to 69 ± 3% of its starting value by 3 ML Ag, and remain constant afterwards. By setting the final reflectivity of thick multilayer Ag to the literature value for bulk Ag of 94%,[37] we estimate that the initial reflectivity for the starting $Fe_3O_4(111)$ thin film was 91 ± 1% using this 31% decrease in absorbed light.

To convert the calorimetrically measured internal energy changes into standard enthalpy changes at the sample temperature (300 K), the excess translational energy of the metal gas atoms at the oven temperature, above that for a 300 K Maxwell-Boltzmann distribution (~0.3 μJ pulse^{-1}), is subtracted, and a small pressure–volume work term (RT per mole) is added, as described elsewhere.[38] The corrected enthalpy of adsorption is thus the standard molar enthalpy of adsorption at 300 K, which at high coverage, where the atoms are adding to bulk-like sites, can be compared directly to the standard heat of sublimation of the metal. The measured heats are expressed as the enthalpy of adsorption on a "per mole of adsorbed Ag" basis by correcting for the sticking probability.

The preparation of the 1 μm-thick Pt(111) single-crystal was described previously.[37,39] The preparation of $Fe_3O_4(111)$ thin films on Pt(111) were based on the recipe of Zscherpel *et al.*[40] In brief, ~1 monolayer of FeO is first grown by depositing ~1 ML of Fe on Pt(111) at 300 K and annealing for 5 min at 900 K in 5×10^{-7} torr O_2. Then ~20 ML of Fe is deposited in 2×10^{-7} torr O_2 at 300 K, and subsequently annealed at 800 K in 5×10^{-7} torr O_2 for 5 min. This led to an $Fe_3O_4(111)$ film of ~6 nm thickness, whose composition, cleanliness and order were verified by AES and LEED.

III. Results

III.1. Silver atom adsorption energies on different oxide surfaces

To assess how different oxides affect the stability of Au nanoparticles, single crystal adsorption calorimetry was used to measure the adsorption energies of Ag atoms onto the MgO(100) surface, slightly-reduced $CeO_2(111)$ surfaces and an $Fe_3O_4(111)$ surface. We used Ag atoms for this study instead of Au, because we do not yet have the ability to produce a high enough flux in a collimated molecular beam of Au atoms to do adsorption calorimetry (due to the much higher sublimation energy of Au), although we expect to have that capability in the near future. Due to the electronic similarities between Ag and Au, we expect the trends with respect to their relative binding strengths to different oxide surfaces to be very similar.

Before presenting the measured heats of Ag adsorption *versus* Ag coverage at 300 K, we first discuss our measurements which determined the Ag film morphology, since it is essential in interpreting the heat data. The Ag film morphology was measured *versus* Ag coverage at 300 K by measuring the AES and He$^+$ ISS peak intensities for the Ag and for the oxide elements (O and/or the oxidized metal) *versus* Ag coverage. For all these oxide surfaces, these AES and ISS data were well fitted by assuming the Ag particles have the shape of hemispherical caps, with a fixed number density that is independent of Ag coverage after the first 2% of a ML. The AES data for MgO(100) and CeO$_2$(111) and their fits to this hemispherical cap model have been presented elsewhere,[41,42] where the fits were shown to give saturation Ag particle densities (N) of 2.5×10^{12} and 4×10^{12} particles cm^{-2}, respectively. This value of N was the same for CeO$_2$(111) independent of whether the surface had a reduced stoichiometry of CeO$_{1.9}$(111) or CeO$_{1.8}$(111) within the probe depth of XPS. For the less reduced CeO$_{1.9}$(111) surface, most of the oxygen vacancies are thought to reside at the step edges.[42] For Fe$_3$O$_4$(111), the Ag particle density that resulted from this fit was 4×10^{12} Ag particles per cm^2. Dividing the Ag coverage (atoms cm^{-2}) by N (particles cm^{-2}) gives the average number of Ag atoms per particle at any given Ag coverage, which can be combined with the bulk density of Ag (5.9×10^{22} atoms cm^{-3}) to give the average particle volume at each coverage. This hemispherical volume then corresponds to an average Ag particle diameter at that coverage. Because diameter varies only as the inverse cube root of N, the factor of ~2 error bar on N makes only a 25% error in diameter, to which the conclusions below are insensitive.

The heat of Ag adsorption was measured *versus* Ag coverage at 300 K for the Fe$_3$O$_4$(111) surface described above. Fig. 1 shows the measured heat of adsorption of a Ag gas atom on Fe$_3$O$_4$(111) plotted *versus* the average Ag particle diameter to which it adds, using the above approach to convert from Ag coverage to Ag particle diameter. As can be seen, the heat of adsorption increases rapidly with coverage (Ag particle size), but reaches a saturation value at large silver particles which is indistinguishable from the bulk heat of Ag sublimation (285 kJ mol^{-1},[43] shown by the dashed horizontal line). The observed increase in adsorption energy with particle size was first observed for Pb adsorption on MgO(100),[44] where it was shown that this can be qualitatively understood from a simple bond additivity model: when a metal atom adds to a very small metal cluster (<2 nm), it makes fewer metal–metal bonds, on average, than when it adds to a large particle.[44,45]

For comparison, we also show in Fig. 1 the heat of adsorption data for Ag on the MgO(100), CeO$_{1.9}$(111) or CeO$_{1.8}$(111) surfaces, which have been presented previously.[41,42] There are large differences between the different surfaces in the stability of metal atoms adding to particles of the same size. Most notably, Ag atoms bind much more strongly to sub-4 nm Ag particles on the Fe$_3$O$_4$(111) and reduced CeO$_2$(111) surfaces than to the same size particles on MgO(100). When metal atoms

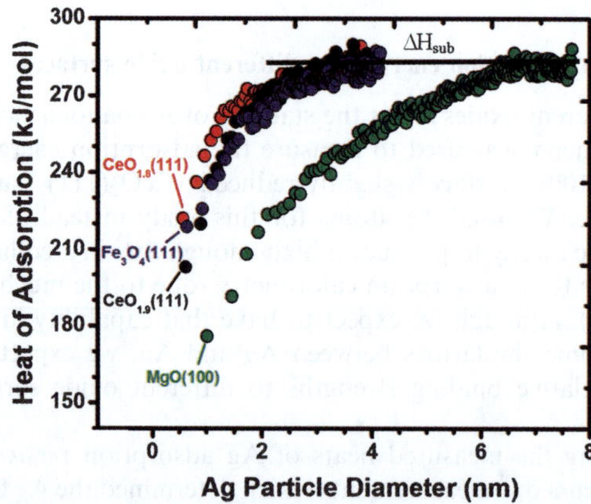

Fig. 1 The measured heat of Ag atom adsorption *versus* the Ag particle diameter to which it adds, for Ag adsorption onto four different surfaces: the Fe$_3$O$_4$(111) thin film and the two CeO$_2$(111) thin films with different extents of surface reduction ($x = \sim0.1$ and 0.2 in CeO$_{2-x}$), all grown on Pt(111) with 4 or 5 nm thickness, and a 4 nm thick MgO(100) film grown on Mo(100). The data for CeO$_2$(111) and MgO(100) are from ref. 42.

add to the particles, on average some add on the tops and sides of the particles but some add to the perimeter, binding to the oxide as well. The difference in average adsorption energy may come from this nearest-neighbor bonding difference, with perhaps a lesser contribution from next-nearest-neighbor sites. Also, the growing Ag lattice may adopt its lattice parameter to that of the oxide surface and thus provide sites for the next Ag atom where the Ag–Ag bond energies are different due to Ag lattice strain effects (with a stretched lattice providing more stable sites for Ag adsorption)

In all four oxide surfaces, the metal particles are probably nucleated at step edges,[42,46] but the particles are big enough in the 1.5–4 nm range that most of the metal atoms at the metal/oxide interface are *not* directly bonding to oxide step atoms, but instead reside at terraces. Thus, these data indicate that the bonding of Ag particles to the oxide is stronger even at the terrace sites of $Fe_3O_4(111)$ and $CeO_2(111)$ than MgO(100). On the MgO(100) surface, Ag atoms bind more strongly atop the oxygen anions than to the metal cation sites (see ref. 41 and refs. therein). On MgO(100) terraces, only half the surface ions are oxygen anions, and half are Mg cations, whereas all of the surface ions on the $Fe_3O_4(111)$ and $CeO_2(111)$ terraces are oxygen anions in a nearly closest-packed array. However, the density of the more favorable oxygen anion sites is 1.3×10^{15} and 7.9×10^{14} anions cm^{-2} on $Fe_3O_4(111)$ and $CeO_2(111)$ terraces, respectively, and 1.1×10^{15} on the MgO(100) terrace. Thus, a higher density of the more favorable anion sites cannot explain the stronger bonding on Ag to these $Fe_3O_4(111)$ and $CeO_2(111)$ terraces than to the MgO(100) terrace. It may instead be due to the differences in electronic structure or oxide lattice parameter and geometry which induce Ag lattice strain differences, which could also affect the heats (see above), or to the presence of cation sites on MgO(100) which are not present on the other two surfaces.

Independent of the exact origin of these differences in Fig. 1, it is clear that Ag atoms in Ag nanoparticles of the same size demonstrate a greater stability on $Fe_3O_4(111)$ and slightly reduced $CeO_2(111)$ surfaces than on MgO(100). This directly maps into a corresponding difference of this same magnitude in the chemical potential of the metal atoms in the particles on different oxides. Kinetic models for the sintering of metal nanoparticles include this chemical potential as a negative term as part of the apparent activation energy for the sintering rate, whether the mechanism for sintering is by Ostwald ripening or particle diffusion/coalescence.[45] Thus the energy differences in Fig. 1 should translate into huge differences in sintering rates for the same size particles on the different oxides: Ag particles of the same size (below 4 nm) should sinter much more slowly on the $Fe_3O_4(111)$ and slightly reduced $CeO_2(111)$ surfaces than on MgO(100). The same effect should also apply to gold nanoparticles. Given that very small nanoparticles (<4 nm) are optimal for gold catalysis (see above), this may explain at least some of the observations that gold catalysts show better activity when dispersed on iron oxide and ceria than when on oxides like MgO.[5–9]

III.2. Variations in metal atom stability with particle size and support affect the stability of adsorbed catalytic intermediates

Fig. 1 clearly shows that the surface metal atoms are bound more weakly to the smallest metal nanoparticles than to large metal particles. One would therefore expect them to bind adsorbed catalytic reaction intermediates more strongly, based on the simple bond energy–bond order conservation concept. Indeed, we have found by temperature programmed desorption (TPD) that oxygen adatoms bind much more strongly to tiny Au nanoparticles than to larger Au nanoparticles on $TiO_2(110)$.[18,47] As shown in Table 1, analysis of their TPD peak temperatures to desorb to make gaseous O_2 ($O_{2,g}$) using simple Redhead analysis gave desorption activation energies that are 36% larger from the smallest Au particles, implying that O adatoms are ~25 kJ mol^{-1} more stable on the smallest Au particles than on bulk Au.

Table 1 Desorption activation energies (E_{des}) for oxygen adatoms from Au nanoparticles of different sizes on $TiO_2(110)$, from refs. 18,47,48. (Note that the same calculation error in ref. 47, which is corrected in an Erratum,[48] also was made in ref. 18, which after correction changes the value of 124.7 kJ mol^{-1} for 1.3-layer particles to 190 kJ mol^{-1}.) The values from 49 for specific sites on bulk Au(211), which has short Au(111) terraces with a periodic array of steps, are also shown for comparison.

Average Au particle thickness (atomic layers)	Desorption peak T/K	E_{des} (kJ per mol $O_{2,g}$)
1.3	740	190
2.3	645	165
6	545	139
Bulk Au(211) step site	540	142
Bulk Au(111) terrace site	515–530	138

In contrast, the opposite effect is observed for CO binding to Pd nanoparticles on $Fe_3O_4(111)$. In a collaboration at the Fritz-Haber Institute with the group of H–J. Freund and S. Schauermann, we found that CO molecules have a lower calorimetric heat of adsorption on tiny Pd nanoparticles than on large Pd particles on $Fe_3O_4(111)$ or bulk Pd(111).[50] The results are summarized in Table 2.

How can one understand these apparently opposite effects of particle size in Tables 1 and 2? The key to this may be in Fig. 1. Note specifically that the smallest particles in Table 2 (1.8 nm) are already within 10 kJ mol^{-1} of their large-particle stability when dispersed on $Fe_3O_4(111)$ in Fig. 1. Thus one would expect these particles to have already nearly reached the large-size limit in the weakness with which they bind adsorbates, if determined by the simple bond energy – bond order conservation model alone. The marked increase in CO adsorption energy in going to larger particles must then be due to some other effect, perhaps associated with the change in lattice strain with particle size (see above). In contrast, if $TiO_2(110)$ behaves more like MgO(100) in Fig. 1 (which still remains to be determined), then the smallest particles in Table 1 are more than 10-fold (>100 kJ mol^{-1}) less stable relative to the large-size limit. This would mean that there should be big changes in adsorption energy for O adatoms in going to the larger particles in Table 1, based on simple bond energy–bond order conservation, just as observed.

Since the surface Ag atoms are bound more strongly to a nanoparticle of the same size on the $Fe_3O_4(111)$ and slightly reduced $CeO_2(111)$ surfaces than on MgO(100) (Fig. 1), one would expect them to bind to adsorbed catalytic reaction intermediates more weakly, if based on the simple bond energy–bond order conservation concept

Table 2 Initial heats of adsorption (ΔH_{ads}) measured by single crystal adsorption calorimetry for CO on Pd nanoparticles of different sizes on $Fe_3O_4(111)$, from ref. 50. The value for bulk Pd (111) is also shown for comparison. Particle thicknesses were calculated assuming hemispherical cap shape and the (111) layer spacing of 0.22 nm.

Average Pd particle diameter (nm)	Average Pd particle thickness (atomic layers)	ΔH_{ads} of CO (kJ per mol)
1.8	2.7	106
4	6	133
8	12	126
Huge	Huge	142
Bulk Pd(111)	Bulk	149

alone. This same effect would probably also occur for gold. This would probably render them less catalytically active for the same particle size, but that of course depends on the details of the reaction and its rate-controlling step(s). Also, this effect is perhaps more than compensated by the fact that the $Fe_3O_4(111)$ and slightly reduced $CeO_2(111)$ surfaces are likely to maintain much smaller gold nanoparticles than MgO(100) at temperatures where sintering can occur (see above). Also, since we really do not yet understand completely what make Au nanoparticles so active in catalysis, it is not clear that this simple bond energy–bond order conservation model is the most important effect to consider, as exemplified by the CO–on–Pd case above. It should be noted, however, that CO–on–Pd may be unusual in *not* following the simple bond energy–bond order conservation, since CO is found by DFT to be not much more stable at step edges than terraces on Pd(111),[51] whereas there is a very large difference between steps and terraces on Pt(111).[52] This is also generally observed in CO TPD from these (111) surfaces, where a high-temperature peak due to step sites is easily seen on Pt(111)[53] but has not been reported on Pd (111), to our knowledge.

III.3.   The stability of the hydroperoxo species (–OOH) on Pt and Au surfaces

Our group has recently reported the heat of formation of adsorbed OD on the Pt (111) surface, measured by adsorption calorimetry.[54] This was determined from the adsorption enthalpy of D_2O gas onto Pt(111) pre-covered with oxygen adatoms, which are well known to react on the surface to produce two surface OD groups (typically stabilized in a 1 : 1 complex with molecularly adsorbed $D_2O^{54,55}$). From the heat of formation of adsorbed OD (-226 kJ mol^{-1}), we also estimated the Pt–OD bond enthalpy which holds OD to the Pt(111) surface to be 263^{54} kJ mol^{-1}. We now will use this bond enthalpy to estimate the Pt–OOH bond energy of the adsorbed hydroperoxo species on Pt(111) and, from that, the Au–OOH bond energy of the adsorbed hydroperoxo species on Au surfaces.

Bercaw and coworkers[56–58] have shown that the strength of metal–ligand bonds in homogeneous organometallic complexes is strongly correlated with the strength of the H-ligand bonds to those same ligands, with a slope very near unity (see Fig. 2). For example, the H-cyclopentyl bond is 29 kJ mol^{-1} weaker than the H-bond to $CH_2CH(CH_3)CH(CH_3)_2$, so the Ir-cyclopentyl bond in the same type of iridium complex is about 26 kJ mol^{-1} weaker then the corresponding Ir bond to CH_2CH

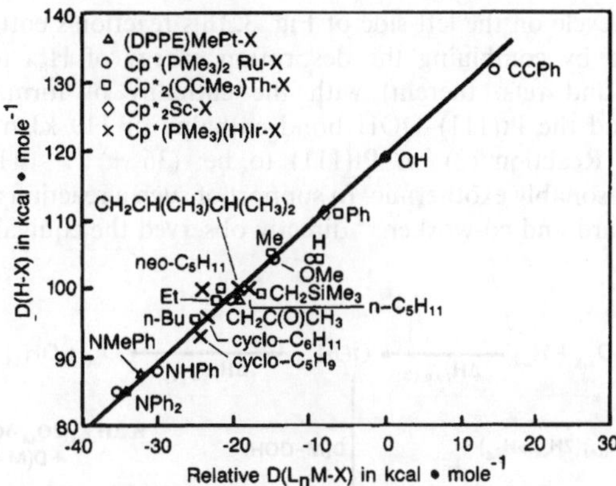

Fig. 2 Plot showing the 1 : 1 correlation between ligand-to-ligand changes in the bond dissociation energy for metal–ligand bonds, $D(L_nM–X)$, in organometallic complexes in solution and that for the corresponding H-ligand bond, $D(H–X)$, taken from ref. 56. Different metals are put on the same plot here by subtracting the M–OH bond energy from all the $D(L_nM–X)$ values for that metal, so these are bond energies relative to that metal's M–OH bond energy.

$(CH_3)CH(CH_3)_2$. We have shown some evidence that this correlation holds for carbon bonds to bismuth surfaces.[59]

Here we focus instead on the strength of metal–O bonds, using our recently-measured value for the Pt(111)–OD bond enthalpy of 263 kJ mol^{-1}[54] and Bercaw's correlation of Fig. 2 to estimate other Pt(111)–OR bond enthalpies. The H–OCH$_3$ bond enthalpy of 437 kJ mol^{-1}[60] is ~63 kJ mol^{-1} weaker than the H–OD bond enthalpy of 500 kJ mol^{-1} (obtained by subtracting the heats of formation of H$_g$ and OD$_g$ from that for HOD$_g$, from 61). According to Bercaw's correlation, M–OCH$_3$ bonds should then be ~63 kJ mol^{-1} weaker than M–OD bonds for the same metal complex. We recently found that the Pt(111)–OCH$_3$ bond is ~64 kJ mol^{-1} weaker than the Pt(111)–OD bond, which is within 2% of this prediction, strongly suggesting that this correlation in Fig. 2 also holds well for the Pt(111) surface.[62] Since the H–OH bond is 1 kJ mol^{-1} weaker than the H–OD bond, the Pt(111)–OH bond enthalpy should then be ~1 kJ mol^{-1} weaker than the Pt(111)–OD bond enthalpy,[61] or ~262 kJ mol^{-1}. Similarly, the H–OOH bond enthalpy of 356 kJ mol^{-1} (obtained by subtracting the heats of formation of H$_g$ and OOH$_g$ from that for H$_2$O$_{2,g}$, from ref 61) is 144 kJ mol^{-1} weaker than the H–OD bond, so this correlation in Fig. 2 predicts that the Pt–OOH bond should be ~144 kJ mol^{-1} weaker than the Pt–OD bond at Pt(111). Subtracting this from our measured value for the Pt(111)–OD bond enthalpy of 263 kJ mol^{-1}[54] gives a Pt(111)–OOH bond enthalpy of just 119 kJ mol^{-1}.

We now show that adsorbed –OOH should readily dissociate on Pt(111) *via*:

$$OOH_{ad} \rightarrow O_{ad} + OH_{ad}. \qquad \text{(Reaction 1)}$$

We estimate this reaction's enthalpy by following the thermodynamic cycle on the right half of Fig. 3: combining this Pt(111)–OOH bond enthalpy of 119 kJ mol^{-1} with the enthalpies of formation of OH$_g$ (39 kJ mol^{-1}) and OOH$_g$ (+2 kJ mol^{-1}),[61] half the enthalpy of O$_2$ adsorption to make two adsorbed oxygen atoms (O$_{ad}$) on Pt(111) (180 kJ mol^{-1})[63] and the Pt(111)–OH bond enthalpy (262 kJ mol^{-1}, see above) gives (119 + 39 − 2 − 180/2 − 262) kJ mol^{-1} = −196 kJ mol^{-1}. This shows that adsorbed –OOH is unstable with respect to dissociation to OH$_{ad}$ + O$_{ad}$ by ~196 kJ mol^{-1} on Pt(111), but it still might be transiently produced.

Consider its production by O$_2$ insertion into the M–H bond of H$_{ad}$:

$$O_{2,g} + H_{ad} \rightarrow OOH_{ad}. \qquad \text{(Reaction 2)}$$

Following the cycle on the left side of Fig. 3, this reaction's enthalpy on Pt(111) can be estimated by combining the desorption energy of H$_{ad}$ to make ½ H$_{2,g}$ (36 kJ mol^{-1}[54] and refs. therein) with the enthalpy of formation of OOH$_g$ (+2 kJ mol^{-1}) and the Pt(111)–OOH bond enthalpy of 119 kJ mol^{-1}. This gives the enthalpy of Reaction (2) on Pt(111) to be: (36 + 2 − 119) kJ mol^{-1} = −81 kJ mol^{-1}, reasonably exothermic. In support of such a reaction and its enthalpy, Goldberg, Goddard and co-workers[64] directly observed the equivalent of Reaction

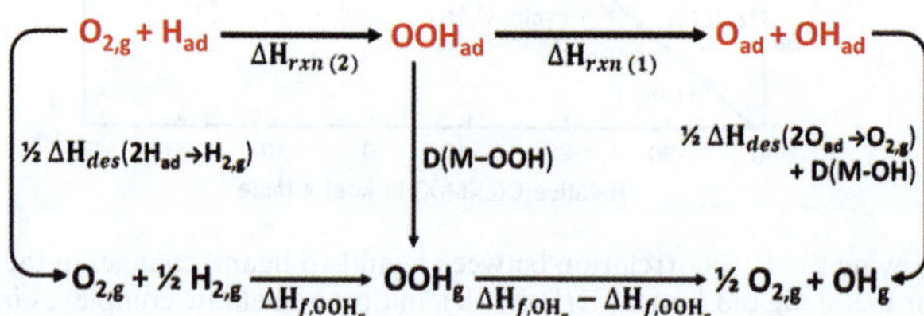

Fig. 3 Thermodynamic cycles involving the surface –OOH species of relevance to gold and platinum catalyst surfaces.

 This journal is © The Royal Society of Chemistry 2011

(2) for a homogeneous Pd complex, and also estimated that reaction to be exothermic by 114 kJ mol^{-1} using DFT with a continuum solvent model. In summary, surface –OOH could be produced by O_2 insertion into the Pt–H bond of a H adatom on Pt(111) *via* Reaction (2), but it would quickly dissociate *via* Reaction (1).

Next we estimate the energetics for Reactions (1) on Au surfaces, based on our above estimates for Pt(111). To do this, we first estimate the Au(111)–OH bond energy. We do this by first recognizing that Abild-Pedersen *et al.*[65,66] showed with DFT estimates of adsorption energies that, for a series of different transition metal surfaces, the differences in their bond energies to –OH scaled as 0.50 times the differences in their bond energies to oxygen adatoms. As noted in Table 1, the desorption energy for O_{ad} to make $O_{2,g}$ on Au(111) terraces is 138 kJ mol^{-1} O_2. On Pt(111), it is 180 kJ mol^{-1} O_2.[63] The difference is $42/2 = 21$ kJ mol^{-1} O, from which the above scaling relationship implies that the Au(111)–OH bond enthalpy is 0.50×21 kJ mol^{-1} $= 11$ kJ mol^{-1} weaker than the Pt(111)–OH bond energy of 262 kJ mol^{-1}. This gives 251 kJ mol^{-1} as an estimate of the Au(111)–OH bond enthalpy. As noted above, the correlation in Fig. 2 implies that a metal–OOH bond should be $\sim$143 kJ mol^{-1} weaker than the corresponding metal–OH bond, giving $251–143 = 108$ kJ mol^{-1} for the Au(111)–OOH bond enthalpy. This and the other enthalpies above can be combined to show that Reaction (1) is exothermic by $\sim$175 kJ mol^{-1} on Au(111). Again, adsorbed –OOH on Au(111) is unstable and would quickly decompose to O_{ad} + OH_{ad}.

We can also estimate the energetics for Reactions (2) on Au surfaces, based on the above enthalpies and the strength of H adatom binding to Au surfaces. We estimate this from the TPD peak temperature for H_{ad} desorbing as H_2 gas from Au (110) $- 2 \times 1$, which is 216 K.[67] This TPD peak temperature gives a desorption energy of 52 kJ mol^{-1} using simple Redhead analysis and assuming the same pre-exponential factor as for H_{ad} on Pt(111) (for which the desorption energy is 72 kJ mol^{-1} and the TPD peak temperature is $\sim$300 K (see ref. 68 and refs. cited therein). Adding half this desorption energy to the heat of formation of OOH gas (2 kJ mol^{-1}, see above) and subtracting the Au(111)–OOH bond enthalpy of 108 kJ mol^{-1} gives the enthalpy for Reaction (2) to be -80 kJ mol^{-1} (*i.e.*, exothermic). This is considerably larger than the computational DFT estimate of -48 kJ mol^{-1},[31] but both estimates give that this reaction is quite exothermic. So again, surface –OOH could be produced by O_2 insertion into the Au–H bond of a H adatom on Au(111), but it would immediately dissociate by Reaction (1) to make O_{ad} + OH_{ad}. Table 1 shows that O_{ad} is 26 kJ mol^{-1} more stable on the smallest Au nanoparticles on TiO_2(110) than on Au(111), implying that –OOH is $0.50 \times 26 = 13$ kJ mol^{-1} more stable on tiny Au nanoparticle catalysts. Thus, Reaction (2) may be exothermic by as much as 93 kJ mol^{-1} on tiny nanoparticle Au catalysts (which is even more exothermic than on Pt(111)). Similarly, Reaction (2) was estimated with DFT calculations to be exothermic by 90 kJ mol^{-1} on free Au_{29} clusters.[31] This suggests that $O_{2,g}$ may be activated on Au catalysts by interaction with H_{ad} *via* Reaction (2), followed by Reaction (1) to make O_{ad} + OH_{ad}. Since O_{ad} on Au nanoparticles is known to react very rapidly with CO gas at room temperature to make CO_2,[18] this seems a plausible mechanism for low-temperature CO oxidation on Au nano-catalysts.

Alternatively, this OOH_{ad} could react with adsorbed propene to make propene epoxide plus OH_{ad} on Au nanoparticles. Propene is known to adsorb very strongly at the edge of tiny Au nanoparticles on TiO_2(110), with a desorption peak at 288 K in TPD, corresponding to an adsorption energy of 72 kJ/mol.[69] Combining the enthalpy of Reaction (1) (-175 kJ mol^{-1}) with that for O_{ad} desorbing to make ½ O_2 ($+95$ kJ mol^{-1} O) and that for the gas-phase reaction ½ $O_{2,g}$ + propene$_g$ $\rightarrow$ propene epoxide$_g$ (-115 kJ mol^{-1}) gives that:

$$OOH_{ad} + propene_g \rightarrow propene\ epoxide_g + OH_{ad} \qquad \text{(Reaction 3)}$$

is exothermic by 195 kJ mol^{-1} on Au nanoparticles. Using the adsorbed propene mentioned above instead of propene gas as the reactant gives the reaction:

$$OOH_{ad} + propene_{ad} \rightarrow propene\ epoxide_g + OH_{ad}, \qquad \text{(Reaction 4)}$$

which is 72 kJ mol^{-1} less exothermic, but still exothermic by 123 kJ mol^{-1}. Alternatively, OOH_{ad} could make HOOH which subsequently reacts with propene to make the epoxide, as proposed by Nijhuis *et al.*[32,34,35]

Note that from the above enthalpies and the heats of formation of H_2O_g and H_g, one can easily calculate that the reaction:

$$H_2O_g \rightarrow H_{ad} + OH_{ad} \qquad \text{(Reaction 5)}$$

is ~10 kJ mol^{-1} exothermic on tiny Au nanoparticles, or more if H_{ad} is more stable on tiny Au than on Au(110).

In summary, reaction mechanisms which involve O_2 insertion into H_{ad} to make OOH_{ad} on Au nanoparticles (Reaction (2)) as a means for activating O_2 are quite plausible from an enthalpic point-of-view, whether that OOH_{ad} is then used to attack an olefin for epoxidation (Reaction (4)) or to make O_{ad} (Reaction (1)) for CO oxidation. Such a mechanism might help explain the promotional role of water and/or H_2 gas in Au-catalyzed reactions. These energetics might be even more favorable at the edges of the Au nanoparticles, where interactions with the oxide support might facilitate water dissociation, H_2 activation or O_2 insertion into the Au–H bond.

Acknowledgements

The authors would like to acknowledge support for this work by the Department of Energy, Office of Basic Energy Sciences, Chemical Sciences Division grant number DE-FG02-96ER14630 and the National Science Foundation under grant number CHE-1010287.

V. References

1 M. Haruta, S. Tsubota, T. Kobayashi, H. Kageyama, M. J. Genet and B. Delmon, Low-Temperature Oxidation of Co over Gold Supported on TiO$_2$, Alpha-Fe$_2$O$_3$, and Co$_3$O$_4$, *J. Catal.*, 1993, **144**, 175–192.

2 M. Haruta, Size-and support-dependency in the catalysis of gold, *Catal. Today*, 1997, **36**, 153.

3 M. Valden, X. Lai and D. W. Goodman, Onset of catalytic activity of gold clusters on titania with the appearance of nonmetallic properties, *Science*, 1998, **281**, 1647.

4 M. Valden, S. Pak, X. Lai and D. W. Goodman, Structure sensitivity of CO oxidation over model Au/TiO$_2$ catalysts, *Catal. Lett.*, 1998, **56**, 7–10.

5 A. S. K. Hashmi and G. J. Hutchings, Gold catalysis, *Angew. Chem., Int. Ed.*, 2006, **45**, 7896–7936.

6 H. H. Kung, M. C. Kung and C. K. Costello, Supported Au catalysts for low temperature CO oxidation, *J. Catal.*, 2003, **216**, 425–432.

7 M. Haruta, Catalysis of gold nanoparticles deposited on metal oxides, *CATTECH*, 2002, **6**, 102–115.

8 M. Haruta and M. Date, Advances in the catalysis of Au nanoparticles, *Appl. Catal., A*, 2001, **222**, 427–437.

9 G. C. Bond and D. T. Thompson, Catalysis by gold, *Catal. Rev. Sci. Eng.*, 1999, **41**, 319–388.

10 C. Della Pina, E. Falletta, L. Prati and M. Rossi, Selective oxidation using gold, *Chem. Soc. Rev.*, 2008, **37**, 2077–2095.

11 L. Kesavan, R. Tiruvalam, M. H. Ab Rahim, M. I. bin Saiman, D. I. Enache, R. L. Jenkins, N. Dimitratos, J. A. Lopez-Sanchez, S. H. Taylor, D. W. Knight, C. J. Kiely and G. J. Hutchings, Solvent-Free Oxidation of Primary Carbon-Hydrogen Bonds in Toluene Using Au–Pd Alloy Nanoparticles, Science, 2011, **331**, 195–199.

12 C. L. Bracey, P. R. Ellis and G. J. Hutchings, Application of copper-gold alloys in catalysis: current status and future perspectives, *Chem. Soc. Rev.*, 2009, **38**, 2231–2243.

13 J. Pritchard, L. Kesavan, M. Piccinini, Q. A. He, R. Tiruvalam, N. Dimitratos, J. A. Lopez-Sanchez, A. F. Carley, J. K. Edwards, C. J. Kiely and G. J. Hutchings, Direct Synthesis of Hydrogen Peroxide and Benzyl Alcohol Oxidation Using Au–Pd Catalysts Prepared by Sol Immobilization, Langmuir, 2010, **26**, 16568–16577.

14 M. S. Chen, D. Kumar, C. W. Yi and D. W. Goodman, The promotional effect of gold in catalysis by palladium-gold, *Science*, 2005, **310**, 291–293.

15 M. S. Chen and D. W. Goodman, The structure of catalytically active gold on titania, *Science*, 2004, **306**, 252–255.

16 M. S. Chen and D. W. Goodman, Catalytically active gold: From nanoparticles to ultrathin films, *Acc. Chem. Res.*, 2006, **39**, 739–746.

17 J. A. Rodriguez, J. Evans, J. Graciani, J. B. Park, P. Liu, J. Hrbek and J. F. Sanz, High Water-Gas Shift Activity in $TiO_2(110)$ Supported Cu and Au Nanoparticles: Role of the Oxide and Metal Particle Size, *J. Phys. Chem. C*, 2009, **113**, 7364–7370.

18 V. A. Bondzie, S. C. Parker and C. T. Campbell, The Kinetics of CO Oxidation by Adsorbed Oxygen on Well-Defined Gold Particles on TiO2(110), *Catal. Lett.*, 1999, **63**, 143–151.

19 C. Lemire, R. Meyer, S. Shaikhutdinov and H. J. Freund, Do quantum size effects control CO adsorption on gold nanoparticles? *Angew. Chem., Int. Ed.*, 2004, **43**, 118–121.

20 N. Lopez, T. V. W. Janssens, B. S. Clausen, Y. Xu, M. Mavrikakis, T. Bligaard and J. K. Norskov, On the origin of the catalytic activity of gold nanoparticles for low-temperature CO oxidation, *J. Catal.*, 2004, **223**, 232–235.

21 R. Zanella, S. Giorgio, C. H. Shin, C. R. Henry and C. Louis, Characterization and reactivity in CO oxidation of gold nanoparticles supported on TiO_2 prepared by deposition-precipitation with NaOH and urea, *J. Catal.*, 2004, **222**, 357–367.

22 A. Sanchez, S. Abbet, U. Heiz, W. D. Schneider, H. Hakkinen, R. N. Barnett and U. Landman, When gold is not noble: Nanoscale gold catalysts, *J. Phys. Chem. A*, 1999, **103**, 9573–9578.

23 J. Guzman and B. C. Gates, Simultaneous presence of cationic and reduced gold in functioning MgO-supported CO oxidation catalysts: Evidence from X-ray absorption spectroscopy, *J. Phys. Chem. B*, 2002, **106**, 7659–7665.

24 Q. Fu, H. Saltsburg and M. Flytzani-Stephanopoulos, Active nonmetallic Au and Pt species on ceria-based water-gas shift catalysts, *Science*, 2003, **301**, 935–938.

25 J. D. Grunwaldt and A. Baiker, Gold/titania interfaces and their role in carbon monoxide oxidation, *J. Phys. Chem. B*, 1999, **103**, 1002–1012.

26 M. M. Schubert, S. Hackenberg, A. C. van Veen, M. Muhler, V. Plzak and R. J. Behm, CO oxidation over supported gold catalysts-"inert" and "active" support materials and their role for oxygen supply during reaction, *J. Catal.*, 2001, **197**, 113–122.

27 L. M. Molina and B. Hammer, Active role of oxide support during CO oxidation at Au/MgO, *Phys. Rev. Lett.*, 2003, **90**, 206102.

28 M. Date and M. Haruta, Moisture effects on CO oxidation over Au/TiO_2 catalyst, *J. Catal.*, 2001, **201**, 221–224.

29 M. Date, M. Okumura, S. Tsubota and M. Haruta, Vital role of moisture in the catalytic activity of supported gold nanoparticles, *Angew. Chem., Int. Ed.*, 2004, **43**, 2129–2132.

30 T. Yan, J. L. Gong, D. W. Flaherty and C. B. Mullins, The Effect of Adsorbed Water in CO Oxidation on $Au/TiO_2(110)$, *J. Phys. Chem. C*, 2011, **115**, 2057–2065.

31 L. Barrio, P. Liu, J. A. Rodriguez, J. M. Campos-Martin and J. L. G. Fierro, Effects of hydrogen on the reactivity of O_2 toward gold nanoparticles and surfaces, *J. Phys. Chem. C*, 2007, **111**, 19001–19008.

32 T. A. R. Nijhuis, T. Visser and B. M. Weckhuysen, The role of gold in gold-titania epoxidation catalysts, *Angew. Chem., Int. Ed.*, 2005, **44**, 1115–1118.

33 M. M. Schubert, V. Plzak, J. Garche and R. J. Behm, Activity, selectivity, and long-term stability of different metal oxide supported gold catalysts for the preferential CO oxidation in H_2-rich gas, *Catal. Lett.*, 2001, **76**, 143–150.

34 T. A. Nijhuis and B. M. Weckhuysen, The direct epoxidation of propene over gold-titania catalysts - A study into the kinetic mechanism and deactivation, *Catal. Today*, 2006, **117**, 84–89.

35 T. A. Nijhuis, T. Visser and B. M. Weckhuysen, Mechanistic study into the direct epoxidation of propene over gold/titania catalysts, *J. Phys. Chem. B*, 2005, **109**, 19309–19319.

36 C. Sivadinarayana, T. V. Choudhary, L. L. Daemen, J. Eckert and D. W. Goodman, The nature of the surface species formed on Au/TiO2 during the reaction of H_2 and O_2: An inelastic neutron scattering study, *J. Am. Chem. Soc.*, 2004, **126**, 38–39.

37 J. A. Farmer, J. H. Baricuatro and C. T. Campbell, Ag Adsorption on Reduced $CeO_2(111)$ Thin Films, *J. Phys. Chem. C*, 2010, **114**, 17166–17172.

38 J. T. Stuckless, N. A. Frei and C. T. Campbell, A Novel Single Crystal Adsorption Calorimeter and Additions forDetermining Metal Adsorption and Adhesion Energies, *Rev. Sci. Instrum.*, 1998, **69**, 2427–2438.

39 H. Ihm, H. M. Ajo, J. M. Gottfried, P. Bera and C. T. Campbell, Calorimetric Measurement of the Heat of Adsorption of Benzene on Pt(111), *J. Phys. Chem. B*, 2004, **108**, 14627–14633.

40 D. Zscherpel, W. Ranke, W. Weiss and R. Schlogl, Energetics and kinetics of ethylbenzene adsorption on epitaxial FeO(111) and Fe_3O_4(111) films studied by thermal desorption and photoelectron spectroscopy, *J. Chem. Phys.*, 1998, **108**, 9506–9515.

41 J. H. Larsen, J. T. Ranney, D. E. Starr, J. E. Musgrove and C. T. Campbell, Adsorption energetics for Ag on MgO(100), *Phys. Rev.*, 2001, **B 63**, 195410.

42 J. A. Farmer and C. T. Campbell, Ceria Maintains Smaller Metal Catalyst Particles by Strong Metal-Support Bonding, *Science*, 2010, **329**, 933–936.

43 D. R. Lide, ed., *CRC Handbook of Chemistry and Physics (Internet Version)*, CRC Press/ Taylor and Francis, Boca Raton, FL, 2011.

44 C. T. Campbell, S. C. Parker and D. E. Starr, The effect of size-dependent nanoparticle energetics on catalyst sintering, *Science*, 2002, **298**, 811–814.

45 S. C. Parker and C. T. Campbell, Kinetic model for sintering of supported metal particles with improved size-dependent energetics and applications to Au on TiO_2(110), *Physical Review B*, 2007, **75**, 035430.

46 C. T. Campbell, Ultrathin Metal Films on Oxide Surfaces: Structural, Electronic and Chemisorptive Properties, *Surf. Sci. Rep.*, 1997, **27**, 1.

47 V. Bondzie, S. C. Parker and C. T. Campbell, Oxygen Adsorption on Well-Defined Gold Particles on TiO2(110), *J. Vac. Sci. Technol. A*, 1999, **17**, 1717–1720.

48 C. T. Campbell, Erratum: Oxygen adsorption on well-defined gold particles on TiO_2(110) (vol 17, pg 1717, 1999), *J. Vac. Sci. Technol., A*, 2008, **26**, 1546–1546.

49 J. Kirn, E. Samano and B. E. Koel, Oxygen adsorption and oxidation reactions on Au(211) surfaces: Exposures using O-2 at high pressures and ozone (O_3) in UHV, *Surf. Sci.*, 2006, **600**, 4622–4632.

50 J. H. Fischer-Wolfarth, J. A. Farmer, J. M. Flores-Camacho, A. Genest, I. V. Yudanov, N. Rosch, C. T. Campbell, S. Schauermann and H. J. Freund, Particle-size dependent heats of adsorption of CO on supported Pd nanoparticles as measured with a single-crystal microcalorimeter, *Phys. Rev. B: Condens. Matter Mater. Phys.*, 2010, **81**, 241416.

51 L. J. Xu and Y. Xu, Effect of Pd surface structure on the activation of methyl acetate, *Catalysis Today*, 2011, **165**, 96–105.

52 B. Hammer, O. H. Nielsen and J. K. Norskov, Structure sensitivity in adsorption: CO interaction with stepped and reconstructed Pt surfaces, *Catal. Lett.*, 1997, **46**, 31–35.

53 C. T. Campbell, G. Ertl, H. Kuipers and J. Segner, A Molecular Beam Study of the Interactions of CO with a Pt(111) Surface, *Surf. Sci.*, 1981, **107**, 207–219.

54 W. Lew, M. C. Crowe, E. Karp, O. Lytken, J. A. Farmer, L. Árnadóttir, C. Schoenbaum and C. T. Campbell, The Energy of Adsorbed Hydroxyl on Pt(111) by Microcalorimetry, *J. Phys. Chem. C*, 2011, **115**, 11586–11594.

55 A. Hodgson and S. Haq, Water adsorption and the wetting of metal surfaces, *Surf. Sci. Rep.*, 2009, **64**, 381–451.

56 H. E. Bryndza, L. K. Fong, R. A. Paciello, W. Tam and J. E. Bercaw, Relative Metal–Hydrogen, –Oxygen, –Nitrogen, and –Carbon Bond Strengths for Organoruthenium and Organoplatinum Compounds; Equilibrium Studies of $Cp^*(PMe_3)_2RuX$ and (DPPE) MePtX Systems, *J. Am. Chem. Soc.*, 1987, **109**, 1444–1456.

57 A. R. Bulls, J. E. Bercaw, J. M. Manriquez and M. E. Thompson, Relative bond-dissociation energies for early transition-metal alkyl, aryl, alkynyl and hydride compounds - equilibration of metallated cyclopentadienyl derivatives of peralkylated hafnocene and scandocene with hydrocarbons and dihydrogen, *Polyhedron*, 1988, **7**, 1409–1428.

58 H. E. Bryndza, P. J. Domaille, W. Tam, L. K. Fong, R. A. Paciello and J. E. Bercaw, Comparison of metal-hydrogen, metal-oxygen, metal-nitrogen and metal-carbon bond strengths and evaluation of functional-group additivity principles for organoruthenium and organoplatinum compounds, *Polyhedron*, 1988, **7**, 1441–1452.

59 M. A. Newton and C. T. Campbell, Benzyl Coupling on Bismuth-Modified Pt{111}, *Catal. Lett.*, 1996, **37**, 15–23.

60 D. F. McMillen and D. M. Golden, Hydrocarbon Bond Dissociation Energies, *Annu. Rev. Phys. Chem.*, 1982, **33**, 493–532.

61 M. W. Chase, Jr, NIST-JANAF Thermochemical Tables, *J. Phys. Chem. Ref. Data, NIST-JANAF Thermochemical Tables, Monograph 9*, 1998, **Monograph 9**, 1–1951 (via NIST Webbook 2011: http://webbook.nist.gov/chemistry/).

62 E. M. Karp, T. L. Silbaugh and C. T. Campbell, in preparation.

63 D. H. Parker, M. E. Bartram and B. E. Koel, Study of high coverages of atomic oxygen on the Pt(111) surface, *Surf. Sci.*, 1989, **217**, 489–510.

64 J. M. Keith, R. P. Muller, R. A. Kemp, K. I. Goldberg, W. A. Goddard and J. Oxgaard, Mechanism of direct molecular oxygen insertion in a palladium(II)-hydride bond, *Inorg. Chem.*, 2006, **45**, 9631–9633.

65 F. Abild-Pedersen, J. Greeley, F. Studt, J. Rossmeisl, T. R. Munter, P. G. Moses, E. Skulason, T. Bligaard and J. K. Norskov, Scaling properties of adsorption energies for hydrogen-containing molecules on transition-metal surfaces, *Phys. Rev. Lett.*, 2007, **99**, 016105.

66 E. M. Fernandez, P. G. Moses, A. Toftelund, H. A. Hansen, J. I. Martinez, F. Abild-Pedersen, J. Kleis, B. Hinnemann, J. Rossmeisl, T. Bligaard and J. K. Norskov, Scaling relationships for adsorption energies on transition metal oxide, sulfide, and nitride surfaces, *Angew. Chem., Int. Ed.*, 2008, **47**, 4683–4686.

67 A. G. Sault, R. J. Madix and C. T. Campbell, Adsorption of Oxygen and Hydrogen on Au (110) − (1 × 2), *Surf. Sci.*, 1986, **169**, 347–356.

68 O. Lytken, W. Lew, J. J. W. Harris, E. K. Vestergaard, J. M. Gottfried and C. T. Campbell, Energetics of cyclohexene adsorption and reaction on Pt(111) by low-temperature microcalorimetry, *J. Am. Chem. Soc.*, 2008, **130**, 10247–10257.

69 H. M. Ajo, V. A. Bondzie and C. T. Campbell, Propene adsorption on gold particles on TiO$_2$(110), *Catal. Lett.*, 2002, **78**, 359–368.

A paradigm for predicting selective oxidation on noble metals: oxidative catalytic coupling of amines and aldehydes on metallic gold

Bingjun Xu,[a] Cynthia M. Friend[ab] and Robert J. Madix[*b]

Received 3rd February 2011, Accepted 10th March 2011
DOI: 10.1039/c1fd00012h

We demonstrate in the present work that relatively straightforward acid/base principles of surface reactivity predict oxygen-assisted amine–aldehyde coupling on metallic gold. Formed *via* the oxygen-assisted (Brønsted acid) N–H bond activation of dimethylamine, $(CH_3)_2N_{(a)}$ acts as a nucleophile to couple with various aldehydes, forming the corresponding amides. At low initial coverages of oxygen on the surface very high selectivities are achieved. The reaction proceeds *via* the surface-bound hemiaminal intermediate, which β-hydride eliminates well below room temperature to form the amide product. On metallic gold desorption of the amide appears to be the rate-limiting step. Under the transient conditions employed in this work oxygen-assisted coupling of the amine with alcohols is limited, suggesting that such reactions must be conducted in the steady state in order to have both the aldehyde and adsorbed $(CH_3)_2N_{(a)}$ present simultaneously.

Introduction

Parravano and Schwank and others long ago reported catalytic activity for metallic gold supported on refractory metal oxides.[1,2] Overall, the activity observed for these catalysts was low, and the studies appeared to be basically "academic" in nature. About a decade later Haruta reported significantly higher activities for some of the same reactions when reducible metal oxides were employed as support materials.[3] One dramatic difference was that CO could be oxidized at temperatures as low as −70 °C. This surprising report echoed a study of the oxidation of CO on single crystal gold, Au(110), that showed CO and adsorbed atomic oxygen react readily at room temperature *via* a Langmiur–Hinshelwood mechanism with an activation energy of only 2.1 ± 1 kcal mol^{-1}.[4] Apparently the reducible metal oxide can supply atomic oxygen to the gold surface in the steady state and promote catalytic oxidation.

The incentive to develop catalytic processes that operate at low temperature with high selectivity has recently given impetus to further development of metallic gold catalysts. In order to rationally design and improve new and existing catalysts, it is important to have a fundamental understanding of the sequence of elementary steps in each reaction; *i.e.*, to know the reaction mechanism. A key step in the mechanistic studies is to find a model system that is simple enough to be studied in detail and representative enough to reflect the reaction mechanism in more

[a]*Department of Chemistry and Chemical Biology, Harvard University, Cambridge, MA, 02138, USA*
[b]*School of Engineering and Applied Sciences, Harvard University, Cambridge, MA, 02138, USA. E-mail: rmadix@seas.harvard.edu; Fax: +617 496 9489; Tel: +617 496 9489*

complicated systems. Excellent correspondence has been established between a model system (Au(111) single crystal surface) under ultra-high vacuum (UHV) conditions[5-9] and nanoporous gold under ambient conditions[10] or supported gold nanoparticles in the liquid phase[11,12] for methanol esterification, which is highly suggestive of a common reaction mechanism in these three different processes. This correspondence is attributed to the following characteristic features of the gold surface: 1) it is inert in the absence of atomic oxygen, preventing the buildup of surface spectator species; 2) it is quite inefficient for dissociating O_2, causing a low steady-state coverage of atomic oxygen in the reacting system and 3) it interacts weakly with water, allowing reactions to be conducted in the liquid phase without interference from the solvent. Therefore, Au(111) appears to be a good model system for fundamental mechanistic studies of partial oxidation reactions, including oxygen-assisted coupling.

A basic conceptualization of oxygen-assisted partial oxidation reactions in terms of acid-base principles was developed for metallic silver and gold surfaces in the early 1980's.[13,14] More recently these concepts have been dramatically extended to much more complex reactions on metallic gold by Friend *et al.*[6-8] Inherent to these partial oxidation reactions is the role of surface adsorbed oxygen, which acts as a Brønsted base and activates the acidic hydrogen in various reactants, BH, as shown in equation 1:

$$BH_{(a)} + O_{(a)} \rightarrow B_{(a)} + OH_{(a)} \tag{1}$$

BH can be, for example, an alcohol, amine, carboxylic acid or unsaturated hydrocarbon; B is the conjugated base of the gas-phase acid, BH. Not only can $B_{(a)}$ react further in unimolecular fashion to create specific products, such as the formation of aldehyde from adsorbed alkoxy, in some cases it may react as a nucleophile, attacking electron-deficient centers in coadsorbed species, *e. g.* adsorbed alkoxy may attack the carbonyl carbon in formaldehyde, to form a surface hemiacetal intermediate.[7,9,15,16] which can undergo β-H elimination leading to ester formation (Eqns. 2 and 3):

$$CH_3O_{(a)} + HC(H){=}O_{(a)} \rightarrow HC(O)(H)OCH_{3(a)} \tag{2}$$

$$2HC(O)(H)OCH_{3(a)} + O_{(a)} \rightarrow 2HC({=}O)OCH_{3(a)} + H_2O_{(a)} \tag{3}$$

Indeed, recent work clearly shows that, in general, adsorbed alkoxys couple with coadsorbed aldehydes on Au(111),[8] forming the corresponding esters with high selectivity (Eqns. 4 and 5):

$$RCH_2O_{(a)} + R'C(H) = O_{(a)} \rightarrow R'C(O)(H)OCH_2R_{(a)}$$
$$R \text{ and } R' = C_nH_{2n+1} \ n = 0, \ 1, \ 2... \tag{4}$$

$$2R'C(O)(H)OCH_2R_{(a)} + O_{(a)} \rightarrow 2RC({=}O)OCH_2R_{(a)} + H_2O_{(a)} \tag{5}$$

When only a single type of alkoxy is available on the surface, a fraction of the alkoxy undergoes a β-H elimination reaction to form the corresponding aldehyde, creating the coadsorbed electrophile *in situ*. The subsequent nucleophilic attack of the remaining alkoxy on the newly formed aldehyde leads to the formation of the self-coupling product (Eqns. 6 and 7):

$$RCH_2O_{(a)} + O_{(a)} \rightarrow RC(H){=}O_{(a)} + OH_{(a)} \tag{6}$$

$$2RCH_2O_{(a)} + 2RC(H){=}O_{(a)} + O_{(a)} \rightarrow 2RC({=}O)OCH_2R_{(a)} + H_2O_{(a)} \tag{7}$$

      This journal is © The Royal Society of Chemistry 2011

The self-coupling reaction of alcohols proceeds well below room temperature on both oxygen-activated Au(111)[7,9] and Au(110).[16] Both experiment and theory show the β-H elimination of the alkoxy is the rate-limiting step on Au(111); this process may be assisted by adsorbed oxygen.[7,8,17] The self-coupling product of methanol, methylformate, is evolved at 215 K in the self-coupling of methanol, 40 K higher than it appears when formaldehyde is reacted with adsorbed methoxy. By supplying formaldehyde to the adsorbed methoxy, the coupling reaction circumvents the rate-limiting β-H elimination of methoxy.[7,8] DFT calculations also suggest that the β-H elimination of methoxy is rate-limiting and that the nucleophilic attack of methoxy on formaldehyde is barrierless.[17]

The mechanistic understanding of the cross-coupling between alkoxys and aldehydes suggests a *generalization* to predict other reaction pathways. By replacing either the alkoxy or aldehyde with another adsorbed nucleophile or electrophile, it is likely that similar nucleophile-electrophile reactions will occur. Adsorbed amide is an attractive candidate for the nucleophile due to its structural similarity with alkoxy. NH_2^- is reported as a strong nucleophile for nucleophilic substitution reactions.[18] The activation of N–H bonds by adsorbed atomic oxygen has been shown on both silver[19,20] and gold surfaces,[21] so that there is a clear route to the selective formation of adsorbed amides. In this paper we present recent results for amide synthesis by the cross-coupling of an amine and an aldehyde.

Aldehydes and ketones react with primary (or secondary) amines in acidic solution to form imines (enamines) *via* a nucleophilic addition-elimination mechanism.[18] Molecular amines nucleophilically attack the electron deficient carbonyl carbon in the aldehyde or ketone, forming the N-protonated carbinolamine intermediate. Imine or enamine is formed after the subsequent water removal and deprotonation.[18] In contrast, the surface bound amide formed *via* the N–H activation of amine by surface oxygen is expected to attack the electron deficient carbonyl carbon in aldehydes on metallic silver and gold to form an adsorbed hemiaminal - structurally analogous to the hemiacetal (Eqns. 4 and 5). The hemiaminal can then β-H eliminate and form the corresponding amide. Though starting with identical reactants, the products from the solution phase and oxygen-assisted surface-mediated reactions are expected to be different.

The use of a secondary amine allows one to create a well defined surface amide, since there is only one N–H bond for activation. Herein we report a detailed study of the cross-coupling reaction between the simplest secondary amine, dimethyl-amine, and different aldehydes as well as alcohols on oxygen-activated Au(111) (O/Au(111)). A combination of temperature programmed reaction spectroscopy (TPRS), X-ray photoelectron spectroscopy (XPS) and high resolution electron energy loss spectroscopy (HREELS) were employed in UHV to obtain a comprehensive picture of reaction pathways analogous to the previously established mechanistic framework for alcohol coupling with aldehydes.

Experimental

All experiments were performed in an ultrahigh vacuum chamber with a base pressure below 2×10^{-10} Torr. The preparation of the clean Au(111) surface has been described elsewhere.[22] The surface was first populated with 0.1 ML O by directly dosing ozone at 200 K. The oxygen atom coverage was calibrated by comparing the amount of O_2 evolved in temperature programmed desorption to that formed for a saturation coverage of oxygen atoms using the same dosing method.[23] Oxidation of the surface in this manner leads to the release of Au atoms to form nanostructures containing Au and O, most of which are smaller than 2 nm in diameter.[22]

Organic molecules were introduced directly onto the O/Au(111) surface at 150 K *via* a dosing tube. Exposures, corrected for dosing enhancement, are given here in terms of Langmuir (L) (1 L = 10^{-6} torr-seconds); the dosing pressure was determined by uncorrected ion gauge pressures.

Temperature programmed reaction was conducted according to well-established protocols, described in detail elsewhere.[22] The heating rate for all reactions was nearly constant at 5 K s^{-1}. The reaction products were identified by quantitative mass spectrometry (Hiden HAL/3F) using authentic samples to obtain fragmentation patterns, which were found to be in general agreement with NIST reference data.[24] HREELS experiments were performed with an LK2000 spectrometer using a primary energy of 7.17 eV at 60° specular geometry. The XPS experiments were conducted in a second UHV chamber, with a base pressure below 5×10^{-10} Torr. XPS spectra were acquired with an analyzer pass energy of 17.9 eV and a multiplier voltage of 3 kV using MgKα X-rays (300 W) as the excitation source. The binding energy (BE) calibration was referenced to the Au $4f_{7/2}$ peak at 83.9 eV. The O(1s) spectra were acquired with 100 scans to enhance the signal-to-noise ratio.

Results

Dimethylamine (DMA) adsorbs and desorbs reversibly on the Au(111) surface, with no detectable amount of N–H bond activation. Only one single N1s peak was observed in the X-ray photoelectron spectrum for adsorbed DMA, with a binding energy of 399.7 eV (Figure 1, bottom trace). When DMA was introduced to the oxgyen covered Au(111) surface ($\theta_O = 0.15$ ML), in addtion to the feature for the molecularly adsorbed dimethylamine, a shoulder at 398.7 eV in the N1s XPS spectrum developed, consistent with the previous assignment to the N atom in $(CH_3)_2N_{(a)}$ on Cu(211),[25] as expected from the reaction pattern previously characterized for reactions of amines with O/Au(111) (Fig. 1, top trace; Eqn. 8)),[26]

$$(CH_3)_2NH_{(a)} + O_{(a)} \rightarrow (CH_3)_2N_{(a)} + OH_{(a)} \tag{8}$$

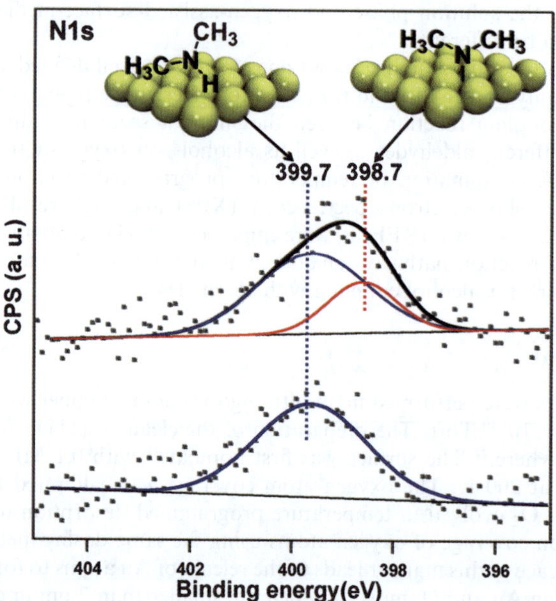

Fig. 1 X-ray photoelectron spectra (N1s) of dimethylamine adsorbed on the Au(111) surface (bottom trace) and on O/Au(111) ($\theta_O = 0.15$ ML, top trace) at 120 K. The N1s peaks at 399.7 eV and 398.7 eV are assigned to N in molecularly adsorbed dimethylamine and the $(CH_3)_2N_{(a)}$ surface intermediate, respectively.

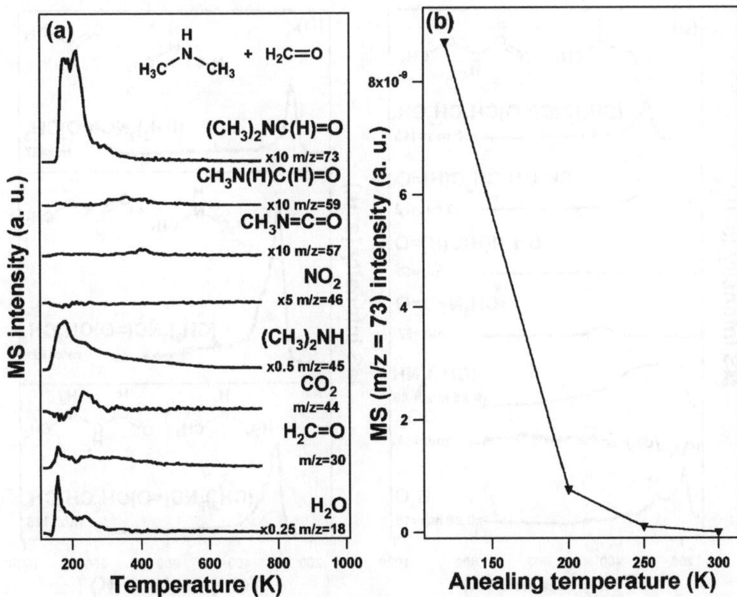

Fig. 2 (a) Reaction of dimethylamine and formaldehyde on the O/Au(111) (θ_O = 0.2 ML) yields the cross-coupling product, dimethylformamide. Dimethylamine and formaldehyde were introduced to the O/Au(111) at 120 K. Surface oxygen was prepared by ozone exposure at 200 K. (b) The amount of the cross-coupling product (dimethylformamide) as a function of the annealing temperature immediately after dimethylamine was introduced, but before formaldehyde was dosed. This procedure leads to secondary oxidation of the adsorbed amide prior to exposure to the aldehyde.

Based on analogy with the coupling reactions of alkoxys and aldehydes, this $(CH_3)_2N_{(a)}$ is expected to nucleophilically attack the electron deficient carbon in aldehydes. Indeed, $(CH_3)_2N_{(a)}$ couples very selectively with formaldehyde forming the surface hemiaminal intermediate, which subsequently β-H eliminates and produces dimethylformamide at around 180 K (Figure 2(a)),[26] as in Eqns. 9 and 10:

$$(CH_3)_2N_{(a)} + H_2C = O \rightarrow (CH_3)_2NC(H_2)O_{(a)} \tag{9}$$

$$2(CH_3)_2NC(H_2)O_{(a)} + O_{(a)} \rightarrow 2(CH_3)_2NC(H)=O + H_2O_{(a)} \tag{10}$$

At low oxygen coverage only a very small amount of the secondary oxidation of dimethylamine occurs, *i.e.* methylformamide and methylisocyanate. The selectivity for cross-coupling product to form dimethylformamide depends strongly on the initial surface oxygen coverage, exhibiting almost 100% selectivity at low oxygen coverage.[26] This efficient cross-coupling reaction illustrates the predictive power of the general mechanistic framework based on acid-base principles.

The thermal stability of the $(CH_3)_2N_{(a)}$ intermediate was determined by annealing experiments. $(CH_3)_2N_{(a)}$ was prepared by dosing dimethylamine on O/Au(111) at 120 K and then quickly heating to a defined temperature and quenching to 120 K before the introduction of formaldehyde. The amount of dimethylformamide formed dropped precipitously as the annealing temperature increased from 120 to 200 K (Figure 2(b)) decreasing to nearly zero at an annealing temperature of 300 K; $(CH_3)_2N_{(a)}$ is not stable close to or above 200 K on O/Au(111), and most of dimethylamide, $(CH_3)_2N_{(a)}$, has already been converted to the precursor $(CH_3NCH_2O_{(a)})$ for secondary oxidation during annealing to 200 K[26] and is lost

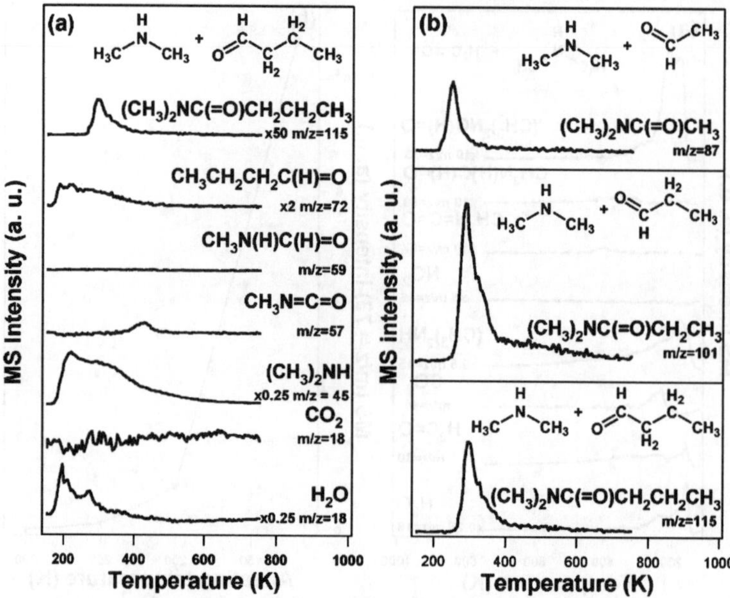

Fig. 3 (a) Reaction of dimethylamine and butanal on O/Au(111) ($\theta_O = 0.1$ ML) yields the cross-coupling product, dimethylbutyramide. (b) Reaction of dimethylamine with acetalde-hyde, propanal and butanal yields dimethylacetamide, dimethylpropanamide and dimethylbu-tyramide, respectively. Dimethylamine and aldehydes were introduced to the O/Au(111) ($\theta_O = 0.1$ ML) at 150 K and the heating rate is 5 K s^{-1}. Surface oxygen was prepared by ozone expo-sure at 200 K.

for initiating nucleophilic attack. These results show that the cross-coupling between $(CH_3)_2N_{(a)}$ and formaldehyde occurs below the temperature at which secondary oxidation becomes significant, which is a major factor leading to the high selectivity for the coupling reaction.

In order to establish the generality of the mechanism of this cross-coupling reac-tion between amine and aldehydes, reactions of the adsorbed dimethylamide with other aldehydes or aldehydes derived from oxy-dehydrogenation of alcohols were examined on O/Au(111), *i.e.* aceldehyde, propanal and butanal. Indeed, $(CH_3)_2N_{(a)}$ couples efficiently with butanal, forming dimethylbutyramide as the main product (Fig. 3(a)). Very little secondary oxidation of dimethylamine (methyl-isocyanate) was observed, suggesting the nucleophilic attack of $(CH_3)_2N_{(a)}$ on buta-nal occurs at sufficiently low temperature to preclude the secondary oxidation reaction. Similar results are found for other aldehydes, with the amides being the primary products (Figure 3(b)).

However, cross-coupling between dimethylamine and primary alkyl alchols (methanol, ethanol, propanol and butanol) was *not* observed under similar condi-tions (Figs. 4(a) and (b)). When dimethylamine and methanol were sequentially ad-sorbed on the O/Au(111) ($\theta_O = 0.1$ ML), no dimethylformamide was produced. Instead, the self-coupling of methanol to yield methylformate and secondary oxida-tion of the adsorbed dimethyamide to methylformamide and methylisocyanate dominated the reaction (Fig. 2(a)). The absence of cross-coupling between methanol and the adsorbed dimethylamide can be attributed to the fact that $(CH_3)_2N_{(a)}$ and formaldehyde never coexist on the surface in significant amounts. β-hydride elimina-tion from methoxy occurs near 215 K, at which secondary oxidation of the adsorbed amide has occurred. Hence, secondary oxidation of the adsorbed amide and self-coupling of methoxy occur independently under the circumstances used here. It is not yet known whether coupling would occur were the excess oxygen removed

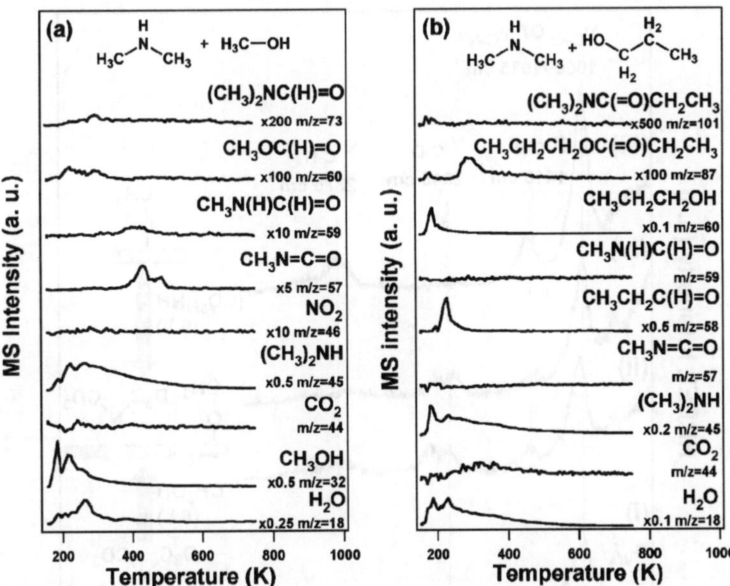

Fig. 4 Reaction of dimethylamine and (a) methanol and (b) propanol on O/Au(111) ($\theta_O = 0.1$ ML). No cross-coupling reaction between dimethylamine and either alcohol was observed Dimethylamine and aldehydes were introduced to the O/Au(111) at 150 K and the heating rate is 5 K s^{-1}. Surface oxygen was prepared by ozone exposure at 200 K.

from the surface. The absence of the cross-coupling reaction between ethanol and dimethylamine can be explained by similar logic.

The absence of cross-coupling between propanol or butanol with dimethylamine has a different cause. Propoxy and butoxy self-couple at or below 150 K, indicating that propanal and butanal form at temperatures below which significant secondary oxidation of the adsorbed dimethylamide occurs. However, no cross-coupling reaction was observed when dimethylamine and propanol were sequentially introduced to O/Au(111) ($\theta_O = 0.1$ ML). Apart from molecular desorption, there is no indication of reaction of dimethylamine at all, even secondary oxidation. Only the partial oxidation and self-coupling products of propanol were detected (Figure 4(b)). This result suggests that propanol completely displaces $(CH_3)_2N_{(a)}$ from the surface (Eqn.11).

$$(CH_3)_2N_{(a)} + CH_3CH_2CH_2OH_{(a)} \rightarrow (CH_3)_2NH + CH_3CH_2CH_2O_{(a)} \qquad (11)$$

Similar displacement reactions have been reported for alkoxy/alcohol interactions Au(111) suface, with longer chain adsorbed alkoxys being more stable.[5] Therefore, the absence of cross-coupling reaction between propanol and dimethylamine is due to the diplacement reaction between propoxy and $(CH_3)_2N_{(a)}$. Similar results were observed in the case of coadsorbed dimethylamine and butanol on the O/Au(111). No cross-coupling product was formed because butoxy can very effectively displace $(CH_3)_2N_{(a)}$. We have not yet achieved a mixed adlayer of dimethylamide and either propoxy or butoxy to determine if cross-coupling will occur. It is very possible that under different circumstances this reaction will proceed.

These matters were investigated in more detail by vibrational spectroscopy (Figure 5). $(CD_3)_2N_{(a)}$ was prepared by reacting $(CD_3)_2NH$ with O/Au(111) ($\theta_O = 0.1$ ML) at 150 K, as evidenced by the characteristic ν_{C-N} and ν_{C-D} absorption bands at 1005 and 2045 cm^{-1}, respectively, in the HREELS spectrum (Figure 5(i)). After the introduction of methanol at 150 K, in addition to the existing bands attributed

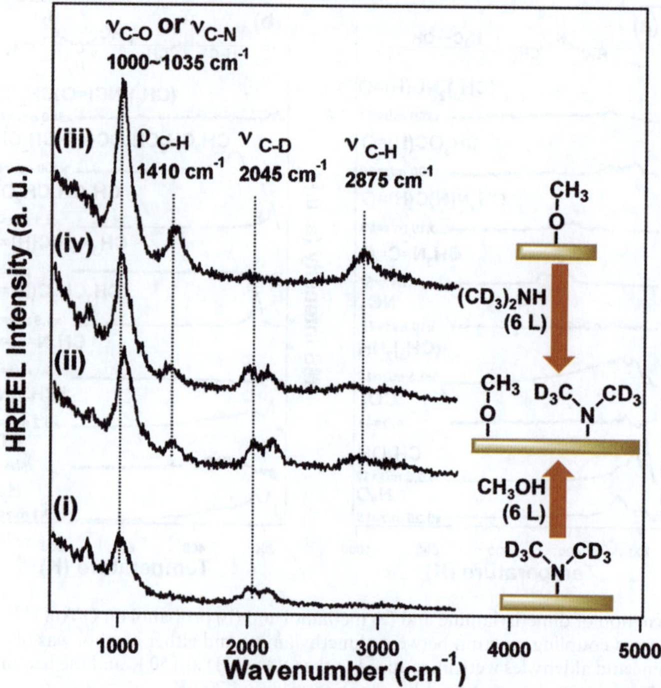

Fig. 5 High-resolution electron energy loss spectra of i) $(CD_3)_2N_{(a)}$, prepared by introducing 6 L of dimethylamine-d_6 to O/Au(111) ($\theta_O = 0.1$ ML) at 150 K; ii) introducing 6 L of methanol to the surface prepared by i); iii) $CH_3O_{(a)}$, prepared by introducing 6 L of methanol to O/Au(111) ($\theta_O = 0.1$ ML) at 150 K; iv) introducing 6 L of dimethylamine-d_6 to the surface prepared by iii).

to $(CD_3)_2N_{(a)}$, new absorption bands characteristic of methoxy appeared at 1410 and 2875 cm^{-1}, corresponding to the ρ_{C-H} and ν_{C-H} vibrational modes (Figure 5 (ii)). Reversing the dosing sequence of $(CD_3)_2NH$ and methanol led to a very similar spectrum (Figure 5(iv)), showing that regardless of dosing sequence, $(CD_3)_2N_{(a)}$ and methoxy can coexist on the surface. Similar results were observed for $(CD_3)_2N_{(a)}$ and ethoxy. Therefore, methoxy, ethoxy and $(CH_3)_2N_{(a)}$ have similar surface stability.

In contrast, propoxy effectively and irreversibly displaces $(CH_3)_2N_{(a)}$. When propanol was introduced to $(CD_3)_2N$-covered O/Au(111) at 150 K (Figure 6(i)), the characteristic vibrational bands of $(CD_3)_2N_{(a)}$ were replaced by modes at 1420 and 2890 cm^{-1} corresponding to the ρ_{C-H} and ν_{C-H} modes of propoxy (Figure 6(ii)). Further introduction of $(CD_3)_2NH$ did not change the spectrum, indicating that adsorbed propoxy is much more stable than $(CD_3)_2N_{(a)}$. These observations are consistent with the TPR results that show the absence of reactions of adsorbed dimethylamide following exposure of $(CD_3)_2N_{(a)}$ to propanol. Similar conclusions follow for butanol.

Discussion

Our results show conclusively that the cross-coupling reaction between amines and aldehydes is fundamentally analogous to the self- and cross-coupling of alcohols and aldehydes, invoking the nucleophilicity of both the adsorbed alkoxy and amide. In these reactions there are three common critical mechanistic steps (Scheme 1): 1) surface oxygen-assisted activation of the acidic hydrogen in alcohol or amine through a Brønsted acid-base reaction; 2) nucleophilic attack of the alkoxy or amide at the electron deficient carbonyl carbon in the aldehydes to form either the surface

     This journal is © The Royal Society of Chemistry 2011

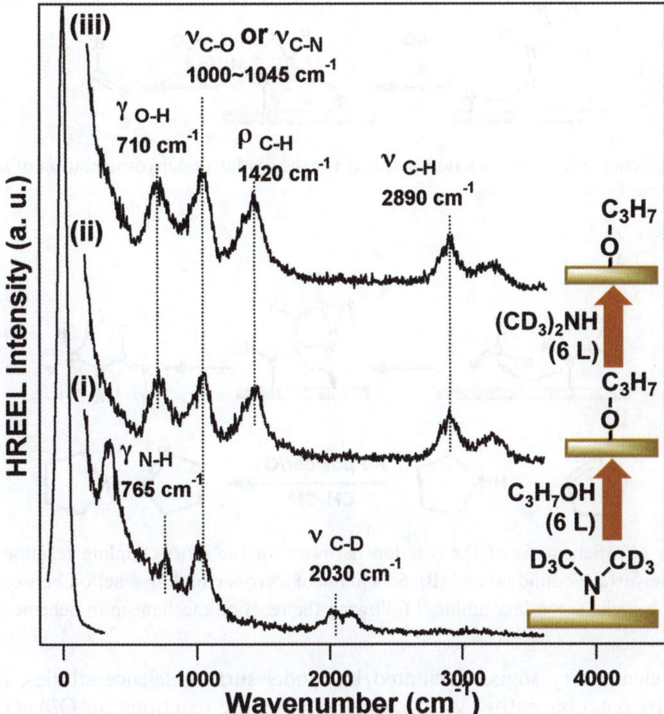

Fig. 6 High-resolution electron energy loss spectra of i) (CD$_3$)$_2$N$_{(a)}$, prepared by introducing 6 L of dimethylamine-d_6 to O/Au(111) ($\theta_O = 0.1$ ML) at 150 K; ii) introducing 6 L of propanol to the surface prepared by i); iii) introducing 6 L of dimethylamine-d_6 to the surface prepared by ii).

hemiacetal or hemiaminal intermediate and 3) β-H elimination from this species to form the ester or amide. In the case of the self-coupling reaction of alcohols, the aldehyde is formed *via* β-H elimination step of the corresponding alkoxy species.

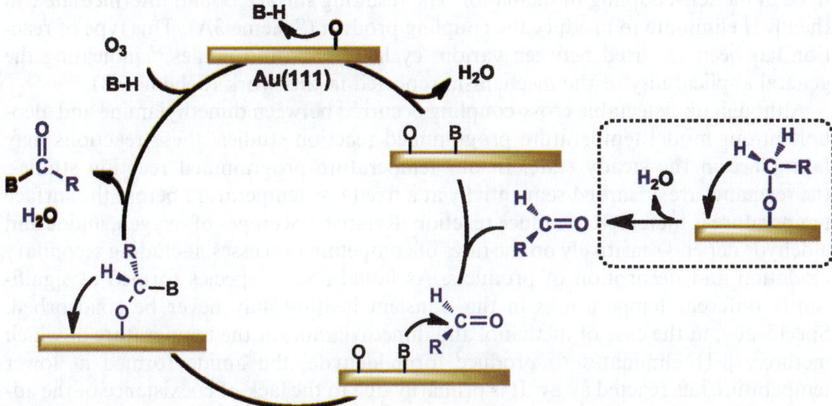

Scheme 1 Schematic of the general mechanism for cross-coupling reactions with its three critical steps: 1) acidic hydrogen activation *via* a Brønsted acid-base reaction; 2) nucleophilic attack to the electron-deficient carbonyl carbon in the aldehyde and 3) β-H elimination of the surface intermediate formed in the previous step. The aldehyde could be either supplied externally or produced *in situ via* β-H elimination of an alkoxy species (the step enclosed in the dashed line).

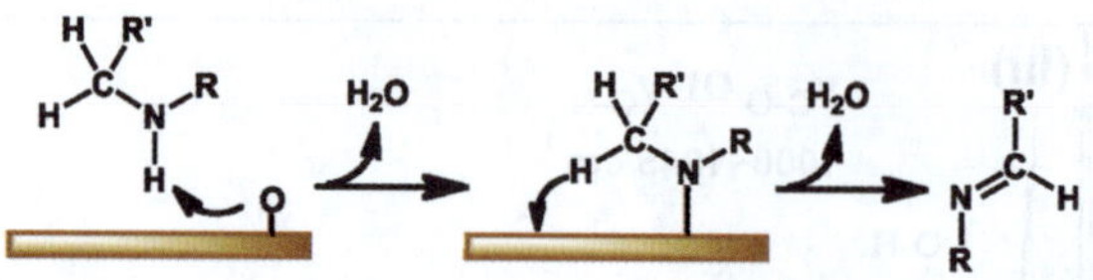

Scheme 2 Schematic of the reaction pathway for the oxidative dehydrogenation of secondary amines.

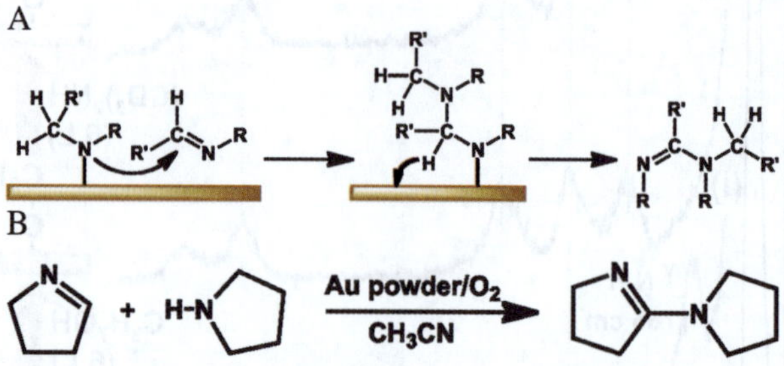

Scheme 3 (**A**). Schematic of the reaction pathway for the cross-coupling reaction between imine and a surface bound amide. (**B**). Schematic of a cross-coupling reaction between a cyclic imine and a cyclic secondary amine,[27] following the reaction mechanism in Scheme 3A.

These elementary steps, elucidated by model surface science studies, not only explain the reaction pathway for oxidative coupling reactions on O/Au(111), but they also illuminate many other gold-mediated reactions in the literature. For example, the recently reported oxidative dehydrogenation of secondary amines to imines[27] on metallic gold most likely occurs by acidic hydrogen activation by adsorbed oxygen to form a surface-bound amide, followed by β-H elimination reaction to yield the corresponding imine (Scheme 2).

Due to the similarity of the electronic structure of imines to aldehydes, with the carbon double-bonded to the nitrogen atom being electron deficient, it seems reasonable that nucleophilic attack of the electron-rich amide, formed in the first step of the Scheme 2, on the imine can occur, similar to the attack of methoxy upon formaldehyde in the self-coupling of methanol. The resulting surface bound intermediate can then β-H eliminate to produce the coupling product (Scheme 3A). This type of reaction has been reported between various cyclic amines and imines,[27] indicating the general applicability of the mechanism reported in this work (Scheme 3B).

Although no detectable cross-coupling occurred between dimethylamine and alcohols in our model temperature programmed reaction studies, these reactions may take place in the steady state. In our temperature programmed reaction studies, the reactants are adsorbed sequentially at a fixed low temperature before the surface temperature is increased to induce reaction. Relative coverages of oxygen, amide and aldehyde depend sensitively on the rates of competing processes, including secondary oxidation and desorption of products. As noted above, species formed at significantly different temperatures in the transient heating may never be coadsorbed. Specifically, in the case of methanol and dimethylamine at the temperature at which methoxy β-H eliminates to produce formaldehyde, the amide formed at lower temperature has reacted away. It is primarily due to the lack of coexistence of the adsorbed amide intermediate and the aldehyde that precludes the nucleophilic reaction from occurring, rather than the feasibility of the reaction itself. However, were the reaction carried out in liquid phase in the steady state at an appropriate temperature and steady state oxygen coverage, it is possible to have amide and aldehyde coexist, leading to the formation of the amide. Indeed, Christensen *et al.* have recently reported the reactions between alkyl alcohols and amines in the liquid phase.[28]

Conclusions

Oxygen-assisted cross-coupling reactions between dimethylamine and various aldehydes to form the corresponding amides occur on gold surfaces in a chemically analogous manner to the coupling reactions of alcohols and aldehydes. Acidic hydrogen abstraction *via* a Brønsted acid-base reaction, nucleophilic attack and oxygen assisted β-H elimination are identified as the critical mechanistic steps. Not only are these steps the corner stones of the mechanism for the coupling reactions on O/Au(111), they also explain similar gold mediated catalytic reactions studied under practical catalytic conditions, including the solution phase, reported in the literature, showing the generality of mechanistic framework.

Comment

Since this paper was submitted for the Discussion further studies have been completed on reactions of alcohols and dimethyl amine. For both methanol and ethanol cross coupling to form amides is observed provided sufficient alcohol is present to remove excess adsorbed oxygen. The results will be reported elsewhere.

Acknowledgements

We gratefully acknowledge the support of the US Department of Energy, Basic Energy Sciences, under Grant No. FG02-84-ER13289 (CMF) and the National Science Foundation, Division of Chemistry, Analytical and Surface Science (RJM) CHE- 0952790. BJX also acknowledges the Harvard University Center for the Environment for support through the Graduate Consortium in Energy and the Environment.

Notes and references

1 J. Schwank, *Gold Bull.*, 1985, **18**, 2–10.
2 J. Schwank, *Gold Bull.*, 1983, **16**, 103–110.
3 M. Haruta, N. Yamada, T. Kobayashi and S. Iijima, *J. Catal.*, 1989, **115**, 301.
4 D. A. Outka and R. J. Madix, *Surf. Sci.*, 1987, **179**, 361.
5 B. Xu, R. J. Madix and C. M. Friend, *J. Am. Chem. Soc.*, 2010, **132**, 16571.
6 B. Xu, J. Haubrich, C. Fryschlag, C. M. Friend and R. J. Madix, *Chem. Sci.*, 2010, **1**, 310.
7 B. Xu, X. Liu, J. Haubrich, R. J. Madix and C. M. Friend, *Angew. Chem., Int. Ed.*, 2009, **48**, 4206.
8 B. Xu, X. Liu, J. Haubrich and C. M. Friend, *Nat. Chem.*, 2009, **2**, 61.
9 X. Y. Liu, B. J. Xu, J. Haubrich, R. J. Madix and C. M. Friend, *J. Am. Chem. Soc.*, 2009, **131**, 5757.
10 A. Wittstock, V. Zielasek, J. Biener, C. M. Friend and M. Baumer, *Science*, 2010, **327**, 319.
11 P. Fristrup, L. B. Johansen and C. H. Christensen, *Catal. Lett.*, 2007, **120**, 184.
12 I. Nielsen, E. Taarning, K. Egeblad, R. Madsen and C. Christensen, *Catal. Lett.*, 2007, **116**, 35.
13 M. A. Barteau, M. Bowker and R. J. Madix, *Surf. Sci.*, 1980, **94**, 303.
14 M. A. Barteau and R. J. Madix, *Surf. Sci.*, 1982, **120**, 262.
15 I. E. Wachs and R. J. Madix, *Surf. Sci.*, 1978, **76**, 531.
16 D. A. Outka and R. J. Madix, *J. Am. Chem. Soc.*, 1987, **109**, 1708.
17 B. Xu, C. M. Friend and K. Efthimios, *J. Phys. Chem. C*, 2011, **115**, 3703–3708.
18 P. Bruice, *Organic Chemistry*, 4th edn., Pearson Education, Inc., Upper Saddle River, 2004.
19 D. M. Thornburg and R. J. Madix, *Surf. Sci.*, 1990, **226**, 61.
20 D. M. Thornburg and R. J. Madix, *Surf. Sci.*, 1989, **220**, 268.
21 X. Y. Deng, T. A. Baker and C. M. Friend, *Angew. Chem., Int. Ed.*, 2006, **45**, 7075.
22 B. K. Min, A. R. Alemozafar, D. Pinnaduwage, X. Deng and C. M. Friend, *J. Phys. Chem. B*, 2006, **110**, 19833.
23 N. Saliba, D. H. Parker and B. E. Koel, *Surf. Sci.*, 1998, **410**, 270.
24 S. E. Stein, P. J. Linstrom, (Ed.) and W. G. Mallard, (Ed.), Institute of Standards and Technology, Gaithersburg MD, 20899, Editon edn.
25 P. R. Davies and J. M. Keel, *Surf. Sci.*, 2000, **469**, 204.

26 B. Xu, L. Zhou, R. J. Madix and C. M. Friend, *Angew. Chem., Int. Ed.*, 2009, **49**, 394.
27 R. J. Angelici, *J. Organomet. Chem.*, 2008, **693**, 847.
28 S. K. Klitgaard, K. Egeblad, U. V. Mentzel, A. G. Popov, T. Jensen, E. Taarning, I. S. Nielsen and C. H. Christensen, *Green Chem.*, 2008, **10**, 419.

Catalytic properties of supported gold nanoparticles: new insights into the size-activity relationship gained from in *operando* measurements

M.-C. Saint-Lager,[*a] I. Laoufi,[a] A. Bailly,[a] O. Robach,[b] S. Garaudée[a] and P. Dolle[a]

Received 2nd March 2011, Accepted 5th April 2011
DOI: 10.1039/c1fd00028d

The relationship between the catalytic activity and the size was studied *in operando* in the case of gold nanoparticles on $TiO_2(110)$ model catalyst during carbon monoxide oxidation. The geometrical parameters, the shape and the dispersion of the particles on the oxide support were examined in detail. The catalytic activity was found optimum for a nanoparticle diameter of about 2 nm and a height of six atomic monolayers. Above the maximum, it fits a power law of the diameter $D^{-2.4 \pm 0.3}$. This indicates that the low-coordinated sites play a major role in the catalytic activity, however such a model still fails to explain the activity maximum. The nanoparticle sintering was also investigated since it is suspected of being responsible for the decrease of the catalyst activity in the course of time. It was clearly observed for particles with a size around the maximum of activity and smaller. At the very beginning of the CO conversion into CO_2, the sintering is strongly activated. The nanoparticles mobility is dependent upon the $TiO_2(110)$ surface direction under consideration: it is higher along the $[001]_{TiO_2}$ than along the $[1{-}10]_{TiO_2}$. Then, the sintering greatly slows down. This could be explained by a nanoparticles' pinning at the step edges. The thermal energy released by the exothermic CO oxidation reaction was evaluated and it suggests that the sintering results from a more complex process than from a reaction-induced local heating.

Introduction

While the bulk surface is known for its chemical inertia gold was revealed to be catalytically very active at the nanometer scale for carbon monoxide (CO) oxidation.[1] Therefore the size dependence of this catalytic activity is one of the key parameter to understand these properties. Haruta was the first to show that, in contrast to platinum, the activity of gold nanoparticles for CO oxidation sharply increases when their diameter decreases below 4 nm and seems to diverge for very small particles with *ca.* 10 atoms.[1,2] He suggested that the reaction takes place at the interfacial perimeter around the gold nanoparticles. Such a scheme was recently supported by the evidence of the oxygen storage capacity at the Au/TiO_2 interface, thus promoting the CO oxidation without additional oxygen in the gas phase.[3]

[a] *Institut Néel, CNRS et Université Joseph Fourier, BP 166, F-38042 Grenoble Cedex 9, France. E-mail: marie-claire.saint-lager@grenoble.cnrs.fr; Fax: +33 (0) 4 76 88 10 38; Tel: +33 (0) 4 76 88 74 15*
[b] *Institut Nanosciences et Cryogénie, Commissariat à l'Energie, Atomique et aux Energies Alternatives, 17 avenue des Martyrs, F-38054 Grenoble Cedex 9, France*

Alternatively, Lopez *et al.* compiled a number of measured CO oxidation activities for different gold catalysts and showed that they are close to follow the same power law of the diameter (D^{-3}).[4] Since decreasing the gold particle size leads to an increase of the relative amount of low-coordinated Au atoms, they proposed that such Au atoms would play a major role in the catalytic activity,[4,5] in line with a previous paper.[6] However, Valden *et al.* highlighted the existence of a maximum of the catalytic activity for diameters around 3 nm.[7] They assigned the catalytic properties to a quantum size effect peaking for two-atom-thick gold clusters of the optimum diameter. More recently, by examining the activity of well-ordered wetting gold films, they claimed that a bilayer feature was critical to explain the catalytic activity.[8,9] In this model, the gold plane at the interface is electron-rich (due to electron transfer from the oxide substrate) and adsorbs the oxygen molecule; while CO is adsorbed on the upper atomic plane consisting of positively charged gold ($Au^{\delta+}$). Thereafter, other studies, also exploring sizes smaller than 2 nm, evidenced a maximum for the catalytic activity.[10,11] However, the interpretation assigning this maximum to a particles' height of two atomic layers remains controversial.[8,9]

In a previous work,[12] we confirmed that the $Au/TiO_2(110)$ catalytic activity for CO oxidation is optimum for a nanoparticle diameter of about 2 nm, as first observed by Valden *et al.*[7] But we also found that the corresponding height is equal to six atomic planes. Below the optimum diameter, the height decreases together with the catalytic activity. These results obtained in *operando* conditions seem to argue against the existence of a bilayer model to explain the high reactivity of the supported gold nanoparticles. This apparent discrepancy could arise from the fact that the result of Valden *et al.* is based on scanning tunneling microscopy (STM) measurements performed in ultra-high vacuum (UHV).[7,9] We also showed that the occurrence of the CO oxidation induces sintering directly correlated to the reaction rate.[12] The sintering is more crucial when the particles are initially small; such particles being more difficult to stabilize. The understanding of nanoparticle sintering is very important since it is suspected of being responsible for the decrease in the catalyst activity in the course of time (deactivation). In this paper, new insight in the catalytic properties of supported gold nanoparticles is gained from a detailed analysis of the relationship between the geometrical parameters of the nanoparticles (size, shape and dispersion on the oxide support) and their catalytic activity in the course of the CO oxidation. This work is based on complementary laboratory and *in operando* measurements in the presence of gases close to atmospheric pressure.

Experimental method

The experiments were performed in a setup consisting of an UHV preparation chamber and a batch reactor running from UHV up to reactive conditions at ambient pressures and allowing both X-ray scattering and catalytic activity measurements in static conditions.[13,14] The setup is operated on a diffractometer at the European Synchrotron Radiation Facility (ESRF) at the beamline BM32, as well as in laboratory. The experimental procedure is the same as in ref. 12; we simply summarize here the main features. The TiO_2 (110) single-crystal was first bombarded with argon ions and then annealed at 1000 K under an oxygen partial pressure of 10^{-3} Pa in order to restore the surface stoichiometry.[15] Gold nanoparticles were grown at 300 K using vapor deposition from a Knudsen cell in UHV (base pressure 5.10^{-8} Pa). As calibrated by a quartz microbalance, the average thickness of the gold deposit was ranging between 0.05 and 3 atomic monolayers (1 ML corresponding to 0.235 nm, equivalent to one Au(111) atomic layer). Once prepared, the Au/TiO_2 (110) samples were transferred under UHV from the preparation chamber into the batch reactor in order to perform grazing incidence small angle X-ray scattering (GISAXS) and catalytic activity measurements during carbon monoxide oxidation. Briefly GISAXS makes it possible to determine *in situ* the morphology (size and

   This journal is © The Royal Society of Chemistry 2011

shape), as well as the spatial distribution of the nanoparticles' assembly on the substrate. The results presented here involve two sets of catalytic activity data. One set was collected at the same time as GISAXS patterns (thus giving access to the particles geometrical features in the course of the CO oxidation reaction) and the second set was recorded at the laboratory in exactly the same experimental conditions (setup, sample preparation, temperature, pressure, and so on) but without any GISAXS data.

In order to establish a correspondence between these two sets of data, the exposure to the gases was always performed according to the same protocol throughout the study since the morphology of the particles strongly depends on the gas environment and on the sample history.[12] Thus the procedure involved three steps; each lasting approximately two to three hours during which gas composition and GISAXS (when applicable) were measured: (1) Au/TiO_2 (110) was inserted in the batch reactor in UHV, (2) then annealed at 473 K in 2.10^3 Pa of oxygen, (3) and finally CO conversion into CO_2 started by adding 20 Pa of CO to the oxygen, while keeping the sample at 473 K. The CO and CO_2 partial pressures in the reactor were respectively

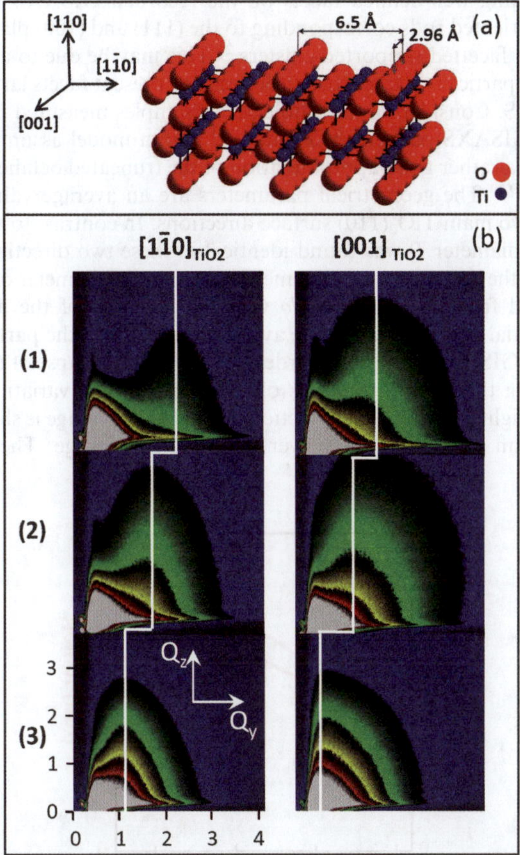

Fig. 1   (a) Ball and stick model of the TiO_2(110) surface. (b) Evolution of the GISAXS patterns recorded for a sample with a 0.1 ML gold coverage during the different steps of gas/temperature as defined in the text (only half of the pattern is shown). The X-ray beam was along [1–10]$_{TiO_2}$ direction (left column) and perpendicular to it (right column). For each image the coordinates are the wavevector transfers Q_y and Q_z that are respectively parallel and perpendicular to the substrate surface. They are expressed in nm^{-1}. The intensity scale is logarithmic and is the same for all the patterns. The white lines mark Q_M, the position of the correlation peak.

deduced from the 28 and 44 ionization currents measured by a mass spectrometer. The reaction rate was defined as the number of CO molecules converted per gold atom per second on the basis of the total number of gold atoms (see ref. 12).

The GISAXS measurements are detailed in ref. 12. The GISAXS patterns were acquired with the X-ray incident beam along the two main $TiO_2(110)$ surface crystallographic directions: [001], along the oxygen dense rows and [1–10], perpendicularly to them[16] (figure 1a). Their evolution during the three different steps of the experimental procedure is illustrated in figure 1b. To a first approximation, the position of the correlation peak at $Q_y = Q_M$ (marked by the white line) can be directly linked to the mean separation distance L between the particles on the substrate by $L = 2\pi/Q_M$.[17] This interparticle distance L measured along the $[001]_{TiO_2}$ and $[1–10]_{TiO_2}$ directions was found different and labelled L_H and L_K, respectively. The mean geometrical parameters of the particles, the diameter D and height H, were deduced from the quantitative analysis of the GISAXS patterns performed using the IsGISAXS software.[18] The model chosen to describe the nanoparticles shape was derived from the images obtained by high resolution transmission electron microscopy (HRTEM).[10,19,20] They evidenced a shape close to truncated octahedrons for particles size in the range of the present work. However, we did not find any trace of signal indicating well-defined facets on the recorded GISAXS patterns, which would present oriented rods corresponding to the (111) and (100) planes as observed for well-ordered facetted supported clusters.[21] This may be due to a misorientation of the gold nanoparticles or to particles too small to present facets large enough to be seen by GISAXS. Considering the whole set of samples measured *in operando*, the best fit of the GISAXS patterns was obtained with a model assuming a truncated sphere; actually a rather good approximation of the truncated octahedrons observed by HRTEM.[10,19,20] The geometrical parameters are an average value of those obtained for the two main $TiO_2(110)$ surface directions. In contrast to the interparticle distance L, the diameter D was found identical in these two directions.

In the case of the laboratory activity measurements, the diameter of the gold particles was derived from the *in operando* GISAXS analysis of the first set of data knowing the initial gold coverage. The average geometry of the particles was determined from the GISAXS patterns recorded during the same lapse of time as the reactivity, *i.e.* during the first ten minutes of the step (3). The variation of the mean diameter and height measured as a function of the gold coverage is shown in figure 2. The diameter can be fitted by a power law of the coverage. The height is then

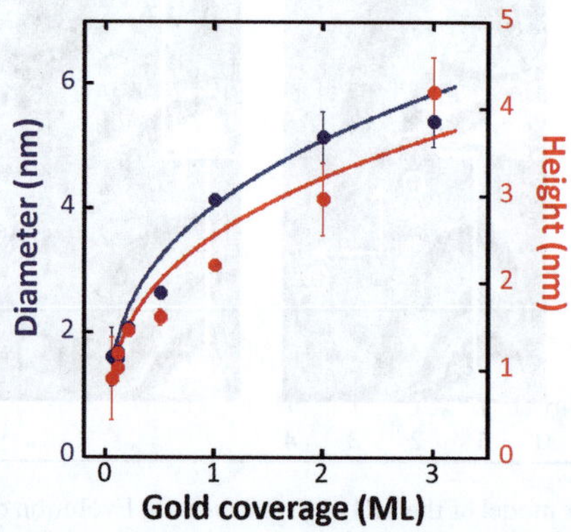

Fig. 2 Variation of the mean diameter D (blue) and height H (red) of the nanoparticles as a function of the deposited gold thickness. The full circles represent the experimental values deduced from the GISAXS patterns, recorded at the beginning of step 3 and fitted with the truncated sphere model. The full lines correspond to the law used to extrapolate D and H for the laboratory measurements.

 This journal is © The Royal Society of Chemistry 2011

deduced using the aspect ratio H/D that was found to be equal to approximately 0.63. Thus it allows us to derive the particles geometrical parameters directly from the gold coverage in the case of the laboratory experiments performed in identical conditions.

Results

Particle size and catalytic activity

The evolution of the catalytic activity is presented in Figure 3a as a function of the gold nanoparticles diameter during CO oxidation. It involves activity data recorded simultaneously with GISAXS patterns and data recorded at the laboratory. The two

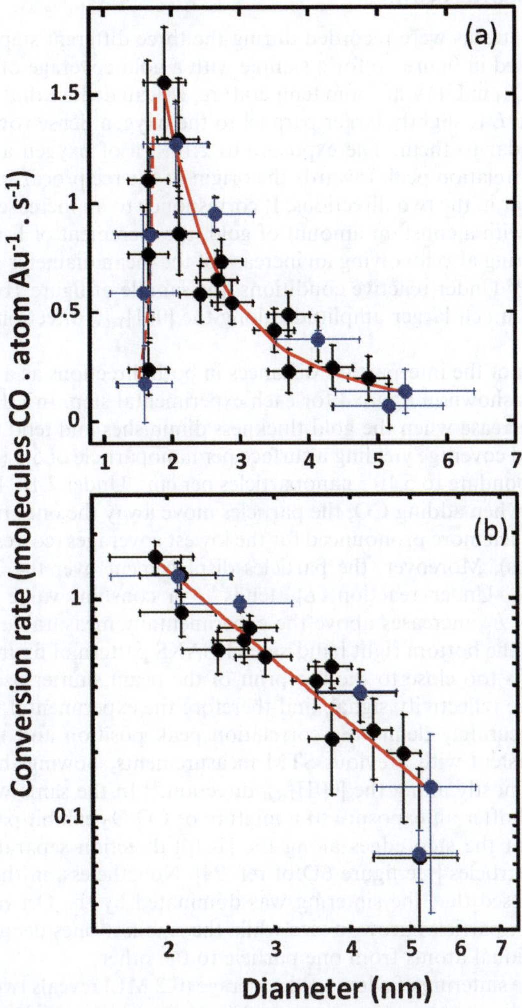

Fig. 3 Reaction rate of the CO conversion into CO_2 per Au atoms per second as a function of the particle diameter: (a) Blue circles: experiments performed *in operando*, that is the catalytic activity and GISAXS patterns were recorded simultaneously during the reaction. Black circles: laboratory measurements; in this case for a given gold amount, the diameter was deduced from the calibration established during the *in operando* measurements. For $D \geq 2$ nm, the red full line represents the power law $D^{-2.4}$ and the dashed one ($D < 2$ nm) is a guide for the eyes. (b) Same as in (a) but for $D \geq 2$ nm and with logarithmic scales; the slope of the red line is -2.4 ± 0.3.

sets of data are consistent, showing that the catalytic activity was not affected by the X-ray beam during the *in operando* measurements. The catalytic activity increases as the diameter of the gold particles decreases down to an optimum diameter of about 2 nm and then, it quickly falls below this optimum. As shown in Figure 3b, the variation of the catalytic activity above the optimum diameter is well accounted for by a power law dependence of $D^{-2.4 \pm 0.3}$. This power law is close to those obtained by extended X-ray absorption fine structure (EXAFS) on dispersed catalysts prepared *via* traditional chemical methods.[22] In this latter case, two exponents were found and assigned to different shape and size distributions depending on the sample preparation procedure. Janssens *et al.* linked these power laws to the major role played by the low-coordinated sites in the gold reactivity.[5]

Sintering and catalytic activity

The GISAXS patterns were recorded during the three different steps of the experiment as illustrated in figure 1b for a sample with a gold coverage of 0.1 ML. From the position of Q_M in UHV at room temperature, we can deduce that the mean interparticle distance L is slightly larger parallel to the oxygen dense rows of $TiO_2(110)$ than perpendicular to them. The exposure to 2.10^3 Pa of oxygen at 473K induces a shift of the correlation peak towards the origin of the reciprocal space; the amplitude being similar in the two directions. It corresponds to an increase of L. Since the system evolves with a constant amount of gold, the increment of L is a clear signature of the sintering also involving an increase of the mean diameter D and height H of the particles.[12] Under reactive conditions, the sample of figure 1b exhibits a new sintering with a much larger amplitude along the $[001]_{TiO_2}$ direction of the oxygen dense rows.

The evolution of the interparticle distances in both directions as a function of the gold coverage is shown in figure 4 for each experimental step. In UHV, L_H and L_K continuously decrease when the gold thickness diminishes and tend to a limit value for very low gold coverage yielding a surface per nanoparticle of 5.5 ($\pm$ 0.5) $\times$ 3.5 ($\pm$ 0.5) nm^2 corresponding to 5.10^{12} nanoparticles per cm^2. Under 2.10^3 Pa of oxygen at 473K, and also when adding CO, the particles move away the ones from the others, with an effect much more pronounced for the lowest coverages (corresponding to the smallest particles). Moreover, the particles displacement over the TiO_2 surface is very anisotropic. Under reaction, L_K tends to a constant value of about 6 $\pm$ 0.5 nm, whereas L_H increases above the experimentally measurable range. Indeed, as illustrated by the bottom right hand-side GISAXS pattern of figure 1b, the correlation peak shifts too close to the footprint of the beam shutter used to mask the extremely intense reflectivity signal, and therefore the experimental sensitivity does not allow to accurately define the correlation peak position and thus L_H. These results are consistent with previous STM measurements showing that the smallest clusters diffuse mostly along the $[001]_{TiO_2}$ direction.[23] In the same way other STM images recorded after an exposure to a mixture of CO/O_2 exhibit particles that are mainly pinned at the step edges along the [1–10] direction separated by terraces almost free of particles [see figure 6D of ref. 24]. Nonetheless, in these two works, the authors stressed that the sintering was dominated by the Ostwald mechanism where the largest particles grow in size while the smallest ones decrease due to the motion of individual atoms from one particle to the other.

Looking at the sintering of a low gold coverage (0.2 ML) reveals two very different behaviors according to the gas environment (under oxygen at 473K or when CO is added while the temperature is kept constant). This is illustrated by the evolution of the interparticle distances measured regularly during the various stages of the experiment (figure 5). Under oxygen at 473K, the increase of L is gentle and somewhat linear. As soon as CO is introduced, L suddenly increases in both directions. The rise is thus clearly associated to the reaction beginning. The increase ΔL_{O_2}, resulting from the three hours of exposure to 2.10^3 Pa of oxygen at 473K as a function of the

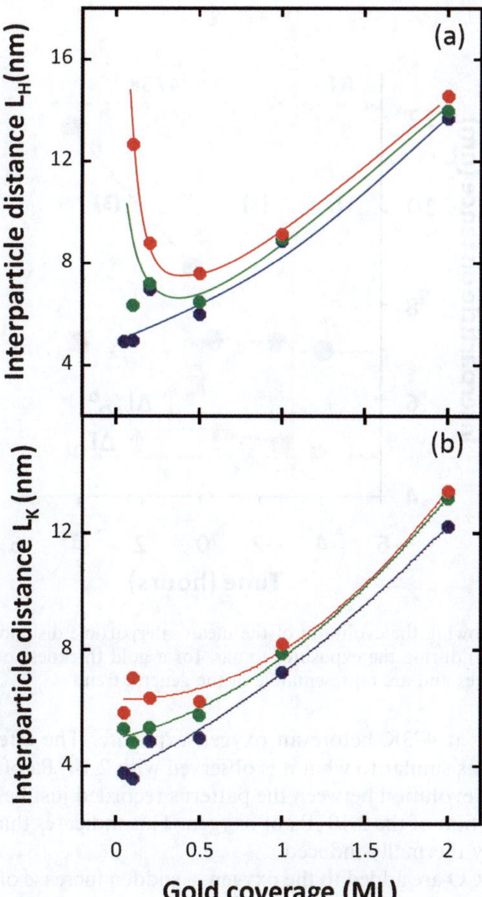

Fig. 4 (a) and (b) Variation of the mean interparticle distances along $[001]_{TiO_2}$ and $[1-10]_{TiO_2}$, respectively, as a function of the gold coverage for each step described in the text: (1) in UHV at RT (blue), (2) under 2.10^3 Pa of oxygen at 473 K (green) and (3) after adding 20 Pa of CO while keeping T = 473 K (red).

gold coverage is reported in figure 6a and the increase ΔL_{CO}^0 at the beginning of the CO oxidation reaction in figure 6b. Their behaviors are quite different. Under oxygen, ΔL_{O_2} tends to diverge for the lowest amounts of gold (corresponding to the smallest particles) as expected for thermally-activated sintering with an energy barrier depending on the particles size. Differently, ΔL_{CO}^0 presents a maximum with a variation as a function of the gold coverage very close to that of the corresponding catalytic activity (figure 6b). It indicates a process directly linked to the reaction itself. This latter being highly exothermic, the sintering might be a consequence of a local heating effect, as already suggested.[25]

The effect of the gas nature and of the temperature on the sintering is shown in figure 7 for coverages around the maximum of activity. For one of the represented samples (0.15 ML), the experimental protocol was identical to the others, except that the oxygen was replaced by argon; the other conditions being exactly the same. The effects during step (2) with 2.10^3 Pa of argon or oxygen at 473K follow the trends described here above with a sintering rate increasing when the coverage diminishes, without noticeable difference according to the gas nature. For another sample, with a gold coverage of 0.2 ML, the protocol was changed and GISAXS patterns were

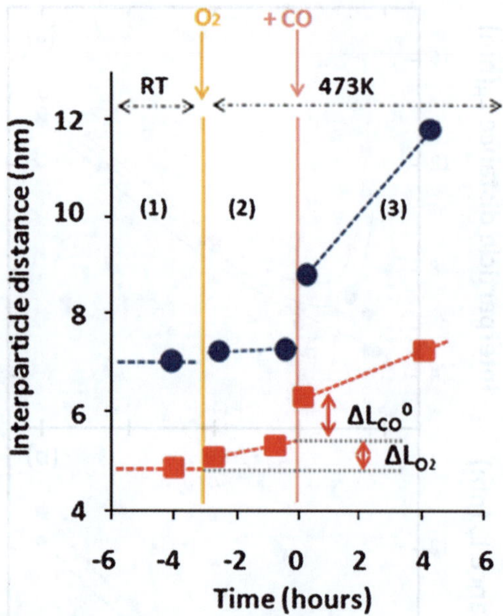

Fig. 5 Diagram showing the evolution of the mean interparticle distances L_K (red squares) and L_H (blue circles) during the exposures to gas, for a gold thickness of 0.2 ML. The lines are guides for the eyes and are representative of the general trend.

recorded in UHV at 473K before an oxygen exposure. The effect of heating the sample in UHV was similar to what it is observed with 2.10^3 Pa of oxygen and there was no significant evolution between the patterns recorded just before and just after a further introduction of the 2.10^3 Pa of oxygen. This indicates that, in this case, the sintering is mainly thermally-induced.

When 20 Pa of CO are added to the oxygen, a sudden increase of the sintering rate occurs. In contrast, no sintering is observed with argon at 473 K. In this latter case, the sintering occurs when the sample is annealed at a higher temperature (step 4 in figure 7). Surprisingly this L rise, activated by a temperature increase of 200 K under neutral gas, is about 0.5 nm which is smaller than the one induced by the reaction itself (~ 1 nm). Therefore the assumption of a sintering activated by a reaction-induced heating implies a very high local temperature augmentation. Since the free energy released by the conversion of one mole of CO into one mole of CO_2 is $\Delta G = -242$ kJ mol^{-1} at 473K, at the maximum of activity this corresponds to a power of about 0.2 mW (the surface sample being 1 cm^2, the number of Au atoms is 2.5×10^{14} and the number of CO_2 moles produced per second is 7×10^{-10}). Compared to the total power of 5 W, necessary to heat the sample up the nominal temperature of 473K, the thermal power supplied by the reaction is negligible. Therefore such drastic effect on the sintering induced by the reaction leads to suspect that the sintering also results from a reactant-induced mass transport mechanism since molecules' adsorption could weaken the Au-Au bound and induce the detachment of Au monomers from supported Au particles. But, as shown while comparing the results when oxygen is replaced by argon in step (2) and (3) of the experimental procedure, it is clear that nor oxygen alone, neither CO alone can explain such effect. We thus have to assume that it is the co-adsorption which provokes the disruption of the Au-Au bound. A more complex mechanism involving hot electrons produced by the CO oxidation reaction was also suggested to explain such reaction-induced sintering.[26] Moreover, figure 7 shows that an additional annealing up to 773K in UHV, after CO oxidation (step 4), does not provoke further sintering. This indicates that the equilibrium reached under reaction conditions is particularly stable.

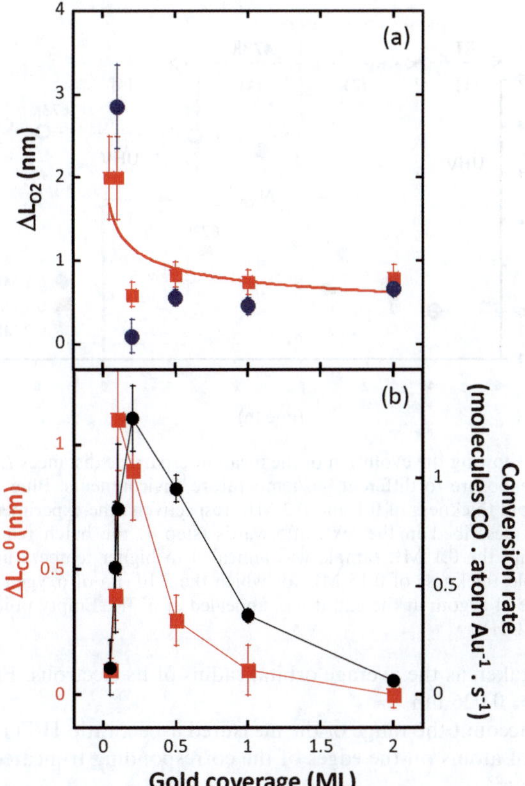

Fig. 6 Variation of the increase of L (a) ΔL_{O_2} under oxygen pressure at 473K (during step 2) along the $[001]_{TiO_2}$ (blue circles) and the $[1–10]_{TiO_2}$ (red squares) directions; (b) red squares: ΔL_{CO}^0 along $[1–10]_{TiO_2}$, induced by the occurrence of the oxidation reaction, black circles: catalytic activity as a function of the gold coverage. Full lines are guides for the eyes.

Low coordinated sites and catalytic activity

In the present work, the relationship between the size and the activity was established by measuring the geometrical parameters in the same time as the catalytic activity. Thus it allows a direct estimation of the statistic of the low-coordinated sites for each particles size. In order to model the gold nanoparticles, truncated octahedrons with (111) facets parallel to the surface were used in line with refs 27 and 28. The number of atomic planes N_P, parallel to the surface, and the number of atoms N_D along the diameter D are related to the mean height H and diameter D (as deduced by GISAXS) by:

$$N_P = (H - 2r_{Au})/d_\perp + 1 \tag{1a}$$

$$N_D = (D - 2r_{Au})/d_{//} + 1 \tag{1b}$$

$d_\perp$ and $d_{//}$ are the interatomic distances perpendicular and parallel to the substrate surface, respectively. Assuming the Au growth being along the (111) direction of its fcc lattice, it leads to $d_\perp = a_0/\sqrt{3}$ and $d_{//} = a_0/\sqrt{2}$ with a_0, the Au bulk lattice parameter equal to 0.408 nm. r_{Au} is the gold atomic radius as seen by X-ray scattering. Since the X-ray scattering form factor of an atom is correlated to its electronic

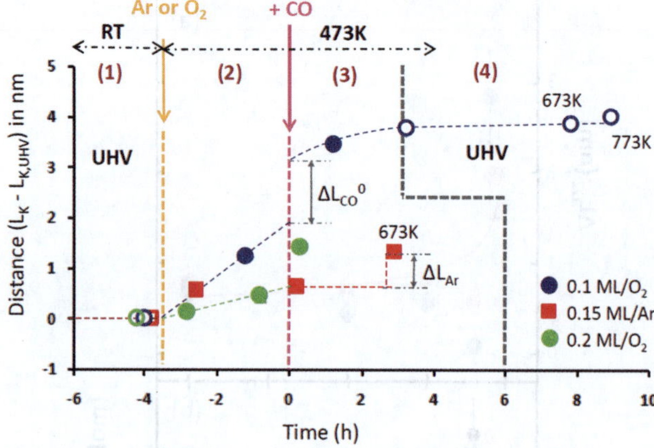

Fig. 7 Diagram showing the evolution of the mean interparticle distances L_K in the course of the time during exposure to different gas/temperature environments. Blue and green circles: samples with a gold thickness of 0.1 and 0.2 ML, respectively; the experimental protocol was first the same as described in the text; afterwards (step 4) the batch reactor was pumped down to UHV and the 0.1 ML sample was annealed to higher temperatures. Red squares: sample with a gold thickness of 0.15 ML for which the 2.10^3 Pa of oxygen were replaced by the same pressure of argon; at the end it was annealed at 673K. Empty points correspond to data recorded in UHV.

density, it was taken as the average orbital radius of its electrons. From ref. 29 and 30, we get $r_{Au} = 0.026$ nm.

Taking into account the range of the measured aspect ratio H/D for the particles, the number m of atoms on the edges of the corresponding truncated octahedron is related to N_D by:[27]

$$N_D \cong 3m - 2 \tag{2}$$

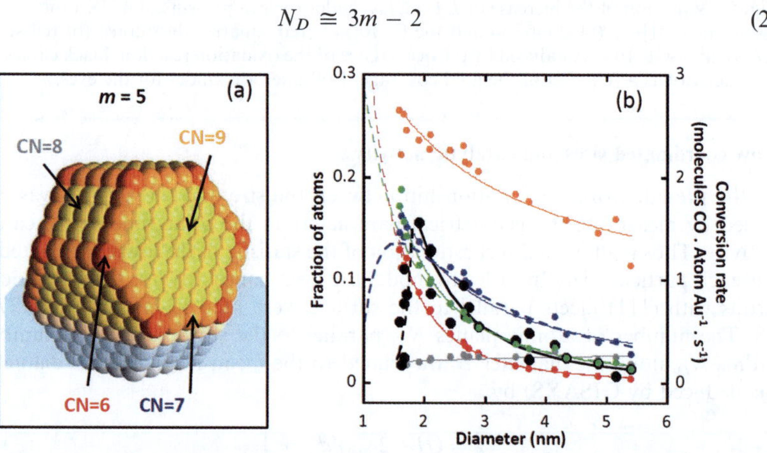

Fig. 8 (a) Schematic representation of a truncated octahedron-shaped gold nanoparticle with a number of atoms along the edges $m = 5$ corresponding to $D = 3.5$ nm, $N_D = 13$ and $N_P = 10$ atomic layers above the substrate (represented as the blue light plane). Five types of gold atoms are identified: perimeter interface, corner, edge not corner, (100) and (111) faces. (b) Comparison of the catalytic activity to the fraction of different types of low-coordinated atoms on the gold particles. The points correspond to the experimental data deduced from the truncated octahedron and lines to a power law fit; black: catalytic activity (*in operando* and laboratory data), red: corners CN = 6 and the power law is $D^{-2.7}$; blue: edge not corner CN = 7, $D^{-1.2}$; grey: (100) surface atoms CN = 8, $D^{0.2}$; orange: (111) surface atoms, CN = 9, $D^{-0.5}$; green: perimeter atoms, $D^{-1.9}$; dashed lines red, blue and green correspond to the variation of extrapolated polyhedron with $N_P = 2m$ for CN= 6, CN = 7 and for perimeter, respectively.

The particles geometry is deduced by taking the N_P top atomic layers of such a truncated octahedron. A schematic representation of the polyhedron deduced from the experimental mean H and D is given in figure 8a. For the data set recorded at the laboratory we used the extrapolated values. Typically $N_P = 2m \pm 2$ over the whole collected data range. In most of the cases, calculated N_P and m were found fractional. These values correspond to a mean polyhedron with a dispersion related to the ones of H and D as deduced from the GISAXS analysis. $\Delta H/H$ being about 0.2, this corresponds to an error bar around the mean N_P value of ± 0.5 atomic plane for the smallest particles up to ± 2 for the largest ones (see equation (1a)). The dispersion $\Delta D/D$ obtained for the experimental diameter is about 0.3 giving an error bar which increases with N_D from ± 1 up to ± 3 atoms. As seen from equation (2), an increment of one unit for m leads to an increment of 3 atoms on the number of atoms N_D. Therefore m cannot be explained only by a mixing of truncated octahedrons (with m integer) but also by polyhedrons with missing atomic planes on one of the octahedron lateral faces, especially for the smallest particles. On the nanoparticles surface, atoms are under coordinated and four types of atoms can be distinguished according to their number of first neighbors (coordination number CN): atoms (i) at corners with CN = 6, (ii) on edge not corner with CN = 7, (iii) on the (100) squared faces (CN = 8) and (iv) on the (111) hexagonal faces (CN = 9). In the bulk of the fcc crystal, CN = 12. The atoms at the interface perimeter form a separate class.

The evolution of the number of the different kinds of low-coordinated surface atoms over the total number of atoms N_v in the particles is compared with the experimental catalytic activity curve as a function of the diameter in Figure 8b. The power law of the activity data above the maximum ($D^{-2.4}$) is between the one of the corners atoms fraction ($D^{-2.7}$) and the one of the atoms at the perimeter ($D^{-1.9}$), but in these two cases no maximum is observed. The only case exhibiting a maximum corresponds to the low-coordinated sites with CN = 7 (edge not corner), disappearing once $m = 2$.

However, on both sides of the maximum, the experimental catalytic activity curve decreases faster than the number of edge not corner sites. This can be explained by the non-isothermal character of the activity curve due to a reaction-induced heating as evidenced by the particle sintering. Since the sintering rate follows the same variation as the reaction rate (figure 6b) this would correspond to a local temperature peak at the activity maximum and may have a consequence on the measured relationship between the activity and the particles size. But we showed that the power supplied by the reaction is very small, even at the maximum, therefore, if this effect exists it has probably a little impact on this relationship. The strong correlation between the fraction of edge not corner sites and the catalytic activity is in line with the DFT calculation of Liu *et al* who found that 7-coordinated steps are more reactive than 6-coordinated kinks despite their higher coordination number.[31] In a slightly different approach, Xu and Mavrikakis have predicted that the best sites for oxygen dissociation are steps presenting a tensile stretch.[32] On the other hand, the ability of *unstrained* stepped surfaces to dissociate the oxygen molecules (the limiting step of the CO oxidation reaction) was recently questioned.[33] The major role of edges on the activity can also be explained by the need of adjacent undercoordinated atoms in the reaction process. Such assumption is consistent with the high activity observed for gold linear structures in the case of the (1 × 3) bilayer Au film.[8]

In the present work we assumed that there is no structural change over the whole size range explored. However, Tai *et al.* have attributed the enhancement of the activity to a structural transition from fcc to icosahedra structure.[11] Indeed, icosahedrons and octahedrons are energetically very close for gold, especially for sizes around 2 nm (*i.e.* 100–300 atoms),[34] that is the sizes corresponding to the maximum in activity. However, an important feature is the apparition of stress in the nanoparticles core. A contraction of the lattice parameter was measured by EXAFS,[11,22] (and references therein) but in contrast to a tensile stretch, it would make gold less active and could rather explain the fall of the activity on the side of the small sizes.[35]

Conclusion

In operando measurements of the relationship between the size and the catalytic activity of supported gold nanoparticles on $TiO_2(110)$ establish the existence of a maximum. They also show that the catalytic activity fits a power law of the nanoparticles diameter, for sizes above the maximum. The counting of sites at the surface of gold clusters indicates that low-coordinated gold atoms play a major role in the catalytic process. However, it is necessary to assume that the enhancement of the activity linked to the increase of the relative amount of these active sites is balanced by other effects affecting the catalytic properties of the gold nanoparticles. Structural changes often evoked in literature can be elucidated by *in operando* X-ray diffraction. The results are presently under investigation and will be the topic of a further publication. Besides, the CO conversion into CO_2 strongly activates the sintering and can impact the relationship between the size and the catalytic activity. It has a drastic effect at the beginning of the reaction. The sintering depends on the $TiO_2(110)$ direction, with a greater atom mobility along $[001]_{TiO_2}$. However it seems that the step edges act as barriers and limit the sintering. Moreover, the present results demonstrate that *in operando* studies on model catalysts offer a promising way to examine the chemistry of supported gold. Grazing incidence X-ray scattering experiments coupled to mass spectrometry are particularly well-adapted for this type of studies since they are non-destructive techniques. This allows to monitor model catalysts from UHV where they are elaborated up to semi-realistic reaction conditions for very low amounts of catalyst and nanoparticles with sizes as small as 1 nm.

Acknowledgements

The technical staffs (SERAS and MCMF) of the Institut Néel-CNRS and of the BM32 CRG beamline at ESRF are acknowledged for their help. We also thank Charles T. Campbell, Alessandro Fortunelli and Mike Bowker for highlighted discussions and comments.

References

1 M. Haruta, N. Yamada, T. Kobayashi and S. Iijima, *J. Catal.*, 1989, **115**, 301.
2 M. Haruta, S. Tsubota, T. Kobayashi, H. Kageyama, M. J. Genet and B. Delmon, *J. Catal.*, 1993, **144**, 175.
3 M. Kotobuki, R. Leppelt, D. A. Hansgen, D. Widmann and R. J. Behm, *J. Catal.*, 2009, **264**, 67.
4 N. Lopez, T. V. W. Janssens, B. S. Clausen, Y. Xu, M. Mavrikakis, T. Bligaard and J. K. Nørskov, *J. Catal.*, 2004, **223**, 232.
5 T. V. W. Janssens, B. S. Clausen, B. Hvolbæk, H. Falsig, C. H. Christensen, T. Bligaard and J. K. Nørskov, *Top. Catal.*, 2007, **44**, 15.
6 M. Mavrikakis, P. Stoltze and J. K. Nørskov, *Catal. Lett.*, 2000, **64**, 101.
7 M. Valden, X. Lai and D. W. Goodman, *Science*, 1998, **281**, 1647.
8 M. S. Chen and D. W. Goodman, *Science*, 2004, **306**, 252.
9 M. S. Chen and D. W. Goodman, *Chem. Soc. Rev.*, 2008, **37**, 1860.
10 R. Zanella, S. Giorgio, C.-H. Shin, C. R. Henry and C. Louis, *J. Catal.*, 2004, **222**, 357.
11 Y. Tai, W. Yamaguchi, K. Tajiri and H. Kageyama, *Appl. Catal., A*, 2009, **364**, 143.
12 I. Laoufi, M.-C. Saint-Lager, R. Lazzari, J. Jupille, O. Robach, S. Garaudée, G. Cabailh, P. Dolle, H. Cruguel and A. Bailly, *J. Phys. Chem. C*, 2011, **115**, 4673.
13 M.-C. Saint-Lager, A. Bailly, P. Dolle, R. Baudoing-Savois, P. Taunier, S. Garaudée, S. Cuccaro, S. Douillet, O. Geaymond, G. Perroux, O. Tissot, J.-S. Micha, O. Ulrich and F. Rieutord, *Rev. Sci. Instrum.*, 2007, **78**, 083902.
14 M.-C. Saint-Lager, A. Bailly, M. Mantilla, S. Garaudée, R. Lazzari, P. Dolle, O. Robach, J. Jupille, I. Laoufi and P. Taunier, *Gold Bull.*, 2008, **41**, 159.

15 J.-M. Pan, B. L. Maschhoff, U. Diebold and T. E. Madey, *J. Vac. Sci. Technol., A*, 1992, **10**, 2470.

16 U. Diebold, *Surf. Sci. Rep.*, 2003, **48**, 53.

17 J. R. Levine, J. B. Cohen and Y. W. Chung, *Surf. Sci.*, 1991, **248**, 215.

18 R. Lazzari, J. Appl. Crystallogr., 2002, 35, 406. http://www.insp.upmc.fr/axe2/Oxydes/IsGISAXS/isgisaxs.htm.

19 S. Giorgio, M. Cabié and C. R. Henry, *Gold Bull.*, 2008, **41**, 167.

20 T. Tanaka, K. Sano, M. Ando, A. Sumiya, H. Sawada, F. Hosokawa, E. Okunishi, Y. Kondo and K. Takayanagi, *Surf. Sci.*, 2010, **604**, L75.

21 M. Rauscher, R. Paniago, T. H. Metzger, Z. Kovats, J. Domke, H. D. Pfannes, J. Schulze and I. J. Eisele, *J. Appl. Phys.*, 1999, **86**, 6763.

22 S. H. Overbury, V. Schwartz, D. R. Mullins, W. Yana and S. Dai, *J. Catal.*, 2006, **241**, 56.

23 C. E. J. Mitchell, A. Howard, M. Carney and R. G. Egdell, *Surf. Sci.*, 2001, **490**, 196.

24 X. Lai and D. W. Goodman, *J. Mol. Catal. A: Chem.*, 2000, **162**, 33.

25 G. M. Veith, A. R. Lupini, S. J. Pennycook, G. W. Ownby and N. J. Dudney, *J. Catal.*, 2005, **231**, 151.

26 F. Yang, M. S. Chen and D. W. Goodman, *J. Phys. Chem. C*, 2009, **113**, 254.

27 A. Carlsson, A. Puig-Molina and T. V. W. Janssens, *J. Phys. Chem. B*, 2006, **110**, 5286.

28 R. Van Hardeveld and F. Hartog, *Surf. Sci.*, 1969, **15**, 189.

29 J. C. Slater, *Quantum Theory of Matter*, McGraw-Hill, New York, 1968.

30 J. T. Waber and D. T. Cromer, *J. Chem. Phys.*, 1965, **42**, 4116.

31 Z.-P. Liu, P. Hu and A. Alavi, *J. Am. Chem. Soc.*, 2002, **124**, 14770.

32 Y. Xu and M. Mavrikakis, *J. Phys. Chem. B*, 2003, **107**, 9298.

33 J. Kim, E. Samano and B. E. Koel, *Surf. Sci.*, 2006, **600**, 4622.

34 C. Mottet, J. Goniakowski, F. Baletto, R. Ferrando and G. Treglia, *Phase Transitions*, 2004, **77**, 101.

35 M. Mavrikakis, B. Hammer and J. K. Nørskov, *Phys. Rev. Lett.*, 1998, **81**, 2819.

Exploring the structure and chemical activity of 2-D gold islands on graphene moiré/Ru(0001)

Ye Xu,[*a] Lymarie Semidey-Flecha,[a] Li Liu,[b] Zihao Zhou[b] and D. Wayne Goodman[*b]

Received 2nd March 2011, Accepted 6th May 2011
DOI: 10.1039/c1fd00030f

Au deposited on Ru(0001)-supported extended, continuous graphene moiré forms large 2-D islands at room temperature that are several nanometers in diameter but only 0.55 nm in height, in the apparent absence of typical binding sites such as defects and adsorbates. These Au islands conform to the corrugation of the underlying graphene and display commensurate moiré patterns. Several extended Au structure models on graphene/Ru(0001) are examined using density functional theory calculations. Close-packed Au overlayers are energetically more stable, but all interact weakly with the support. Preliminary tests found the Au islands/graphene/Ru(0001) surface to be active for CO oxidation at cryogenic temperature, which suggests that the Au itself is the locus of catalytic activity.

Introduction

Since the seminal discovery of Haruta and co-workers that Au nanoparticles are catalytically active toward CO oxidation at sub-zero temperatures,[1] research efforts have increased exponentially to explore the potential of Au as catalysts.[2–5] To date, Au has been found as an effective catalyst for a wide range of reactions, including low temperature CO oxidation; water gas shift;[6,7] selective oxidation including epoxidation;[8–11] hydrogen peroxide synthesis;[12] and C–O coupling,[13] among others, in many cases offering more environmentally friendly "green" pathways than those currently in practice for industrially relevant reactions. The performance of Au catalysts depends on many variables,[14] but the chief factor is the particle size which needs to be maintained in the desired nanometer range at least in some dimensions.[15,16]

The size of the Au particles is intimately related to the interaction of Au with support materials. The choice of the support material affects not only the size of Au particles, but also their geometric and electronic structures, state of oxidation, and stability, all of which can affect the catalytic properties of Au. Strong interaction of Au with reactive oxide surfaces is generally thought to be necessary for stabilizing and dispersing Au nano-structures,[16,17] but the binding sites for and interfacial structures of Au NPs are compositionally and structurally diverse and can change with the reaction environment and dynamics,[18–20] so they remain a matter of intense research. On the other hand, relatively inert support materials such as carbon tend to result in Au agglomeration into large particles,[21,22] which is detrimental to CO oxidation activity.

When viewed in this light, graphene moirés present some intriguing possibilities. Large-scale, continuous graphene monolayers that grow over steps without

[a]Center for Nanophase Materials Sciences, Oak Ridge National Laboratory, P.O. Box 2008, Oak Ridge, TN, 37831, USA. E-mail: xuy2@ornl.gov.; Fax: +1-865-574-1753; Tel: +1-864-574-9761
[b]Department of Chemistry, Texas A&M University, P.O. Box 30012, College Station, TX, 77842, USA. E-mail: goodman@mail.chem.tamu.edu.; Fax: +1-979-845-6822; Tel: +1-979-845-0214

interruption have been synthesized on several single-crystal metal surfaces.[23] On some of these surfaces, including Ir(111), Rh(111), and Ru(0001),[23–26] graphene develops long-range moiré patterns because of small mismatch between the graphene and metal lattices. Moieties of C atoms either properly or improperly aligned with underlying metal atoms for bonding repeat periodically, with regions in which C atoms are bonded with the metal becoming locally reactive and functioning as nucleation sites for a second deposited metal. Pioneering work by N'Diaye *et al.* has demonstrated that graphene moiré on Ir(111) can be used as a template for forming superlattices of monodisperse nanoclusters of Ir, Pt, W, Re.[25,27–31] The nanoclusters nucleate exclusively in the so-called hcp regions and are confined therein until well in excess of 1 ML of coverage of the second metal. Therefore, graphene moiré offers a relatively inert support that is unlikely to adsorb reactants but is nonetheless capable of forming well dispersed metal NPs; is homogenous in composition and structure, which can help simplify the determination of metal cluster-support interaction and the nature of the active sites; and is thermally and chemically stable under moderate conditions.[32–34]

To explore these possibilities, we have deposited varying amounts of Au on graphene/Ru(0001), and observed the formation of large 2-D Au islands at room temperature that extend over several graphene moiré unit cells and conform to the corrugation of the graphene moiré, in the absence of typical binding sites such as defects and adsorbates in the graphene. The Au islands/graphene/Ru(0001) surface shows activity toward CO oxidation at cryogenic temperature. Density functional theory (DFT) calculations have been performed to assess several different Au overlayer structures, which all turn out to be energetically metastable with respect to bulk Au and anionic. Of these, close-packed overlayers are energetically more stable but do not agree closely with the observed height of the 2-D Au islands. A highly textured overlayer exhibits a maximum height that agrees closely with the observed height but is less stable than the close-packed overlayers.

Methods

Experimental

Experiments were carried out in an ultrahigh vacuum chamber (UHV) analysis chamber (base pressure $\sim 1 \times 10^{-10}$ Torr) equipped with a double-pass cylindrical mirror analyzer for Auger electron spectroscopy (AES), low energy electron diffraction (LEED) optics, and STM (Omicron STM 1). The Ru(0001) substrate was cleaned using a standard approach that includes cycles of Ar ion sputtering, annealing in 4.0×10^{-8} Torr of O_2 at ~ 1100 K, and flashing in vacuum to ~ 2000 K. Sample cleanliness was confirmed with AES, LEED, and STM. Single-layer graphene was formed by first dissociating methane (at 700 K) or ethylene (at 300 K) on the clean Ru(0001) and then annealing to ~ 1400 K. This process can be repeated several times to generate graphene that fully covers the Ru substrate. Au deposition was achieved by evaporating Au metal from a heated source. Deposition (in monolayers, ML) was performed at a substrate temperature of room temperature and quantified using Auger break points of the ratios of the most intensive Auger transitions of the target metals and Mo. All STM images were scanned in constant-current mode with an electrochemically etched tungsten tip. The bias voltages are reported with reference to the sample.

Theoretical

Periodic DFT calculations were performed using the Vienna *Ab Initio* Simulation Package (VASP)[35–37] within the generalized gradient approximation (GGA-PBE).[38] The core electrons were described by the PAW method,[39] while the Kohn–Sham valence states expanded in plane wave basis sets up to a cut-off energy of 400 eV. A first-order Methfessel-Paxton scheme[40] was used to smear the electronic

   This journal is © The Royal Society of Chemistry 2011

states with a width of 0.1 eV. The optimized lattice constants for Ru ($a = 0.273$ nm and $c = 0.404$ nm), Au ($a = 0.417$ nm) and graphite ($a = 0.246$ nm) are in close agreement with the corresponding experimental values (Ru: 0.273/0.431 nm; Au: 0.408 nm;[41] graphite: 0.246 nm[42]) The cohesive energy of Au is calculated to be 2.99 eV at 0 K, under-predicting the experimental value of 3.81 eV.[43] This is a well-known error of the GGA despite the inclusion of relativistic effects,[44] although the reactivity of Au surfaces is adequately described by the GGA.

The graphene/Ru(0001) surface was represented by a single (12 × 12) graphene layer laid on top of a three-layer, (11 × 11) unit cell of Ru(0001), for a total of 288 C and 363 Ru atoms. This (12 × 12)-graphene/(11 × 11)-Ru(0001) model formally represents a 1.58% expansion of the graphene lattice in the xy plane. A vacuum space of 2.05 nm separated neighbouring Ru slabs in the z direction. The surface Brillouin zone was sampled at the Γ point only. Energetic and structural convergence with respect to the k-point sampling density was verified using a 2 × 2 × 1 Monkhorst–Pack k-point mesh for the one of the Au models (Model A; see Table 1). Sampling the surface Brillouin zone at the Γ-point gave very reasonable energetic and structural results compared to the much more expensive higher sampling density. All the C, Au, and top-layer Ru atoms were fully relaxed and their positions optimized until no force in relaxed degrees of freedom was greater than 0.05 eV Å^{-1}. The electrostatic potential was adjusted accordingly[45] to account for the fact that the graphene and Au were present on only one side of the Ru slab. Bader charge analysis was performed with the code developed by Henkelman and coworkers.[46]

The adsorption energy of an Au structure was calculated as: $E_{\text{ads}} = E_{\text{total}} - E_{\text{graphene/Ru}} - nE_{\text{Au}_x}$, where E_{total} and $E_{\text{graphene/Ru}}$ are the DFT total energies of the overall system (including the Au structure) and graphene/Ru(0001) only, and E_{Au_x} is the DFT total energy of a single Au atom in the reference state, which can be either the gas phase, bulk fcc phase, or a free monolayer/bilayer, as indicated.

Results and discussion

Formation of graphene moiré and 2-D Au islands

The preparation of graphene/Ru(0001) has been discussed elsewhere[24,34] and will be omitted here for brevity. Fig. 1a displays a typical STM topographic image of the graphene moiré on Ru(0001), which can be grown to hundreds of nm and over

Table 1 E_{ads} (in eV, per Au atom) and maximum height (in nm) of the Au structures on graphene/Ru(0001), and total Bader charge of the Au structures normalized by unit cell area (in e^-/nm^2)[a]

model	w.r.t. $n\text{Au}_{(g)}$	w.r.t. $n\text{Au}_{(s)}$	w.r.t. $n\text{Au}_L$	Max. height[b]	Bader charge
A – (11 × 11) Au monolayer	−2.76	+0.23	−0.08	0.40	−0.17
(A, 2 × 2 × 1 k-point mesh)	−2.74	+0.25	−0.05	0.41	−0.48
B – (11 × 11) Au bilayer	−2.77	+0.22	−0.04	0.70	−0.35
C – (12 × 12) Au monolayer	−2.68	+0.31	—	0.68	−0.40
D – sparse Au "trilayer"	−2.66	+0.33	—	0.59	−0.35
Au adatom	−1.41	+1.58	—	0.22	—

[a] $\text{Au}_{(g)}$, $\text{Au}_{(s)}$, and Au_L refer to an Au atom in the gas-phase; in the bulk fcc phase; and in a free Au monolayer/bilayer. [b] Relative to maximum height of the mound region of clean graphene moiré/Ru(0001) except for Au adatom, for which the height is the distance between the adatom and the C atom directly beneath it.

Ru steps without signs of defect or disuption. As indicated by the marked moiré unit cell, three different regions, which appear bright, medium dark, and dark, are observed. A typical LEED pattern is shown as an inset where the satellite points surrounding the (1 × 1) spots of the Ru(0001) substrate demonstrates the formation of a graphene overlayer.[24,47] The bright, medium dark, and dark regions displayed in Fig. 1a are designated as mound, fcc, and hcp regions, respectively, based on the DFT-optimized (12 × 12)-graphene/(11 × 11)-Ru(0001) model (Fig. 1b), in which the graphene develops a distinct moiré pattern.[48,49] Relative to the surface layer of Ru atoms, the highest C atom, which is in the mound region, has a height of 0.37 nm while the lowest C atom, located in the fcc region, remains at 0.22 nm, thus for a total corrugation of 0.15 nm. Our DFT results[50] are in excellent agreement with previously reported DFT calculations by Wang *et al.*[48,49] and corroborate our and others' LEED and STM results.[47,48,51] This model is a good approximation and the real commensurate superstructure is more complex.[48,52–56]

Fig. 2 shows STM images of 0.2 and 0.6 ML Au deposited on graphene/Ru (0001).[34] At the onset of deposition, small 2-D Au cluster seeds reside exclusively in fcc regions; no nucleation in the mound or hcp regions is observed. DFT calculations confirm the fcc region to contain the most reactive sites (in contrast to graphene/Ir(111), where the hcp region contains the most reactive sites), as with respect to *e.g.* the adsorption of most 3d,[57] 4d and 5d[50] metal adatoms, including Au. With an increase in coverage, a handful of large islands form instead of more small clusters, each of which are more than 10 nm in diameter spanning several moiré unit cells. These large Au islands also display moiré patterns, with the distance between the neighbouring bright spots being exactly the same as that in the moiré of the underlying graphene. The Au islands show no preference for areas above Ru step edges.

We take a closer look at the Au islands and the graphene/Ru support by performing line scan on different surface motifs with 0.25 ML of Au deposited at room temperature (Fig. 3).[58] The step height of graphene on Ru(0001) is measured to be 0.21 nm, in agreement with the height of a mono-atomic step on Ru(0001). The Au islands in fact conform to the morphology of the graphene moiré. The height of a typical Au island is 0.55 nm, which is twice of the height of a mono-atomic

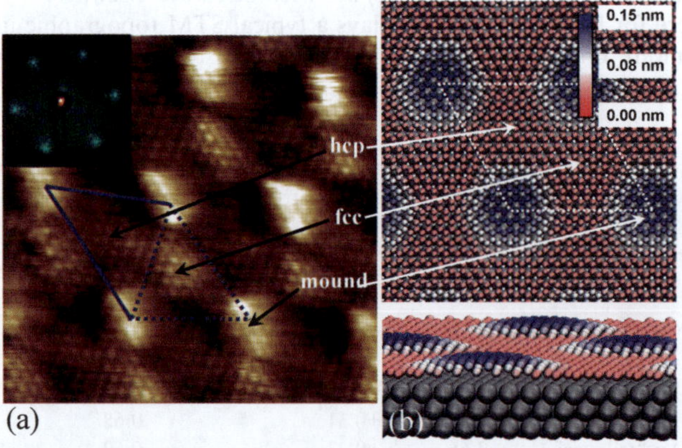

Fig. 1 (a) A typical room temperature STM image (8 nm × 8 nm, $V_b = -0.3$ V, $I_t = 1.0$ nA) of the graphene moiré structure on Ru(0001). The highlighted moiré unit cell and typical LEED pattern (as an inset) are also shown. Thermal drafts at room temperature cause the periodic arrangement of mounds deviate from perfect hexagons. (b) Top and tilted side views of DFT-optimized (12 × 12)-graphene/(11 × 11)-Ru(0001) unit cell. The underlying graphene moiré/Ru(0001) unit cell is outlined. The mound, fcc, and hcp regions are indicated. Gray spheres represent Ru atoms; C atoms are colour-coded according to height (lowest C = 0 nm).

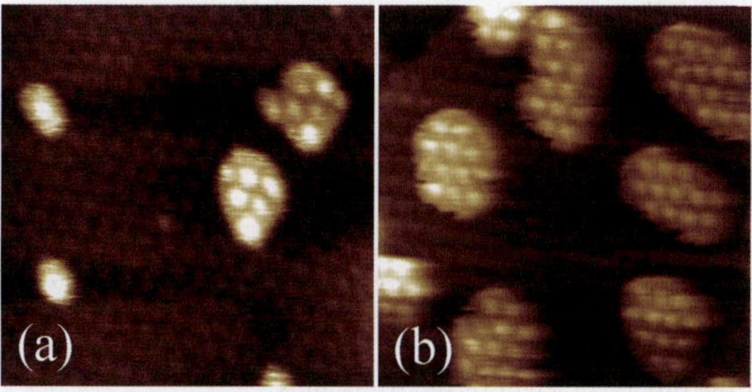

Fig. 2 STM images (50 nm × 50 nm, $V_b = +1.0$ V, $I_t = 0.1$ nA) of (a) 0.2 ML and (b) 0.6 ML of Au deposited on graphene/Ru(0001) at room temperature.[34]

step on Au(111) (0.25 nm) and appears to suggest a bilayer structure. The distance between the graphene and the first layer of Au islands is not necessarily the height of a step on Au(111), but this distance would not be accessible to STM if the Au were indeed bilayer. No bias-dependent change in the image and the apparent height of the Au islands is observed when the tip bias voltage is varied from -1.5 V to 1.5 V. Therefore, the measured height should correspond to the geometric height of the Au islands.

Structure of 2-D Au islands

The simplest model for the Au islands conceivably involves a continuous layer of Au adsorbed on the graphene moiré. In a freestanding Au monolayer and bilayer, the equilibrium in-plane Au–Au distance is calculated to be 0.276 nm in GGA-PBE,

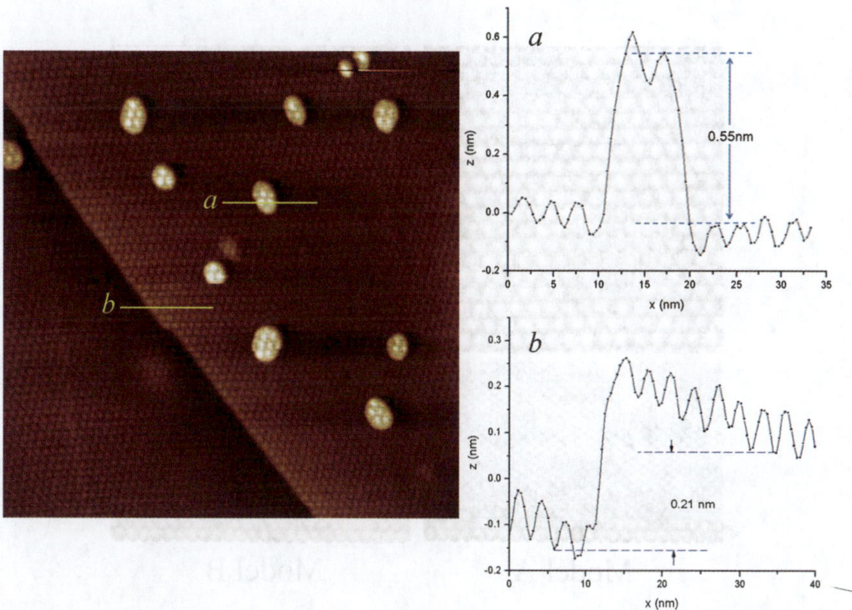

Fig. 3 STM image (200 nm × 200 nm, $V_b = +1.0$ V, $I_t = 0.1$ nA) of 0.25 ML of Au deposited on graphene/Ru(0001) at room temperature.[58] Insets show two line scan profiles: (a) across an Au island; (b) across a Ru step.

which is significantly contracted compared to the bulk Au–Au distance of 0.288 nm because of quantum size effect. 0.276 nm is only *ca.* 1% greater than the in-plane lattice constant of Ru (0.273 nm), which suggests that a layer of Au should adopt nearly the same atomic density of Ru(0001) in the weak adsorption limit. Therefore we begin by examining a close-packed (11 × 11) monolayer (Model A) and bilayer (Model B) of Au, which contain 121 and 242 Au atoms per moiré unit cell, respectively.

Fig. 4 shows the optimized structures for Models A and B. In both structures, the Au layers prefer to conform to the corrugation of the underlying graphene and possess a moiré pattern with a similar corrugation (0.19 and 0.16 nm, respectively). The maximum height of neither model (0.40 and 0.70 nm, respectively), however, agrees with the observed height of the Au islands.

It is instructive to compare the (11 × 11) Au monolayer to an isolated Au adatom adsorbed on graphene/Ru(0001) (Fig. 5a). The E_{ads} for the Au adatom is −1.41 eV (Table 1), which is in agreement with the results of Wang *et al.*[49] and similar to the adsorption energy of an Au adatom on *e.g.* rutile $TiO_2(110)$[59] or supported TiO_x film.[60] The Au adatom preferentially adsorbs at 0.22 nm directly above a C atom that is located above a Ru atom in the subsurface layer, and is 0.29 nm from the next-nearest C atoms in the graphene. The charge density difference plot indicates through-C bonding between Au and Ru (Fig. 5b). On the other hand, the minimum separation between the Au and the graphene is *ca.* 0.37 nm in Models A and B, considerably greater than the bond between the Au adatom and graphene. The calculated E_{ads} for the Au structures in Models A and B is −2.76 and −2.77 eV/Au with respect to individual Au atoms (Table 1), considerably more negative than that of the Au adatom. This is attributed to the formation of strong Au–Au bonds that out-competes bonding with the surface. With respect to a free Au monolayer or bilayer the E_{ads} is quite weak, at −0.08 and −0.03 eV/Au, respectively, consistent with the significant separation between the Au layers and graphene. The weak adhesion of large Au clusters has been noted before: Lopez *et al.*, for

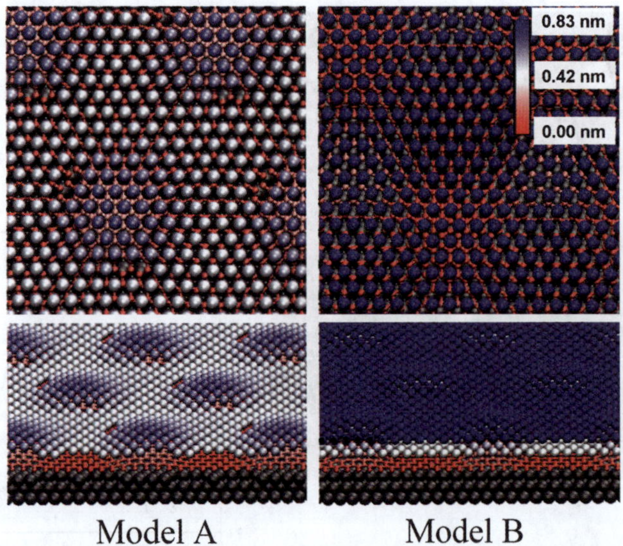

Fig. 4 Top (upper) and side (lower) views of DFT-optimized geometries for: Model A: (11 × 11)-Au monolayer, and Model B: (11 × 11)-Au bilayer (AA-stacking), adsorbed on graphene/Ru(0001). Gray spheres represent Ru atoms; C and Au atoms are colour-coded according to height (lowest C = 0 nm). For clarity, C atoms are represented as ball-and-stick.

   This journal is © The Royal Society of Chemistry 2011

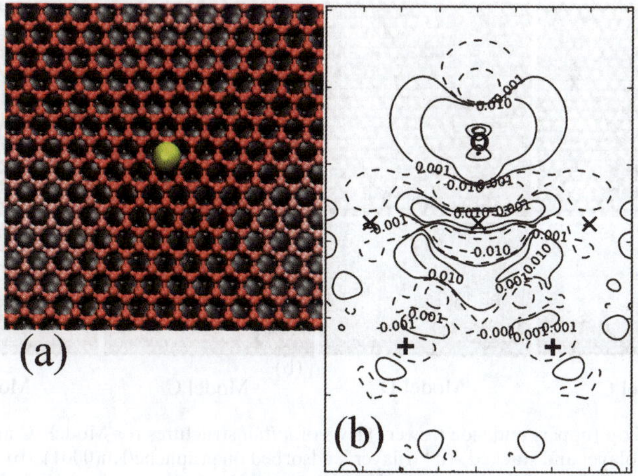

Fig. 5 (a) Geometry and (b) charge density difference contours ($\Delta\rho = \rho_{total} - \rho_{graphene/Ru} - \rho_{Au_x}$) for an isolated Au adatom on graphene/Ru(0001) to illustrate Au–C–Ru bonding. The plane is cut through the Au adatom along the (01$\bar{1}$0) direction of Ru(0001). The contours included are ±0.1, ±0.01, and ±0.001 e/Å^3. Solid and dashed lines represent positive (density gain) and negative (density loss) values, respectively. "O" indicates Au atoms, "×" indicate C atoms, and "+" indicate Ru atoms.

example, reported that an Au monolayer on defect-free $TiO_2(110)$ has an adsorption energy of +0.11 eV/Au (GGA-RPBE) relative to a free Au monolayer.[59]

Given that the commensurate Au monolayer and bilayer do not agree with the experimentally observed height of the 2-D Au islands, we calculate two additional Au structures with more than 1 but less than 2 ML of Au atoms: a (12 × 12)-Au monolayer (Model C; Fig. 6); and a stacked "trilayer" structure (Model D; Fig. 6) that is based on the placement of Au atoms, at three different relative heights, all above C_6 ring sites on the graphene. Models C and D each contain 144 Au atoms.

The initial structure of Model C involves the packing of 144 Au atoms into a smooth, continuous monolayer (Fig. 6a), in which the Au–Au distance is compressed substantially by *ca.* 7%. In the optimized structure, this compressive strain is relieved by displacing the Au atoms above the mound region significantly outward (Fig. 6b). In the optimized Model D (Fig. 6b), a nearly complete Au ML forms, with the excess Au atoms squeezed out forming strings of Au adatoms. Although the optimized structures of Models C and D are quite different, they are both less stable than the close-packed Models A and B. The bulging mound regions make the Au in Model C significantly corrugated (maximum height is 0.68 nm relative to the mound of the clean graphene moiré), whereas the maximum height of Model D is the closest of the structures examined so far to the observed height of the Au islands in STM (Table 1).

In our experiments there is no visible defect in the graphene moiré or detectable foreign species such as oxygen, so the enhancement of Au-support adhesion due to defects or oxidation[19,20,61–63] can be ruled out, leading us to conclude that there is no special anchoring site besides the native structure of the graphene moiré on Ru(0001) itself. Our DFT calculations suggest that Au is able to form metastable, extended structures that conform to the corrugation of the graphene moiré, in the absence of direct Au–C bonding. Van der Waals contribution to the adsorption energy, which is not captured by standard GGA formulation of the DFT, is estimated using Grimme's DFT-D2 formulation[64,65] for Model A to be a sizable −0.48 eV/Au. Nonetheless, the geometries of these Au structures develop as the result of energy minimization without van der Waals interaction being taken into account, which suggests chemical interaction possibly of an Au d–C π nature and

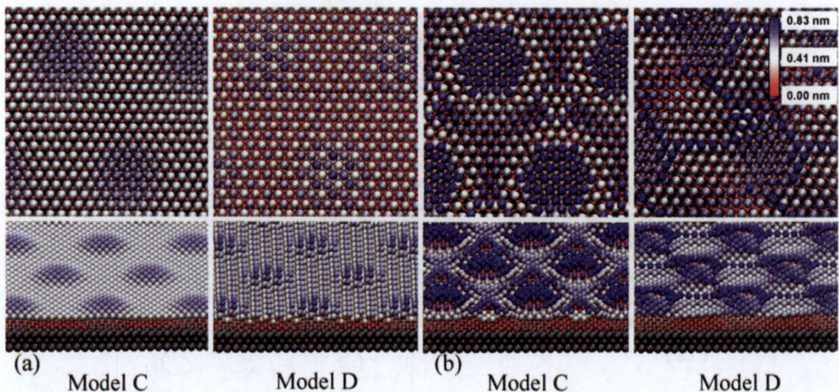

(a) Model C   Model D   (b) Model C   Model D

Fig. 6 (a) Top (upper) and side (lower) views of *initial* structures for Models C and D, (12 × 12)-Au monolayer and stacked Au "trilayer", adsorbed on graphene/Ru(0001). (b) Top (upper) and side (lower) views of DFT-*optimized* Models C and D. Gray spheres represent Ru atoms; C and Au atoms are colour-coded according to height (lowest C = 0 nm). For clarity, C atoms are represented as ball-and-stick.

reflects the fluxional nature of Au structures. With respect to bulk Au metal they are all metastable (Table 1), with the close-packed Au monolayer and bilayer being energetically more favourable than the sparse, textured structures.

Catalytic activity of 2-D Au islands

Our preliminary results show that the Au islands/graphene/Ru(0001) surface adsorbs CO at 85 K, and the pre-adsorbed CO can be titrated by O_2 at the same cryogenic temperature.[58] Consistent with the chemical inertness of graphene, the bare graphene/Ru(0001) surface does not adsorb any detectable amount of CO. Therefore there is less ambiguity in this case that the Au islands themselves are the locus of CO and O_2 adsorption and catalytic activity, than in the cases where functional support materials capture and store reactants.[66]

The textured surfaces of Models C and D present many under-coordinated Au atoms, which have been shown to be vital to *e.g.* CO oxidation catalysis by enhancing reactant adsorption and activation.[67–72] The structure of the periphery of the Au islands may also be important because of the presence of under-coordinated Au atoms and sites formed by Au and the support.[14,73,74] In addition, negative charge on Au has been suggested as another factor that can enhance CO oxidation,[75–77] and Bader charge analyses of our Au structure models indicate that all of them are slightly anionic (Table 1). Detailed catalytic behaviour and conclusive identification of the structure, anchoring mechanism, and active site of the 2-D Au islands are the subject of ongoing experimental and theoretical investigations.

Conclusions

Large 2-D Au islands formed upon the vapour deposition of Au on graphene moiré supported on Ru(0001) at room temperature. These Au islands each span several moiré unit cells but are only *ca.* 0.55 nm high relative to the graphene as appears in STM. They conform to the corrugation of the graphene structurally and display commensurate moiré patterns. Several Au overlayer models on graphene/Ru(0001) are examined using density functional theory calculations. Smooth, close-packed Au monolayer and bilayer (Models A and B) are energetically more stable but do not agree with the observed height of the 2-D Au islands. A textured structure (Model D) does exhibit a maximum height that agrees closely with the observed height of the Au islands but is less stable than the close-packed structures. All the overlayer models examined are, however, metastable with respect to bulk Au metal. Preliminary tests find the Au islands/graphene/Ru(0001) surface to be active for CO oxidation at cryogenic temperature, which suggests that the Au itself is the locus of catalytic activity.

Acknowledgements

This material is based upon work supported as part of the Center for Atomic-Level Catalyst Design, an Energy Frontier Research Center funded by the Office of Basic Energy Sciences, the Office of Science of the U.S. Department of Energy under Award Number DE-SC0001058. This research used computing resources of the Oak Ridge Leadership Facility at the Oak Ridge National Laboratory, which is supported by the Office of Science of the U.S. Department of Energy under Contract DE-AC05-00OR22725.

References

1 M. Haruta, T. Kobayashi, H. Sano and N. Yamada, *Chem. Lett.*, 1987, 405.
2 G. C. Bond and D. T. Thompson, *Catal. Rev. Sci. Eng.*, 1999, **41**, 319.
3 M. Haruta, *Chem. Rec.*, 2003, **3**, 75.
4 G. J. Hutchings, *Gold Bull.*, 2004, **37**, 3.
5 R. Meyer, C. Lemire, S. K. Shaikhutdinov and H. Freund, *Gold Bull.*, 2004, **37**, 72.
6 D. Andreeva, V. Idakiev, T. Tabakova, A. Andreev and R. Giovanoli, *Appl. Catal., A*, 1996, **134**, 275.
7 R. Si and M. Flytzani-Stephanopoulos, *Angew. Chem., Int. Ed.*, 2008, **47**, 2884.
8 B. N. Zope, D. D. Hibbitts, M. Neurock and R. J. Davis, *Science*, 2010, **330**, 74.
9 A. Grirrane, A. Corma and H. Garcia, *Science*, 2008, **322**, 1661.
10 X. Y. Deng and C. M. Friend, *J. Am. Chem. Soc.*, 2005, **127**, 17178.
11 E. Sacaliuc, A. M. Beale, B. M. Weckhuysen and T. A. Nijhuis, *J. Catal.*, 2007, **248**, 235.
12 J. K. Edwards, B. Solsona, E.N. N, A. F. Carley, A. A. Herzing, C. J. Kiely and G. J. Hutchings, *Science*, 2009, **323**, 1037.
13 B. J. Xu, X. Y. Liu, J. Haubrich, R. J. Madix and C. M. Friend, *Angew. Chem., Int. Ed.*, 2009, **48**, 4206.
14 M. Haruta, *Catal. Today*, 1997, **36**, 153.
15 M. Valden, X. Lai and D. W. Goodman, *Science*, 1998, **281**, 1647.
16 M. S. Chen and D. W. Goodman, *Science*, 2004, **306**, 252.
17 M. S. Chen and D. W. Goodman, *Acc. Chem. Res.*, 2006, **39**, 739.
18 S. Laursen and S. Linic, *Phys. Rev. Lett.*, 2006, **97**, 026101.
19 J. G. Wang and B. Hammer, *Top. Catal.*, 2007, **44**, 49.
20 P. Frondelius, H. Hakkinen and K. Honkala, *Angew. Chem., Int. Ed.*, 2010, **49**, 7913.
21 W. C. Ketchie, M. Murayama and R. J. Davis, *Top. Catal.*, 2007, **44**, 307.
22 Z. Ma, C. D. Liang, S. H. Overbury and S. Dai, *J. Catal.*, 2007, **252**, 119.
23 J. Wintterlin and M. L. Bocquet, *Surf. Sci.*, 2009, **603**, 1841.
24 M.-C. Wu, Q. Xu and D. W. Goodman, *J. Phys. Chem.*, 1994, **98**, 5104.
25 J. Coraux, A. T. N'Diaye, C. Busse and T. Michely, *Nano Lett.*, 2008, **8**, 565.
26 J. Coraux, A. T. N'Diaye, M. Engler, C. Busse, D. Wall, N. Buckanie, F. Heringdorf, R. van Gastei, B. Poelsema and T. Michely, *New J. Phys.*, 2009, **11**, 023006.
27 A. T. N'Diaye, J. Coraux, T. N. Plasa, C. Busse and T. Michely, *New J. Phys.*, 2008, **10**, 043033.
28 A. T. N'Diaye, S. Bleikamp, P. J. Feibelman and T. Michely, *Phys. Rev. Lett.*, 2006, **97**, 215501.
29 A. T. N'Diaye, T. Gerber, C. Busse, J. Myslivecek, J. Coraux and T. Michely, *New J. Phys.*, 2009, **11**, 103045.
30 P. J. Feibelman, *Phys. Rev. B: Condens. Matter Mater. Phys.*, 2008, **77**, 165419.
31 P. J. Feibelman, *Phys. Rev. B: Condens. Matter Mater. Phys.*, 2009, **80**, 085412.
32 B. Borca, F. Calleja, J. J. Hinarejos, A. L. V. de Parga and R. Miranda, *J. Phys.: Condens. Matter*, 2009, **21**, 134002.
33 H. Zhang, Q. Fu, Y. Cui, D. L. Tan and X. H. Bao, *J. Phys. Chem. C*, 2009, **113**, 8296.
34 Z. Zhou, F. Gao and D. W. Goodman, *Surf. Sci.*, 2010, **604**, L31.
35 G. Kresse and J. Furthmuller, *Comput. Mater. Sci.*, 1996, **6**, 15.
36 G. Kresse and J. Furthmuller, *Phys. Rev. B: Condens. Matter*, 1996, **54**, 11169.

37 G. Kresse and J. Hafner, *Phys. Rev. B: Condens. Matter*, 1994, **49**, 14251.
38 J. J. Perdew, K. Burke and M. Ernzerhof, *Phys. Rev. Lett.*, 1996, **77**, 3865.
39 G. Kresse and J. Joubert, *Phys. Rev. B: Condens. Matter Mater. Phys.*, 1999, **59**, 1758.
40 M. Methfessel and A. T. Paxton, *Phys. Rev. B*, 1989, **40**, 3616.
41 N. W. Ashcroft and N. D. Mermin, *Solid State Physics*, Saunders College, Orlando, FL, 1976.
42 R. Nicklow, N. Wakabayashi and H. G. Smith, *Phys. Rev. B: Solid State*, 1972, **5**, 4951.
43 C. Kittel, *Introduction to Solid State Physics*, Wiley, New York, 1996.
44 P. H. T. Philipsen and E. J. Baerends, *Phys. Rev. B: Condens. Matter*, 2000, **61**, 1773.
45 J. Neugebauer and M. Scheffler, *Phys. Rev. B: Condens. Matter*, 1992, **46**, 16967.
46 W. Tang, E. Sanville and G. Henkelman, *J. Phys.: Condens. Matter*, 2009, **21**, 084204.
47 S. Marchini, S. Gunther and J. Wintterlin, *Phys. Rev. B: Condens. Matter Mater. Phys.*, 2007, **76**, 075429.
48 B. Wang, M. L. Bocquet, S. Marchini, S. Gunther and J. Wintterlin, *Phys. Chem. Chem. Phys.*, 2008, **10**, 3530.
49 B. Wang, S. Gunther, J. Wintterlin and M. L. Bocquet, *New J. Phys.*, 2010, **12**, 15.
50 L. Semidey-Flecha, D. Teng, D. S. Sholl and Y. Xu, submitted for publication.
51 W. Moritz, B. Wang, M. L. Bocquet, T. Brugger, T. Greber, J. Wintterlin and S. Gunther, *Phys. Rev. Lett.*, 2010, **104**, 136102.
52 D. Martoccia, P. R. Willmott, T. Brugger, M. Bjorck, S. Gunther, C. M. Schleputz, A. Cervellino, S. A. Pauli, B. D. Patterson, S. Marchini, J. Wintterlin, W. Moritz and T. Greber, *Phys. Rev. Lett.*, 2008, **101**, 126102.
53 A. L. V. de Parga, F. Calleja, B. Borca, M. C. G. Passeggi, J. J. Hinarejos, F. Guinea and R. Miranda, *Phys. Rev. Lett.*, 2008, **100**, 056807.
54 Y. Pan, H. G. Zhang, D. X. Shi, J. T. Sun, S. X. Du, F. Liu and H. J. Gao, *Adv. Mater.*, 2009, **21**, 2777.
55 D. Martoccia, M. Bjorck, C. M. Schleputz, T. Brugger, S. A. Pauli, B. D. Patterson, T. Greber and P. R. Willmott, *New J. Phys.*, 2010, **12**, 043028.
56 B. Borca, S. Barja, M. Garnica, M. Minniti, A. Politano, J. M. Rodriguez-Garcia, J. J. Hinarejos, D. Farias, A. L. V. de Parga and R. Miranda, *New J. Phys.*, 2010, **12**, 093018.
57 O. V. Yazyev and A. Pasquarello, *Phys. Rev. B: Condens. Matter Mater. Phys.*, 2010, **82**, 045407.
58 L. Liu, Z. Zhou, Q. Guo, Z. Yan, Y. Yao and D. W. Goodman, *Surf. Sci.*, 2011, DOI: 10.1016/j.susc.2011.04.040.
59 N. Lopez and J. K. Nørskov, *Surf. Sci.*, 2002, **515**, 175.
60 Y. Zhang, L. Giordano and G. Pacchioni, *J. Phys. Chem. C*, 2008, **112**, 191.
61 M. Sterrer, M. Yulikov, E. Fischbach, M. Heyde, H. P. Rust, G. Pacchioni, T. Risse and H. J. Freund, *Angew. Chem., Int. Ed.*, 2006, **45**, 2630.
62 M. S. Chen and D. W. Goodman, *Top. Catal.*, 2007, **44**, 41.
63 D. Matthey, J. G. Wang, S. Wendt, J. Matthiesen, R. Schaub, E. Lægsgaard, B. Hammer and F. Besenbacher, *Science*, 2007, **315**, 1692.
64 X. Wu, M. C. Vargas, S. Nayak, V. Lotrich and G. Scoles, *J. Chem. Phys.*, 2001, **115**, 8748.
65 S. Grimme, *J. Comput. Chem.*, 2006, **27**, 1787.
66 L. M. Liu, B. McAllister, H. Q. Ye and P. Hu, *J. Am. Chem. Soc.*, 2006, **128**, 4017.
67 M. Mavrikakis, P. Stoltze and J. K. Nørskov, *Catal. Lett.*, 2000, **64**, 101.
68 Y. Xu and M. Mavrikakis, *J. Phys. Chem. B*, 2003, **107**, 9298.
69 N. Lopez, T. V. W. Janssens, B. S. Clausen, Y. Xu, M. Mavrikakis, T. Bligaard and J. K. Nørskov, *J. Catal.*, 2004, **223**, 232.
70 I. N. Remediakis, N. Lopez and J. K. Nørskov, *Angew. Chem., Int. Ed.*, 2005, **44**, 1824.
71 S. H. Overbury, V. Schwartz, D. R. Mullim, W. F. Yan and S. Dai, *J. Catal.*, 2006, **241**, 56.
72 W. D. Williams, M. Shekhar, W. S. Lee, V. Kispersky, W. N. Delgass, F. H. Ribeiro, S. M. Kim, E. A. Stach, J. T. Miller and L. F. Allard, *J. Am. Chem. Soc.*, 2010, **132**, 14018.
73 L. M. Molina and B. Hammer, *Phys. Rev. Lett.*, 2003, 90.
74 Z. P. Liu, P. Hu and A. Alavi, *J. Am. Chem. Soc.*, 2002, **124**, 14770.
75 M. S. Chen and D. W. Goodman, *Catal. Today*, 2006, **111**, 22.
76 L. M. Molina and B. Hammer, *Appl. Catal., A*, 2005, **291**, 21.
77 U. Landman, B. Yoon, C. Zhang, U. Heiz and M. Arenz, *Top. Catal.*, 2007, **44**, 145.

The effect of the metal to non-metal transition on the activity of gold catalysts

Geoffrey C. Bond*

Received 21st January 2011, Accepted 23rd February 2011
DOI: 10.1039/c1fd00010a

The problem of obtaining exact information on the particle size dependence of the rate of gold-catalysed reactions is surveyed, and it is shown to be possible that over a range of *mean* sizes the activity for CO oxidation is predominantly due to the fraction of particles smaller than about 2.5 nm. This size coincides with the point at which transition from metallic to non-metallic or 'molecular' behaviour is expected, and means of describing the electronic structure of the microparticles is considered. This transition may be the cause of enhanced and activated chemisorption of the CO molecule. Analysis of results for the rate dependence on size for CO oxidation on Au/TiO_2 catalysts by the compensation relation confirms the importance of the transition. It is also held responsible for the ability of the PtAu system to form small homogeneous alloy particles.

1.   Introduction

Progress in any branch of science depends on the formation of a consensus among its principal practioners, and this must cover both experimentally determined features and the associated theoretical explanation. Reading the literature on such a formally simple system as the gold-catalysed oxidation of CO is therefore a humbling experience, since the number of things on which there is universal agreement is small, and expressible only in very general terms. This system has been a happy hunting ground for theoreticians,[1-3] but the chemist in the lab will sympathise with Omar Khayyam, who it will be recalled

 "….when young did eagerly frequent

 Doctor and saint, and heard great argument

 About it and about…."

It may therefore be salutary to examine briefly the reasons for the lack of wider agreement on the origin of gold's activity and for the plethora of explanations offered for it. To do this, it is necessary to re-visit and perhaps to re-analyse some older work, and to remind ourselves of sundry pieces of information that will help to plot the way forward. Some emphasis will also have to be placed on what we do not know, as well as on what we do.

In preparation for submitting an abstract for GOLD2009, the late David Thompson and I undertook to survey the then available literature in order to identify and quantify the many features bearing on gold's activity. Although it was possible to recognise their existence,[4] attempts to assess their exact importance were frustrated by the very varied conditions under which activities were reported. Even for a single system (*e.g.* Au/TiO_2), rates when plotted on an Arrhenius diagram were widely scattered, and at a fixed temperature varied by more than two orders of magnitude. The only two factors on the importance of which there was anything approaching unanimity were[4] (i) the gold's particle size, and (ii) the choice of support. High

9 Townfield, Rickmansworth, WD3 7DD, UK. E-mail: geoffrey10bond@aol.com; Tel: +44 1923 774156

---

activity demanded particle sizes well below 5 nm, while best performance was usually obtained with reducible supports such as TiO_2 and Fe_2O_3, although subsequently it appeared[4] that ceramic oxides (Al_2O_3 and SiO_2) could also be acceptable supports.

2.   Problems in precisely measuring catalytic activity

A sound understanding of the true nature of a catalytic process first demands accurate measurement of its rate under specified conditions. This is essential when trying to evaluate the effects of process variables and of alterations to structure, composition and preparative procedure. It presents problems in all catalysed reactions, and the gold-catalysed oxidation of CO is no exception; indeed there are some unique difficulties. Extracting quantitative conclusions from the literature is complicated by the use of different O_2/CO ratios, choice depending on the foreseen end use: most commonly used is either 1% CO in air or the stoichiometric ratio of 1:2, but *without knowing orders of reaction there is no way in which the two sets of results can be harmonised.* Then many gold catalysts experience a more or less rapid deactivation, due primarily to the formation of inactivating side-products such as CO_3^{2-} and HCO_3^-; *unfortunately the time-on-stream at which the rate is obtained is not always specified.*[5] Next, *the probable occurrence of mass-transport limitation at high conversions is frequently ignored;* it is impossible to derive a meaningful rate from the temperature at which conversion attains 100%. Finally, values of temperatures at which conversion reaches 50% or any other value are of little help unless the information is provided to convert them into rates.

Even if a well-defined rate under kinetic control is measured, we have only reached the end of the beginning, because for supported catalysts we need to know the gold content; *it is sometimes assumed without evidence that all the metal used in the preparation ends up on the support.* If the loading is measured, we should then have a rate in units such as mol CO reacted per mol Au in unit time; this is a *specific rate*, but with well characterised materials it should be possible to estimate the metal's exposed surface area, and hence to get a rate as mol CO per m^2 Au in unit time, or with certain assumptions as mol CO per mol Au surface atoms in unit time; the latter is termed *turnover frequency*. We must now see whether to do so is advantageous.

3.   The size of gold particles

Supported gold catalysts made by chemical routes inevitable contain particles covering a range of sizes. Characterisation by TEM provides a size distribution curve (or histogram) from which the mean and most probable sizes are deduced; these may be expressed as either a length (diameter) or with assumptions concerning shape as an area or a volume. Very small particles may however escape detection; they may be seen by aberration-corrected STEM (see below), but this technique is not yet widely available. An XAFS measurement gives a mean coordination number,[6] but its conversion to a size depends on the model selected. Now it is usual to relate activity to mean size, but *in fact it represents the integrated activity over the whole size range*, and the true dependence of activity on size can only be deduced by the use of catalysts containing only monodisperse particles. Let us suppose for the sake of argument that all particles smaller than 2.5 nm are 150 times more active than all those that are larger (this factor is validated below); then if we take typical size distributions having maxima at 2, 4 and 6 nm (Figure 1A), the net activity will be dominated by the fraction of the smaller particles (Figure 1B). The commonly observed activity *vs.* size curve is readily accounted for in this way.

We cannot therefore be absolutely certain of the exact form of the size-activity relation based on information as usually obtained; this consideration limits the usefulness of trying to account for activity in terms of any specific observable or calculated parameter, a matter to which we now turn.

     This journal is © The Royal Society of Chemistry 2011

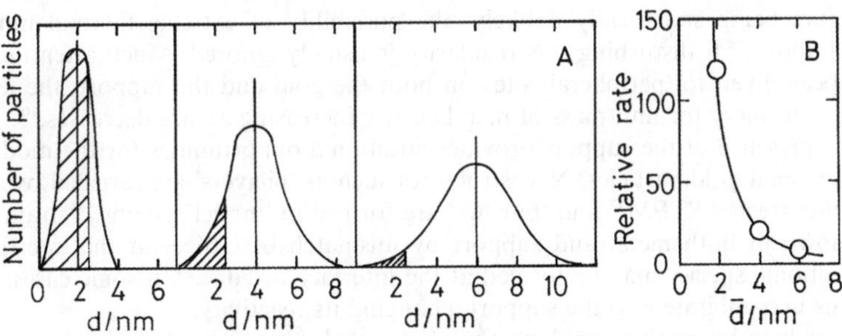

Fig. 1 (A) Schematic diagram showing how for size distributions having maxima at 2, 4 and 6 nm the fraction of particles smaller than 2.5 nm may vary. (B) Dependence of relative rate (a. u.) on size, assuming that all particles below 2.5 nm are 150 times more active than all that are larger.

4. Variation of physical properties with particle size

Whatever uncertainties surround efforts to obtain exact knowledge of how activity varies with particle size, the fact remains that it does so very dramatically, in a way that is not seen with other metals.[7] Microparticulate gold owes its activity for CO oxidation to its ability to chemisorb it to a sufficient extent, but unlike platinum not so strongly as to render it unreactive. However the commonly observed fractional positive orders of reaction for both reactants[5,8,9] show that the optimum coverage is not reached at normally used pressures, in which case some increase in adsorption strength should prove beneficial.

The origin of this exceptional activity has been much discussed,[4,10,11] and the changes in physicochemical character with particle size which may be responsible have been extensively researched.[10] Some of the same features will be observed regardless of whether the particle is supported or not, although the support will introduce additional factors needing attention. Observable factors have been classified as (i) structural, (ii) energetic, and (iii) electronic; the term 'optoelectronic' embraces the last two,[8] and mutual interaction between all three is to be expected. It is unnecessary to rehearse the experimental facts in detail, but one or two pertinent changes may be mentioned. Under (i) we note that some particles formed by vapour deposition of atoms sometimes have pentagonal symmetry rather than the fcc structure,[12] while the bond length in supported particles of cubic symmetry decreases from the bulk value of 2.88 nm to 2.72 nm as size becomes smaller.[13] Under (ii) the melting temperature decreases, and the vibrational amplitude of surface atoms and their mobility becomes greater, as size goes down.[8] It therefore seems possible that a small catalysing gold particle may be quaking like a jelly,[14] and indeed a 2 nm particle has been observed to undergo extensive structural transformation on a time-scale of 1 s in an electron microscope.[15] Under (iii) it is well established that the binding energy of the Au $4f_{7/2}$ level decreases by about 1 eV at sizes below about 5 nm, due probably to a combination of initial and final state effects.[8] The inner potential starts to rise at about 5 nm, and increases very rapidly below 2 nm.[15] The changes in optoelectronic properties are a consequence of the increasing fraction of surface atoms, and thus to the uses that are found for the electrons not required for bonding; the mean coordination number is a simple way of assessing this.[12] These are all manifestations of a significant alteration in electronic structure, a matter to which we will return.

The variation of activity with size below 5 nm has been attributed at one time or another to most if not all the physical features that show size-dependence.[8,11] The first and most obvious of these is the rising fraction of surface atoms, especially those having very low coordination numbers at edges and corners. This correlation is frequently based on perfect crystal forms having a complete outer layer, and therefore being statistically unlikely; the possibility of extreme thermal motion, noted above,[14,15] disturbing this regularity is usually ignored. Much attention has also been given to 'peripheral' sites on both the gold and the support, the length of the periphery for unit mass of metal clearly increasing as size decreases.[10]

The presence of the support provides additional opportunities for the modification of small gold particles. New structures such as 'bilayers' are revealed by aberration-corrected STEM[17] and 'bi-rows' are formed in 'model'systems,[18] strain may be shown in both metal and support by mismatch of lattices at the interface,[12] and cationic species may be formed at the interface or edge;[19] in some cases, gold cations may migrate into the support, changing its reactivity.

It needs to be emphasised, however, that *establishing a correlation does not automatically explain its cause.* Where so many characteristics demonstrate similar size-dependence, all related to the number of atoms per particle and the surface/volume ratio, it is pointless to elevate one above the rest. It is futile to dispute whether steric/geometric factors are more important than electronic, since they are mutually dependent. One does not argue whether one's left leg is more important than the right; both are essential to the proper functioning of the body, and catalytic activity is the result of the totality of relevant factors, acting in concert.

5. The physical properties of very small gold particles: the metal to non-metal transition

Whatever the importance of microstructural factors, the electronic constitution of very small particles cannot be neglected. There is much physical evidence to show that below a certain size they become non-metallic and non-conducting, although the point at which this occurs seems to depend on the technique used, on the criterion applied, on the support if any, and on the preparation method.[8,20] In consequence the critical point is located somewhere between 100 and 800 atoms, *i.e.* between about 2 and 3 nm. This question was considered in several papers given at a Faraday Discussion on 'Small Metallic Particles' in 1991. It is suggested that conduction will start to fall when the particle size approaches that of the electron's characteristic de Broglie wavelength, and indeed microwave absorption measurements indicate that conduction in 4 nm particles is 10^7 times smaller than for the bulk.[19] Such particles resemble giant molecules, and may be termed 'molecular gold'. So for example the ligand-coated Au_{55} cluster (d = $\sim$1.5 nm) is judged to be non-metallic (band gap = 0.17 eV) or on the verge of becoming metallic.[8,20] Perhaps the clearest evidence for an emerging molecular state is provided by scanning-tunnelling microscopy (STS), which for model Au/TiO_2 particles reveals the appearance of a band gap at 3 nm, this rising to above 1.5 eV for quasi-two dimensional particles 1.5 nm in diameter and two atom layers thick.[8,18] Further evidence for the anomalous electronic structure of extremely small gold particles is shown by the detection of an esr signal at g = 2.07–2.12 only with low-loaded Au/SiO_2 (%Au [21].(%0.05 $\leq$ This was attributed to spin-orbit coupling between localised atomic orbitals, as no signal was seen at higher loadings. It is strange that this observation appears never to have been confirmed or exploited. Using X-ray magnetic circular dichroism, gold particles 1.9 $\pm$ 0.2 nm in size exhibit ferromagnetic spin polarisation,[22] and a further sign of major change in electronic properties is the rapid rise in redox potential with decreasing size.[8]

There is therefore a *prima facie* case for associating high catalytic activity with non-metallic character,[16,18,23] since whatever other criteria are needed all gold particles smaller than about 2.5 nm will have this property. At this size, about half the atoms are superficial, having coordination numbers below 12; most of the rest are in contact with one or more surface atoms and are somewhat modified by this. The problem of devising what electronic structure is responsible for the observed behaviour therefore becomes essentially that of defining a local density of states

(LDOS) for surface atoms. It is worth stressing that *surface atoms on large particles are much less active than those on small particles*, so the mere fact of being on the surface is not enough; what goes on below is clearly relevant.

Before addressing the interpretation of these findings, we should note that many of them have of necessity been obtained using 'model' preparations and UHV conditions that do not equate to those used in practical catalysis.[24] The transfer of information from these studies to the real world needs to be treated cautiously, and in particular gold particles made by atom deposition onto evacuated oxide surfaces certainly differ significantly from those made chemically.[4] This reservation must be kept in mind in what follows.

6.   An interpretation of the metal to non-metal transition

The problem we now face is to find a suitable way of portraying the electronic structure of gold particles comprising 10 to 100 or so atoms. Although DFT calculations are usually made with clusters of up to only 13 atoms,[2] the point of transition from non-metal to metal falls in the range (1–10 nm) that is currently inaccessible to computation.[25] We can however do two things: (i) we can approach the behaviour of very small particles from that of clusters of a few atoms, and (ii) we can employ a simple argument based on electrostatics to explore their electronic state.

Representation of the electronic structure of very small gold particles presents certain problems. Figure 2 is an attempt to show how on passing from atom to small particle to large particle the valence energy levels coalesce to form bands; the location on the energy scale follows that used by Phala and van Steen.[25] The band structure of a 20 nm particle is fairly straightforward: the 5*d*, 6*s* and 6*p* bands overlap, their proximity being helped by relativistic effects. The Fermi energy E_F falls just below the top of the *d*-band to account for the appearance of a weak white line, so this structure is clearly metallic (Figure 3). As particle size falls and approaches

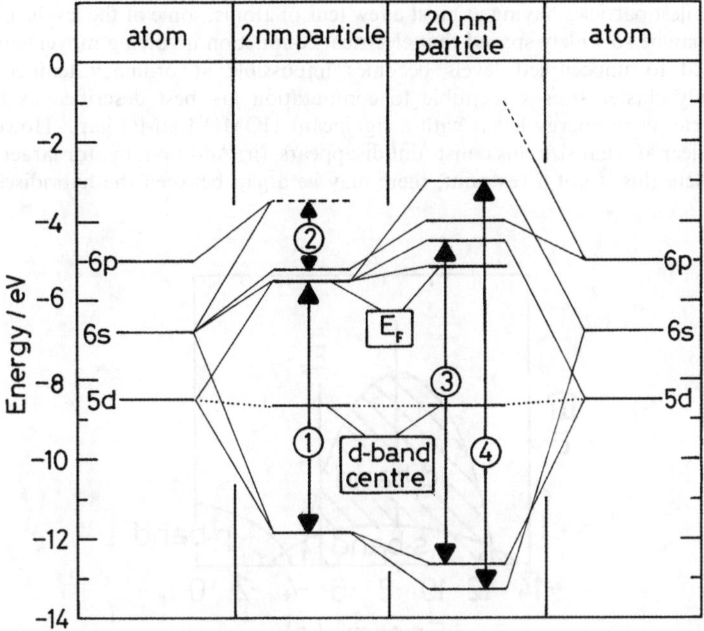

Fig. 2   Illustration of the suggested electronic band structures of 2 nm and 20 nm particles and their relation to the energy levels in free atoms. (1) Filled *ds*-band. (2) Empty *sp*-band. (3) Almost filled *d*-band. (4) Partially filled *s*-band.

the transition size, the bands become narrower and the spacing between the energy levels starts to increase; E_F is also not a fixed quantity, because as the highest accessible level it falls somewhat with particle size (Figure 2).[13,19,24,25] These changes are accompanied by the decrease in the mean interatomic spacing, noted above.[13]

It is sometimes stated that the transition size can be inferred by considering the point at which the spacing of the energy levels exceeds the thermal energy kT;[19,26] at room temperature this is 2.5 kJ mol^{-1}. This spacing is then related approximately to E_F/n where n is the number of atoms in the particle. While this gives a sensible answer for gold ($E_F = 5.22$ eV $= 504$ kJ mol^{-1}), *i.e.* $n = \sim200$, at the transition point, it is doubtful whether it ought to apply in systems where there is more than one electron in the conduction band, and its use is therefore best avoided.

A number of physical techniques (STS, Mössbauer spectroscopy, FEM, XPS *etc.*) report observing a 'band gap' in very small gold particles, but they are infuriatingly coy in specifying the bands or levels between which this occurs. This makes the task of devising a band structure diagram for a non-metallic gold particle, analogous to Figure 3, somewhat difficult. Some help is provided by an XPS study of 1.6 nm Au/diamond,[27] which found a valence band maximum at 1.5 eV below E_F, with *sp* states just below E_F, and by an analysis[28] of the optical absorption spectrum of 1.4–3.2 nm particles. Intraband structures originated in a broad conduction $6s^1p$ band, giving a surface plasmon frequency at 5 eV, and the first interband transition which started 1.6–1.7 eV was assigned to $5d^{10}$ excitation to unoccupied states above E_F. The $d{\rightarrow}p$ character results in strong absorption, and the characteristic colour. Most interestingly however was the observation with 1.4 nm particles of a step-like feature at intervals of about 0.3 eV between 1.6 and 2.4 eV, thought to be due to transitions into higher unoccupied levels. This is related to near-IR luminescence shown by ligand-stabilised particles between 1 and 2 nm in size; its intensity improves with decreasing size and is much greater (x 10^5) than with bulk gold.[29] These authors[28] believe that "the optoelectronic properties of nano-gold molecules cannot be understood by extrapolation of the bulk properties of fcc gold, but require new concepts."

There would appear to be two possible ways of regarding this energy gap. (i) With the smallest particles having at most a few tens of atoms, some of the levels in the *ds* band may be so widely spaced that electrical conduction involving movement from occupied to unoccupied levels becomes impossible at ordinary temperatures. Certainly cluster sizes susceptible to computation are best described as having a discrete set of energy levels with a significant HOMO-LUMO gap.[2] However it is not clear at what size this constraint disappears. (ii) Additionally, for larger particles where this is not a restraint, there may be a gap between the hybridised $d^{10}s^1$

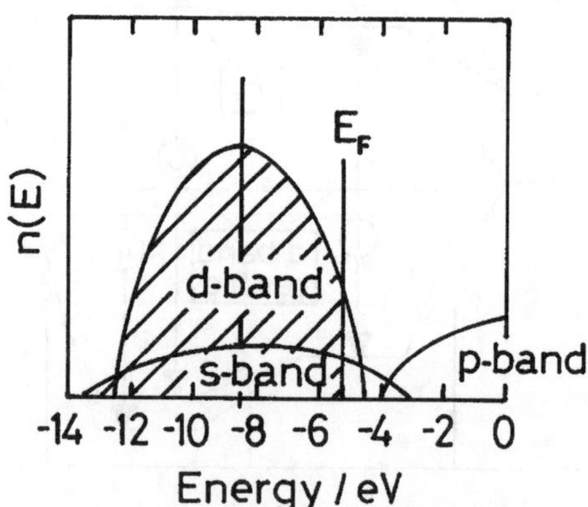

Fig. 3 Distribution of energy levels for a large (20 nm) particle; note the E_F lies below the top of the *d*-band, accounting for the appearance of a small white line.

band and an empty *sp* band. This assignment is supported by calculations employing EHT methodology on an Au_{63} cubic fcc cluster,[30] with an edge length of 1.1 nm; this showed molecular orbitals below E_F to be linear combinations of atomic 5*d*- and 6*s*-orbitals, while the LUMO and higher molecular orbitals are linear combinations of 6*s*- and 6*p*-orbitals. The HOMO-LUMO band gap is just under 0.3 eV. Helped by this information, Figure 4 translates the band structure of Figure 2 into a density-of states diagram for a typical 'molecular' particle.

7. DFT calculations on small particles

A possible approach to understanding the behaviour of very small gold particles is to examine the trends that occur in clusters of a few atoms. Short has carried out such calculations on Au_2, Au_4 and Au_8 clusters,[30] with the Au_4 being both square planar and the stabler[31] planar rhombus, and the Au_8 being both an element of fcc structure and the stabler[31] planar form. Values for E_F are somewhat lower than the bulk value, while the *d*-band centre remains fixed and is similar to the bulk value.; the band width increases with the number of atoms, as expected (Figure 5). Both the Au_4 structures give closely similar results, but E_F for the stabler Au_8 is somewhat more negative, and there is a difference of about 2.7 eV between the HOMO and LUMO levels. It is surprising and gratifying to find how closely the values for the fcc structures resemble those chosen for the 2.0 nm particle in Figure 2 (repeated in Figure 5), arrived at in quite different ways. The orbitals of the $6s^1$ electron hybridise with those of the $5d^{10}$ electrons, but appear close to E_F, as we have assumed above for very small particles.

The shape of the Au_8 cluster affects the width of the *d*-band, but not the location of its centre (Figure 5). We cannot at this time extend the calculations to cover the types of bi-layer[17] and bi-row[18] that are claimed to show exceptional activity, while accepting that shape is likely to have a profound effect on a particle's electronic composition.[24] Theoretical work[32] on the bi-layer structure does not illuminate the source of its activity. Clearly all the atoms in both structures will be superficial or nearly so, and extension in two dimensions should not alter this, but it remains to be seen whether the development of energy bands in one or two dimensions is of any consequence.

These changes in electronic constitution are not unique to gold; in particular the decrease of E_F with decreasing size has been demonstrated in other cases (copper and iron);[33] measurements on small vanadium clusters reveal a transition from molecule-like structure to a two-band spectrum at $n = 13$–16.[34]

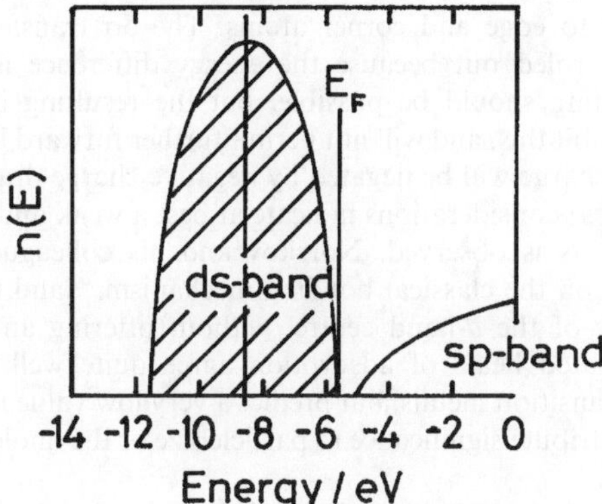

Fig. 4 Distribution of energy levels for a very small (2 nm) particle; there is a small band gap above the Fermi energy.

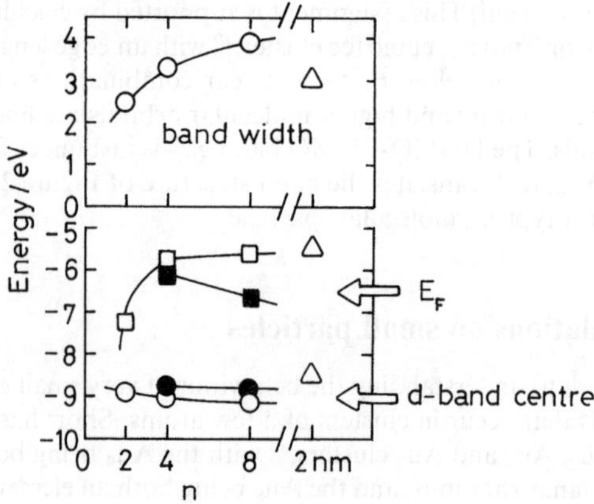

Fig. 5 Calculated values of E_F, location of the d-band centre, and the band width, for Au$_2$, Au$_4$ and Au$_8$ clusters: open points, fcc structures; filled points, stablest structures;[31] values for a 2 nm particle (Figure 2) shown for comparison.

8.   The effect of molecular character on chemisorption and catalysis

It is now necessary to explore how alteration of electronic structure with decreasing size may affect the mode of chemisorption of the reactants, and so suggest an explanation for the activity increase. We must confine our attention to the CO molecule, since there is general agreement that it at least has to chemisorb on the gold, whereas the way in which O$_2$ enters the reaction is still uncertain. The location of its chemisorption almost certainly depends on the type of support, and theory suggests a number of possibilities.[2] The much studied bonding of CO to metal surfaces however provides a firm basis for examining its interaction with gold particles.

Now the classic description of the coordination or chemisorption of CO at a metal centre involves two mutually reinforcing processes: (i) a forward donation of charge from the CO 5σ orbital to the metal's vacant σ orbitals, and (ii) a back-bonding from the metal's $d\pi$ orbitals into the CO $2\pi^*$ antibonding orbitals.[8] A first attempt at a molecular description of CO chemisorption on the low Miller index planes of fcc metals was made in an earlier Faraday Discussion.[35] These synergistic processes tend towards electroneutrality, and a very stable bond results.

With massive gold or large gold particles, however, the possibility of forward donation would be limited, because the d-band vacancy is small, and adsorption is probably limited to edge and corner atoms. The 5σ transfer into the unfilled s-band is probably ruled out because the energy difference is too large ($5\sigma = -7$ eV). Back-bonding should be possible, but the resulting increase in nuclear charge will soon inhibit this, and will not permit further forward bonding. Moreover any rise in nuclear charge will be negated by negative charge drawn from the many adjacent atoms. These considerations indicate at best a weak and not very extensive CO chemisorption, as is observed. Nørskov and his colleagues have developed a theoretical model on the classical bonding mechanism,[36] and this assigns importance to the energy of the d-band centre, without offering an easily understood explanation. Calculated heats of adsorption agree quite well with experimental values for several transition metals, but predict a very low value for gold. His model does not however attribute significance to particle size or the 'molecular' character of very small particles.

The situation with very small 'molecular' particles may well be substantially different from that pertaining to massive metal and large particles. One approach has been advocated by Phala and van Steen,[25] who employ classical electrostatics

   This journal is © The Royal Society of Chemistry 2011

to describe the effect of size on electronic structure. According to Wood,[37] the work function of a spherical particle exceeds that at a planar surface as

$$W_r = W_\infty + 3e^2/8r$$

where r = particle size, W_r and W_∞ the work functions at radius r and infinity respectively. This effect finds its origins in the increased localisation of frontier orbitals, and a narrower energy band. The change in E_F (note $E_F = -W$) predicted by this relation[25] is shown in Figure 6, the values at 2 and 20 nm being those shown in Figure 2. This trend, which is sometimes mentioned in a qualitative way,[13,19,24,25] is supported by the results reported in Section 7, and may be taken to apply approximately to particles that are not strictly spherical. Phala and van Steen[25] then suppose that the energy of the d-band centre is invariant, as is also indicated by our DFT calculations (Figure 5), and that the quantity that determines chemisorption energy is the difference between it and E_F, but the justification for doing this is unclear. Application of the Nørskov model[3,36] however leads to a rapid increase in the heat of adsorption when size falls below 3–4 nm. This conclusion, which is nevertheless supported by experimental and calculated values of the heat, is interpreted as requiring a third bonding mechanism, involving a "metal d-adsorbate – sp interaction" that becomes operative when E_F is sufficiently close to the d-band centre. While such a mechanism finds support in the literature,[38] its relevance to the present problem is uncertain.

An alternative approach to finding the cause of the rapid rise in activity, due probably to more extensive and favourable CO chemisorption, therefore needs to be explored, based on the occurrence of a band gap in very small particles, the associated increases in energy level spacing, and a lower E_F caused by the narrowing of the valence band (Figures 2 and 6). In order to increase the strength of adsorption, and hence surface coverage, while at the same time weakening the C–O bond, it is necessary to invoke the synergistic bonding mode. We may suppose that back-bonding from the top of the d-band is still possible; this accounts for the observed electron depletion observed by XANES in small (2 nm) particles,[13] and causes a further reduction in E_F. This allows some forward bonding to take place, and due to the smaller number of neighbours to the adsorption site and the larger spacing of the energy levels the increased nuclear charge is not so easily restored by electron transfer from other atoms. Most importantly however the band narrowing and lower E_F now allows forward bonding from the 5σ level into the bottom of the sp conduction band (see Figure 4), a possibility already mooted,[8] so that bonding more characteristic of that at individual atoms takes place. The XANES results [13] require the

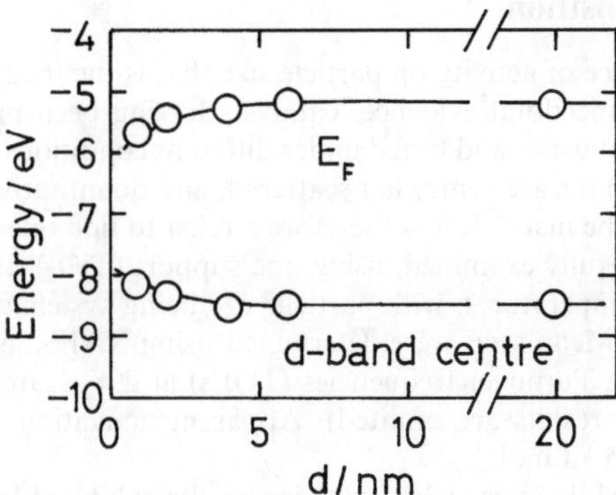

Fig. 6 Variation of Fermi energy according to Wood's equation,[37] and of the energy of the d-band centre,[25] with particle size.

back-bonding mode to dominate, with consequence weakening of the C–O bond. The net effect however must be a type of bonding in which the Au–C bond is sufficiently weak to be reactive, and not as strong as the Pt–C bond, which renders platinum a relatively inefficient catalyst for CO oxidation. Incidentally with this metal there is no indication that passage into the 'molecular' state is in any way beneficial.

A DFT calculation[30] has examined the effect of attaching a CO molecule to one of the two-coordinate gold atoms in the planar Au_8 cluster. There are surprising and extensive alterations to the charge carried by all the atoms; that on the atom carrying the CO decreases very substantially (from -0.19 to -0.06 eV), while that on the other two-coordinate atoms increases, but there is a net transfer from the cluster to the CO. This is in line with the XANES results showing that the metal suffers electron loss.[13] The electronic consequences are therefore not limited to the adsorption site, and the idea of a charge redistribution proposed above receives some support. The heat of adsorption is 86.5 kJ mol^{-1}; this is close to the average of values for Au_5 to Au_{13} neutral clusters (60–125 kJ mol^{-1}) quoted in the review by Willock *et al.*,[2] obtained in various ways. Direct measurements on 'model' catalysts have given values of 50–80 kJ mol^{-1} for 2–3 nm particles.[39] Vannice's much lower value (11 kJ mol^{-1}),[40] derived from the temperature coefficient of the adsorption constant under reaction conditions with $\sim$30 nm Au/TiO$_2$ particles, fits the trend shown in ref. 25.

It is tragic that simple and straightforward measurements that might confirm the above analysis and cast light on the reaction mechanisms have not yet been made. Factors affecting 'activity' need to be expressed as rate constants and activation energies. The dependence of rate upon reactant pressures at several temperatures when analysed by Langmuir-Hinshelwood formalism can afford estimates of heats of adsorption *under reaction conditions.*[40–42] It is a clear consequence of the foregoing arguments that a decease in the orders of reaction for CO with decreasing particle size is to be expected. Such kinetic measurements as have been performed have almost invariably reported in Power Rate Law formalism,[5] that is to say, as the exponent of a logarithmic relation into which they have been forced, but this gives at best a qualitative indication of how coverages are affected by pressures. The literature[5] provides faint support for very low CO orders (<0.05), suggesting that strong adsorption is indeed found on small particles (2 nm Au/TiO$_2$; 1.2 nm Au/Co$_3$O$_4$).[43] The only very thorough kinetic study[39] was performed early on in the history of gold catalysis using poorly dispersed Au/TiO$_2$, but the results pointed clearly to a non-competitive mechanism. Reluctance to apply classical kinetic methodology suggests that it is probably no longer taught.

9. Application of the compensation phenomenon to explore the metal to non-metal transition

The steep dependence of activity on particle size that is the focus of this paper rests largely on almost anecdotal evidence, catalysts having been prepared on various supports, in different ways, and tested under different conditions.[5] On the frequently shown plot,[19] the points are somewhat scattered, and quantitative assessment of the trend cannot easily be made. It was therefore a relief to find one study[6] in which the effect had been carefully examined, using one support (TiO$_2$), two gold concentrations, a range of temperatures, with particle size being systematically increased by thermal treatment. Mean sizes were determined using XAFS, assuming a hemi-cubooctahedral shape. Turnover frequencies (TOFs) at 298 K are shown in Figure 7 ('anomalously low' results are omitted). Apparent activation energies E_a ranged from about 10 to 35 kJ mol^{-1}.

It seemed worthwhile to see whether these results exhibited 'compensation'. It is often seen[44] that for a set of related reactions the Arrhenius parameters E_a and ln A_a (A = pre-exponential factor) vary sympathetically, as

    This journal is © The Royal Society of Chemistry 2011

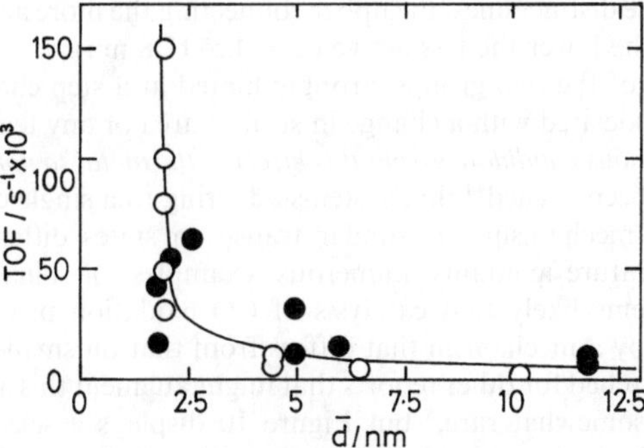

Fig. 7 Dependence of TOF (s^{-1}) at 298 K for CO oxidation over Au/TiO$_2$ on mean particle size;[6] open points, 7.2% Au; filled points, 4.5% Au.

$$\ln A_a = mE_a + c$$

so a high E_a is *compensated* by a high $\ln A_a$. Lest it be thought that the application of this relation to gold catalysis is the product of an over-heated imagination, attention is drawn to the work of Eley[45] on the decomposition of H$_2$O$_2$ on Pd-Au alloy wires; the Arrhenius parameters show excellent 'compensation' with E_a values between 8 and 92 kJ mol^{-1}. These were produced by varying alloy composition and pre-treatment, and the proximity of all points to the line makes this one of the most impressive examples in the literature. A primary cause of this relation is now seen to be the use of *apparent* rather than *true* Arrhenius parameters, the variation in E_a being due to differences in heats of adsorption according to the Temkin Equation.[44]

Rates obtained in the study mentioned above were expressed as TOFs, *i.e.* as rates corrected for the expected differences in metal areas, and also as 'atomic rates', *i.e.* as mol CO (mol Au$_{total}$ s)$^{-1}$. Figure 8 shows the compensation plot using values of $\ln A_a$ obtained from the TOFs;[46] there is a general sense of compensation, but the points are widely scattered, and the literature contains much more convincing plots for other systems. Nothing daunted, the compensation plot based on the atomic rates was then tried (Figure 9), and surprisingly the points for both the gold loadings

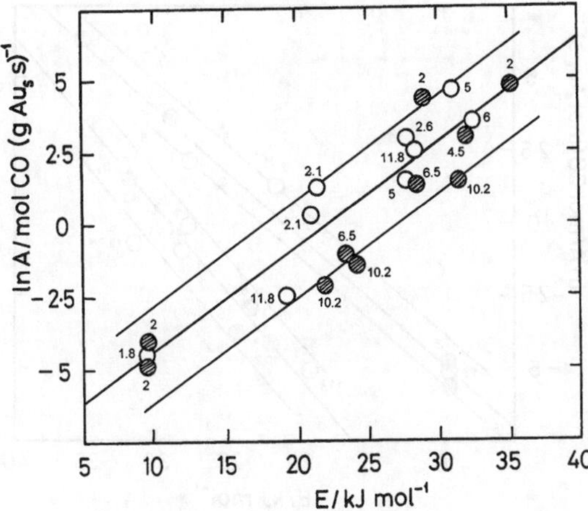

Fig. 8 Compensation plot (E_a *vs.* $\ln A_a$) corresponding to the results shown in Figure 7; the numbers are the mean particle sizes.

fell about two quite distinct lines, the upper connecting the more active catalysts (d = 1.8–2.6 nm) and the lower the less active (d = 4.5–11.8 nm).

The separation of the two groups strongly hinted at a step change in activity at about 3 nm, unassociated with a change in surface area or any feature proportional to it. *The most obvious candidate giving this effect is the metal to non-metal transition.* It has previously been argued[44] that systems adhering to a single compensation line share a common mechanism, *i.e.* similar transition states differing only in their energy. The literature contains numerous examples of binary compensation plots,[44,47] so it seems likely that catalysis of CO oxidation proceeds on particles larger than 3 nm by a mechanism that differs from that on smaller ones. The literature was then searched for other reports that might augment this analysis. Measurements of E_a are somewhat rare,[5] but Figure 10 displays a selection of available values; a few of the higher rates from Figure 9 are included, and these lie close to the line through most of the remainder. High values of E_a (>40 kJ mol^{-1}) are often observed with catalysts containing an 'active' oxide component (*e.g.* Fe_2O_3, SnO_2). A few further systems showing low rates surround the broken line taken from the lower line of Figure 9. If we compare rates calculated from points at E_a = 20 kJ mol^{-1}, the faster rate exceeds the slower by about 150 times; this justifies the assumption made earlier.

The degree of compensation along the major line is not complete; there is about 100-fold difference in rates between top and bottom. Arrhenius plots giving points on a compensation line necessarily intersect at a common temperature T_i, where RT_i equals m^{-1}. T_i for the major line is 227 K, so most of the data have been collected above this temperature, so that high rates are associated with high E_a. Classical Langmuir-Hinshelwood analysis shows that the true activation energy E_t should exceed E_a by $\sum f_i q_i$, where f_i are the orders of reaction and q_i the heats of adsorption of the reactants; this is the Temkin Equation.[44] This is probably not the case here, because E_t is unlikely to be above 70 kJ mol^{-1}. The only available value from the old work of Vannice is 41 kJ mol^{-1}, so lower values may well be inferior to their E_t by reason of the decrease in reactant coverage as temperature rises. Activation energies above the likely E_t have previously been ascribed to the existence of a pre-equilibrium that moves favourably with increasing temperature, but this idea cannot be pursued here.

This analysis suggests that where the mean size exceeds about 5 nm there are few if any of the smallest highly active particles present, so distributions may in fact be narrower than those shown in Figure 1.

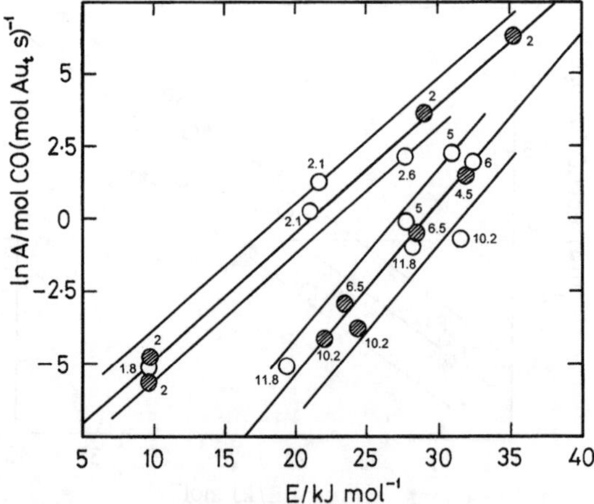

Fig. 9 Compensation plot (E_a *vs.* ln A_a) for the same system but A_a now derived from 'atomic rates' (mol CO (mol Au $_{total}$ s)$^{-1}$).

     This journal is © The Royal Society of Chemistry 2011

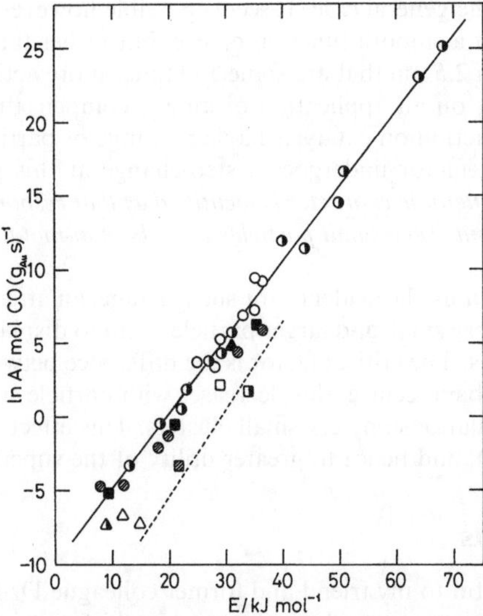

Fig. 10 Compensation plot for CO oxidation on various supported gold catalysts; the broken line is the same as the lower line in Figure 9. Open points, TiO_2; hatched points, Al_2O_3; half-filled points, other supports. ▲ World Gold Council Au/TiO_2; ■ points from Figure 8 for small particle catalysts; other squares, catalysts made by cluster beam deposition; △ other catalysts having large particles. For references to the source of these and other points, see ref. 46.

10. Homogeneous platinum-gold bimetallic particles

A further significant symptom of the altered chemistry of very small gold particles lies in their apparent ability to from homogeneous alloys with platinum.[48,49] This runs counter to expectation based on the macroscopic phase diagram, which shows a broad miscibility gap between about 20 and 95% gold at ambient temperature. A variety of preparative methods afford homogeneous alloy particles, having sizes generally below 5 nm; there is little doubt about the conclusion, since two X-ray determinations of the lattice parameters of PtAu colloids give results conforming very well to Vegard's Law.[47] These observations do not appear to have been exposed to rigorous theoretical treatment, but is seems that stable bonds can be formed using localised valence orbitals where particles are too small for there to be coherent energy bands.

A little light is cast on this problem by comparing DFT results[30] for Au_2 and PtAu dimers. In the latter the gold orbitals lie considerably below the corresponding platinum orbitals, so that electrons in bonding MOs will be mainly on the gold, but those in antibonding MOs will congregate on the platinum. The Au_2 bond energy is 1.87 eV, and that for PtAu is 1.5 eV, so that quite strong bonding between the components is indicated in small bimetallic particles.

11. Conclusions

Accurate assessment of the effect of particle size on activity for CO oxidation demands the availability of monodisperse particles, which at present can only be formed in 'model' systems having little catalytic utility. The transfer of knowledge thus obtained to 'real' catalysts prepared by chemical routes is risky,[4] because of the different states of the support surface when the particle is formed. Even the elegant work[17] on Au/Fe_2O_3 that detected the superior ability of bilayer particles does not constitute a route to determining the effect of size *variation*, so it appears difficult, perhaps impossible, ever to validate a quantitative monotonic change of activity with size in the general case. It seems possible however that activity for this reaction is not in fact a smooth function of size, but rather that it chiefly resides in particles smaller than 2.5 nm that are some 150 times more active than those larger. This conclusion rests on an application of kinetic compensation to the Arrhenius parameters of the reaction on catalysts having a range of particle sizes, and implies that the determining factor undergoes a step-change at this point. *The transition from metallic to non-metallic character is identified as that responsible; whatever other factors may be relevant, very small particles must be non-metallic and 'molecular' in character.*

After briefly examining the evidence for such change, an attempt is made to define the energy levels in very small and larger particles, and to display schematic forms of band energy diagrams. The critical factor is the difference between the Fermi energy E_F and that of the d-band centre; this decreases with particle size, a conclusion supported by DFT calculations on very small clusters. This effect may lead to stronger chemisorption of CO, and hence to greater utility of the superficial gold atoms.

Acknowledgements

I am extremely grateful to my friend and former colleague Dr Eric Short for extensive electronic discussion of the ideas presented in this paper, and for carrying out the DFT calculations referred to above.

References

1 P. Pyykkö, *Chem. Soc. Rev.*, 2008, **37**, 1967.
2 R. Coquet, K. L. Howard and D. J. Willock, *Chem. Soc. Rev.*, 2008, **37**, 2046.
3 J. J. Nørskov and B. Hammer, *Surf. Sci.*, 1995, **343**, 211.
4 G. C. Bond and D. T. Thompson, *Gold Bull.*, 2009, **42**, 247.
5 V. Aguilar-Guerrero and B. C. Gates, *Catal. Lett.*, 2009, **130**, 108.
6 S. H. Overbury, V. Schwartz, D. R. Mullins, W. F. Yan and S. Dai, *J. Catal.*, 2006, **241**, 56.
7 N. Lopez, T. V. W. Janssens, B. S. Clausen, Y. Xu, M. Mavrimakis, T. Bliggard and J. K. Nørskov, *J. Catal.*, 2004, **223**, 232.
8 G. C. Bond, C. Louis and D. T. Thompson, *Catalysis by Gold*, I C Press, London, 2006.
9 G. C. Bond and D. T. Thompson, *Catal. Rev. Sci. Eng.*, 1999, **41**, 319.
10 G. C. Bond and D. T. Thompson, *Gold Bull.*, 2000, **33**, 41.
11 C. Louis, in *Nanoparticles and Catalysis*, ed. D. Astruc, Wiley-VCH, Weinheim, 2007, p. 477.
12 G. C. Bond, *Metal-Catalysed Reactions of Hydrocarbons*, Springer, New York, 2005, chapter 2.
13 J. T. Miller, A. J. Kropf, Y. Zha, J. R. Regalbuto, L. Delannoy, C. Louis, E. Bus and J. A. van Bokhoven, *J. Catal.*, 2006, **240**, 222.
14 K. Ueda, T. Kawasaki, H. Hasegawa, T. Tanji and M. Ichihashi, *Surf. Interface Anal.*, 2008, **40**, 1725.
15 A. Howie, *Faraday Discuss.*, 1991, **92**, 1.
16 K. Okazaki, S. Ichikawa, Y. Maeda, M. Haruta and M. Kohyama, *Appl. Catal., A*, 2005, **291**, 45.
17 A. A. Herzing, C. J. Kiely, A. F. Carley, P. Landon and G. J. Hutchings, *Science*, 2008, **321**, 1331.
18 M.-S. Chen and D. W. Goodman, *Chem. Soc. Rev.*, 2008, **37**, 1747.
19 P. P. Edwards and J. M. Thomas, *Angew. Chem., Int. Ed.*, 2007, **46**, 5480.
20 R. Meyer, C. Lemire, Sh. K. Shaikhutdinov and H.-J. Freund, *Gold Bull.*, 2004, **37**, 72.
21 P. A. Sermon, G. C. Bond and P. B. Wells, *J. Chem. Soc., Faraday Trans. 1*, 1979, **75**, 385.
22 Y. Yamamoto, T. Miura, M. Suzuki, N. Kawamura, H. Miyagawa, T. Nakamura, K. Kobayashi, T. Teranishi and H. Hori, *Phys. Rev. Lett.*, 2004, **93**, 116801.
23 D. W. Goodman, *Topics Catal.*, 2001, **14**, 71.
24 D. Uzio and G. Berhault, *Catal. Rev.*, 2010, **52**, 106.
25 N. Phala and E. van Steen, *Gold Bull.*, 2007, **40**, 150.
26 W. Romanowski, *Highly Dispersed Metals*, Ellis Horwood, Chichester, 1987.
27 M. E. Lin, R. Reifenberger and R. P. Andres, *Phys. Rev. B*, 1992, **65**, 075412.
28 M. M. Alvarez, J. T. Khoury, T. G. Schaaff, M. N. Shafigullin, I. Vezmar and R. L. Whetteen, *J. Phys. Chem. B*, 1997, **101**, 3706.

29 T. P. Bigioni, R. L. Whetten and Ö. Dag, *J. Phys. Chem. B*, 2000, **104**, 6983.

30 E. L. Short, unpublished work. The DFT calculations were carried out with the B3LYP exchange and correlation functional and the LanL2DZ basis set.

31 H. Häkkinen, Bokwon Yoon, U. Landmann, Xi Li, H.-J. Zhai and L. S. Wang, *J. Phys. Chem. A*, 2003, **107**, 6168.

32 N. C. Hernández, J. F. Sanz and J. A. Rodriguez, *J. Am. Chem. Soc.*, 2006, **128**, 15600.

33 V. Ponec and G. C. Bond, *Catalysis by Alloys*, Elsevier, Amsterdam, 1995, ch. 5.

34 H.-B. Wu, S. R. Desai and L.-S. Wang, *Phys. Rev. Lett.*, 1996, **77**, 2436.

35 G. C. Bond, *Discuss. Faraday Soc.*, 1966, **41**, 200.

36 B. Hammer, Y. Morikawa and J. K. Nørskov, *Phys. Rev. Lett.*, 1996, **76**, 2141.

37 D. M. Wood, *Phys. Rev. Lett.*, 1981, **46**, 749.

38 R. Hoffmann, *Solids and Surfaces: a Chemist's View of Bonding in Extended Structures*, VCH, Weinheim/New York, 1988.

39 D. C. Meier and D. W. Goodman, *J. Am. Chem. Soc.*, 2004, **126**, 1892.

40 M. A. Bollinger and M. A. Vannice, *Appl. Catal. B: Env.*, 1996, **8**, 417.

41 G. C. Bond, *Catal. Rev.*, 2008, **50**, 532.

42 G. C. Bond, F. Rosa C and E. L. Short, *Appl. Catal., A*, 2007, **329**, 46.

43 M. Haruta, S. Tsubota, T. Kobayashi, H. Kageyama, M. J. Genet and B. Delmon, *J. Catal.*, 1993, **144**, 175.

44 G. C. Bond, M. A. Keane, H. Kral and J. A. Lercher, *Catal. Rev. Sci. Eng.*, 2000, **42**, 323.

45 D. D. Eley and D. M. MacMahon, *J. Colloid Interface Sci.*, 1972, **38**, 502.

46 G. C. Bond, *Gold Bull.*, 2010, **43**, 88.

47 G. C. Bond, *Zeit. Phys. Chem. NF*, 1985, **144**, 21.

48 G. C. Bond, *Platinum Met. Rev.*, 2007, **51**, 21.

49 G. De and C. N. R. Rao, *J. Mater. Chem.*, 2005, **15**, 891.

General discussion

Dr Nijhuis opened the discussion of the paper by Professor Campbell: I would like to ask a question in relation to the stability of OOH. In your paper you mention that OOH decomposes directly very fast. What I have observed, as well as others in literature, is that the decomposition of peroxide in the presence of hydrogen is very fast and primarily occurs by means of hydrogenation of the peroxide. Can you comment on how you think the OOH decomposition occurs in the presence of hydrogen? Would this also occur *via* a direct decomposition first, or *via* an alternative route?

Professor Campbell responded: In the paper, we wrote: "This and the other enthalpies above can be combined to show that reaction (1) is exothermic by $\sim$175 kJ mol^{-1} on Au(111). Again, adsorbed –OOH on Au(111) is unstable and would quickly decompose to O_{ad} + OH_{ad}." We were assuming here that this large exothermicity would give rise to a fast rate. But you make a good point: Fast is relative and so it could hydrogenate even faster with enough H present. Note that the low stability of the Au–OOH bond should also make it very reactive with other adsorbed species like H (or propene).

Dr Whiston said: What temperature is conceivable on nanoparticles during oxidationreactions due to the exothermic nature of the reactions and the insulating character of the supports in general? What measurements or information are available on this point.

Professor Campbell responded: The nature of the bonding within the metal nanoparticles and at their metal/oxide interfaces is not so different compared to bulk materials, so I expect that the termal conductivities involved in dissipating the reaction energy are at least as big as in the bulk of the oxide support. Given that, and the low turnover frequencies usually involved in gold catalysis, I suspect that this is usually not a big effect. However, it is not uncommon for fast and highly exothermic oxidation reactions to have temperature runaway. Due to the importance of such effects, some chemical engineers have become expert in calculating temperature gradients in reactor beds, given the reaction rate per unit area *versus* temperature, the thermal conductivity of the solid and its structure (porosity, *etc.*), and the gas composition and flow rates. So there are experts out there who could give you a far better answer than I can, and are probably easy to locate *via* the Web of Science. They would probably need to first prove that the typical particle–particle separations across the oxide surface are so small that the thermal diffusion time between them is small compared to the time between reaction events per particle. In that case, it is safe to assume that the reaction heat is dissipated uniformly across the oxide surface, which will make for a far simpler calculation.

Professor Bowker commented: I guess the Ag nanoparticles that you grow nucleate on defects in the surface. I think we would expect the binding energy at such sites to be high—is that the case and is there a chance you can measure it?

Professor Campbell responded: You are right, Mike. Such late transition metals are well known to nucleate mainly at the step sites (which are the dominant defects) on all three of the oxide surfaces I discussed. So the nanoparticle energetics I report here are for nanoparticles mainly decorating oxide step edges. Note, however, that for a 2 nm diameter particle, most of its metal atoms adjacent to the oxide surface are already at oxide terraces, due to the finite width of step edges and their large average separations). Indeed, it is thought that nucleation occurs at defects because these defect sites bind such metals more strongly than terrace sites on these oxide supports, at least according to DFT.[1] We have not really measured that experimentaly yet, but now we have the capability to do so, and expect to do just that very soon. Doing so just requires measuring the heats of metal adsorption, like we reported here with the surface at 300 K, but instead with the surface held at low temperature so that all metal atoms diffuse quickly across terraces to step sites but stop once they get there. We expect this will be the case at ~100 K, which we can do now.[2]

1. C. T. Campbell and O. Lytken, *Surf. Sci.*, 2009, **603**, 1365–1372.
2. W. Lew, O. Lytken, J. A. Farmer, M. C. Crowe, and C. T. Campbell, *Rev. Sci. Instrum.*, 2010, **024102**, 9.

Professor Friend asked: Au and Ag have different eletronegativities? what is the basis for your assumption that Au and Ag will follow similar trends?

Professor Campbell answered: This is simply because studies of the growth and morphology of late transition metals on oxide surfaces shows nothing unusual for Au compared to what is reported for Ag, Cu and Pd *etc.*, except for what is to be expected from the fact that Au has a high sublimation energy, which we have shown in earlier studies is a very important parameter. (The adsorption energies tend to scale with bulk sublimation energy.)

Professor Friend said: I did not see data showing determination of the particle sizes. It seems that your measurement averages the adsorption energy over the surface. From where are your particle sizes derived and could the change in the heat of adsorption be related to defects at low concentration?

Professor Campbell answered: We estimated particle sizes by knowing very accurately the average particle thickness (from knowing the Ag coverage and the fraction of the surface covered by Ag islands). We then used this thickness and a hemispherical particle shape (approximately known from STM studies on related systems) to estimate the particle diameter. To further confirm this, both the ISS and AES signals *versus* Ag coverage are quantitatively consistent with approximately hemispherical particle shape.

Details are given in our paper about Ag on ceria cited in our manuscript.

Dr Jupille commented: This question is about the contribution of the interfacial energy to the heat of adsorption of silver particles supported on oxide surfaces (Figure 3). In the case of Ag/MgO(100), we have observed changes in the average parameter of silver by extended X-ray absorption spectroscopy during the growth of the silver clusters.[1] The parameter of bulk silver (0.409 nm) is smaller than that of MgO (4.21 nm). Silver grows with a cube-on-cube epitaxy on MgO(100). In the case of very small clusters of ≈ 1.5 nm in size, the silver parameter (3.95 nm) is contracted below the bulk value. This is assigned to surface stress. As particles increase in size (2-3 nm), the silver parameter increases because of the misfit strain up to nearly match the parameter of MgO. Finally, as particles keep growing, the value of the silver parameter progressively decreases toward that of the bulk metal. These data suggest that the interface energy may change in a non-monotonous way during the first stage of the growth of a weakly bound particle. I wish you to comment on the possibility that such behavior might produce some breaks in slope in figure 3 where the heat of adsorption of silver is reported as a function of the particle size.

1. P. Lagarde, S. Colonna, A.-M. Flank and J. Jupille, *Surf. Sci.*, 2003, **524**, 102.

Professor Campbell answered: I fully agree that that such lattice-parameter effects might produce some breaks in the slope in Figure 3, as so beautifully shown by your elegant structural characterizations of growing nanoparticles. However, because we have a very broad distribution of particle sizes in Fig. 3, such effects would be hidden there, I am afraid. Someday we may be able to do such calorimetry measurements with nearly monodisperse particles, but we are still far from that dream.

Professor Hutchings remarked: I am interested in the way in which you consider the central OOH intermediate is formed. In your paper you involve Au–H, a gold hybride; but in the mechanistic proposals of Bond and Thompson they have suggested that this reaction species could be formed using a hydroxyl group at a defect site adjacent to the periphery of the gold nanoparticles. Can you comment on this and is it possible to test between the two possibilities?

Professor Campbell replied: I suppose that the Au-H bond is much weaker than the O-H bond in the surface species Bond proposed for the following reasons. As noted in our paper, the TPD peak temperature for H adatoms from Au(110) of 216 K gives a desorption energy of 52 kJ mol^{-1} H$_2$, which implies a Au-H bond energy of only $\sim$244 kJ mol^{-1} (as an upper limit). This is much weaker than O-H bond energies, which are typically above 350 kJ mol^{-1}. So it will be easier for O$_2$ to react with H on Au sites. Note however that there will be more H in the form of OH on oxide sites (due to its greater stability), so it is possible that its abundance compensates. Note that the H on Au sites may be produced by reverse spillover from OH on oxide sites, in which case these two steps in my mechanism may be combined into a single step in Bond's mechanism, and thus quite difficult to distinguish experimentally. I suspect that DFT might be able to tell us which process is most feasible, if the calculated activation barrier differences are large enough to trust within the errors of DFT.

Professor Madix asked: Do you believe that this OOH species would be a good proton acceptor from a reasonably strong gas phase acid such as methanol?

Professor Campbell answered: Excellent question, Bob! Methanol is a weaker gas-phase acid than HOOH by $\sim$21 kJ mol^{-1} (see Taft *et al.*).[1] As you have so beautifully shown in your earlier work with other adsorbed species, this difference means that if co-adsorbed –OOH and methoxy should compete for surface hydrogen (adsorbed –H or –OH), then methoxy would win. This also means that it is $\sim$21 kJ mol^{-1} uphill for methanol gas to adsorb and donate its acid H to adsorbed –OOH to make HOOH gas.

1. R. W. Taft, I. A. Koppel, R. D. Topsom, and F. Anvia, *J. Am. Chem. Soc.*, 1990, **112**, 2047.

Professor Madix responded: I agree completely, Charlie, that based on these numbers, hydrogen peroxide is a stronger gas phase acid than methanol. Thus, as you say, it is energetically uphill for methanol to transfer its alcoholic proton to HO$_{2-}$ in the gas phase. Taking the general guideline for predicting surface displacement reactions from their relative gas phase acidities, one would come to the same conclusion for the reaction of the adsorbed species.[1] However, for comparison, the relative ΔH_{acid} for ethanol and methanol differ by 3 kcal mol^{-1}, yet each will partially displace the other on Au(111), provided relative concentrations are used as a driving force. That is, if adsorbed ethoxy is exposed to sufficient methanol in ultrahigh vacuum, a mixed layer of ethoxy and methoxy is formed, and conversely. In fact, cross-coupling of these alcohols to form methylacetate on metallic gold via coadsorbed methoxy and ethoxy requires excess methanol in the reactant mixture to compensate for its lower gas phase acidity.[2] Thus, it would appear possible that in excess methanol, adsorbed OOH could act as a proton acceptor, leading to the formation of hydrogen peroxide and adsorbed methoxy. One could thus easily imagine that dilute hydrogen peroxide in methanol solution could promote the ester-ification of methanol over supported gold catalysts were it possible to form adsorbed HO_{2-} on the gold surface. This displacement reaction between HO_{2-} and other alcohols to form their adsorbed alkoxy species should become even more facile for higher molecular weight alcohols, such as butanol or glycerol.

1. M. A. Barteau and R. J. Madix, *Surf. Sci.*, 1982, **120**, 262.
2. B. Xu, R. J. Madix and C. M. Friend, *J. Am. Chem. Soc.*, 2010, **132**, 16571.

Professor Meyerstein commented: It is worthwhile noting that in aqueous suspensions, which clearly differ from solid gas phase interfaces, $CH_3O_2\cdot$ radicals, that are very similar to $HO_2\cdot$ radicals, oxidize the Au^0 NPs in a three electron oxidation process as derived from the analysis of the organic products.[1]

1. R. Bar-Ziv, I. Zilbermann, T. Zidki, H. Cohen and D. Meyerstein, *J. Phys. Chem. C*, 2009, **113**, 3281–3286.

Professor Campbell answered: That is a very good point, and not inconsistent with our results, I think. Of course, exactly what happens in producing those observed organic products is probably not so obvious.

Professor Poliakoff said: Do you have any idea why one needs water to be present to get catalytic oxidation to occur on Au?

Professor Campbell responded: Yes. We proved in earlier work that CO reacts very rapidly with O adatoms on Au surfaces to make CO_2, once they have been produced. The slow step is to get O adatoms on Au from O_2 gas. The implication of the energetics we estimated in our paper here is that O_2 insertion into the H–Au bond on Au surfaces provides a potential route to activate O_2, producing Au–OOH which then decomposes to produce the necessary O adatom. Thus, I suggest that the promotional roles of both water vapor and hydrogen gas may be to produce a low equilibrium concentration of Au-H species, which serve as catalysts to activate O_2 but that get regenerated in the net reaction cycle.

Prof Dr Baeumer said: In Figure 1 of your paper, it seems that the heat of adsorption as a function of Ag particle size exhibits 2 different slopes. Is it possible that in the range of small particle sizes oxidation of the Ag (or charge transfer in general) plays a role, resulting in a different slope than in the range of larger metallic aggregates?

Professor Campbell replied: Indeed, there is some charge transfer at low Ag coverages. We have observed with synchrotron-based photoemission combined with Auger parameter analysis that the $CeO_2(111)$ surface is slightly reduced upon Ag deposition, which our evidence indicates must be due to the reverse spillover of oxygen atoms from the oxide onto the Ag nanoparticles' surfaces.[1] However, I would hesitate to analyze the Fig. 1 data in terms of two slopes, given the fact that each point is surely an average over a fairly broad size distribution (based on beautiful STM studies of related systems by yourself and others). It just occured to me that this oxygen spillover from ceria to silver may have implications with respect to how oxygen adatoms get on the surface of gold nanoparticles as well. Thanks for the question.

1. D. Kong, G. Wang, Y. Pan, S. Hu, J. Hou, H. Pan, C. T. Campbell and J. Zhu, *J. Phys. Chem. C*, 2011, **115**, 6715–6725.

Professor Bowker opened the discussion of the paper by Professor Madix: How important is the surface structure for these reactions? Presumably Au atom removal

     This journal is © The Royal Society of Chemistry 2011

from the surface plane to make Au–O nanoparticles can occur on other surface planes too.

Professor Madix answered: This is an excellent question, because reaction intermediates are usually stabilized by added metal atoms on the group 1B metal surfaces. For example, Cu, Ag and Au metal adatoms are enlisted to stabilize adsorbed atomic oxygen on the (110) surfaces, and there is evidence that a similar effect occurs on the close-packed (111) surface as well. Since there is now ample evidence from scanning tunneling microscopy that oxygen-containing intermediates such as formate, sulfite, carbonate, etc enlist added metal atoms to form local "complexes", it must be anticipated that gold will also do the same. The kinetics of reactions of these complexes must differ from that of the isolated intermediate. These factors are important in theoretical computations of the rates of elementary steps for microkinetic modeling.

Professor Hutchings commented: In the reaction when an aldehyde RCHO forms the hemi acetyl, have you considered substituent effects as this might help to probe the reaction mechanism. For example are benzylic substituents more readily reacted? Our work on toluene oxidation also accesses a hemi acetyl intermediate to from benzyl benzoate. In our case the addition of Pd to Au significantly enhances the activity; do you think this is related to an enhanced formation of the surface oxygen species? In our case we are using sol-immobilised nanoparticles which will have polyvinylalcohol ligands present. Perhaps these play a role in restricting access to the surface sites and help control the selectivity we observe.

Professor Madix replied: In general the reactions of aldehydes with adsorbed alkoxy species to form the hemiacetal are quite facile with little energy barrier, occurring below 200 K. For the higher molecular weight aldehydes, for example, benzaldehyde, the rate-limiting step for ester formation is desorption, so it is difficult to compare the ease of formation of the hemiacetal quantitatively. Prof. Friend has studied these effects in more detail and perhaps she can add further comments.

With regard to alloying with Pd, it is quite possible that Pd does enhance the activation of oxygen and that the polyvinyl ligands restrict access of the phenyl group to the surface, enhancing selectivity for partial oxidation. We know, for example, that the phenyl group interact strongly with clean Au(111) leading to C–H bond cleavage in the ring.

Professor Friend addressed Professor Madix and Professor Hutchings :We have investigated the effect of phenyl rings and they are oxidized on the Au. There are two effects that are the consequence of the ring interacting with the surface. First, the ring-surface interaction makes it more subject to attack by O adsorbed on the surface. Second, the ring-surface interaction increases the bond strength so that when coupling products are formed–for example benzaldehyde reaction with methoxy–it stays on the surface to higher temperature, making it more subject to attack by O. Overall, the ring leads to more nonselective reaction compared to straight-chain alcohols.

Professor Bowker commented: There is some commonality of Au reactivity patterns with Cu and Ag—can you please comment on these patterns?

Professor Madix answered: The analogous reactivity of group IB metals has been discussed in previous papers, some recently.[1-3] Briefly, the patterns of reactivity observed on silver surfaces appear to provide a foundation for anticipating partial oxidation behavior on gold. However, subtle differences between gold and silver in the activation energies for specific elementary steps can produce significant differences in the selectivities observed for these complex reactions. Generally, gold appears to facilitate partial oxidation at lower temperatures with greater selectivity. This difference may be due to the greater difficulty of activating oxygen on gold, which restricts the surface oxygen concentration and mitigates against secondary oxidation.

1. R. J. Madix, C. M. Friend and X. Liu, *J. Catal.*, 2008, **258**, 410–413.
2. X. Liu, R. J. Madix and C. M. Friend, *Chem. Soc. Rev.*, 2008, **137**, 2243–2261.
3. C. Freyschlag and R. J. Madix, *Mater. Today*, 2011, **14**, 134.

Professor Pyykkö asked: At atomic and quantum-mechanical level, only one of the three coinage metals is *normal* and that is silver. Copper is anomalous having a compact 3d shell not having any radial nodes (for this point, see ref. 1, chapter 4.1.). This gives a strong electron–electron repulsion inside the 3d shell and possibly explains the brown colour of copper and the divalency of Cu(ii). Similarly, gold is anomalous having strong relativistic effects that stabilise the 6s shell and destabilise the 5d shell. For the record, element 111 or roentgenium (Rg) is very strongly relativistic and even chooses a $7s^2 6d^9$ electron configuration in its atomic ground state.

Silver already has an inner (3d) shell inside its 4d shell and is not yet so strongly relativistic, the valence-shell relativistic effects increasing roughly as Z^2, where Z is the full nuclear charge. Therefore *silver is the only normal coinage metal*.

1. P. Pyykkö, *Phys. Chem. Chem. Phys.*, 2011, **13**, 161–168.

Professor Friend responded: These are interesting points and suggest that differences in the behavior of the three coinage metals—Cu, Ag, and Au—are to be expected. It is interesting that there are many qualitative similarities in their reactivity, but there are many quantitative differences. Thank you for this insight.

Prof Dr Bhargava responded: Explanation accepted

Professor Campbell asked: Excellent data! Do you know the difference in energy barriers for beta hydride elimination for the same molecule on Au *vs.* Ag and Cu, with all three surfaces at the same unreacted O coverage (preferably with NO excess O in all cases)?

Professor Madix replied: The best studied example for this comparison is the C–H bond cleavage in the formate intermediate on the (110) surfaces of Cu, Ag and Au.[1–3] A valid comparison of the activation energies for the three metals can be obtained from the temperatures at which the formate decomposes to CO_2 and adsorbed hydrogen. This temperature is 475 K, 410 K and 340 K for Cu(110), Ag(110) and Au(110), respectively. Apparent activation energies compute to be 29.5, 25.0 and 21.2, respectively. For both Cu(110) and Ag(110) more detailed studies of the kinetics have been made; the preexponential factors for this unimolecular decomposition were 10^{16} s^{-1}, with activation energies of 32 and 29 kcal mol^{-1}, respectively. The activation energy for the reaction on Au(110) was not precisely determined, but can be estimated to be approximately 4 kcal mol^{-1} lower than that for Ag(110). In the case of both copper and silver, CO_2 and H_2 are simultaneously evolved, giving confidence that the reaction occurs in the absence of excess adsorbed oxygen. However, on Au (110) both water and hydrogen are evolved simulataneously, indicating that adsorbed oxygen may influence the C-H bond cleavage. Work in underway in our laboratory to investigate this effect.

1. D. H. S Ying and R. J. Madix, *J. Catal.*, 1980, **61**, 48.
2. M. A. Barteau and R. J. Madix, *Surf. Sci.*, 1980, **94**, 303.
3. D. A. Outka and R. J. Madix, *Surf. Sci.*, 1987, **179**, 361.

Professor Friend commented: There is still an open question about whether O, OH, other oxygenates or the Au can promote the loss of H from the methoxy

     This journal is © The Royal Society of Chemistry 2011

intermediates. It is not clear if hydride elimination can occur on the metallic Au. DFT suggests that it does not occur; however this is still an open question.

Professor Bowker opened the discussion of the paper by Dr Saint-Lager: Regarding the heating of nanoparticles—you have dismissed this as a possibility. Maybe this should not have been eliminated, since, although the power is low, it is focussed on the nanoparticles which are of very small total mass. The extent of particle heating also depends on thermal conductivity of the support and the time-scale of the experiment.

Dr Saint-Lager answered: I agree and it is very difficult to evaluate the effect of the reaction-induced heating since many unknown parameters are involved for determining the heat dissipation not only *via* the TiO_2 substrate but also *via* the surrounding gas environment. This is the reason why we have not completely excluded the role of the heating induced by the oxidation reaction.

Professor Bond remarked: The form of dependence of rate on particles size shown in Figure 3A differs significantly from that shown in Figure 7 of my paper for chemically-made particles, in that the ratio of rates at 2 and 5 nm in your case is about 7, whereas in my case it is at least 70.[1] One can only speculate about how the population of low coordination sites varies with size in your particles, but I suggest that measurement of rates at several temperatures with derivation of Arrhenius parameteres would provide additional insight. As to the activity decrease below 2 nm, is it possible that your particles may become encapsulated by the support during the annealing step? Extensive work on the Strong Metal-Support Interaction showed how readily TiO_x species can migrate to achieve this.[2]

1. G C Bond, *Faraday Discuss.*, DOI: 10.1039/c1fd00010a .
2. G C Bond, Metal-Catalysed Reactions of Hydrocarbons, Springer, New York, 2005, p. 137.

Dr Saint-Lager answered: Actually the discrepancy between figure 3 of our paper and figure 7 of your paper is less important, if we take into account the error bar (fig 3B shows a factor of 20 between 2 and 5 nm). However this dependence is different since it is sensitive to the shape of the particles which could change according to the sample preparation. Nevertheless, it is clear that, as you suggest, measuring the reaction rate at several temperatures with derivation of the Arrhenius parameters would provide additional insights. As concern the migration of TiO_x species which could encapsulate the smallest gold clusters, is an important issue. This potential effect could be evidenced with GISAXS measurements through (1) change in the roughness of the substrate and (2) more subtle changes in the NPs GISAXS signal, linked to the difference of electronic density between gold and TiO_x species. For the effect (1), we did not observe any significant change in the roughness signal of the GISAXS patterns during the whole experimental protocol from UHV to exposure under reactive gases. The effect (2) is more difficult to evidence and requires a deeper quantitative GISAXS analysis.

Professor Friend asked: Your TiO_2 is bulk reduced, based on the description of its preparation and the fact that you stated it was blue. Subsurface defects in titania can promote reactions on the surface, *e.g.* deoxygenation, and also might affect the binding of the Au. Did you investigate the role of reduction on reaction rate and on sintering?

Dr Saint-Lager answered: For X-ray measurements, it is not necessary to use a conducting sample, in contrast to scanning tunnelling microscopy and more generally, electron-based techniques. Consequently, as explained in our paper, during the sample preparation, the TiO2(110) substrate was annealed under oxygen pressure in order to restore the stoichiometry after Ar^+ sputtering. The TiO_2 crystal kept a white color all along the experience at ESRF under X-ray beam, when the study on sintering was performed. Actually, some tests on "blue" crystal were only performed during preparation run at the laboratory. They revealed that the residual reactivity on clean TiO2 was higher than on the 'white' one.

Professor Bowker said: You think the decrease in activity at the smallest particle sizes is due to sintering?

Dr Saint-Lager answered: This work does not make it possible to draw such conclusion.

Professor Fortunelli remarked: You have observed increased sintering of Au nano-particles in reaction conditions, *i.e.*, when CO and O_2 are simultaneously present. What we have seen in our simulations is that the co-adsorption of O_2 (dissociated) and CO on an oxide-supported Au3 cluster favors facile breaking of the complex into fragments, possibly leading to diffusion of the fragments and so to some forms of Ostwald ripening. Could this be a mechanism explaining sintering observed in your experiments?

Dr Saint-Lager responded: This can be an explanation of the sintering observed when CO and CO_2 are presents. These predictions could be valid as far as the results on such small clusters can be extrapolated to bigger gold nanoparticles involved in our work (hundreds of atoms).

Professor Campbell remarked: It is well known that when you add oxygen or chlorine gas to Pt/alumina, sintering occurs more rapidly. Almost any adsorbate could facilitate Ostwald ripening by the following mechanism:

The apparent activation barrier for Ostwald ripening typically includes the energy to detach a metal atom from a metal cluster and move it to the support surface (see Campbell *et al.*).[1]

Since this metal atom has more nearest neighbors in the initial state than the fnal state for this step, it will bind gas molelules more strongly in the final state. In the presence of such an adsorbed gas, this step will therefore be less uphill in energy, and therefore faster according to Brønsted relations.

1. S. C. Parker and C. T. Campbell, *Phys. Rev. B*, 2007, **75**, 035430.

Dr Saint-Lager replied: Thanks for this detailed comment. If adsorbed gases make easier the detachment of Au monomers from Au Clusters, we have to assume that it is co-adsorption of CO and oxygen molecules which could provoke such a detachment. This seems to be possible as predicted by A. Fortunelli *et al.* from their simulations on an oxide-supported Au_3 cluster (see his question here above).

Professor Bowker remarked: Regarding the mobility of surface species. It really depends, I think, on the nature of the surface. For instance you often cannot resolve the surface structure of clean metal single crystals at ambient temperature because of the streaking caused by metal atom diffusion across the surface (*e.g.* Pd(110)). Yet you can get atomic resolution at 500C in STM when there is a surface oxide monolayer there (see Bowker *et al.*).[1] Perhaps the oxides are more stable due to a Madelung potential and there is intrinsically more diffusion in the metallic state (*i.e.*, in your experiments, in a reducing environment like CO) and hence more sintering?

1. R. Bennett, S. Poulston, I. Jones and M. Bowker, *Surf. Sci.*, 1998, **401**, 72–81.

Dr Saint-Lager responded: The reducing environment cannot be the explanation since the enhancement of the sintering is observed only when CO is added to oxygen, but not when it is added to argon.

Prof Dr Baeumer asked: Do your data provide indication that small Au particles on titania are stabilized by oxidized gold at the perimeter of the particles (or at the Au–titania interface)?

Dr Saint-Lager responded: Unfortunately we cannot get such informations from the present X-ray measurements.

Prof Dr Baeumer opened the discussion of the paper by Professor Madix: What is the relative importance of OH as a Broensted base as compared to $O_{(ad)}$ (reaction (1) in the paper)?

Professor Madix responded: Excellent point. Based on comparable studies on Cu and Ag single crystals, transfer of the acidic proton to OH appears facile as well. For methanol and formic acid, for example, all of the initially adsorbed oxygen can be reacted away by exposure to these acids. This requires that the hydroxyl groups receive the acidic hydrogen from the reactants. Since the surface chemistry of Cu, Ag and Au appears similar, we expect this reaction to proceed on Au surfaces as well. However, to date, direct confirmation of this hypothesis has proved elusive, as adsorbed OH species are quite transient and therefore difficult to directly titrate.

Professor Friend answered: We have not successfully isolated adsorbed OH on Au (111). It forms transiently but is unstable with respect to disproportianation to water and adsorbed O. Therefore, we have not specifically investigated the possible role of OH. We do anticipate that OH will activate OH and C–H bonds on Au, though, since methoxy can activate these bonds.

Professor Selvam opened the discussion of the paper by Professor Campbell: How about the geometrical constraint while addressing the adsorption on gold nanoparticles of various sizes and the orientation of different facets?

Professor Campbell answered: I am sure there are geometrical effects, like particle shape and exposed facets, that must play a role. Unfortunately, the best we have been able to do thus far experimentally (with adsorption calorimetry) is to look at average particle size effects. Once people publish measurements with techniques such as STM, AFM or X-ray scattering which reveal particle growth conditions where the shape can be controlled, we should also be able to study shape effects on particle stability or molecule adsorption energies on particles. Of course there have been other experimental measurements such as catalytic activity and TPD which have revealed that shape does play a role.

Professor Rotello opened the discussion of the paper by Dr Xu: Do you have a proposed rationale for the formation of bilayer gold structures? In biological and polymer systems bilayers are formed from directional building blocks such as lipids. Is it possible that directionality in your system could be generated by polarization by either the graphene or air/vacuum interface?

Dr Xu replied: We are continuing to our investigation of the bonding mechanism of the Au islands. This system appears to rest on a delicate balance of weak interactions, of which polarization could be a factor.

Dr Cadete Santos Aires asked: If I understand correctly in your STM images the periodic contrast is due to a Moiré (graphene/Re(0001). You also have a Moiré on the Au islands standing on the graphene. Is there a relationship between the two?

Dr Xu responded: Yes, the two do appear to be related. A close look of our STM images (*cf.* Figs. 2 and 3) suggests that the moire pattern in the Au islands, where it is exhibited, has nearly the same periodicity as the moire pattern in graphene/Ru (0001).

Dr Cadete Santos Aires asked: So in that case, since both have the same parameters, could it be that you are sensing the LDOS of graphene/Ru(0001) through the Au (bi)layer (its thickness is consistent with this assumption)?

Dr Xu replied: Within 1 eV of the Fermi level Au has substantial DOS, so what is responsible for the tunneling current above the Au islands cannot be solely due to graphene or Ru. If the DOS of the different components of this surface substantially overlap spatially as well as energetically in this energy range, it would imply chemical interaction, in which case it would be difficult to distinguish the contribution to the tunneling current from each component. However, there is currently no evidence to support direct chemical bonding between the Au islands and g/Ru, which leads us to conclude that what STM sees above the Au islands are mainly due to Au DOS.

Professor Bowker said: What happens when you put down more gold? Do the islands grow further layers, and in what mode of growth?

Dr Xu replied: We have tried putting down 2 ML equivalent of Au on graphene/Ru(0001). About 60% of the Au clusters that formed were 2-D, and the remaining were 3-D. Both the 2-D Au islands and the 3-D Au clusters displayed similar lateral size distributions.

Professor Campbell said: Very nice results. It is curious that your Au layer seems to prefer 2 layers rather than one or thicker. It is easy to understand an enegy preference for one-layer structures over thicker structures, but a bilayer being most stable is harder to explain. To my knowledge, the only good explanation for this is the buildup of strain with thickness when the overlayer adapts a different lattice constant due to lattice mismatch with the substrate. This same type argument was invented by Jerry Tersoff to explain why a growing SiGe film on Si(100) suddenly dewets to form quantum dots at ~5 monolayer thickness. Do you care to comment.

Dr Xu replied: The preference of the Au for the bi-layer structure is clearly a consequence of a peculiar growth process. Further confirmation of the bilayer structure and an understanding of the growth process are the goals of on-going investigations. It is unclear what role strain plays in this system. In the absence of strong chemical bonding between the Au and graphene, there should be nothing to force the Au lattice to expand or contract. It should therefore adopt the equilibrium lattice size of a free Au bilayer. This we calculated to be 0.276 nm, similar to the in-plane lattice constant of Ru (calculated to be 0.273 nm), thus giving a free Au bilayer the same surface atomic density as Ru (*i.e.*, the equilibrium density is approximately (11 x 11) Au on (11 x 11) Ru). According to our DFT calculations, an Au monolayer with a greater atomic density (Model C) develops significant buckling, whereas an Au structure with a less atomic density (Model D) collapses into a (11 x 11) monolayer, both of which are consistent with the Au islands being nearly freestanding structures.

Professor Bowker said: How accurately can you measure a height difference when there may well be local work function differences, for instance, between graphene and the gold islands?

Dr Xu answered: The tunneling current in STM has an exponential dependence on the square root of the work function, which needs to be taken into account when interpreting STM results. When the tip and the sample have different work functions, with a voltage applied on top of that, the work function becomes a function of the distance between the tip and the sample. A typical approximation to make is that the effective work function is equal to:

$$\varphi = (\varphi_{tip} + \varphi_{sample} - e|V|) / 2.$$

We note that 1) the work functions of free graphene and bulk Au are 4.6 and 5.1 eV respectively; 2) no change is observed in the apparent height of the 2-D Au islands relative to the graphene as the bias voltage is varied from -1.5 V to $+1.5$ V, over which $\sqrt{(\varphi_A/\varphi_B)}$ varies from 1.025 to 1.030 (A, B being the Au and graphene regions of the surface). Since in the constant current mode, $W_A \times \sqrt{\varphi_A} = W_B \times \sqrt{\varphi_B}$ (W being the tip-surface distance), we conclude that W_B/W_A varies by less than 1% with the bias voltage (between 0 and ±1.5 V). Given the typical tip-surface separation of 10 Å, 1% amounts to 0.1 Å, meaning that W_A and W_B differ only by a small amount between 0 and ±1.5 V. This, in turn, suggests that the observed height difference between the Au and graphene (0.55 nm) is predominantly geometric, not electronic.

Professor Hutchings asked: Can you comment further on how the CO is titrated off of the surface? Can you comment on how the oxygen is activated?

Dr Xu responded: What we have done is dosing O_2 on CO-precovered Au/graphene/Ru(0001) at 85 K. PM-IRAS spectra show that the intensity of the CO peak (starting at 2095 cm^{-1}) decreases significantly with increasing O_2 dosage (20–110 L), and that the CO peak broadens and slightly blue-shifts. These findings suggest the co-existence of CO and O_2 on the surface and the loss of some CO from the surface. Given that O_2 adsorbs more weakly than CO, we hypothesize that the disappearance of CO is due to CO reacting with O_2. Further experiments are being carried out for confirmation.

Professor Bowker remarked: Do you think that electronic communication between the metal film and the substrate Ru may be important in the reactivity of these systems? Clearly there can be electron exchange, because STM can image through thin film insulator surfaces (*e.g.* SiO$_2$)

Dr Xu answered: Insofar as electronic conductivity is concerned, all three components of this system— Au islands, extended graphene sheet, and Ru—are conductors, so the composite surface is also a conductor, but that in itself should not have a direct bearing on the reactivity of the Au. An analogous system of a long Au wire connecting two carbon or Ru electrodes illustrates this point.

Strong interaction between thin ad-layers and substrates (*e.g.* one or a few monolayers of metal A grown pseudomophically on metal B) is known to modify the surface reactivity of the ad-layers strongly in many cases. The fact that no CO adsorption was observed on the graphene moiré covered Ru(0001) in PM-IRAS suggests that the graphene is not significantly modified by the Ru substrate, or conversely, the Ru is well shielded by the graphene. It seems unlikely that the Au islands, which are located on top of the graphene, are electronically modified by the Ru or the graphene. No chemical bonding between the Au and graphene as reported by our DFT calculations is consistent with this view. We are conducting further investigations of the surface reactivity of the Au islands to determine how it differs from bulk Au surfaces.

Professor Golunski opened the discussion of the paper by Professor Bond: When using conventional methods to prepare supported precious metal catalysts, the total metal area can be increased by using higher metal loadings. However, the average metal particle size also increases. How can we get round this problem when trying to produce highly active and durable gold catalysts?

Professor Bond responded: I do not think there is really a problem here. Moreau's work[1] showed that deposition-precipitation is capable of of creating small particles up to 2% loading, while that of Overbury *et al.*[2] cited in my paper achieves this with loadings of 7.2%. I doubt there is much call for catalysts containing more Au than this. An alternative approach is to prepare colloidal Au particles of the desired size and then put them on a support, but I am not sure what is the maximum loading obtainable by this method.

1. F. Moreau, G. C. Bond and A. O. Taylor, *J. Catal.*, 2005, 235, 105.
2. S. H. Overbury, V. Schwartz, D. R. Mullins, W. F. Yan and S. Dai, *J. Catal.*, 2006, **241**, 56.

Dr Willock said: Your paper discusses compensation plots in which the natural logarithm of the pre-exponential factor is plotted against the activation energy. In transition state theory we view the pre-exponential factor as containing the entropy of the reaction process including the number of active sites and the energy barrier as the energy required to achieve the transition state. It is remarkable that these two factors should show the observed linear relation over large ranges of data. Do you have any physical interpretation of the occurance of the compensation effect?

Professor Bond replied: Some years ago, an informal scientific meeting was held, at which several of those fascinated by the compensation phenomenon met to discuss its cause. The result was a substantial review article (ref. 44 of my paper), in which we concluded, at least for the cases we considered, that compensation was simply a consequence of using apparent rather than true kinetic aprameters. Observed variations in apparent activation energy are therefore due solely to differences in the heats of adsorption of the reactants. As I have several times remarked, the only determintion of the true parameters for CO oxidation was reported in 1996 (ref. 40 of my paper). Proof that the normally reported kinetic parameters are in fact apparent would be provided by observing an effect of changing reactant concentrations on activation energy. Discussion of possible correlations between entropy and enthalpy of adsorption has to be based on true kinetic parameters, which unfortunately are not available.

Professor Haruta asked: I appreciate very much for your instructive paper. Concerning the dependency of reaction rate on the size of Au particles, Dr. Fujitani has shown an equation "Rate $= k \cdot d^{-2}$, where k is rate constant and d is the diameter of Au particles.[1] They prepared catalyst samples by using single crystal of rutile TiO_2 and by cathodic arc plasma method. Each catalyst specimen has the same metal loading but has different sizes. The reaction rate was in proportion to the inverse second power of the diameter, indicating the perifery is the site for reaction.

1. T. Fujitani *et al.*, *Angew. Chem. Int. Ed.*, in press.

Professor Bond responded: I shall be very interested to read the paper by Dr Fujitani when it appears; it sounds as if his results closely resemble those of Dr Saint-Lager *et al.* presented in paper 15. In a question I put to her I noted that the ratio of rates shown by particles of 2.5 and 5 nm is very much greater with those made by a chemical route than with those by vapour deposition.

I think we have to be extermely careful how we compare these two classes of particle and how we try to transfer information from one to the other. I believe

    This journal is © The Royal Society of Chemistry 2011

the fractional charge carried by the particles is not the same in each case and that this may account for the very different dependence of rate on size that they show.

Professor Hutchings opened the discussion of the paper by Professor Bond: I am impressed by your very careful analysis of the work by Overbury as you are able to indicate different compensation effects for particles >3nm and <3nm in diameter. What would you expect the effect of the support to be, could it change this markedly?

Professor Bond responded: Provided there is no electronic connection between the gold particles and the support, I would expect to see a similar effect with other supports. Unfortunately, as I have said in another comment, confidence is so often placed in the results of measuring activity chnages at a single temperature, so that my expectation cannot be tested experimentally.

Professor Campbell commented: Regarding this issue of metallic versus non-metallic behavior and its relationship to particle size effects in Au catalysis, I do not expect this to be important for the following reason:

The reaction mechanisms and reaction rates of small molecules with clean Si(100) and clean other semiconductor surfaces (as studied with UHV surface analysis tools) have not been reported to depend on whether the bulk is doped n-type or p-type nor on the extent of such doping. I conclude that chemisorption (and therefore catalysis) depend much more on the local electronic structure, and not on whether the sample has more or less long-range conductivity. Of course there may be some indirect correlations between reactivity and conductivity when comparing different elements, but only because they both are related to the electronic energy levels.

Professor Bond responded: You are I think correct to stress the importance of the local electronic structure (LDOS) of surface atoms for catalytic activity; in fact I said as much in Section 5 of my paper. Now in a 2 nm spherical particle containing about 230 atoms, some 60% of them are superficial, and most of the rest are in contact with them. My submission is that surface atoms sense what is going on beneath them, and therefore that high coordination atoms on large particles will have an LDOS predicated on the underlying band structure. Edge and corner atoms, including those at defects, show greater independence, and will have an LDOS more adept to adsorption and catalysis. With the 2 nm particle, the surface atoms overlay a 'molecular' insulating interior, and the LDOS of all of them is appropriate to catalysis. The transition to the non-metallic state is inevitable; what is lacking is sound experimental evidence for the effect this has on the reactivity of surface atoms.

Dr Willock said: We have seen in periodic DFT calculations on ethene adsorption to Pt surfaces that the adsorption energy can be thought of as a combination of local attractive interactions with the d-states of the surface atoms and repulsive inteactions between the molecule and electron density in the sp band states. It could be that in small clusters the loss of the band structure due to the small particle size removed the repulsive interactions and so increases the adsorption energy of molecules on the surface.

Professor Bowker remarked: I'd like you to pin your flag to the mast - what is your view of what dictates the maximum in CO oxidation activity as a function of size (occurring at $\sim$ 3nm size)? Do you think there is a particular size which is best, perhaps dictated by the 'band gap' in such small particles?

Professor Bond replied: Some care is needed to answer this, because the assertion it contains, namely, that there is a rate maximum at about 3 nm, is open to question. Such a maximum is observed, so far as I am aware, only with catalysts made by vacuum deposition under UHV conditions.[1,2] Those made by a chemical route, showing mean sizes of less than 2.5 nm, give TOFs at 298 K of 0.05 to 0.15 s^{-1} (see Figure 7), and those less than 2 nm[3] TOF values at 273-300 K of 0.28 to 0.45 s^{-1}. True, some much lower values are also found[3], but they cannot be caused by a too small size. Particles formed under UHV conditions differ significantly from those made chemically, because in the first case the surface is oxygen-deficient and Au-Ti bonds may be formed, whereas in the second case the surface is fully hydroxylated and Au-O bonds are formed at the interface. This difference affects the charge on the particles[3]. Another significant point of difference is the ratio of rates at 2.5 and 5 nm; for 'model' particles this ratio is about 7[1] or less[2], while for 'real' particles it is nearer 100 (see Figure 7 and ref. 2) Discussion should therefore centre on the reason for the activity decrease shown by small 'model' particles; for 'real' particles the question becomes, what is the smallest number of Au atoms that will give high CO oxidation activity? A maximum must exist, but there is as yet no evidence for its location.

1. M-C. Saint-Lager, I. Laoufi, A. Bailly, O. Robach, S. Garaudee and P. Dolle, *Faraday Discuss.*, DOI: 10.1039/c1fd00028d.
2. M. Valden, X. Lai and D. W. Goodman, *Science*, 1998, **281**, 1647.
3. G. C. Bond and D. T. Thompson, *Gold Bull.*, 2009, **42**, 247.

Professor Hutchings commented: It should be noted that real catalysts comprise a broad range of different particle sizes and morphologies. I feel there is general agreement that there will be a maximum in the relationship between the activity for CO oxidation and the gold particle size; after all we do not consider individual gold atoms to be active. The affect, however, may be very support dependent.

Professor Friend responded: There is clearly a correlation between activity for CO oxidation and gold particle size and I also agree that the support is very important. In my view, these effects at least partly increase the supply of O as a reactant. There may be other effects as well. Ideally, we will establish a better understanding of particle size and support.

Prof Dr Bhargava responded: Agreed.

  This journal is © The Royal Society of Chemistry 2011

Oxidative coupling of alcohols on gold: Insights from experiments and theory

Bingjun Xu[a] and Cynthia M. Friend[*ab]

Received 10th February 2011, Accepted 22nd March 2011
DOI: 10.1039/c1fd00015b

Molecular level understanding of the mechanism of oxidative coupling of alcohols on metallic Au(111) activated by oxygen is achieved through a combination of experiments and theoretical calculations. The facility of the β–H elimination of the alkoxys, which increases with the length of the alkyl chain, is identified to be critical in determining the product distributions. Dioxymethylene serves as a formaldehyde reservoir in the cross-coupling reaction between methanol and formaldehyde through its reversible formation and decomposition, contributing to the high selectivity for the coupling products.

Introduction

Much research effort has been devoted to gold-based catalysts since Haruta discovered that gold nanoparticles supported on a reducible metal oxide can catalyze CO oxidation at low temperature.[1] More recently, gold-based catalysts have been shown to mediate complex oxidative coupling reactions at low temperature with high selectivity[2–11]—a potential replacement for the traditional stoichiometric chromium- or manganese-based synthesis and its accompanying toxic reactants.[12] Gold-based catalysts have thus emerged as promising candidates as "green" catalysts.

There is considerable debate about the origin of catalytic activity of gold-based catalysts for oxidative reactions. Proposals include quantum size effects of the nanoparticles, edge effects on the crystallites and effects originating from the support. Indeed, a particle size effect has been claimed for many different reactions—gold particles with diameter between 1∼10 nm[13,14] being most reactive. However, micron-scale particles of gold powder catalyze complex coupling reactions in solution,[15] nanoporous gold is a catalyst for several oxidation processes—selective self-coupling of methanol,[16] selective oxidation of glucose,[17] and CO oxidation.[18] Furthermore, single crystal gold surfaces promote a wide variety of partial oxidation reactions.[19–21] Neither a particle size effect nor influence of the support can account for these phenomena.

The nature and mode of transport of the active oxygen species onto gold surface are also debated. It is generally agreed that bulk gold is inert toward the activation and dissociation of molecular oxygen.[22,23] However, gold nanoparticles themselves and the gold-metal oxide interface have been proposed to activate O_2.[23–25] Reactive species, such as atomic oxygen, may migrate across the interface between oxide support onto the gold particles. Adsorbed peroxidic dioxygen has also been suggested as the active form of oxygen.[26] In strongly basic solution, adsorbed hydroxyl groups may initiate surface-mediated coupling reactions.[27] These various possibilities notwithstanding, *atomic* oxygen bound to metallic Au unquestionably activates

[a]Department of Chemistry and Chemical Biology, Harvard University, Cambridge, MA, 02138, USA
[b]School of Engineering and Applied Sciences, Harvard University, Cambridge, MA, 02138, USA.
E-mail: cfriend@seas.harvard.edu; Tel: +617 496 9489

numerous oxidative coupling reactions of alcohols, aldehydes and amines. The mechanism for these reactions follows a well-defined pattern of oxygen-assisted reactivity found with both gold powder and supported gold in solution.

From both experimental and computational studies[20,28] we have found that atomic oxygen adsorbed on Au(111)—not clean gold surface itself—activates the hydroxyl hydrogen in the alcohols to form surface bound alkoxys at low temperatures (*e. g.* ≤ 150 K for methanol,[20] eqn (1)):

$$CH_3OH_{(a)} + O_{(a)} \rightarrow CH_3O_{(a)} + OH_{(a)} \tag{1}$$

In the catalytic steady state, or during heating in temperature programmed reaction, the methoxy undergoes β–H elimination to form formaldehyde (eqn (2)), which in turn can react with remaining adsorbed methoxy to yield methylformate. All of these reactions occur sequentially on the surface without readsorption of any of the products. This mechanism provides a general construct for oxidative coupling processes mediated by gold in which a nucleophile, *e.g.* methoxy, attacks the electron-deficient carbonyl carbon in formaldehyde to form the alkoxy-hemiacetal surface intermediate (eqn (3)), which subsequently eliminates another hydrogen to yield methylformate (eqn (4)). The β–H elimination of the alkoxy-hemiacetal intermediate can also be mediated by the Au(111) surface alone, by losing one β–H to the surface, as in eqn (5). This surface assisted process may play an important role in the steady-state reaction with very low surface oxygen coverages due to the low oxygen dissociation efficiency of the gold surface.

$$CH_3O_{(a)} + O_{(a)} \rightarrow HC(=O)H_{(a)} + OH_{(a)} \tag{2}$$

$$CH_3O_{(a)} + HC(=O)H_{(a)} \rightarrow CH_3OC(=O)H_{2(a)} \tag{3}$$

$$CH_3OC(=O)H_{2(a)} + O_{(a)} \rightarrow CH_3OC(=O)H_{(a)} + OH_{(a)} \tag{4}$$

$$CH_3OC(=O)H_{2(a)} \rightarrow CH_3OC(=O)H_{(a)} + H_{(a)} \tag{5}$$

The self-coupling mechanism established from model studies on Au(111) under low pressure was used to predict the catalytic behavior of nanoporous gold at ambient pressure in the self-coupling of methanol with O_2 as the oxidant.[16] It also accounts completely for the self-coupling of ethanol in solution at high pressure.[5,21,29] The strong parallel between the studies carried out on the model Au (111) surface and on gold catalysts under working catalytic conditions strongly suggests that the mechanistic framework on oxidative coupling reactions established using well defined single crystal surfaces bridges the so-called "pressure gap". We attribute the wide applicability of the mechanistic framework to several key features of gold surface: 1) it is inert in the absence of adsorbed atomic oxygen, preventing spectator species from building up and blocking active sites; 2) surface hydroxyl and adsorbed water are unstable and 3) steady state atomic oxygen coverages are low.

The general mechanistic principles found for the specific case of self-coupling of methanol are applicable to self-coupling reactions of other alcohols.[21] The first step in the process is activation of the alcohol by adsorbed atomic oxygen. Thereafter, many cross-coupling reactions can be envisioned, including coupling of methanol with various aldehydes,[30] and cross coupling of different alcohols.[31,32] The product distribution can be tailored by the educated choice of the relative facility of the β–H elimination of different alkoxys and their relative surface concentrations.[31]

  This journal is © The Royal Society of Chemistry 2011

The concept of nucleophilic attack by the adsorbed alkoxy can be generalized to other possible adsorbed nucleophiles, also generated by oxygen activation of specific bonds of a select reactant. For example, we have demonstrated the oxygen-assisted coupling of amines and aldehydes on Au(111), for example, the coupling of dimethyl amine and formaldehyde to afford dimethylformamide.[19] Strong parallels can also be drawn between our mechanistic work and such reactions in solution-phase by oxygen-assisted gold mediated coupling reactions.[33,34]

In this paper we report a combined study of experiments and DFT calculations on key factors controlling the selectivity towards oxidative coupling reactions *vs.* combustion on atomic oxygen covered Au(111) surface. DFT results predict that methoxy is able to activate C–H bond in a nearby methoxy, which is confirmed by the isotopic labeling experiments. Partial oxidation and coupling reactions compete favorably with combustion at low temperature, suggesting the facility of β–H elimination of alkoxys is critical in determining the product distributions of the self-coupling of alcohols. We further show that dioxymethylene is a reservoir for formaldehyde *via* reversible formation and decomposition. This pathway for supplying formaldehyde to the surface contributes to the relatively high selectivity for coupling of methanol and formaldehyde.

Experimental

All experiments were performed in an ultrahigh vacuum (UHV) chamber with a base pressure below 2×10^{-10} Torr. The preparation of the clean Au(111) surface has been described elsewhere.[35] The surface was first populated with 0.1 monolayer (ML) of atomic oxygen (O/Au(111)) by introducing an appropriate amount of ozone at 200 K. The oxygen atom coverage was calibrated by comparing the amount of O_2 evolved in temperature programmed desorption to that formed for a saturation coverage of oxygen atoms.[36] Oxidation of the surface in this manner leads to the release of Au atoms to form nanostructures containing Au and O, most of which are smaller than 2 nm in diameter.[35] This oxygen-covered Au surface is referred to here as O/Au(111).

Organic molecules were introduced onto the O/Au(111) surface at 150 K *via* a dosing tube. Exposures, corrected for dosing enhancement, are given here in terms of Langmuir (L) (1 L = 10^{-6} torr-seconds); the dosing pressure was determined by uncorrected ion gauge pressures.

Temperature programmed reaction—a method for determining the product distributions and for obtaining kinetic information—was conducted according to well-established protocols, described in detail elsewhere.[35] The heating rate for all reactions was nearly constant at 5 K s^{-1}. The reaction products were identified by quantitative mass spectrometry (Hiden HAL/3F) using fragmentation patterns obtained from authentic samples, which were found to be in general agreement with NIST reference data.[37] XPS spectra were acquired with an analyzer passing energy of 17.9 eV and a multiplier voltage of 3 kV using MgKa X-rays (300 W) as the excitation source. The binding energy (BE) calibration was referenced to the Au 4f$_{7/2}$ peak at 83.9 eV. The O(1s) spectra were acquired with 100 scans to enhance the signal-to-noise ratio.

All DFT calculations in this work were preformed using the VASP code[38,39] with the GGA-PW91[40] functional to describe electron exchange and correlation. We employed the Projector Augmented Wave (PAW) function method[41] with plane wave basis sets (energy cut-off: 400 eV). For reciprocal space we used a $3 \times 3 \times 1$ Monkhorst–Pack k-point grid. We tested a higher density Γ-centered $4 \times 4 \times 1$ k-point in several cases and found no significant differences in either adsorption energies or in activation barriers (typical changes in these quantities were ~ 0.02 eV). The Au(111) surface was modeled by a 3-layer slab in the (111) direction, a $p(3 \times 3)$ unit cell in the lateral directions, and a vacuum of 15 Å between slabs; the 2 upper layers were allowed to relax, with the atoms in the bottom layer fixed at the ideal bulk positions. The bulk gold positions of the bottom layer were taken from the calculated lattice constant of 4.17 Å, which is in good agreement with the experimental value 4.08 Å.[42] Since insufficient convergence of the slab thickness with respect to properties such as adsorption and reaction energies can lead to significant errors, we also tested a $p(3 \times 3)$ 4-layer slab model in several cases with the bottom two layers kept frozen, and a $3 \times 3 \times 1$ Monkhorst–Pack k-point grid. The systematic errors in adsorption energies and energy barriers between the 3-layer and 4-layer slabs were smaller than 0.02 eV. As a result, we conducted the DFT analysis on the 3-layer models for computational efficiency. The dipole correction scheme[43] was tested in the z-direction, which is perpendicular to the surface, in many cases and produced similar deviations ($\Delta E < 0.02$ eV). Thus, no spurious electrostatic interactions between the periodic images are present in the calculations. The electronic structure was converged to within 10^{-4} eV, and the geometries optimized until the forces were smaller in magnitude than 0.02 eV Å^{-1}. Both spin polarized and unpolarized calculations were performed when a single H atom is involved in a reaction, and the spin polarization has a negligible impact on the energetics and activation barriers.

We obtained the reaction barriers using the climbing nudged elastic band method (cNEB)[44–46] at a reduced force threshold of 0.05 eV Å^{-1}, with at least three images in-between the fixed starting and ending points. In several cases, we employed as many as eight images in order to test the convergence of the energy barriers. From the above convergence studies, as well as from previous experience with similar systems, our best estimate for the error bars of activation energies is <0.1 eV. All transition state structures have been subjected to the vibration analysis and only one imaginary frequency was found along the reaction coordinate. The activation barriers E_B are defined as the energy difference between the transition state (E_{TS}) and the initial state (E_i), and the thermodynamic barriers E_{Th} are the energy difference between the final state (E_f) and the initial state. We do not include any further corrections for coverage and thermodynamic effects due to finite temperature or partial pressure.

Results and discussion

The nature of O on Au(111)

Metallic gold is activated by atomic oxygen for a variety of selective reactions, including oxidative coupling of alcohols. Unfortunately, the rate of O_2 dissociation on extended single crystal studies is undetectable under the low pressure conditions used for model studies.[22] Hence, other means of delivering O on the surface is necessary. In the studies described herein, ozone was used as a source of oxygen on Au (111). Previous studies have established that ozone efficiently delivers atomic oxygen to the surface and that a wide range of coverages, up to ~1.2 monolayer, can be achieved even at low surface temperature.[36] While the rate of O atom production in a catalytic process will be important in determining the overall rate of reaction, our model studies provide insight into the reactive steps that ensue.

The adsorption of oxygen on Au is complex, leading to release of gold atoms from the surface and creation of O-containing Au islands on the surface, as is evident in scanning tunneling microscopy experiments (Fig. 1). Release of Au atoms from the (111) surface is induced by other electronegative atoms, $e.g.$ S and Cl. Although we specifically investigated the Au(111) surface because it is the thermodynamically most stable surface, analogous release of Au has been reported for, $e.g.$ stepped Au[47] and Au(110).[48,49]

The distribution of Au and O on the surface depends on the conditions for oxidation—the temperature and rate at which O is delivered to the surface—and on the total amount of O deposited. For the purposes of this paper, we focus on the low coverage limit because it provides the optimum selectivity and also because it is

    This journal is © The Royal Society of Chemistry 2011

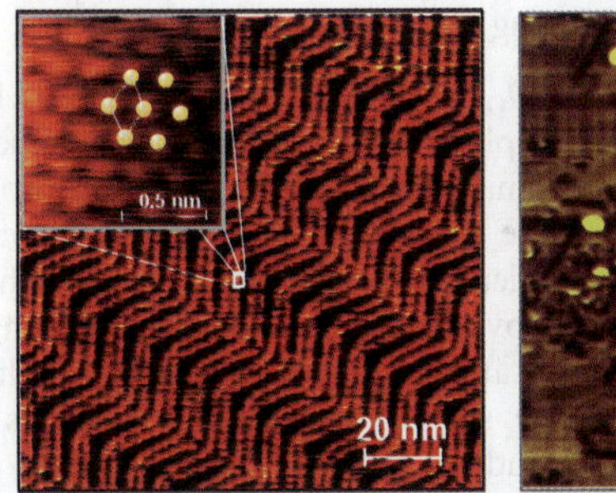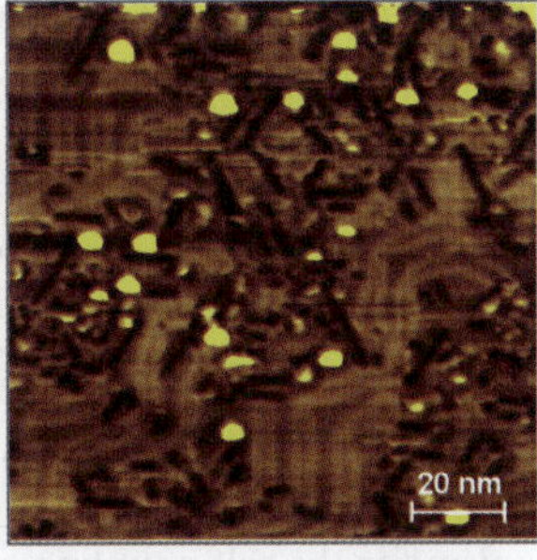

Fig. 1 Scanning tunneling microscope images show the formation of O-containing Au islands upon adsorption of atomic oxygen. The nanoparticles that form contain both O and Au[11] and are very reactive toward alcohols and other organic molecules. Left: Clean Au(111) surface with herringbone reconstruction at 200 K. Right: Au(111) surface with an atomic oxygen coverage of 0.2 ML at 200 K.

most relevant to catalytic conditions. The surface structures used in our reactivity studies are small Au–O particles on the Au crystal, 85% of which are ∼2 nm in diameter (Fig. 1).[35] These structures are metastable and disordered, but also provide the highest activity and selectivity. These nanoparticles remain metallic since they are supported by the bulk Au substrate.

Most oxygen is bound in local 3-fold coordination sites on Au(111) at low coverage and using a low temperature of oxidation—the conditions used for the selective oxidative coupling. A combination of density functional theory (DFT) and spectroscopic measurements show that O is predominantly bound in 3-fold fcc sites even in the presence of defects, such as steps, adatoms or vacancies (Fig. 2).[50] Defects were specifically studied since the release of Au atoms from the surface to make the nanoparticles necessarily creates defects. The bond strength of O in the three-fold sites is affected slightly by the presence of defects; however, direct bonding of the O to the defects themselves is never favored over the 3-fold site. There are also other forms of O bonding to Au(111) that are present at higher coverage and higher temperature.[51] These other forms of O on Au are less selective and less reactive and, therefore, are not considered in detail herein.

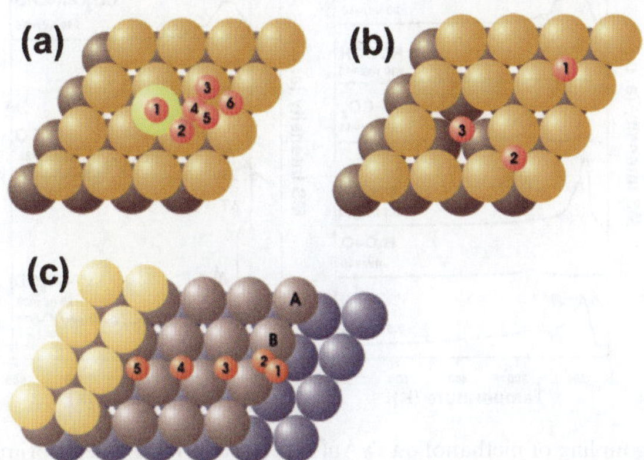

Fig. 2 Three-fold fcc sites are the preferred adsorption sites for defective gold surfaces. Possible oxygen adsorption sites on Au(111) surface with (a) adatom (1/9 ML) (b) vacancy (1/9 ML) and (c) stepped Au(533) surface. The most stable atomic oxygen adsorption sites are site 5 and 2 in (a) and (b), respectively. Site 1, 2 and 3 have almost the same oxygen adsorption energy in (c), more stable than site 4 and 5. Reprint with permission from ref. 50.

Mechanistic insight into methanol coupling on O/Au(111)

Methanol self-couples on O/Au(111) ($\theta_O = 0.1$ ML) to yield methylformate at $\sim$215 K (Fig. 3(a));[20] however, the coupling reaction occurs in competition with other processes. A small amount of formaldehyde and combustion products (CO_2 and water) were also formed under these conditions; no formic acid was detected. Importantly, no reaction is detected on clean Au(111) either with or without surface defects. A defective surface was created by first forming the O-containing Au nanoparticles through ozone decomposition and subsequently removing all O *via* CO oxidation. STM results show that the surface morphology remains essentially unchanged for temperatures of 300 K and below.[52]

The mechanism for the self-coupling of methanol also applies to coupling of two dissimilar alcohols introduced simultaneously to O/Au(111), which results in the formation of a mixture of two alkoxys.[31,32] Analogous to methoxy, the alkoxys will then undergo β–H elimination and form the corresponding aldehydes, which leads to the oxidative coupling reaction between alkoxys and aldehydes. The product distributions depend critically on two factors: 1) the relative facility of β–H elimination of the dissimilar alkoxys and 2) the relative concentration of alkoxys on the surface.[31]

A key part of designing catalytic reactions is to understand how to control both activity and selectivity. Given that atomic oxygen is required for molecular activation of alcohols, it is important to understand how it participates in other elementary steps. We have used isotopic labeling in combination with investigation of the dependence in the product distributions on the initial concentration of O on the surface to further probe how oxygen participates in various reactive steps.

Reaction of CD_3OH on O/Au(111) ($\theta_O = 0.05$ ML) demonstrates there are at least two likely pathways for the β–H elimination of methoxy. The labeling experiment shows that molecular desorption of methanol (130–200 K)[53] (Fig. 3(a)) occurs well below methanol-d_4, which must be formed from reaction of a methyl D of CD_3O with another adsorbed methoxy. In addition to the self-coupling product, methylformate-d_6, and the partial oxidation product, formaldehyde-d_2,

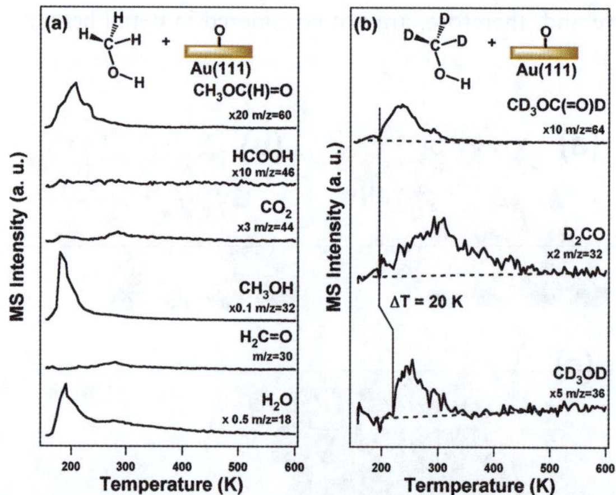

Fig. 3 (a) Self-coupling of methanol on O/Au(111) ($\theta_O = 0.1$ ML) yields primarily methylformate. (b) The onset of the perdeuterated methanol (CD_3OD) evolution is 20 K higher than those of perdeuterated methylformate and formaldehyde in the self-coupling reaction of CD_3OH on O/Au(111) ($\theta_O = 0.05$ ML). The contribution from fragmentation of methanol (CH_3OH in (a) and CD_3OH in (b)) was subtracted for clarity. Methanol (normal and deuterated) was introduced to the O/Au(111) at 150 K and the heating rate is 5 K s^{-1}. Surface oxygen was prepared by ozone exposure at 200 K.

methanol-d_4 was also observed around 240 K (Fig. 3(b)). The peak evolution of $D_2C=O$ and $CD_3O-C(D)=O$ at the same temperature (Fig. 3(b)) is in agreement with the β–H elimination reaction from the methoxy being the rate-limiting step in the self-coupling reaction of methanol (Scheme 1).

When examined in more detail, however, the temperature programmed reaction results suggest more than one pathway for the formation of formaldehyde and methylformate. The onset of the formation of both $D_2C=O$ and perdeutero-methylformate is 20 K lower than that of CD_3OD, which indicates a pathway(s) with a lower activation barrier than that of methoxy disproportionation (eqn (6)). DFT calculations suggest that the oxygen assisted β–H elimination reaction (eqn (2)) indeed has a lower activation barrier than the methoxy assisted pathway (eqn (6)) by 0.17 eV (Table 1).[28] This pathway may result from the presence of some excess adsorbed oxygen. The overall mechanistic framework of the self-coupling reaction of methanol is summarized in Scheme 1. All steps thereafter are rapid.

There are several possible pathways for the reformation of methanol, in this case, CD_3OD (eqn 6, 7, and 8).

$$CD_3O_{(a)} + CD_3O_{(a)} \rightarrow D_2C=O_{(g)} + CD_3OD_{(g)} \tag{6}$$

$$CD_3O_{(a)} + OD_{(a)} \rightarrow O_{(a)} + CD_3OD_{(g)} \tag{7}$$

$$CD_3O_{(a)} + D_{(a)} \rightarrow CD_3OD_{(g)} \tag{8}$$

We used density functional theory (DFT) and a detailed analysis of our reactivity data to probe the three pathways considered and concluded that methoxy-assisted formaldehyde production accounts for the production of methanol at 240 K (Table 1). The extremely low surface life-time of hydroxyl and water at temperatures above 200 K renders pathway 7 unlikely.[54] According to several calculations, atomic hydrogen is not bound on Au(111) surface,[55] rendering abstraction of a methyl D by Au endothermic.[28] However, atomic hydrogen recombination has been reported to yield H_2 at 200 K from Au(110) in temperature programmed desorption experiments. It is possible that the binding of H to Au is very weak and therefore not described well by DFT calculations. Hydrogen evolution has not been observed in various coupling reactions studied on Au(111),[19–21,30] suggesting that atomic

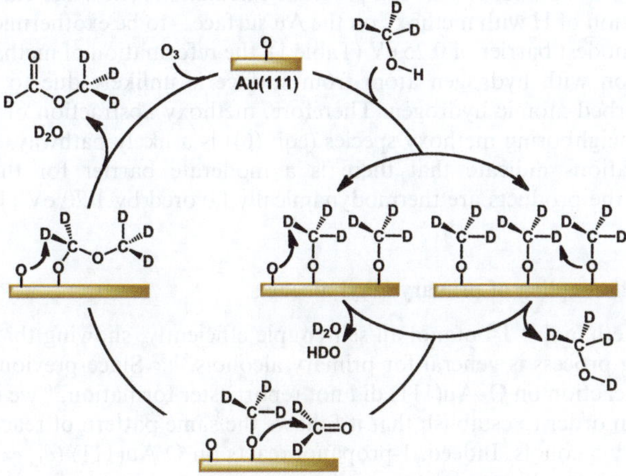

Scheme 1   Schematic of the mechanism for the self-coupling of methanol.

Table 1 Calculated activation and thermodynamic barriers for possible reaction pathways for methanol activation

Reaction pathway	Activation barrier[a] (eV)	Thermodynamic energy[b](eV)
Activation of methanol to methoxy		
1[c] $CH_3OH_{(a)} + O_{(a)} \rightarrow CH_3O_{(a)} + OH_{(a)}$	0.41	0.27
β–H elimination of methoxy to formaldehyde		
2[c] $CH_3O_{(a)} + O_{(a)} \rightarrow H_2C{=}O_{(a)} + OH_{(a)}$	0.49	−0.98
Nucleophilic attack of methoxy to formaldehyde to form alkoxy hemiacetal		
3[c] $CH_3O_{(a)} + HC{=}O)H_{(a)} \rightarrow CH_3OC({=}O)H_{2(a)}$	0	−0.75
β–H elimination of alkoxy hemiacetal to methylformate		
4[c] $CH_3OC{=}O)H_{2(a)} + O_{(a)} \rightarrow$ $CH_3OC({=}O)H_{(a)} + OH_{(a)}$	0	−2.18
5[c] $CH_3OC({=}O)H_{2(a)} \rightarrow CH_3OC({=}O)H_{(a)} + H_{(a)}$	0.22	−0.59
Reformation of methanol		
6[c] $CH_3O_{(a)} + CH_3O_{(a)} \rightarrow H_2C{=}O_{(a)} + CH_3OH_{(a)}$	0.66	−1.20
7[c] $CH_3O_{(a)} + OH_{(a)} \rightarrow CH_3OH_{(a)} + O_{(a)}$	0.14	−0.27
8[c] $CH_3O_{(a)} + H_{(a)} \rightarrow CH_3OH_{(a)}$	0.25	−1.33
Dioxymethylene formation		
9 $H_2C{=}O_{(a)} + O_{(a)} \leftrightarrow H_2CO_{2(a)}$	0	−0.54
Formate formation		
10 $H_2C{=}O_{2(a)} + O_{(a)} \rightarrow HCO_{2(a)} + OH_{(a)}$	0.12	−2.87
CO₂ formation		
11 $HCO_{2(a)} + OH_{(a)} \rightarrow CO_{2(g)} + OH_{(a)}$	1.19	−2.50

[a] Activation barrier (E_B) is defined as the difference between the energy of the transition state (E_{TS}) and the energy of the initial state (E_i), i. e., $E_B = E_{TS} - E_i$. [b] Thermodynamic energy (E_{Th}) is defined as the difference between the energy of the final state (E_f) and the energy of the initial state (E_i), i. e., $E_{Th} = E_f - E_i$. [c] adopted from ref. 28.

hydrogen is not released directly to the surface. An alternative explanation for the absence of hydrogen evolution could be that the instability of atomic hydrogen on Au(111) causes it to be consumed rapidly by reactions such as eqn (8). However, to the best of our knowledge, no existing experimental or computational evidence supports this explanation. In the following discussion, very low surface hydrogen concentration is assumed. Although previous calculations show the energy of reaction 8—reaction of H with methoxy on the Au surface—to be exothermic by 1.33 eV with only a modest barrier of 0.25 eV (Table 1), the reformation of methanol *via* methoxy reaction with hydrogen atom from surface is unlikely due to the lack of surface adsorbed atomic hydrogen. Therefore, methoxy abstraction of a methyl H (D) from a neighboring methoxy species (eqn (6)) is a likely pathway. Indeed, our DFT calculations indicate that there is a moderate barrier for this reaction, 0.66 eV and the products are thermodynamically favored by 1.20 eV (Table 1).

Oxidative self-coupling of primary alkyl alcohols

Methanol,[20] ethanol,[21] 1-butanol all self-couple efficiently, showing that the oxidative coupling process is general for primary alcohols.[31,32] Since previous studies of 1-propanol reaction on O–Au(111) did not report ester formation,[56] we investigated 1-propanol in order to establish that it follows the same pattern of reactivity of the other primary alcohols. Indeed, 1-propanol reacts on O/Au(111) ($\theta_O = 0.1$ ML) at 150 K to form propyl propanoate in temperature programmed reaction experiments (Fig. 4(a)). Propanal, formed *via* partial oxidation of 1-propanol, is the primary product under these conditions. No propanoic acid was formed and only minor amounts of combustion products (CO_2 and water) were observed.

The product distributions for primary alcohol oxidation depend on the alkoxy chain length (Fig. 4(b)), which we attribute to variation in the facility for β–H elimination. Despite the relatively large error bars, which arise from the subtraction of contributions from fragments of different products, it is clear that the propensity toward aldehydes grows significantly, from ∼10% in the case of methanol to ∼70% for 1-butanol, reflecting the increasing facility of β–H elimination reaction of alkoxys with the length of the alkyl chains.[31] The selectivity towards the self-coupling products increases slightly from methanol to ethanol (∼55% to ∼70%) and decreases substantially for 1-propanol and 1-butanol (70% to 30%).

The yield of the self-coupling product is influenced by the surface coverage of the alkoxy, aldehyde and adsorbed oxygen. The higher rate of β–H elimination apparently shifts the balance of aldehyde and alkoxy on the surface during temperature programmed reaction, reducing the rate of formation of the alkoxy-hemiacetal. Therefore, it is not reasonable that the more aldehyde forms at the expense of the self-coupling products.

The limited combustion of the higher molecular weight alcohols (Fig. 4(b)) is apparently due to preferred reaction of the aldehyde with the alkoxy, rather than with residual oxygen. The combustion in the self-coupling of methanol occurs *via* the successive secondary-oxidation steps of formaldehyde on both Ag[57] and Au[20] surfaces (eqn (9)–(11)).

$$H_2C{=}O_{(a)} + O_{(a)} \leftrightarrow H_2CO_{2(a)} \tag{9}$$

$$H_2CO_{2(a)} + O_{(a)} \rightarrow HCO_{2(a)} + OH_{(a)} \tag{10}$$

$$HCO_{2(a)} + O_{(a)} \rightarrow CO_{2(g)} + OH_{(a)} \tag{11}$$

The reduction in combustion of the higher molecular weight alcohols may be due to formation of the corresponding aldehydes at temperatures significantly lower than that at which formaldehyde forms from methanol. It is also possible that less excess oxygen remains on the surface after exposure to these alcohols.

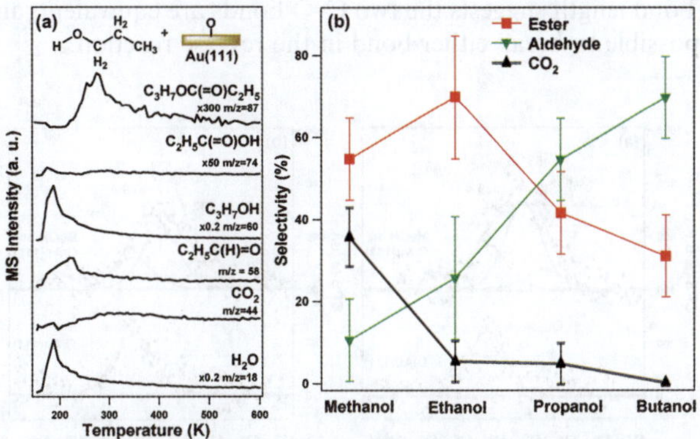

Fig. 4 (a) Self-coupling of 1-propanol on O/Au(111) ($\theta_O = 0.1$ ML). The contribution from fragmentation of 1-propanol was subtracted for clarity. Methanol (normal and deuterated) was introduced to the O/Au(111) at 150 K and the heat rate is 5 K s^{-1}. Surface oxygen was prepared by ozone exposure at 200 K. (b) The trend in selectivity for products (ester, aldehyde and CO_2) for self-coupling of methanol, ethanol, 1-propanol and 1-butanol on O/Au(111) ($\theta_O = 0.1$ ML). The error is primarily due to the corrections for overlapping m/z for different products.

To address this matter the self-coupling reaction of methanol and 1-butanol were examined with X-ray photoelectron spectroscopy and found that the amount of adsorbed oxygen consumed from reaction with 1-butanol is more than that for reaction of methanol (Fig. 5). Adsorbed oxygen ($\theta_O = 0.1$ ML) exhibits a single peak at 529.4 eV in the O1s spectrum (Fig. 5(a), bottom trace).[19] After the introduction of methanol at 150 K, two additional peaks at 531.5 and 532.7 eV appear (Fig. 5(a), top trace). The 532.7 eV peak is assigned to the oxygen in molecular methanol, based on a control experiment where methanol was condensed on the Au(111) surface. The peak at 531.5 eV is assigned to the methoxy species, according to the previous vibrational studies on the same system[16,18] and is also in agreement with previous studies of methoxy on Cu(110) and Cu(111).[58] Most of the adsorbed O (529.4 eV), ~70%, remains after methanol is introduced to the surface. The low signal to noise ratio is due primarily to the low surface coverage of the species under study, and the error bar for the peak fitting is expected to be ~15%. Similar experiments were performed for 1-butanol on O/Au(111), and the 531.0 eV and 532.9 eV peaks are assigned to 1-butoxy and molecular 1-butanol, respectively. When 1-butanol was introduced to O/Au(111) at 150 K, only 50% of the oxygen was left, even though a smaller flux for 1-butanol (0.4 L) was used (methanol (1.5 L)) to prevent the buildup of condensed 1-butanol. Evidently, 1-butanol reacts more efficiently with surface oxygen and less combustion is expected for 1-butanol relative to methanol.

Selectivity control and combustion

Identification of surface intermediates leading to combustion is an important part of fully understand the oxidation of alcohols on gold. We used a combination of DFT calculations and experimental measurements to probe possible pathways. The reaction of formaldehyde and adsorbed oxygen to form dioxymethylene (eqn (9)) was calculated to be spontaneous and exothermic by 0.54 eV. The starting and ending configurations are shown in Fig. 6(a–d). Atomic oxygen prefers to adsorb at a 3-fold FCC site, with a adsorption energy of over 3 eV,[50] whereas formaldehyde adsorbs on the Au(111) surface only very weakly ($E_{ads} < 0.1$ eV), and does not have a preferred adsorption site. When formaldehyde approaches the adsorbed surface oxygen, they spontaneously react and form dioxymethylene, with each oxygen atom binding close to a 2-fold bridging site. The length of both C–O bonds in dioxymethylene is 1.40 Å, very close to the calculated length of the C–O single bond in methanol (1.43 Å), indicating that they are single bonds in nature. Further, the identical bond length suggests the two C–O bonds are equivalent, and therefore, it is equally possible to break either bond in the reverse reaction.

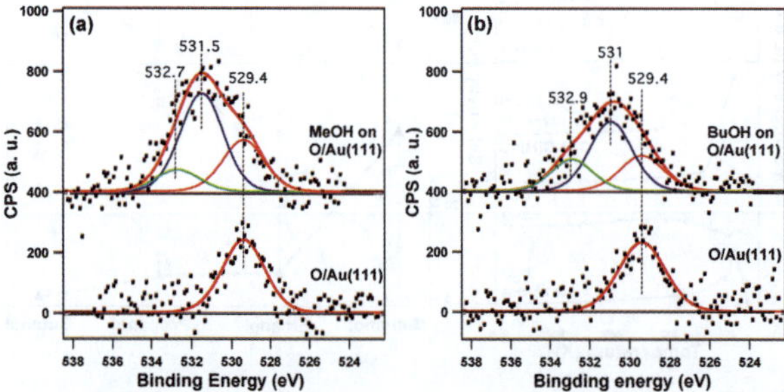

Fig. 5 X-ray photoelectron spectra show the consumption of atomic O by reaction with methanol and 1-butanol: (a) O/Au(111) ($\theta_O = 0.1$ ML, bottom trace) followed by exposure to methanol (top trace) at 150 K. (b) O/Au(111) ($\theta_O = 0.1$ ML, bottom trace) and 1-butanol on O/Au (111) (top trace) at 150 K. Methanol (1.5 L) and 1-butanol (0.4 L) were introduced to the O/Au (111) at 150 K. Surface oxygen was prepared by ozone exposure at 200 K. All spectra were collected at 150 K.

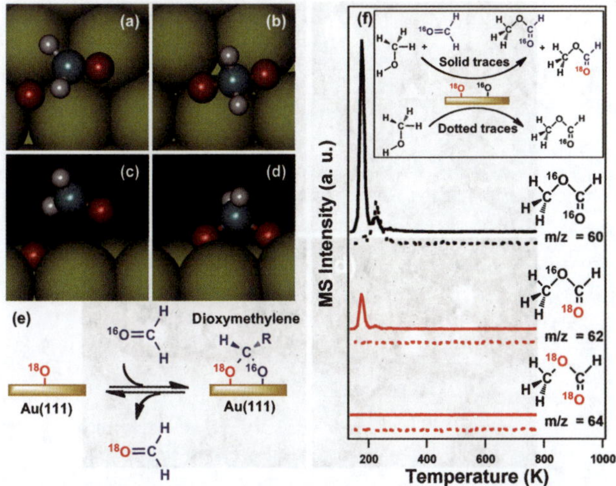

Fig. 6 Reversible formation of dioxymethylene leads to oxygen exchange between formaldehyde and surface oxygen. Top view ((a) and (b)) and side view ((c) and (d)) of the starting and ending configurations of the formation of dioxymethylene on the Au(111) surface. Large yellow spheres, red spheres, blue spheres and small while spheres represent Au, O, C and H atoms, respectively. (e) Schematics of the oxygen exchange of formaldehyde with surface adsorbed ^{18}O. (f) Coupling between methanol and formaldehyde (solid traces) and self-coupling of methanol (dotted traces) on the ^{16}O- and ^{18}O-covered Au(111) surface ($\theta_O = 0.05$ ML, ^{16}O:^{18}O $\approx$ 1 : 1). Methanol (normal and deuterated) was introduced to the O/Au(111) at 150 K and the heating rate is 5 K s^{-1}. Surface oxygen was prepared by ozone exposure at 200 K. ^{18}O was prepared by exposing ^{16}O/Au(111) to 6 L of H$_2$^{18}O at 250 K.

Experiments also provide support for formation of dioxymethylene from formaldehyde and O on Au (111). When H$_2$C=O is adsorbed on the ^{18}O-covered Au(111) surface, H$_2$C=^{18}O was detected in the subsequent temperature programmed reaction experiments, with a H$_2$C=^{18}O/H$_2$C=^{18}O ratio around unity, schematically shown in Fig. 6(e) (data not shown). Similar results on oxygen exchange have been reported for 2-butanone,[59] most likely *via* a similar bidentate intermediate (CH$_3$CH$_2$C(O)(O)CH$_3$).

When methanol and formaldehyde are coadsorbed on a mixed ^{16}O/^{18}O layer on Au(111) ($\theta_O = 0.05$ ML, ^{16}O:^{18}O $\approx$ 1 : 1), both CH$_3$OC(H)=^{16}O ($m/z = 60$) and CH$_3$OC(H)=^{18}O ($m/z = 62$) (Fig. 5(f), solid traces) are formed in a ratio of $\sim$6 : 1. This clearly indicates that a significant amount of the formaldehyde which reacts to the ester passes through the dioxymethylene intermediate (Fig. 6(e)). No CH$_3$^{18}OC(H)=^{18}O ($m/z = 64$) was detected, a clear indication that the oxygen in methoxy does not exchange with surface oxygen. Not all of the formaldehyde exchanges with surface O because of the low initial oxygen coverage ($\theta_O = 0.05$ ML) used. More importantly, over half (4/7) of the formaldehyde that participates in the cross-coupling reaction comes from the dissociation of dioxymethylene. The calculated thermodynamic barrier for dissociation of dioxymethylene to formaldehyde and adsorbed O is 0.54 eV, in agreement with the product evolution temperature (180 K).[30] The reversible formation and decomposition of dioxymethylene is a reservoir of formaldehyde, *via* which formaldehyde can be stored or released for reaction with methoxy.

In contrast, only CH$_3$OC(H)=^{16}O ($m/z = 60$) was detected when methanol alone was introduced to the mixed ^{16}O/^{18}O overlayer on Au(111) ($\theta_O = 0.05$ ML, ^{16}O:^{18}O $\approx$ 1 : 1) (Fig. 6(f), dotted traces). In this case, any formaldehyde formed does not exchange with adsorbed ^{18}O. Since the rate-limiting step for methanol self coupling is β–H elimination of methoxy[20,28] to formaldehyde becomes significant only at

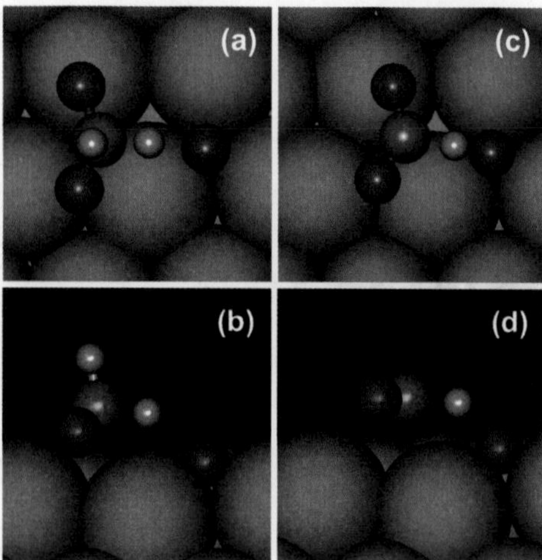

Fig. 7 Transition state structures for transfer of a proton of dioxymethylene to adsorbed oxygen forming formate and hydroxyl, showing (a) top and (b) side views. Transition state structures for transfer of a proton of formate to adsorbed oxygen CO_2 and hydroxyl, showing (c) top and (d) side views. The calculations used coverages of 1/9 ML for O, dioxymethylene and formate.

temperatures above 200 K (Fig. 3(b)), the rapid secondary oxidation steps (eqn 10 and 11) of dioxymethylene apparently make the reformation of formaldehyde less likely. These observations are also consistent with the fact that the methylformate: CO_2 ratio from coupling of methoxy with formaldehyde is higher than that in self-coupling of methanol by a factor of 7.[30]

Consequently, the subsequent steps (eqn 10 and 11) in the combustion reaction were studied by DFT calculations. Dioxymethylene loses a proton to a nearby adsorbed oxygen atom with an activation barrier of 0.12 eV. The proton is shared by the carbon atom in the dioxymethylene and the adsorbed oxygen (Fig. 7(a–b)). Although the activation barrier for this process is low, the rate of the reaction may be limited by the low oxygen coverage and low diffusion rate of surface species at low temperature, leading to the competitive advantage of the coupling reaction against combustion at low temperature (<200 K).

The last step in the combustion process—transfer of a proton from formate to surface oxygen forming CO_2 and hydroxyl—has an activation barrier of 1.19 eV. The formate bends towards the adsorbed oxygen in the transition state structure, and the proton is shared by both oxygen atoms (Fig. 7(c–d)). In the starting configuration the O–C–O plane is perpendicular to the surface and both C–O bonds are 1.27 Å in length, between the length of C–O single and double bonds. The relatively high activation barrier is likely due to the substantial bending required for the O–C–O plane to get H close enough to the surface oxygen. The calculated barrier agrees well with the experimental observation that CO_2 is produced around 300 K in the self-coupling of methanol, well above the desorption temperature of methylformate.

Conclusions

A general mechanistic framework for oxidative coupling reactions of alcohols, amines and aldehydes has been established through a combination of experimental and theoretical investigations. The amount of combustion reduces in the self-coupling reaction of alcohols as the length of the alkyl chain in the alcohol increases. It is attributed to the increasing facility of the β–H elimination reaction of alkoxys as the length of the alkyl chain grows, leading to more efficient partial oxidation and coupling reactions at low temperature. Dioxymethylene is shown to be able to act as a reservoir of formaldehyde in the cross-coupling reaction between methanol and formaldehyde, enhancing the selectivity for the coupling products.

Acknowledgements

We gratefully acknowledge the support of the U.S. Department of Energy, Basic Energy Sciences, under Grant No. FG02-84-ER13289 (CMF). BJX also acknowledges the Harvard University Center for the Environment for support through the Graduate Consortium in Energy and the Environment. STM images in Fig. 1 are by courtesy of Dr Ling Zhou and Dr Weiwei Gao.

References

1 M. Haruta, N. Yamada, T. Kobayashi and S. Iijima, *J. Catal.*, 1989, **115**, 301–301.
2 P. Fristrup, L. B. Johansen and C. H. Christensen, *Catal. Lett.*, 2008, **120**, 184–190.
3 A. Abad, A. Corma and H. Garcia, *Chem.–Eur. J.*, 2008, **14**, 212–222.
4 I. Nielsen, E. Taarning, K. Egeblad, R. Madsen and C. Christensen, *Catal. Lett.*, 2007, **116**, 35–35.
5 B. Jorgensen, S. E. Christiansen, M. L. D. Thomsen and C. H. Christensen, *J. Catal.*, 2007, **251**, 332–332.
6 D. I. Enache, D. Barker, J. K. Edwards, S. H. Taylor, D. W. Knight, A. F. Carley and G. J. Hutchings, *Catal. Today*, 2007, **122**, 407–411.
7 A. Abad, C. Almela, A. Corma and H. Garcia, *Tetrahedron*, 2006, **62**, 6666–6672.
8 A. Abad, P. Concepcion, A. Corma and H. Garcia, *Angew. Chem., Int. Ed.*, 2005, **44**, 4066–4069.
9 S. Carrettin, P. McMorn, P. Johnston, K. Griffin, C. J. Kiely and G. J. Hutchings, *Phys. Chem. Chem. Phys.*, 2003, **5**, 1329–1336.
10 S. Biella and M. Rossi, *Chem. Commun.*, 2003, 378–379.
11 B. K. Min, X. Y. Deng, X. Y. Li, C. M. Friend and A. R. Alemozafar, *ChemCatChem*, 2009, **1**, 116–121.
12 A. J. Mancuso, S. L. Huang and D. Swern, *J. Org. Chem.*, 1978, **43**, 2480–2482.
13 M. Haruta, *Catal. Today*, 1997, **36**, 153–166.
14 M. S. Chen and D. W. Goodman, *Catal. Today*, 2006, **111**, 22–33.
15 M. Lazar, B. L. Zhu and R. J. Angelici, *J. Phys. Chem. C*, 2007, **111**, 4074–4076.
16 A. Wittstock, V. Zielasek, J. Biener, C. M. Friend and M. Baumer, *Science*, 2010, **327**, 319–322.
17 H. M. Yin, C. Q. Zhou, C. X. Xu, P. P. Liu, X. H. Xu and Y. Ding, *J. Phys. Chem. C*, 2008, **112**, 9673–9678.
18 A. Wittstock, B. Neumann, A. Schaefer, K. Dumbuya, C. Kubel, M. M. Biener, V. Zielasek, H. P. Steinruck, J. M. Gottfried, J. Biener, A. Hamza and M. Baumer, *J. Phys. Chem. C*, 2009, **113**, 5593–5600.
19 B. Xu, L. Zhou, R. J. Madix and C. M. Friend, *Angew. Chem., Int. Ed.*, 2009, **49**, 394–398.
20 B. Xu, X. Liu, J. Haubrich, R. J. Madix and C. M. Friend, *Angew. Chem., Int. Ed.*, 2009, **48**, 4206–4209.
21 X. Y. Liu, B. J. Xu, J. Haubrich, R. J. Madix and C. M. Friend, *J. Am. Chem. Soc.*, 2009, **131**, 5757–5759.
22 X. Y. Deng, B. K. Min, A. Guloy and C. M. Friend, *J. Am. Chem. Soc.*, 2005, **127**, 9267–9270.
23 Y. Xu and M. Mavrikakis, *J. Phys. Chem. B*, 2003, **107**, 9298–9307.
24 J. A. Rodriguez, S. Ma, P. Liu, J. Hrbek, J. Evans and M. Perez, *Science*, 2007, **318**, 1757–1760.
25 L. M. Molina and B. Hammer, *Appl. Catal., A*, 2005, **291**, 21–31.
26 A. Bongiorno and U. Landman, *Phys. Rev. Lett.*, 2005, **95**, 106102.
27 B. N. Zope, D. D. Hibbitts, M. Neurock and R. J. Davis, *Science,* 330, pp. 74–78.
28 B. Xu, J. Habrich, T. A. Baker, C. M. Friend and K. Efthimios, *J. Phys. Chem. C*, 2010, In press.
29 R. J. Madix, C. M. Friend and X. Y. Liu, *J. Catal.*, 2008, **258**, 410–413.
30 B. Xu, X. Liu, J. Haubrich and C. M. Friend, *Nat. Chem.*, 2009, **2**, 61–65.
31 B. Xu, R. J. Madix and C. M. Friend, *J. Am. Chem. Soc.*, 2010, **132**, 16571–16580.
32 B. Xu, J. Haubrich, C. Fryschlag, C. M. Friend and R. J. Madix, *Chem. Sci.*, 2010, **1**, 310–314.

33 S. K. Klitgaard, K. Egeblad, U. V. Mentzel, A. G. Popov, T. Jensen, E. Taarning, I. S. Nielsen and C. H. Christensen, *Green Chem.*, 2008, **10**, 419–423.

34 C. Marsden, E. Taarning, D. Hansen, L. Johansen, S. K. Klitgaard, K. Egeblad and C. H. Christensen, *Green Chem.*, 2008, **10**, 168–170.

35 B. K. Min, A. R. Alemozafar, D. Pinnaduwage, X. Deng and C. M. Friend, *J. Phys. Chem. B*, 2006, **110**, 19833–19838.

36 N. Saliba, D. H. Parker and B. E. Koel, *Surf. Sci.*, 1998, **410**, 270–282.

37 S. E. Stein, P. J. Linstrom, (Ed.) and W. G. Mallard, (Ed.), Institute of Standards and Technology, Gaithersburg MD, 20899, Editon edn.

38 G. Kresse and J. Hafner, *Phys. Rev. B: Condens. Matter*, 1993, **48**, 13115–13118.

39 G. Kresse and J. Hafner, *Phys. Rev. B: Condens. Matter*, 1993, **47**, 558–561.

40 J. P. Perdew and Y. Wang, *Phys. Rev. B: Condens. Matter*, 1992, **45**, 13244–13249.

41 G. Kresse and D. Joubert, *Phys. Rev. B: Condens. Matter Mater. Phys.*, 1999, **59**, 1758–1775.

42 D. R. Lide, *CRC Handbook of Chemistry and Physics*, CRC Press, New York, 1996.

43 G. Makov and M. C. Payne, *Phys. Rev. B: Condens. Matter*, 1995, **51**, 4014–4022.

44 D. Sheppard, R. Terrell and G. Henkelman, *J. Chem. Phys.*, 2008, **128**, 134106.

45 G. Henkelman, B. P. Uberuaga and H. Jonsson, *J. Chem. Phys.*, 2000, **113**, 9901–9904.

46 G. Henkelman and H. Jonsson, *J. Chem. Phys.*, 2000, **113**, 9978–9985.

47 J. Kirn, E. Samano and B. E. Koel, *Surf. Sci.*, 2006, **600**, 4622–4632.

48 J. M. Gottfried, K. J. Schmidt, S. L. M. Schroeder and K. Christmann, *Surf. Sci.*, 2003, **525**, 184–196.

49 J. M. Gottfried, K. J. Schmidt, S. L. M. Schroeder and K. Christmann, *Surf. Sci.*, 2003, **525**, 197–206.

50 T. A. Baker, C. M. Friend and E. Kaxiras, *J. Phys. Chem. C*, 2009, **113**, 3232–3238.

51 T. A. Baker, B. J. Xu, X. Y. Liu, E. Kaxiras and C. A. Friend, *J. Phys. Chem. C*, 2009, **113**, 16561–16564.

52 T. A. Baker, X. Y. Liu and C. M. Friend, *Phys. Chem. Chem. Phys.*, 2010, **13**, 34–46.

53 J. Gong, D. W. Flaherty, R. A. Ojifinni, J. M. White and C. B. Mullins, *J. Phys. Chem. C*, 2008, **112**, 5501–5509.

54 R. G. Quiller, T. A. Baker, X. Deng, M. E. Colling, B. K. Min and C. M. Friend, *J. Chem. Phys.*, 2008, **129**, 064702.

55 B. Hammer and J. K. Norskov, *Nature*, 1995, **376**, 238–240.

56 J. L. Gong, D. W. Flaherty, T. Yan and C. B. Mullins, *ChemPhysChem*, 2008, **9**, 2461–2466.

57 I. E. Wachs and R. J. Madix, *Surf. Sci.*, 1978, **76**, 531–558.

58 A. F. Carley, A. W. Owens, M. K. Rajumon, M. W. Roberts and S. D. Jackson, *Catal. Lett.*, 1996, **37**, 79–87.

59 T. Yan, J. Gong and C. B. Mullins, *J. Am. Chem. Soc.*, 2009, **131**, 16189–16194.

Kinetic study of propylene epoxidation with H_2 and O_2 over Au/Ti–SiO$_2$ in the explosive regime†

Jiaqi Chen, Sander J. A. Halin, Jaap C. Schouten and T. Alexander Nijhuis*

Received 10th February 2011, Accepted 3rd March 2011
DOI: 10.1039/c1fd00014d

A kinetic study of propene epoxidation with hydrogen and oxygen over a Au/Ti–SiO$_2$ catalyst has been performed in a wide range of reactant concentrations including the explosive region in a micro reactor. The observed rate dependency on the reactants for the epoxidation and the competing direct water formation is discussed in relation to the current mechanistic insights in the literature. The formation rate of propene oxide is most dependent on the hydrogen concentration, in which the formation of an active peroxo species on the gold nanoparticles is the rate determining step. Deactivation is mainly caused by consecutive oxidation of propene oxide. Oxygen favours the regeneration of the deactivated catalytic sites. Water formation and propene epoxidation are strongly correlated. Water is formed *via* two routes: through the active peroxo intermediate responsible for epoxidation and from direct formation without involving this active intermediate. Improving the hydrogen efficiency should distinguish between these two routes of water formation. The active peroxo intermediate in epoxidation is competitively consumed by hydrogenation and epoxidation. The active gold site is blocked during deactivation.

1. Introduction

The direct epoxidation of propylene to propene oxide (PO) using the co-reactants hydrogen and oxygen, demonstrated for the first time by Haruta *et al.*,[1] is highly selective over gold-titania based catalysts and has gained considerable attention in the past decade.[2–10] Although this catalytic reaction has been investigated extensively, the mechanism is still not well understood. The most important steps in the reaction mechanism proposed in the literature include the formation of a hydroperoxy species on the gold nanoparticles, a reactive adsorption of propene on titania sites adjacent to gold, and the consecutive oxidation of propene by the hydroperoxy species to form propene oxide.[11–13] The highlights of this catalytic reaction are its exceptionally high selectivity towards propene oxide and the mild operating conditions. However, the main obstacles to its industrial application originate from the stability of the catalyst, the relatively low activity, and the insufficient hydrogen efficiency. Encouraging progress has been made over the past years concerning the three key factors that determine the catalytic performance,

Laboratory of Chemical Reactor Engineering, Department of Chemical Engineering and Chemistry, Eindhoven University of Technology, P.O. Box 513, 5600 MB Eindhoven, The Netherlands. E-mail: t.a.nijhuis@tue.nl; Fax: +31-40-2446653; Tel: +31-40-2473671

† Electronic Supplementary Information (ESI) available: the characterization of the catalyst support and the derivation for the calculation of activation energy difference between water formation and propene epoxidation. See DOI: 10.1039/c1fd00014d/

i.e. the size (or the morphology) of the gold nanoparticles,[10,14,15] the state of Ti in the support,[3,8,16–20] and the preparation method of the catalysts which plays an important role in the selective dispersion of the gold and its interaction with the Ti sites.[6,15,21–23] Very high productivity of propene oxide has been reported by different groups and relatively stable performance has been achieved.[5,6,8,10,15,17] However, this is usually at relatively high temperatures (150–200 °C), where hydrogen efficiency is less. While the hydrogen efficiency of a certain catalyst can be mainly attributed to the synergy between the gold nanoparticles and the titanium sites in proximity, promoters[5,8] and surface modification[24] may additionally depress the hydrogen consumption (or hydroperoxy decomposition) to some extent by adjusting the surface acidity. Further improvement in hydrogen efficiency seems difficult. To evade the issue of hydrogen efficiency in the direct epoxidation of propene, the Au-clusters–$C_3H_6/O_2/H_2O$ system has come into view albeit the low activity and the selectivity.[25–27] The reported H_2 efficiency of almost 100% on a Au/TS-1 catalyst with a very low gold loading[23] reveals the potential in enhancing the H_2 efficiency. The highest hydrogen efficiency (47%) reported recently over the Au/TS-1 catalyst without any post-treatment, which still preserves a high conversion of propene, opens the door to a further step in catalyst development.[15]

A kinetic study of propylene epoxidation using H_2 and O_2 over a gold–titania based catalyst is important in revealing the reaction mechanism. However, two aspects are complicating: short-term deactivation due to build-up of carbonate-carboxylate species on active Ti-sites[11,28,29] and possible long-term changes in activity mainly caused by the sintering of the gold nanoparticles.[30,31] The short term deactivation is reversible by burning the strongly adsorbed species and can be solved in a sense by choosing a relatively stable catalyst where Ti is highly dispersed and by operating at elevated temperatures. The sintering of gold particles is irreversible and the effect can be minimized by performing the catalytic testing in a period as short as possible or preparing catalysts which are stable (low Cl content, surface-stabilized Au nanoparticles). Another factor constraining a kinetic study is the explosive nature of the reactant mixture. Most reported studies were performed within the non-explosive region.[12,13] An important exception is the work performed by the group of Oyama, who reported propene epoxidation in a packed-bed catalytic membrane reactor.[32] Feeding (part of) the hydrogen through a membrane allowed them to work safely with much higher hydrogen and oxygen concentrations (up to 40% each), which gave them a stable propene conversion of 10% at 80% propene oxide selectivity at 453 K. Their results showed fractional orders in a power-rate-law expression similar to what had been published previously.[12,13] Although corresponding relations between hydrogen consumption and product selectivities as well as the formation rate of propene oxide can be observed among earlier studies,[24,33–35] further quantitative investigations have not been reported.

In this work, a microreactor system is utilized to perform the propene epoxidation over a gold on titania–silica catalyst in an extended range of reactant concentrations including the explosive region. Different from previously published kinetic studies, the formation rate of propene oxide in this work is decoupled from the short-term deactivation by the methodology described in our previous paper,[36] which uses the initial activity of the catalyst extrapolated by using a simplified deactivation model based on the mechanism proposed by Nijhuis *et al.*[11] In this study, quantitative description of the deactivation in a wider range of reactant concentrations will help to gain further insight into the mechanism of the propene epoxidation in hydrogen and oxygen. The relationship between the catalyst stability, the water formation during the epoxidation, and the propene epoxidation itself is thoroughly examined, which may provide mechanistic implications for improving the hydrogen efficiency in the future development of the catalyst.

2. Experimental

2.1 Catalyst preparation and characterization

A catalyst consisting of 1 wt% gold on titania dispersed on silica was used, prepared in a manner similar to the method described by Nijhuis *et al.*[28] A total of 15 g of dry silica support (Davisil 643, Aldrich, 300 m^2 g^{-1}, pore size 150 Å, pore volume 1.15 cm^3 g^{-1}) was dispersed in 250 mL of anhydrous isopropanol (Aldrich, 99.5%) under a nitrogen atmosphere in a glove box. The slurry was stirred for 10 min and afterwards 0.70 mL of tetraethylorthotitanate (TEOT) (Aldrich, 97%) was added. The amount of TEOT added was determined by calculating a theoretical titania coverage of 5% monolayer (Ti/Si atom ratio) on the silica surface. The slurry was stirred for 30 min. The isopropanol was slowly evaporated at 318 K and 140 mbar in a rotary evaporator under nitrogen. After the isopropanol was removed, the powder was dried overnight at 353 K and subsequently calcined first at 393 K (5 K min^{-1} heating) for 2 h and then at 873 K (10 K min^{-1} heating) for 4 h.

Gold was deposited on the catalyst support by a deposition–precipitation method using aurochloric acid and ammonia. A total of 10 g of support was dispersed in 100 mL of water. The pH of the slurry was adjusted to 9.5 by dropwise adding ammonia (2.5 wt%). A total of 575 mg of an acidic 30 wt% HAuCl$_4$ solution (Aldrich) was diluted in 20 mL of demineralized water and was added dropwise to the support slurry over a 15 min period. While HAuCl$_4$ solution was being added, the pH was kept at 9.5 using aqueous ammonia. After the addition of the gold solution, the slurry was stirred for one hour. The slurry was filtered and washed 3 times using 200 mL of water. The catalyst was dried overnight at 353 K and calcined first at 393 K (5 K min^{-1} heating) for 2 h and afterwards at 673 K (10 K min^{-1} heating) for 4 h. Drying and calcination of the support and the catalyst were performed under atmospheric pressure in stationary air.

ICP analysis showed that the gold loading was 0.91 wt%, which is close to the target loading of 1 wt%. The titanium content on the catalyst was 1.29 wt%. TEM analysis showed a narrow particle size distribution for the gold nanoparticles, centered at 4.5 nm. A deconvoluted UV–vis spectrum of the support showed that Ti is highly dispersed mainly in tetrahedral coordination, while a minor amount of penta- or hexacoordinatd Ti was also observed. No adsorption band was found above 300 nm (see ESI).†

2.2 Catalyst testing

Catalytic tests were performed using a microreactor system. The microreactor consisted of a stainless steel capillary tube (0.9 mm inner diameter), which was loaded with 20 mg of catalyst (sieved, 50–60 μm) and was operated at a gas feed rate of 3.33 NmL min^{-1}. The oxygen was mixed with the other two reactants shortly before the catalyst bed. Immediately after the catalyst bed, additional helium was added to the gas stream, diluting the gases to a composition outside the explosive region. In this manner, the total volume of a potentially explosive gas mixture was minimized.

The catalytic experiments were performed in cycles. Prior to each test the catalyst was (re)activated by heating it to 573 K (10 K min^{-1}) in a 10% oxygen in helium stream for 1 h. After cooling to the desired reaction temperature, the gas feed was switched to the desired composition and a 5 h catalytic test was performed. Experiments were performed at varying feed concentrations ranging from 2 to 80 vol% for hydrogen and oxygen and 2 to 40 vol% for propene, with the remaining part being helium. The "standard" feed concentration is 10 vol% for each reactant (hydrogen, oxygen and propene) with helium as balance. In the experiments in which the composition was varied, two reactants were kept at the concentration of 10 vol% with only the concentration of the third reactant being changed. A kinetic study over the full range of reactant concentrations was performed at 388 K, 403 K and 418 K. The long-term stability was examined before the kinetic study and was found to be as excellent at 388 K as in our previous paper.[36] At the higher temperatures, however, a minor activity loss of 5–10% compared with the initial activity was observed over a period of 20 days (around 50 cycles). In the kinetic study, the catalyst was replaced every week in order to minimize the long-term change in catalyst activity and the activity was monitored by repeating the reaction at the "standard" condition at the end of each series of catalytic cycles. Due to the similarity in trend of the results at different temperatures, we mainly report and discuss the result at 403 K in detail. The pressure at the reactor outlet is atmospheric pressure and the pressure drop along the catalyst bed (including a particle filter afterwards) is at the level of 0.7 bar at 403 K.

Product analyses were performed using a fast Interscience Compact GC system, equipped with a Porabond Q column and a Molsieve 5A column, capable of analyzing all products in 4 min.

3. Results and discussion

3.1 Product formation

In Figure 1, the formation rates of propene oxide and water in a single catalytic cycle at the chosen standard conditions are shown. Both formation rates show a relatively fast decrease within the first two hours. The rates continue to decrease, much slower, in the second half of the catalytic cycle. The performance between 150 min and 270 min is used for calculating the averaged H_2 efficiency (defined as r_{PO}/r_{H2O}) and C_3H_6 conversion for a single reaction cycle, which will be used for evaluating the catalyst performance. Different from the H_2 efficiency and C_3H_6 conversion, the PO selectivity is unchanged since the very beginning of a reaction cycle.

The breakthrough curve (green dash) at the beginning of the reaction cycle shown in Figure 1 for the PO formation is due to the strong adsorption of PO on the catalyst surface.[37] This breakthrough is more obvious when the initial activity of PO formation is low, but it may be undetectable when the initial activity is high. The concomitant decrease in the formation rate of water with that of PO along time is commonly observed in literature for gold–titania catalysts.[12,24,33–35] The formation rate of water is the overall rate unless specified otherwise. The strong correlation

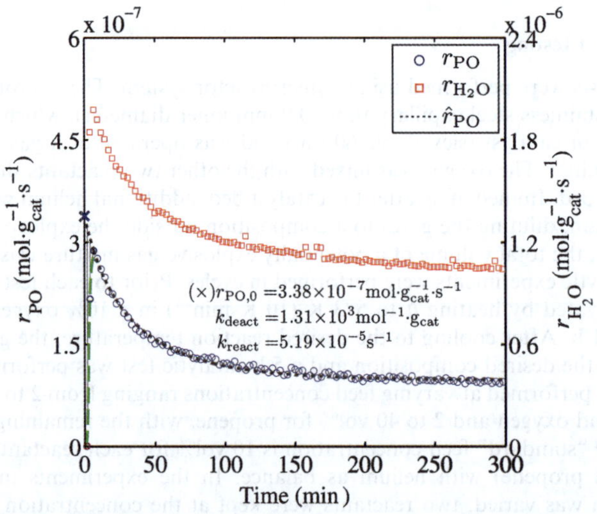

Fig. 1 Formation rates of propene oxide and water as a function of time: O, propene oxide; □, water; ······, fitted rate of propene oxide by Equation 5 (1 wt% Au on Ti − SiO$_2$, gas feed 10 vol% hydrogen, 10 vol% oxygen, 10 vol% propene in helium, 403 K, GHSV 10000 mL g$_{cat}^{-1}$ h^{-1}).

between water formation and propene epoxidation during deactivation indicates that a shared site or a common reactive intermediate exists between these two reactions. Most likely, the deactivating species blocks this shared site or a site producing the common intermediate. Based on the current insights in literature on the mechanisms for both the epoxidation and the water formation,[9,12,13,38,39] it is most likely that a peroxo species is this common intermediate.

3.2 Propene epoxidation

In the previous work,[36] a model was developed to describe the time-dependent propene oxide formation in the catalytic cycles on the Au/Ti–SiO$_2$ catalyst. This model is based on the reaction mechanism proposed previously.[11,28] The model was able to accurately describe the propene oxide formation rate in time, using three parameters:

- $r_{PO, 0}$: the extrapolated propene oxide formation rate at $t = 0$, *i.e.* the activity the catalyst would have had in the absence of deactivation;
- k_{deact}: the deactivation rate constant;
- k_{react}: the reactivation rate constant.

The model is based on the following assumptions:

1. The propene oxide formation rate is proportional to the initial rate $r_{PO, 0}$ (without deactivation) multiplied with the fraction of active sites available on the catalyst (*i.e.*, $1 - \theta_d$, where θ_d is the fraction of the deactivated sites), according to Equation 1.

2. The catalyst deactivation proceeds *via* a consecutive reaction of propene oxide (or an intermediate directly correlated to the propene oxide formation rate) adsorbed on an active site, forming a deactivated site. The rate of this reaction is according to Equation 2.

3. The catalyst reactivation occurs *via* the desorption of the deactivating species from a deactivated site. The rate of this reaction is according to Equation 3.

$$r_{PO} = r_{PO,0}(1 - \theta_d) \tag{1}$$

$$r_{deact} = \left(\frac{d\theta_d}{dt}\right)_{deactivating} = k_{deact} r_{PO}(1 - \theta_d) \tag{2}$$

$$r_{react} = \left(\frac{d\theta_d}{dt}\right)_{regenerating} = -k_{react}\theta_d \tag{3}$$

The rate of formation of deactivated sites can be determined from a site balance for the deactivated sites:

$$\frac{d\theta_d}{dt} = k_{deact} r_{PO}(1 - \theta_d) - k_{react}\theta_d \tag{4}$$

By combining equations 1 and 4 and taking the initial conditions of $r_{PO} = r_{PO, 0}$ and $\theta_d = 0$ at $t = 0$, the equations can be solved analytically. The analytical solution is provided by Equation 5:

$$r_{PO} = r_{PO,0}\left(1 - \frac{1 - \exp(-At)}{a - \frac{1}{a}\exp(-At)}\right) \tag{5a}$$

with

$$A = \left(a - \frac{1}{a}\right) k_{\text{deact}} r_{\text{PO,0}} \qquad (5b)$$

and

$$\frac{k_{\text{react}}}{k_{\text{deact}} \cdot r_{\text{PO,0}}} = a + \frac{1}{a} - 2 \quad (a > 1) \qquad (5c)$$

This 3-parameter model was fitted to the experimental cycles at different reaction conditions. An example of such a fit is shown in Figure 1. The model describes the time-dependent experimental results well.

This model, however, is not yet a reaction mechanism based on the elementary reaction steps, but rather on a simplified mechanism describing the observed reaction rates well with parameters which lump a number of elementary steps. For this reason, the two rate constants are not only dependent on temperature, but also on the concentrations of the three reactants, *viz.*, hydrogen, oxygen, and propene. The dependency of the three parameters on the reactant concentrations is shown in Figure 2.

At low concentrations, the propene oxide formation rate is dependent on the concentrations of all three reactants, while at higher concentrations, the rate is only proportional to the hydrogen concentration (Figure 2(a)). This indicates that at higher reactant concentrations, only hydrogen is involved in the rate determining

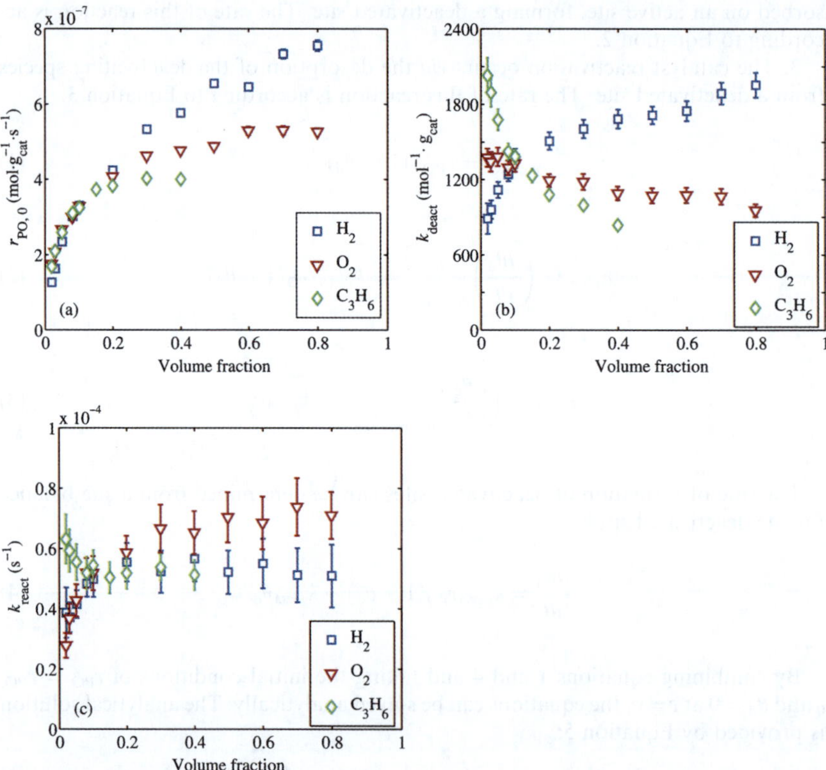

Fig. 2 Initial activity $r_{\text{PO, 0}}$ (a), deactivation rate constant k_{deact} (b), and reactivation rate constant k_{react} (c) fitted from the deactivation model (Equation 5) at different reactant concentrations (1 wt% Au on Ti–SiO$_2$, 403 K, GHSV 10000 mL g$_{\text{cat}}^{-1}$ h^{-1}, the concentration of one reactant is varied while the other two are fixed at 10 vol%).

       This journal is © The Royal Society of Chemistry 2011

step. This observation is in agreement with the most common assumption in the literature[2,11–13] that the formation of peroxo species (the active oxidizing species) on the gold nanoparticles is the rate determining step. For this reaction a dissociative adsorption of hydrogen is required.

In Figure 2(b), it can be seen that the deactivation rate constant increases with the hydrogen concentration, while it decreases with the propene concentration and is only weakly dependent on the oxygen concentration. This is explained by propene oxide (or a precursor thereof) being oxidized further by the same peroxo species as is the active oxidant in propene epoxidation. If the propene concentration is higher, the concentration of this species on the catalyst is lower, reducing the rate of the unwanted consecutive oxidation. If the hydrogen concentration is higher, this peroxo species is produced at a higher rate. The slightly mitigated deactivation at higher oxygen concentration will be discussed in Section 3.5.

In Figure 2(c), it can be seen that the catalyst reactivation rate is only dependent on the oxygen concentration. Nijhuis *et al.*[28] previously determined that the deactivating species are consecutive oxidation products of propene oxide, which can only leave the surface once they are completely combusted to produce CO_2. Our current results indicate that this consecutive oxidation to produce carbon dioxide is not by the hydroperoxy species, but rather by oxygen.

3.3 Water formation and link to epoxidation

In Figure 1 it can be seen that propene oxide production and water formation show a similar pattern. This can be considered as an indication for a strong correlation between the water formation and the propene epoxidation. Since the amount of water produced is far larger than the stoichiometric amount produced from the epoxidation reaction, this similarity in the rate of formation cannot be explained by the fact that water and propene oxide would be the co-products of a single reaction. To examine this correlation further, the water formation rate as a function of the propene oxide formation rate for a number of catalytic cycles is plotted in Figure 3.

It can be seen that the correlation between the water production and the propene oxide production is almost perfectly linear. The intercept of the linear correlation is not at the origin, which indicates that water is produced *via* two different routes, one correlated to the propene oxide formation, and one which is not. The water formation, which is directly correlated to the propene oxide formation, is tentatively assigned to the water produced out of a peroxo species (*i.e.* the species which is also responsible for the epoxidation of propene). The water formation which is not correlated to the propene oxide formation, is assigned to the direct water formation from the reaction between hydrogen and oxygen, not being formed *via* a peroxo species. These two routes for water formation are depicted in Scheme 1, which is resembling the scheme presented previously by Edwards *et al.*[40] for the direct formation of hydrogen peroxide over Au–Pd catalysts. It should be mentioned that even though the hydroperoxy species is denoted as HOOH in Scheme 1, this species might also be OOH + H adsorbed in some other form.[41,42]

To more closely examine both the direct water formation and the water formation from the peroxo species, which is also responsible for the epoxidation, the linear correlation is used to fit each of the data sets according to Equation 6:

$$r_{H_2O} = cr_{PO} + d \qquad (6)$$

The coefficient c represents the ratio between the change in the overall water formation rate and the loss in the PO formation rate on the active Au–Ti centers. The loss of activity for both water formation and propene epoxidation can only be attributed to a blockage on this active Au–Ti center. Occupation of Ti alone by the deactivating species does not necessarily result in this synchronous decrease

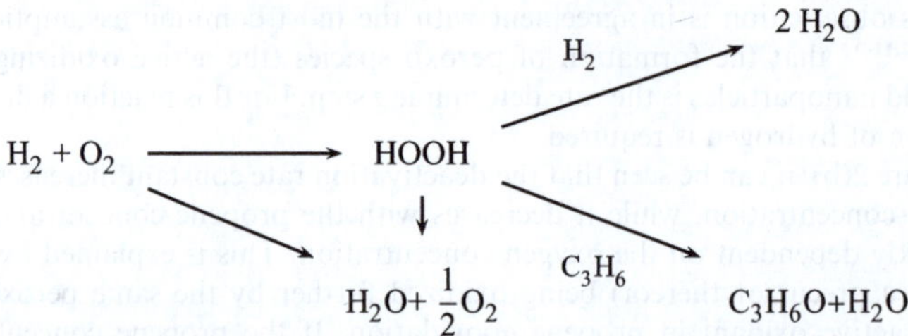

Scheme 1 Routes for water formation in propene epoxidation using hydrogen and oxygen.

in rates in the context of the sequential mechanism, *i.e.* HOOH forms first on gold and then spills over to Ti, since HOOH can decompose to water without Ti. Either gold is occupied in the deactivation, or the strong metal support interaction[43,44] is so weakened by the deactivation of Ti sites that the formation of the active peroxo species is interrupted. The constant d represents the direct formation rate of water. Direct water formation can occur at gold sites near Ti *via* a route that does not involve a peroxo species responsible for epoxidation, but it can also occur at gold sites not in proximity to Ti.

In Figure 4, the values of c and d are shown as a function of the feed concentrations of hydrogen, oxygen, and propene.

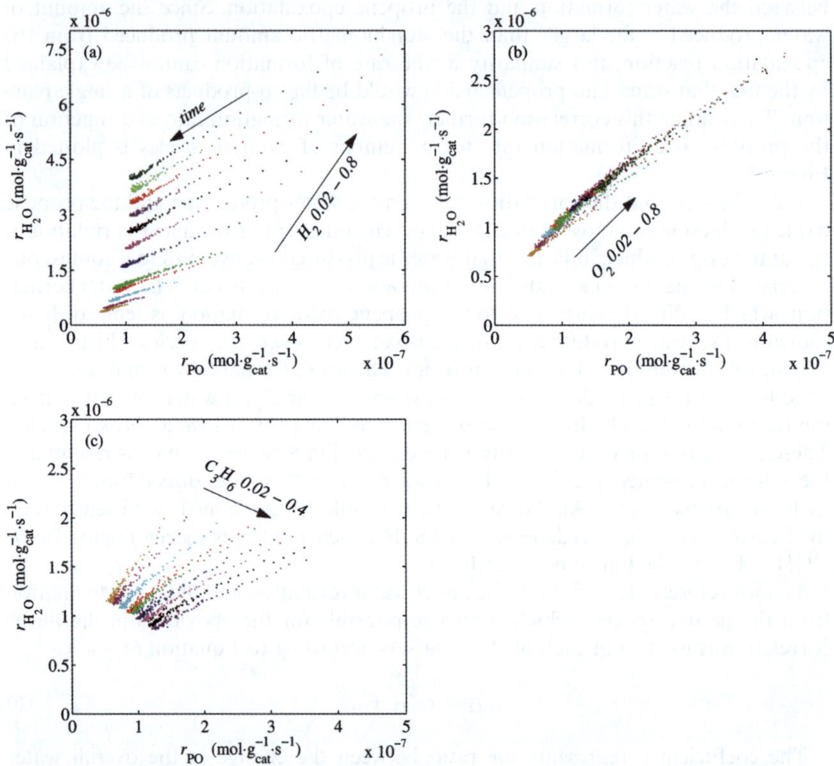

Fig. 3 Formation rate of water as a function of PO formation rate within each catalytic cycle when only the concentration of (a) hydrogen, (b) oxygen, (c) propene is changed while the other two are fixed at 10 vol% (1 wt% Au on Ti–SiO₂, 403 K, GHSV 10000 mL g$_{cat}^{-1}$ h^{-1}, balance in helium).

   This journal is © The Royal Society of Chemistry 2011

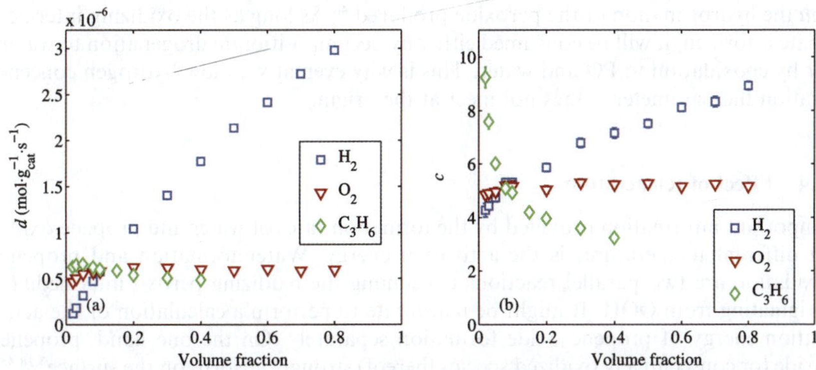

Fig. 4 Direct formation rate of water not involved in epoxidation (a) and the ratio between the over all water formation rate and PO formation rate on the active Au–Ti sites (b) at different reactant concentrations (1 wt% Au on Ti–SiO$_2$, 403 K, GHSV 10000 mL g$_{cat}^{-1}$ h^{-1}, balance in helium).

In Figure 4(a), d is mainly a function of the hydrogen concentration in the feed and not dependent on propene and oxygen. The fact that parameter d is almost first order dependent on the hydrogen concentration and not dependent on the oxygen concentration (except for only mildly at very low oxygen concentration) suggests that direct water formation is rate limited by either the hydrogen adsorption or the hydrogen dissociation on the gold nanoparticles. Interestingly, at lower oxygen concentrations the value of d seems to originate from a non-zero value, which is different for the situation for hydrogen. One possible explanation is that oxygen adsorption on gold is so much stronger than (dissociative) hydrogen adsorption, that only at much lower oxygen concentrations (beyond our current experimental capabilities) it would become limiting. The effect by propene on the parameter d is trivial, since the parameter d represents the direct water formation out of hydrogen and oxygen.

In Figure 4(b), it can be seen that parameter c is not dependent on the oxygen concentration, but positively dependent on the hydrogen concentration, and inversely dependent on the propene concentration. In Section 3.2, it has been discussed that propene oxide formation is directly proportional to the formation of a peroxo species. The parameter c here is not representing the formation of a peroxo species on the catalyst, but the way this peroxo species is reacting further. Scheme 1 is showing the possible reactions for this peroxo species (for simplicity simply shown as HOOH).

In Figure 4(b), it can be seen that oxygen does not play a role in the reaction or decomposition of the peroxo species on the gold nanoparticles. This is in line with the expectation that oxygen is not directly involved in the rate determining step. The observed inverse proportionality of c on the propene concentration is in agreement with the reaction route of HOOH as shown in Scheme 1. Parameter c is actually inversely proportional to the efficiency with which the peroxo species is used for the desired epoxidation. A large value of c implies that a lot of water is produced compared to the amount of propene oxide produced. The inverse proportionality of c to the propene concentration can be easily explained by the fact that as the propene concentration is increasing, a larger fraction of the peroxo species will react with propene to produce propene oxide, and less will be lost by decomposition. The positive dependency of c on the hydrogen concentration can be explained by the fact that excessive hydrogen speeds up the loss of the peroxo species by hydrogenation to form two water molecules. This is in agreement with the literature, where it is indeed concluded that in the direct hydrogen peroxide formation out of hydrogen and oxygen over gold nanoparticles, the largest loss of peroxide occurs *via* the hydrogenation of the peroxide produced.[40] As long as the oxidizing intermediate is formed, it will be consumed either by decomposition/hydrogenation to water or by epoxidation to PO and water. This is why even at very low hydrogen concentration the parameter c does not meet at the origin.

3.4 Effect of temperature

Important information provided by the formation rates of water and propene oxide at different temperatures is the activation energy. Water formation and propene oxidation are two parallel reactions consuming the oxidizing peroxo intermediate originating from OOH. It might be inaccurate to perform a calculation of the activation energy of propene oxide formation separately. On the one hand, propene oxide (or consecutively oxidized species thereof) strongly adsorbs on the surface[3,28,45] resulting in an activation energy influenced by adsorption enthalpies and surface coverages. On the other hand, the competitive consumption of a reactive peroxo intermediate by epoxidation and water formation will cause the uneven distribution of this reactive intermediate towards one of these two reactions when the temperature increases, *i.e.* in $r = A \exp(- E_a/RT)f(C)$ the $f(C)$ may change when T changes. The latter situation can be observed by a saturation and even a decrease in one of the two reaction rates when the effect of the strong product adsorption is not obvious.

Table 1 gives the general performance of the catalyst at different temperatures. The time-on-stream performance at different temperatures is shown in Figures 5(a) and 5(b). The main side products are isomers of propene oxide, most probably, due to ring opening of propene oxide on acidic Ti^{4+} sites.[3,5,45] Propionaldehyde is primary among these side products, which is consistent with the study on the isomerization of propene oxide by Namuangruk *et al.*[46] Since significant formation of CO_2 appears at temperatures higher than 453 K, we consider these side products as formed by consecutive reaction of propene oxide at temperatures below (and including) 445 K. The overall formation rate of carbon oxygenates at $t = 0$ is denoted as $r_{C3,\,0}$. The relation between the activation energy of water formation and that of epoxidation on the Au–Ti site is given by Equation 9, which is explained in more detail in the ESI,†

$$\ln\left(\frac{dr_{H_2O}}{dr_{PO}/Sel.} - 1\right) = -\frac{E_{a,H_2O}^{obs} - E_{a,PO}^{obs}}{RT} + \text{const.} \tag{7}$$

where *Sel* is the selectivity towards propene oxide.

Table 1 General performance of 1 wt% Au on Ti–SiO$_2$ in direct propene epoxidation at different temperatures and standard reactant concentrations[a]

T/K	C_3H_6 conversion (%)	Selectivity (%)					H_2 efficiency (%)
		PO	EA	PA	AC	CO_2	
373	0.56	>99	0	0	0	0	20
394	0.73	99	0.7	0.1	0	0	12
407	0.91	94	1.8	3.8	0.3	0	9.1
415	1.06	91	2.0	6.2	1.1	0	8.2
425	1.26	84	2.5	9.8	3.9	0	6.6
435	1.45	77	3.2	13	5.7	1.8	5.2
445	1.62	69	4.2	17	7.7	1.2	4.0
453	2.05	48	4.9	28	12	6.9	2.1
473	3.13	17	8.4	42	17	16	0.6

[a] PO, propene oxide; EA, acetaldehyde; PA, propionaldehyde; AC, acetone.

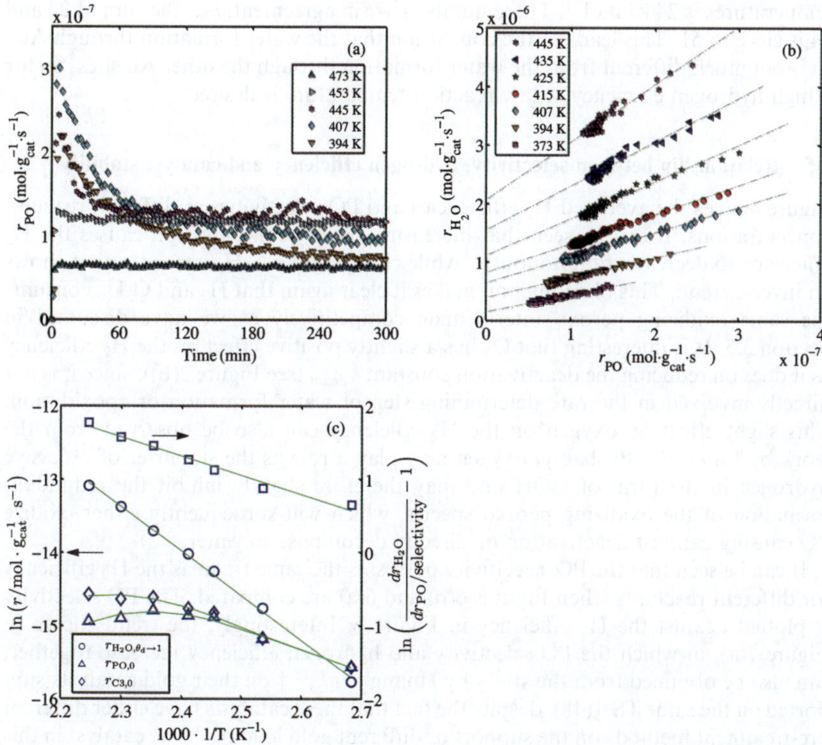

Fig. 5 Formation rate of propene oxide as a function of time (a) and water formation rate as a function of PO formation rate (b) at different temperatures and (c) Arrhenius plots of the direct water formation rate, initial PO formation rate, modified initial rate of PO, and the ratio between water formation rate and modified PO formation rate at the active Au–Ti sites (1 wt% Au on Ti–SiO$_2$, gas feed 10 vol% hydrogen, 10 vol% oxygen, 10 vol% propene in helium, GHSV 10000 mL g$_{cat}^{-1}$ h^{-1}).

The Arrhenius plots of $r_{H_2O, \theta_d \to 1}$ (the parameter d in Equation 8), $r_{PO, 0}$, and $r_{C3, 0}$ are presented in Figure 5(c). The difference in overall activation energy between water formation and epoxidation at the Au–Ti site given by Figure 5(c) is 22 ± 3 kJ mol^{-1}, while the overall activation energy of the direct water formation is 51 ± 5 kJ mol^{-1}. The activation energy estimated by $r_{PO, 0}$ at the three lowest

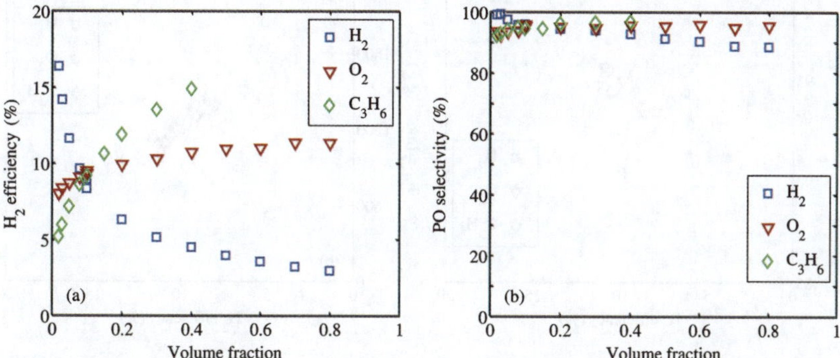

Fig. 6 Hydrogen efficiency (a) and selectivity to propene oxide (b) at different reactant concentrations (1 wt% Au on Ti–SiO$_2$, 403 K, GHSV 10000 mL g$_{cat}^{-1}$ h^{-1}, the concentration of one reactant is varied while the other two are fixed at 10 vol%).

temperatures is 24 kJ mol^{-1}. These numbers are in agreement, *i.e.*, the sum of 22 and 24 is close to 51. This leads to the conclusion that the water formation through Au–Ti is not much different from the water formation through the other Au sites. So for a high hydrogen efficiency, a low reaction temperature is desired.

3.5 Relationship between selectivity, hydrogen efficiency and catalyst stability

Figure 6 gives the averaged H$_2$ efficiencies and PO selectivities at different reactant concentrations. It can be seen that increasing the H$_2$ concentration causes the H$_2$ efficiency to decrease monotonously, while changing the C$_3$H$_6$ concentration shows an inverse trend. This phenomenon makes it clear again that H$_2$ and C$_3$H$_6$ consume the same oxidizing peroxo intermediate competitively as we have discussed in Section 3.3. It is interesting that O$_2$ has a slightly positive effect on the H$_2$ efficiency as it does on reducing the deactivation constant k_{deact} (see Figure 2(b)), since it is not directly involved in the rate determining step of water formation or epoxidation. This slight effect by oxygen on the H$_2$ efficiency can also be observed from the work by Lu *et al.*[13] Probably, oxygen may play a role as the stabilizer of excessive hydrogen in the form of OOH and may therefore slightly inhibit the redundant formation of the oxidizing peroxo species, which will subsequently either oxidize PO causing catalyst deactivation or directly decompose to water.

It can be seen that the PO selecitivity possesses the same trend as the H$_2$ efficiency for different reactants when Figures 6(b) and 6(a) are compared. The PO selectivity is plotted against the H$_2$ efficiency in Figure 6. Interestingly, the trend shown in Figure 7(a), in which the PO selectivity and hydrogen efficiency increase together, can also be obtained from the study by Huang *et al.*,[15] † on their gold catalysts supported on the same TS-1(48), despite the fact that these catalysts have either different pre-treatment methods on the support or different gold loadings. The catalyst in this study has a titanium density of 2.7×10^{-4} mol g$_{\text{cat}}^{-1}$, which is far more than the amount of Ti that functions in epoxidation.† Propene oxide can easily react with these Lewis acidic sites and form mainly its isomers. A higher formation rate of water may increase the extent of hydrolysis of Ti–O–Si bonds, which yields a stronger Brønsted acidity on the catalyst surface favouring the ring opening of propene oxide.

In Figure 7(b), the deactivation rate constant k_{deact} is plotted against the H$_2$ efficiency at different reactant concentrations. It can be clearly seen that a higher H$_2$ efficiency corresponds to a lower deactivation rate indicating that the deactivation process is closely linked to water formation, or more precisely, to the way how the (excessive) oxidizing peroxo species is used. A catalyst with a higher hydrogen efficiency should have a more stable performance and a better PO selectivity. A

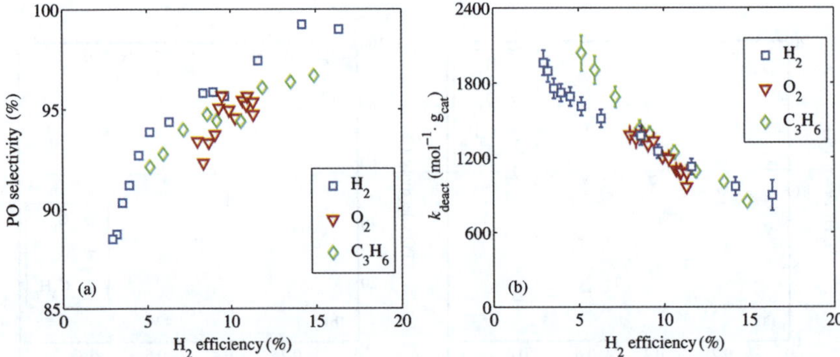

Fig. 7 Selectivity towards propene oxide (a) and deactivation rate constant (b) as a function of hydrogen efficiency at different reactant concentrations (1 wt% Au on Ti–SiO$_2$, 403 K, GHSV 10000 mL g$_{\text{cat}}^{-1}$ h^{-1}, the concentration of one reactant is varied while the other two are fixed at 10 vol%).

  This journal is © The Royal Society of Chemistry 2011

very important conclusion that can be drawn from Figure 7 is that a stable, selective and hydrogen efficient catalyst should be possible.

4. Summarizing discussion

Deactivation on the catalyst in this study is mainly caused by consecutive oxidation of propene oxide as proposed by Ruiz et al.,[29] while the water productivity reflects the formation rate of the oxidizing peroxo species. The deactivating species in propene epoxidation was found to be carbonatecarboxylate adsorbed on active Ti sites.[28] Building up of carbonates on gold is also the cause of deactivation in CO oxidation on gold catalysts.[47,49] The concurrent decrease in water formation and epoxidation observed in this sutdy indicates that deactivating species block the active Au–Ti sites.

The competing roles of propene and hydrogen in consuming the active peroxo intermediate indicate that a moderate hydrogen concentration is preferred for an acceptable propene conversion without much loss in hydrogen efficiency. Including the decomposition of the active peroxo intermediate into the rate expression based on a real mechanism[13] can well explain the saturation of PO formation at higher propene concentrations and the fractional order on propene in the power-rate-law expression, which is close to zero and normally within the range of 0.18–0.35.[12,13,32] Most likely, hydrogen speeds up this decomposition by increasing surface coverage of dissociated hydrogen. This results in a rate expression for PO formation in the following form,

$$r_{PO} = k_{HOOH}\theta_{OOH}\frac{P_{H_2}}{1+\sqrt{K_{H_2}P_{H_2}}}\frac{k_{PO}P_{C_3H_6}}{k_{PO}P_{C_3H_6}+k_{H_2O}\dfrac{\sqrt{K_{H_2}P_{H_2}}}{1+\sqrt{K_{H_2}P_{H_2}}}} \tag{8}$$

which has no essential difference from the rate expression proposed by Lu et al.,[13] but may explain the generally observed lower order (0.55–0.60) on hydrogen in propene epoxidation than that in hydrogen oxidation (0.7–0.8) on gold–titania catalysts.[2,12,13,32,38]

Due to the parallel consumption of the active peroxo species by hydrogen and propene, a higher propene concentration will increase the utilization efficiency of this active intermediate towards propene oxide and suppresses the unwanted water formation, which accordingly enhances the catalyst stability. Higher oxygen concentrations favour the regeneration of the deactivated sites and might stabilize the excessive hydrogen alleviating the deactivation process.

Further improvement in hydrogen efficiency will place a premium on the theoretical investigation into the pathway of water formation on gold. Barton and Podkolzin[38] proposed the HOOH pathway through which the O–O bond cleaves and two hydroxyls form. The study by Ford et al.[42] reinforced this perspective and suggested another energetically competitive pathway on Au(111) facets, in which OOHH is formed and O–O bond scission occurs leaving an oxygen atom on gold. Similar hydrogen-induced OOH dissociation on gold surface is also proposed in hydrogen-promoted CO oxidation[41] and in direct H_2O_2 synthesis.[50] A recent DFT study by Li et al.[51] on propene epoxidation in oxygen and water on gold clusters suggested a pathway in which the scission of the oxygen bond in OOH on gold surface is preferred and the oxygen atom left epooxidizes propene. Considering our low activation energy of propene epoxidation and the proposal by Joshi et al.[52] that there is an extra energy barrier for HOOH attacking Ti–OH, the reaction route on our catalyst might not be the sequential mechanism involving H_2O_2 transfer on Au/titanosilicate catalysts.[9] In general, a lower reaction temperature is preferred in our system for a higher hydrogen efficiency, but this makes catalyst regeneration more difficult. An efficient activation of hydrogen on gold nanoparitles/clusters[53–56] as well as the synergy between Au and Ti is the key issue for a desirable performance.

5. Conclusions

A kinetic study of propene epoxidation with hydrogen and oxygen over the Au/Ti–SiO$_2$ catalyst has been performed over a wide range of reactant concentrations including the explosive region by utilizing a micro reactor system. Analysis of the dynamic deactivation process at different reactant concentrations showed that the formation rate of propene oxide is most dependent on the hydrogen concentration and that the formation of an active peroxo species on the gold nanoparticles is the rate determining step. Deactivation is mainly caused by the consecutive oxidation of propene oxide (or a precursor thereof). Higher hydrogen concentrations speed up the deactivation by increasingly forming the oxidizing peroxo species. When the propene concentration is higher, the concentration of this oxidizing species is lower by epoxidizing propene to form the desired propene oxide and therefore the deactivation is mitigated. Oxygen favours the regeneration of the deactivated sites.

Water formation and epoxidation are strongly correlated. It can be concluded from our results that there are two routes for water formation, *i.e.* the water formation on the Au–Ti center through the active peroxo intermediate which is also responsible for epoxidation, and the direct water formation not related to epoxidation. Water formation and propene epoxidation on the active Au–Ti sites are two parallel reactions competitively consuming the same active peroxo intermediate. When the hydrogen concentration is higher, the hydrogenation of the peroxo intermediate to form water is more dominant instead of its consumption by epoxidation. Higher propene concentrations are preferred for the efficiency of utilizing this peroxo intermediate to form propene oxide. Oxygen has no influence on the direct water formation and does not affect the ratio of the water productivity and epoxidation rate on Au–Ti centers indicating that OOH is the true intermediate and that its reaction with hydrogen forming the active peroxo species is the rate determining step. Catalyst deactivation is caused by the blockage of the active Au–Ti center.

Saturation of propene oxide formation is observed, which can be attributed to the nature of the parallel water formation and propene epoxidation consuming a common intermediate. The activation energy of propene epoxidation is 22 kJ mol^{-1} lower than the water formation on the Au–Ti center suggesting a low reaction temperature for propene epoxidation is favoured. A moderate hydrogen concentration combined with high propene and oxygen concentrations is preferred for a desirable performance of the catalyst, *i.e.* a higher hydrogen efficiency and consequently better stability and selectivity.

Acknowledgements

The Netherlands Organization for Scientific Research (NWO) is kindly acknowledged for providing an ECHO grant (700.57.044).

References

1 T. Hayashi, K. Tanaka and M. Haruta, *J. Catal.*, 1998, **178**, 566–575.
2 E. E. Stangland, K. B. Stavens, R. P. Andres and W. N. Delgass, *J. Catal.*, 2000, **191**, 332–347.
3 G. Mul, A. Zwijnenburg, B. Van der Linden, M. Makkee and J. A. Moulijn, *J. Catal.*, 2001, **201**, 128–137.
4 A. Zwijnenburg, A. Goossens, W. G. Sloof, M. W. J. Crajé, A. M. Van der Kraan, L. J. De Jongh, M. Makkee and J. A. Moulijn, *J. Phys. Chem. B*, 2002, **106**, 9853–9862.
5 B. Chowdhury, J. J. Bravo-Suárez, M. Dat, S. Tsubota and M. Haruta, *Angew. Chem., Int. Ed.*, 2006, **45**, 412–415.
6 L. Cumaranatunge and W. N. Delgass, *J. Catal.*, 2005, **232**, 38–42.
7 T. A. Nijhuis, T. Visser and B. M. Weckhuysen, *Angew. Chem., Int. Ed.*, 2005, **44**, 1115–1118.

8   A. K. Sinha, S. Seelan, S. Tsubota and M. Haruta, *Angew. Chem., Int. Ed.*, 2004, **43**, 1546–1548.

9   J. J. Bravo-Suárez, K. K. Bando, J. Lu, M. Haruta, T. Fujitani and S. T. Oyama, *J. Phys. Chem. C*, 2008, **112**, 1115–1123.

10   B. Taylor, J. Lauterbach and W. N. Delgass, *Appl. Catal., A*, 2005, **291**, 188–198.

11   T. A. Nijhuis and B. M. Weckhuysen, *Catal. Today*, 2006, **117**, 84–89.

12   B. Taylor, J. Lauterbach, G. E. Blau and W. N. Delgass, *J. Catal.*, 2006, **242**, 142–152.

13   J. Lu, X. Zhang, J. J. Bravo-Suárez, S. Tsubota, J. Gaudet and S. T. Oyama, *Catal. Today*, 2007, **123**, 189–197.

14   J. Lu, X. Zhang, J. J. Bravo-Suárez, K. K. Bando, T. Fujitani and S. T. Oyama, *J. Catal.*, 2007, **250**, 350–359.

15   J. Huang, T. Takei, T. Akita, H. Ohashi and M. Haruta, *Appl. Catal., B*, 2010, **95**, 430–438.

16   T. A. Nijhuis, B. J. Huizinga, M. Makkee and J. A. Moulijn, *Ind. Eng. Chem. Res.*, 1999, **38**, 884–891.

17   B. Taylor, J. Lauterbach and W. N. Delgass, *Catal. Today*, 2007, **123**, 50–58.

18   T. Liu, P. Hacarlioglu, S. T. Oyama, M. Luo, X. Pan and J. Lu, *J. Catal.*, 2009, **267**, 202–206.

19   H. Yang, D. Tang, X. Lu and Y. Yuan, *J. Phys. Chem. C*, 2009, **113**, 8186–8193.

20   Y. Liu, H. Yu, X. Zhang and J. Suo, *Acta. Phys. Chim. Sin.*, 2010, **26**, 1585–1592.

21   E. E. Stangland, B. Taylor, R. P. Andres and W. N. Delgass, *J. Phys. Chem. B*, 2005, **109**, 2321–2330.

22   E. Sacaliuc-Parvulescu, H. Friedrich, R. Palkovits, B. M. Weckhuysen and T. A. Nijhuis, *J. Catal.*, 2008, **259**, 43–53.

23   J. Lu, X. Zhang, J. J. Bravo-Suárez, T. Fujitani and S. T. Oyama, *Catal. Today*, 2009, **147**, 186–195.

24   B. S. Uphade, T. Akita, T. Nakamura and M. Haruta, *J. Catal.*, 2002, **209**, 331–340.

25   M. Ojeda and E. Iglesia, *Chem. Commun.*, 2009, 352–354.

26   J. Huang, T. Akita, J. Faye, T. Fujitani, T. Takei and M. Haruta, *Angew. Chem., Int. Ed.*, 2009, **48**, 7862–7866.

27   S. Lee, L. M. Molina, M. J. López, J. A. Alonso, B. Hammer, B. Lee, S. Seifert, R. E. Winans, J. W. Elam, M. J. Pellin and S. Vajda, *Angew. Chem., Int. Ed.*, 2009, **48**, 1467–1471.

28   T. A. Nijhuis, T. Visser and B. M. Weckhuysen, *J. Phys. Chem. B*, 2005, **109**, 19309–19319.

29   A. Ruiz, B. van der Linden, M. Makkee and G. Mul, *J. Catal.*, 2009, **266**, 286–290.

30   N. Yap, R. P. Andres and W. N. Delgass, *J. Catal.*, 2004, **226**, 156–170.

31   S. C. Parker and C. T. Campbell, *Top. Catal.*, 2007, **44**, 3–13.

32   S. T. Oyama, X. Zhang, J. Lu, Y. Gu and T. Fujitani, *J. Catal.*, 2008, **257**, 1–4.

33   C. Qi, M. Okumura, T. Akita and M. Haruta, *Appl. Catal., A*, 2004, **263**, 19–26.

34   A. K. Sinha, S. Seelan, M. Okumura, T. Akita, S. Tsubota and M. Haruta, *J. Phys. Chem. B*, 2005, **109**, 3956–3965.

35   T. A. Nijhuis, E. Sacaliuc, A. M. Beale, A. M. J. van der Eerden, J. C. Schouten and B. M. Weckhuysen, *J. Catal.*, 2008, **258**, 256–264.

36   T. A. Nijhuis, J. Chen, S. M. A. Kriescher and J. C. Schouten, *Ind. Eng. Chem. Res.*, 2010, **49**, 10479–10485.

37   T. A. Nijhuis, T. Q. Gardner and B. M. Weckhuysen, *J. Catal.*, 2005, **236**, 153–163.

38   D. G. Barton and S. G. Podkolzin, *J. Phys. Chem. B*, 2005, **109**, 2262–2274.

39   T. A. Nijhuis, M. Makkee, J. A. Moulijn and B. M. Weckhuysen, *Ind. Eng. Chem. Res.*, 2006, **45**, 3447–3459.

40   J. K. Edwards, A. F. Carley, A. A. Herzing, C. J. Kiely and G. J. Hutchings, *Faraday Discuss.*, 2008, **138**, 225–239.

41   E. Quinet, L. Piccolo, F. Morfin, P. Avenier, F. Diehl, V. Caps and J.-L. Rousset, *J. Catal.*, 2009, **268**, 384–389.

42   D. C. Ford, A. U. Nilekar, Y. Xu and M. Mavrikakis, *Surf. Sci.*, 2010, **604**, 1565–1575.

43   D. W. Goodman, *Catal. Lett.*, 2005, **99**, 1–4.

44   M. Boronat and A. Corma, *Dalton Trans.*, 2010, **39**, 8538–8546.

45   F. Sun and S. Zhong, *J. Nat. Gas Chem.*, 2006, **15**, 45–51.

46   S. Namuangruk, P. Khongpracha, P. Pantu and J. Limtrakul, *J. Phys. Chem. B*, 2006, **110**, 25950–25957.

47   M. M. Schubert, A. Venugopal, M. J. Kahlich, V. Plzak and R. J. Behm, *J. Catal.*, 2004, **222**, 32–40.

48   Y. Denkwitz, B. Schumacher, G. Kučerová and R. J. Behm, *J. Catal.*, 2009, **267**, 78–88.

49   F. Gao, T. E. Wood and D. W. Goodman, *Catal. Lett.*, 2010, **134**, 9–12.

50   R. Todorovic and R. J. Meyer, *Catal. Today*, 2011, **160**, 242–248.

51   C. Chang, Y. Wang and J. Li, *Nano Res.*, 2011, **4**, 131–142.

52   A. M. Joshi, W. N. Delgass and K. T. Thomson, *J. Phys. Chem. C*, 2007, **111**, 7841–7844.

53 A. G. Sault, R. J. Madix and C. T. Campbell, *Surf. Sci.*, 1986, **169**, 347–356.
54 T. V. Choudhary and D. W. Goodman, *Appl. Catal., A*, 2005, **291**, 32–36.
55 A. Corma, M. Boronat, S. González and F. Illas, *Chem. Commun.*, 2007, 3371–3373.
56 M. Boronat, F. Illas and A. Corma, *J. Phys. Chem. A*, 2009, **113**, 3750–3757.

   This journal is © The Royal Society of Chemistry 2011

Methane activation and partial oxidation on free gold and palladium clusters: Mechanistic insights into cooperative and highly selective cluster catalysis

Sandra M. Lang and Thorsten M. Bernhardt*

Received 24th February 2011, Accepted 10th March 2011

DOI: 10.1039/c1fd00025j

The catalytic activation, dehydrogenation, and direct oxidative conversion of methane into more valuable products such as larger hydrocarbons, alcohols, aldehydes *etc.* are of considerable industrial interest. To investigate the energetics and kinetics of elementary bond-breaking and bond-formation processes, free metal clusters can serve as versatile catalytic model systems. In this context, temperature dependent reactions of small cationic gold clusters Au_x^+ with methane as well as a mixture of methane and molecular oxygen have been performed in an octopole ion trap under multi-collision conditions and compared with the corresponding reactions on palladium clusters Pd_x^+ ($x = 2$–4). Binding energies of methane to all investigated cluster cations are determined from kinetic measurements *via* statistical analysis. Furthermore, among the gold clusters, the dimer Au_2^+ is found to be able to dehydrogenate methane and to convert it into ethylene in a highly selective catalytic reaction. In contrast, all investigated palladium clusters activate methane under non-selective formation of a variety of dehydrogenated products. Most interestingly, methane dehydrogenation is observed for Pd_x^+ and Au_2^+ only, if a cluster specific 'critical number' of methane molecules is pre-adsorbed. This emphasizes the importance of cooperative coadsorption effects in the dehydrogenation process on these clusters. Finally, the reaction between Au_2^+ and both O_2 and CH_4 yields a low temperature product of the stoichiometry $Au_2(C_3H_8O_2)^+$ that clearly contains activated O_2 and dehydrogenated methane indicating a possible C–O bond formation process. The palladium dimer Pd_2^+ on the other hand exhibits only the mere coadsorption of molecular oxygen and non-dehydrogenated methane.

1. Introduction

The utilization of methane as energy source as well as feedstock for the chemical production of more valuable products such as larger hydrocarbons, alcohols, and aldehydes is of growing industrial and economical interest due to its large abundance in fossil and biogenic resources. Industrially, the conversion of methane mainly proceeds *via* the generation of syngas, a mixture of carbon monoxide and hydrogen that is then further processed in the Fischer–Tropsch synthesis.[1] However, with regard to a 'green', *i.e.* sustainable and economically sensitive processing the direct and selective conversion of CH_4 is highly desirable. In particular, extensive research has been devoted to the partial selective

Institute of Surface Chemistry and Catalysis, University of Ulm, Albert-Einstein-Allee 47, 89069 Ulm, Germany. E-mail: thorsten.bernhardt@uni-ulm.de; Fax: +49-731-50-25452; Tel: +49-731-5025455

oxidation of methane under a wide range of pressure and temperature conditions, though only poor selectivity and product yield were achieved.[1,2] The activation of the stable C–H bond of methane represents a general long-standing problem and usually is the rate determining reaction step in a catalytic process which requires highly active catalysts and/or high temperatures. This demand then often results in little product selectivity or complete combustion of the products.[2,3]

In this context, we aim to gain fundamental mechanistic insights into the C–H bond activation as well as C–C and C–O bond formation mechanisms by investigating small gas-phase gold and palladium clusters as catalytic model systems. The particular importance of small gold clusters in real heterogeneous catalysis has recently been demonstrated by the discovery that those oxide supported gold particles containing only about ten atoms are the actual catalytically active species in the CO oxidation reaction.[4] This observation complements earlier model studies on mass selected and deposited gold clusters Au_x ($x \leq 20$) which found the CO oxidation to be initiated by gold particles containing eight or more atoms.[5]

Although gas-phase studies might never account for the complex mechanisms proceeding in real catalysis under industrially realized reaction conditions,[6] they could be shown to provide a powerful means to elucidate elementary mechanistic details, such as cooperative processes and activation barriers,[7-10] that might guide the future conceptional development of new catalytically active and selective materials. Furthermore, gas-phase studies cannot only be performed under well defined reaction conditions (partial pressure, temperature, reaction time) but also allow for the exact characterization of the clusters regarding their size, charge state, composition, and number of active adsorption sites which represent important parameters for tuning the reactivity and catalytic performance.

Gas-phase studies concerning the activation and oxidation of methane on gold and palladium clusters are scarce. Both cationic atoms, Au^+ an Pd^+, as well as small cationic gold clusters Au_x^+ ($x = 2$–4) were shown to be unreactive toward CH_4 under single collision conditions.[11] Yet, under multi-collision conditions several stable gold-methane complexes were observed in a fast flow reactor and in an octopole ion trap.[9,12,13] The activation and dehydrogenation of CH_4 was, however, only observed to proceed on Au_2^+.[9,10] Neutral Pd_x ($x = 1$–24) were studied and found to react with CD_4 with the exception of Pd, Pd_3, and Pd_4 while, however, no details on the reaction products and the possibility of methane activation were given.[14] Recently, the room temperature chemical kinetics for the catalytic oxidation of methane with ozone on Pd^+ has been investigated in a joined experimental and theoretical study demonstrating the simultaneous formation of methanol (main reaction channel) and formaldehyde (side reaction) in an overall exothermic reaction.[15] Oxidation reactions on gas-phase palladium clusters were not performed so far.

In the present contribution, we report on detailed temperature dependent reactivity and reaction kinetic studies comparing the activation and dehydrogenation of methane on small cationic gold and palladium clusters with two to four atoms as well as on the partial oxidation of CH_4 (CD_4) facilitated by dimers Au_2^+ and Pd_2^+ in an octopole ion trap. All investigated palladium clusters as well as the gold dimer are found to dehydrogenate methane under the cooperative action of multiple methane molecules. Kinetic studies of the dimer reactions demonstrate the highly selective formation of ethylene on Au_2^+ in a full catalytic cycle while several hydrocarbons are simultaneously formed on Pd_2^+. The addition of molecular oxygen does not give evidence for the catalytic formation of alcohols or aldehydes on Pd_2^+. In contrast, a low temperature product is observed to be formed on the gold dimer indicating the simultaneous activation of CH_4 and O_2 and the subsequent C–O bond formation.

2. Methods

A. Experimental setup

The experimental setup to study methane activation mechanisms and catalytic reactions mediated by small metal cluster cations consists of a variable temperature radio frequency (rf) octopole ion trap inserted into a tandem quadrupole mass spectrometer. Details of the experimental layout are described in detail elsewhere.[16]

The metal cluster cations are produced by a CORDIS (cold reflex discharge ion source)[17] sputtering source. Clusters are mass-selected in a first quadrupole filter. The cluster ion beam containing only clusters of one specific mass then enters the octopole ion trap which is prefilled with about 1 Pa partial pressure of helium buffer gas and a small, well defined fraction of reactants (CH_4 and CD_4, respectively, or a mixture of CH_4 (CD_4) and O_2). The ion trap enclosure is attached to a helium cryostat that allows for temperature adjustment in the range between 20 K and 300 K. At the applied pressures, thermal equilibration of the clusters with the buffer gas is achieved within a few milliseconds,[16] whereas the cluster ions are stored in the ion trap typically for several seconds. The absolute pressure inside of the ion trap is measured by a Baratron gauge (MKS, Typ 627B) attached to the ion trap *via* a 1 mm inner diameter teflon tube. For exact pressure determination inside of the ion trap, the effect of thermal transpiration has to be taken into account arising from the temperature difference between the Baratron capacitance manometer (stabilized at 318 K) and the temperature variable ion trap.[18,19] The correction has been applied for quantitative data evaluation to determine reaction rate constants and binding energies while the pressures given in the figure captions below represent uncorrected values.

After a chosen reaction time (storage time in the ion trap) t_R, all ionic reactants, intermediates, and products are extracted from the ion trap, and the ion distribution is analyzed *via* a second quadrupole mass filter. By recording all ion intensities as a function of the reaction time (reaction kinetics), the rates of the reaction at a well defined reaction temperature can be studied.

Please note, due to the broad natural isotope distribution of palladium, reactivity measurements of Pd_x^+ were performed with deuterated methane CD_4. In contrast, Au_x^+ clusters were studied in the presence of CH_4 and labeling experiments with CD_4 were only performed to confirm the peak assignments and to exclude interferences from background gases.

B. Data analysis

The data evaluation procedure applied to analyze the temperature dependent kinetic traces and the corresponding error analysis has been described in great detail elsewhere[19] and will only be briefly summarized here. The normalized kinetic traces are evaluated by fitting the integrated rate equations of proposed potential reaction mechanisms to the experimental data by using the *'Detmech'* software[20] yielding pseudo-first order rate constants k.

Since the experiments were performed in the kinetic low pressure regime, the details of simple association reactions between a metal cluster cation M_x^+ and a neutral molecule L ($= CH_4$, O_2)

$$M_x^+ + L \xrightarrow{k} M_xL^+ \tag{1}$$

can be described by the Lindemann energy transfer model for association reactions[21]

$$M_x^+ + L \underset{k_d}{\overset{k_a}{\rightleftharpoons}} M_xL^+ \tag{2a}$$

$$(M_xL^+)* + He \xrightarrow{k_s} M_xL^+ + He^* \tag{2b}$$

According to this model, the charged metal cluster M_x^+ reacts with L forming the energized intermediate $(M_x^+)^*$ with a rate constant k_a. This intermediate either decomposes unimolecularly back to the reactants (rate constant k_d) or is stabilized by an energy-transfer collision with helium buffer gas (rate constant k_s). Consequently, the overall reaction (1) depends on the helium buffer gas pressure and becomes of third order. In the low pressure limit, the corresponding termolecular rate constant is

$$k^{(3)} = \frac{k_a k_s}{k_d}. \tag{3}$$

The ion-molecule association rate constant k_a and the stabilization rate constant k_s are approximated by ion-molecule collision rate constants as specified by Langevin theory.[21,22] Since CH_4, O_2, and He, are nonpolar molecules, the rate constants k_a and k_s are temperature independent.[21] Therefore, any observed temperature dependence of reaction (1) must be contained in the unimolecular decomposition rate constant k_d.

Combining eqn (2)–(3) thus allows for the determination of an experimental unimolecular decomposition rate constant k_d which is also well described by RRKM (Rice-Ramsperger-Kassel-Marcus) theory.[23,24] RRKM theory is usually applied to determine k_d from the known binding energy of the ligand to the cluster ion under assumption of the dissociation to proceed *via* a transition state (TS) $(M_xL^+)^\ddagger$. In our experimental approach k_d is the quantity derived from the experimental data, which consequently allows for the determination of the binding energy. In order to obtain RRKM binding energies E_0, the experimental decomposition rate constants k_d have been simulated using the software package '*MassKinetics*' developed by Drahos *et al.*[25] In the simulations assumptions concerning the TS structure have to be made. If the TS structure is presumed to resemble the energized complex (reactant) structure, a "tight" TS model with vibrational frequencies similar to the energized complex is applied. In contrast, if the TS structure is rather close to the products, the TS frequencies are derived from the energized complex by scaling with a factor f ("loose" transition state). The latter TS model is more realistic for the simple bond dissociation reactions considered here.

3. Results

A. Methane activation and C–C coupling

1. Room temperature reactivity and binding energies. Fig. 1 displays ion mass distributions of Au_x^+ ($x = 2$–4, traces a–c) and Pd_x^+ ($x = 2$–4, traces d–f) clusters exposed to methane for a reaction time of 0.1 s at room temperature. Under such experimental conditions the gold dimer Au_2^+ adsorbs a first methane molecule yielding $Au_2(CH_4)^+$.[12] Upon reaction with a second CH_4, a product is formed that can be assigned to the stoichiometry $Au_2(C_2H_4)^+$ clearly indicating the dehydrogenation of CH_4 under elimination of two H_2.[7] In contrast, Au_3^+ sequentially adsorbs two methane molecules while Au_4^+ is not able to form any stable products at such short reaction time (methane adsorption on Au_4^+ is only observed at longer reaction times).[12] Similarly, both Pd_2^+ and Pd_3^+ are able to sequentially adsorb two CD_4 molecules and Pd_4^+ only adsorbs one CD_4. From the recorded mass spectra, there is no direct evidence for the dehydrogenation of methane molecules on Au_x^+ ($x > 2$) and all investigated Pd_x^+. This is already a first indication for the outstanding reaction behavior of the gold dimer toward methane. Except for Au_4^+, no additional products are observed at longer reaction times.

The corresponding kinetic data (as well as kinetic data at lower reaction temperatures) are analyzed by employing the Lindemann energy transfer model for association reactions in conjunction with statistical RRKM theory to obtain binding

  This journal is © The Royal Society of Chemistry 2011

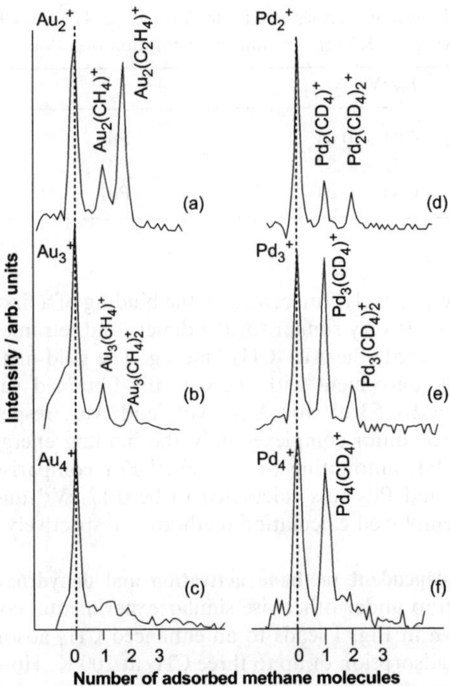

Fig. 1 Cluster size dependent ion mass distributions of (a–c) Au_x^+ ($x = 2$–4)[9,12,13] and (d–f) Pd_x^+ ($x = 2$–4) obtained after the clusters were trapped at 300 K for $t_R = 0.1$ s inside the octopole ion trap filled with helium buffer gas and CH_4 (Au_x^+) and CD_4 (Pd_x^+), respectively.

energies E_0 for a first adsorbed CH_4 (Au_x^+) and CD_4 (Pd_x^+), respectively (for details see Section 2.A. and References 13, 19). For the determination of E_0 utilizing the software package '*MassKinetics*'[25] the vibrational frequencies of the energized complex $(M_xCH_4^+)^*$ $((M_xCD_4^+)^*)$ and the transition state $(M_xCH_4^+)^‡$ $((M_xCD_4^+)^‡)$ must be specified. The gold cluster as well as Pd_2^+ and Pd_3^+ metal–metal vibrations are adapted from Reference 26, while the Pd_4^+ vibrations are assumed to be distributed according to the Debye model of phonon frequency dispersion as described by Jarrold *et al.*[27] The nine CH_4 and CD_4 vibrations, respectively, are taken from Shimanouchi,[28] the M_x^+–CH_4 (M_x^+–CD_4) stretching frequency along the reaction coordinate[13,29] is estimated to be 434 cm⁻¹, and the unknown bending vibrations[13,19] are chosen to be 50 cm⁻¹. Furthermore, a 'loose' TS model is employed that was found to be suitable to model simple cluster–molecule bond cleavage reactions.[13,19] In this model, the transition state is described by the same vibrational modes as the energized molecule minus the M_x^+–CH_4 (M_x^+–CD_4) stretching vibration that is treated as internal translation along the reaction coordinate. Furthermore, the low frequency bending vibrations (50 cm⁻¹) are additionally scaled by a factor $f = 0.5$.[13,19,27] Adiabatic rotations are taken into account by considering a 'rotational barrier'[13,19,24,25] E_{RB}, amounting to 0.09 eV for Au_2^+ and Au_3^+ as well as 0.08 eV for Au_4^+ and Pd_x^+ ($x = 2$–4).[13]

The resulting cluster size dependent RRKM binding energies are summarized in Table 1. For both small gold and palladium clusters the binding energies decrease with increasing cluster size with the largest value of 0.91 ± 0.04 eV for the gold dimer and of 0.94 ± 0.20 eV for the palladium dimer. The binding energies of all the here investigated small clusters clearly exceed the experimental values for the binding of CH_4 to extended Au(111) as well as Pd(111) surfaces which were reported in literature to amount to 0.15 eV[30] (Au) and 0.17 eV[31,32] (Pd, theoretical values:

Table 1 Experimental binding energies of CH_4 to Au_x^+ ($x = 2$–4)[12] and CD_4 to Pd_n^+ ($n = 2$–4) as determined by employing RRKM theory under assumption of a 'loose' TS

	E_0/eV		E_0/eV
Au_2^+	0.91 ± 0.04	Pd_2^+	0.94 ± 0.20
Au_3^+	0.72 ± 0.07	Pd_3^+	0.86 ± 0.12
Au_4^+	0.64 ± 0.04	Pd_4^+	0.69 ± 0.11

0.02–0.15 eV[32,33]), respectively. Interestingly, the binding of a first methane molecule to gold and palladium is very similar for the dimers and tetramers, merely the Pd_3^+–CD_4 value slightly exceeds the Au_3^+–CH_4 binding. The gold-methane binding energies are in favorable agreement with theoretically obtained values amounting to 0.88 eV, 0.83 eV, and 0.53 eV for Au_2^+, Au_3^+, and Au_4^+, respectively,[13] while for cationic methane–palladium complexes only the binding energy to the atom has been calculated so far, amounting to 0.74 eV.[34] For comparison, the binding of CH_4 to neutral Pd and Pd_2 was calculated to be 0.17 eV[35] and 0.17 eV–1.13 eV (depending on the employed calculation method),[36] respectively.

2. Cluster size dependent methane activation and dehydrogenation. Successive cooling of the ion trap under otherwise similar experimental conditions as for the measurements shown in Fig. 1 leads to an enhanced CH_4 adsorption on Au_3^+ and Au_4^+ resulting in an adsorption of up to three CH_4 at 200 K. However, also at lower reaction temperatures the detected ion mass distributions do not indicate the dehydrogenation of methane on these clusters.

In contrast, cooling of the ion trap uncovers completely new features in the product mass distributions of Au_2^+ and of all palladium clusters Pd_x^+. Fig. 2 illustrates the temperature dependent reactivity of Au_2^+ (panel a), Pd_2^+ (panel b), Pd_3^+ (panel c), and Pd_4^+ (panel d) toward CD_4 in 2D contour representations of the mass spectra obtained at $t_R = 0.1$ s and a series of different reaction temperatures. For the gold dimer, in addition to the room temperature complexes $Au_2(CD_4)^+$ and $Au_2(C_2D_4)^+$, a new mass peak is detected at lower temperatures that can be assigned to $Au_2(CD_4)_2^+$ containing two not dehydrogenated methane-d_4 molecules. The intensity of this new product increases on the cost of $Au_2(C_2D_4)^+$ with decreasing

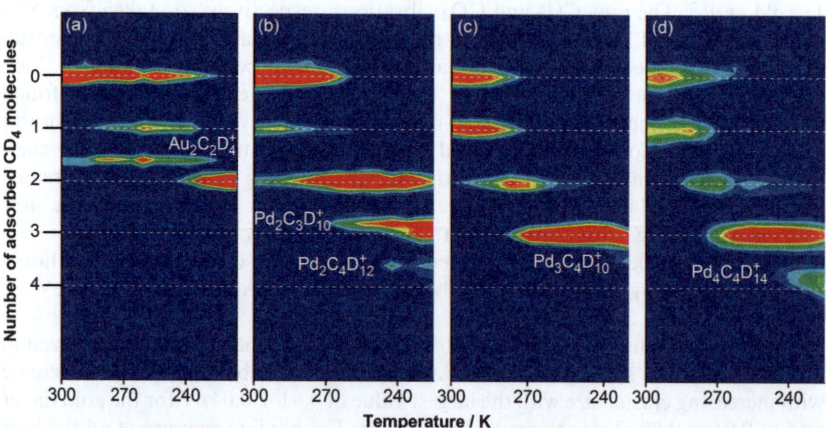

Fig. 2 2D contour plots of the temperature dependent product ion mass distributions after trapping (a) Au_2^+, (b) Pd_2^+, (c) Pd_3^+, and (d) Pd_4^+ for 0.1 s in the presence of methane-d_4. The relative ion intensity is color coded ranging from blue (0%) to red (100%).

temperature. Furthermore, a low intensity product, $Au_2(C_2D_4)(CD_4)^+$, is detected at intermediate temperatures,[7] which is, however, not apparent under the experimental counditions of Fig. 2a, but will be discussed below.

Depending on the reaction temperature the palladium clusters are able to adsorb several methane-d_4 molecules yielding the product complexes $Pd_2(CD_4)^+$, $Pd_2(CD_4)_2^+$, $Pd_2(CD_4)_3^+$, $Pd_2(CD_4)_4^+$, $Pd_3(CD_4)^+$, $Pd_3(CD_4)_2^+$, $Pd_3(CD_4)_3^+$, $Pd_4(CD_4)^+$, $Pd_4(CD_4)_2^+$, $Pd_4(CD_4)_3^+$, and $Pd_4(CD_4)_4^+$. As can be seen from Fig. 2b–d, besides these products containing not dehydrogenated methane-d_4, additional new products corresponding to $Pd_2C_3D_{10}^+$, $Pd_2C_4D_{12}^+$, $Pd_3C_4D_{10}^+$, and $Pd_4C_4D_{14}^+$ appear. The stoichiometries of these complexes provide direct evidence that all investigated palladium clusters are able to activate CD_4 under liberation of hydrogen. The dehydrogenation process on Pd_3^+ and Pd_4^+ is even more pronounced at longer reaction times. Studies performed at about 225 K and $t_R = 1.0$ s (not shown here) reveal additional dehydrogenated products $Pd_3C_4D_{14}^+$, $Pd_3C_5D_{12}^+$, and $Pd_3C_5D_{16}^+$, $Pd_4C_5D_{16}^+$ and $Pd_4C_5D_{18}^+$ apart from $Pd_3(CD_4)_4^+$, $Pd_3(CD_4)_5^+$, and $Pd_4(CD_4)_5^+$. In contrast, for the palladium dimer only the relative product intensities change with reaction time but no extra products appear.

3. Catalytic C–C bond formation on dimers. The large number of Pd_3^+ and Pd_4^+ product complexes that contain dehydrogenated methane molecules and only appear at low temperatures and long reaction times renders the exact determination of a reaction mechanism *via* kinetic measurements nearly impossible.

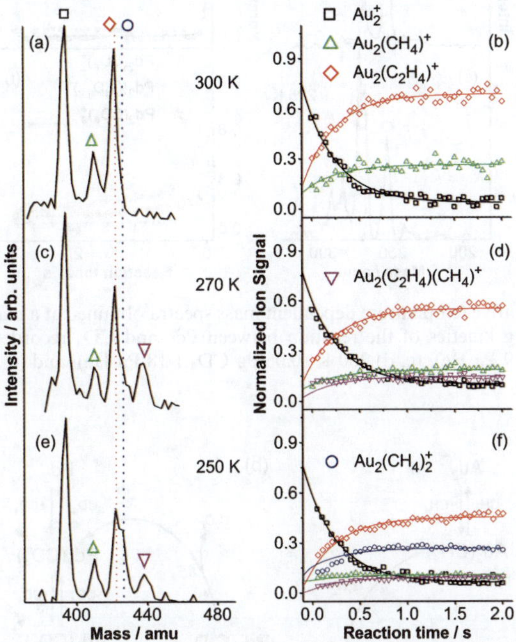

Fig. 3 Representative temperature dependent mass spectra obtained at a reaction time of 0.1 s and corresponding kinetics of the reaction between Au_2^+ and CH_4 recorded at (a, b) 300 K (0.08 Pa CH_4; 0.96 Pa He), (c, d) 270 K (0.04 Pa CH_4; 1.0 Pa He), and (e, f) 250 K (0.04 Pa CH_4; 1.02 Pa He).[9] The open symbols in the kinetics represent experimental data, normalized to the total ion concentration in the ion trap. The solid lines are obtained by fitting the integrated rate equations of the proposed reaction mechanism (*cf.* Fig. 5a) to the experimental data. Please note that a fragment signal, $Au(C_2H_4)(CH_4)^+$, which is formed to a minor extent above 250 K, is omitted here for the sake of clarity.[9] The effect of the fragment on the kinetics is a slow overall decrease of all ion signal intensities with increasing reaction time.

In contrast, Au_2^+ and Pd_2^+ exhibit only one and two dehydrogenated complexes, respectively, that appear over a larger temperature range. To deduce a reaction mechanism in these cases it is necessary to additionally consider the temperature dependent kinetic data. Such kinetic traces together with representative mass spectra at three selected temperatures are displayed in Fig. 3 and 4. The reaction mechanisms that best fit theses kinetics for Au_2^+ and Pd_2^+ at all studied temperatures are shown in Fig. 5. The fits of the integrated rate equations of these mechanisms to the experimental data are represented by the solid lines in the kinetic data of Fig. 3 and 4.

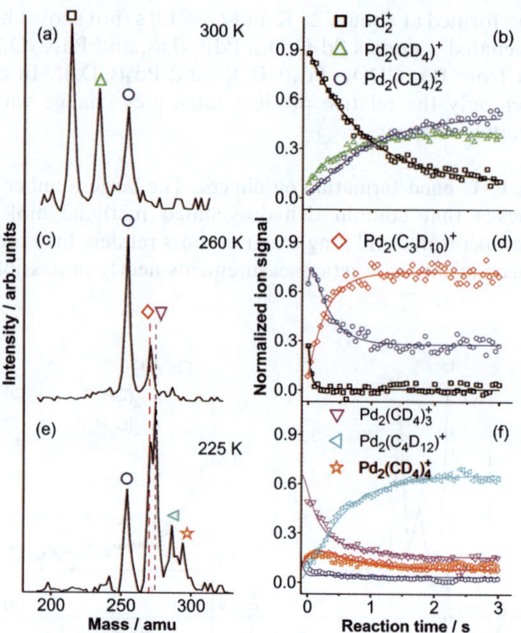

Fig. 4 Representative temperature dependent mass spectra obtained at a reaction time of 0.1 s and corresponding kinetics of the reaction between Pd_2^+ and CD_4 recorded at (a, b) 300 K (0.05 Pa CD_4; 0.89 Pa He), (c, d) 260 K (0.06 Pa CD_4; 1.18 Pa He), and (e, f) 225 K (0.05 Pa CD_4; 1.03 Pa He).

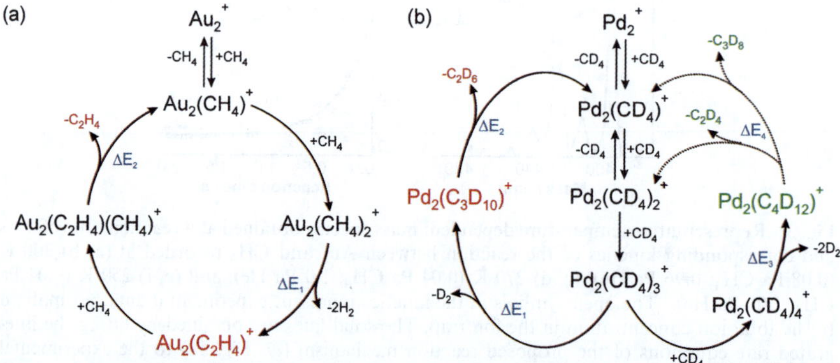

Fig. 5 Proposed reaction mechanisms for the reaction of (a) Au_2^{+9} and (b) Pd_2^+ with methane yielding the best fit of the experimental kinetic data shown in Fig. 3 and 4.

   This journal is © The Royal Society of Chemistry 2011

In the case of Au_2^+, the first reaction step is represented by the reversible adsorption of one methane molecule yielding $Au_2(CH_4)^+$. At 300 K a further product, $Au_2(C_2H_4)^+$, is observed that must result from the adsorption of a second CH_4. The binding of this second methane molecule then facilitates the subsequent formation of ethylene under elimination of 2 H_2 while the not dehydrogenated product $Au_2(CH_4)_2^+$ is, however, not detected at this temperature (*cf.* Fig. 3a and b). Especially striking is the equilibrium concentration between the bare clusters Au_2^+ and the products $Au_2(CH_4)^+$ and $Au_2(C_2H_4)^+$ that are formed at room temperature, as can be seen from Fig. 3b. Such kinetic traces can only be modeled by either an equilibrium reaction mechanism in which all products are connected with each other by reversible reactions or by a catalytic cycle.[21] However, due to the appearance of $Au_2(C_2H_4)^+$ that is obviously formed through activation and dehydrogenation of methane, an equilibrium reaction is physically not reasonable. Thus, ethylene must be formed on Au_2^+ as part of a catalytic reaction cycle. The ethylene-complex then further reacts with a third CH_4 forming the intermediate $Au_2(C_2H_4)(CH_4)^+$ (*cf.* Fig. 3c–f) which then enables the liberation of C_2H_4 under re-formation of $Au_2(CH_4)^+$. An alternative cycle closing resulting in the reformation of Au_2^+ instead of $Au_2(CH_4)^+$ can be excluded on the basis of the kinetic traces, consequently, the reversible adsorption of the first CH_4 is not part of the cycle.[9]

In the case of Pd_2^+, the first reaction steps, already proceeding at room temperature, are represented by the subsequent adsorption of two methane-d_4 yielding the products $Pd_2(CD_4)^+$ and $Pd_2(CD_4)_2^+$ (*cf.* Fig. 4a and b). At an intermediate temperature (*cf.* Fig. 4c and d) the additional product $Pd_2C_3D_{10}^+$ appears that must be formed from $Pd_2(CD_4)_3^+$. The corresponding kinetic traces show that Pd_2^+, $Pd_2(CH_4)_2^+$, and $Pd_2C_3D_{10}^+$ are in equilibrium with each other which can only be explained by a catalytic reaction cycle. The cycle is then closed by the elimination of an ethane molecule C_2D_6 under re-formation of $Pd_2(CD_4)^+$. An alternative closing of the cycle is the re-formation of $Pd_2(CD_4)_2^+$ upon loss of CD_2 which cannot be excluded on the basis of the experimental data. However, the formation and especially the liberation of the highly reactive carbene CD_2 should be energetically unfavorable and is thus unlikely to occur.

At an even lower temperature (*cf.* Fig. 4e and f) a further dehydrogenation product, $Pd_2C_4D_{12}^+$, is observed which must be formed from $Pd_2(CD_4)_4^+$. Consequently, the previously described formation of $Pd_2(C_3D_{10})^+$ from $Pd_2(CD_4)_3^+$ competes with the adsorption of a fourth CD_4. Since all products are again in equilibrium, the kinetic data can only be described by a combination of two catalytic cycles as displayed in Fig. 5b. The cycle on the right hand side is either closed by the reaction step $Pd_2C_4D_{12}^+ \rightarrow Pd_2(CD_4)_2^+$ under elimination of ethylene, C_2D_4, or by the reaction step $Pd_2C_4D_{12}^+ \rightarrow Pd_2(CD_4)^+$ under elimination of butane C_3D_8. On the basis of the experimental kinetic data only, it cannot be decided which of these reaction steps occurs or if both ethylene and butane are formed simultaneously. But detailed theoretical investigations of the energetics would be desirable in this regard, which are, however, beyond the scope of the present contribution.

B. Methane oxidation on dimers Au_2^+ and Pd_2^+

Since both dimers Au_2^+ and Pd_2^+ have been shown to activate and dehydrogenate methane in full thermal catalytic cycles, next the reaction of these dimers with a mixture of CH_4 (CD_4) and molecular oxygen have been investigated.

At room temperature, the gold dimer Au_2^+ only adsorbs and dehydrogenates methane yielding the products $Au_2(CH_4)^+$ and $Au_2(C_2H_4)^+$ identical to the ion mass distribution observed in the presence of methane only (*cf.* Fig. 2a and 3a).[9,10] Consequently, the room temperature reaction can be described by the catalytic reaction mechanism shown in Fig. 5a yielding ethylene. However, cooling the ion trap below 270 K considerably changes the detected ion mass distribution. Fig. 6 displays, as an example, a mass spectrum (black line) and the corresponding kinetic traces obtained after reacting Au_2^+ with $CH_4/^{16}O_2$ at 210 K. Apart from the bare cluster and the non-dehydrogenated $Au_2(CH_4)_2^+$ two new, oxygen containing products, $Au_2(CH_4)_2O_2^+$ and $Au_2(C_3H_8O_2)^+$, are detected, while the 300 K products $Au_2(CH_4)^+$ and $Au_2(C_2H_4)^+$ are not apparent any more. For comparison, an additional mass spectrum (red line) illustrating the reaction of Au_2^+ with $CH_4/^{18}O_2$ is shown. The labeling experiments with monoisotopic $^{18}O_2$ result in a mass shift by 4 amu of the products $Au_2(CH_4)_2O_2^+$ and $Au_2(C_3H_8O_2)^+$ while the bare cluster signal and the product $Au_2(CH_4)_2^+$ remain unaffected which confirms the peak assignment. Further confirmation was obtained by additional labeling with deuterated methane instead of CH_4. While $Au_2(CH_4)_2O_2^+$ might be a simple coadsorption product of two methane and one oxygen molecules, $Au_2(C_3H_8O_2)^+$ clearly contains dehydrogenated methane and possibly activated O_2 indicating a potential C–O coupling reaction.

The corresponding kinetic traces (*cf.* Fig. 6b) display that, most interestingly, the dehydrogenated $Au_2(C_3H_8O_2)^+$ is in equilibrium with $Au_2(CH_4)_2^+$ and $Au_2(CH_4)_2O_2^+$ while the latter represents the precursor for dehydrogenation. As discussed above, such kinetic data suggest the formation of $Au_2(C_3H_8O_2)^+$ as part of a catalytic cycle. Stoichiometrically, $Au_2(C_3H_8O_2)^+$ might be the result of formaldehyde $Au_2(CH_2O)_2(CH_4)^+$ or carbon dioxide $Au_2(CO_2)(CH_4)_2^+$ formation on Au_2^+. However, based on the mass spectra and the kinetic data alone the structure of this complex cannot be determined. Details of the structure and the formation process in conjunction with theoretical simulations of the reaction path will be addressed in a separate contribution.[8]

At intermediate temperatures a smooth crossover of the dehydrogenated products $Au_2(C_2H_4)^+$ and $Au_2(C_3H_8O_2)^+$ is observed suggesting a highly temperature dependent competition between the ethylene cycle (*cf.* Fig. 5a) and the catalytic cycle that leads to $Au_2(C_3H_8O_2)^+$.

In contrast to Au_2^+ that only forms oxygen complexes at low temperatures, the palladium dimer Pd_2^+ is significantly more reactive toward molecular oxygen. Fig. 7 demonstrates the temperature dependent reactivity of Pd_2^+ in a 2D contour plot representation. Depending on the reaction temperature the methane products $Pd_2(CD_4)^+$, $Pd_2(CD_4)_2^+$, $Pd_2(CD_4)_3^+$, as well as the dehydrogenated products

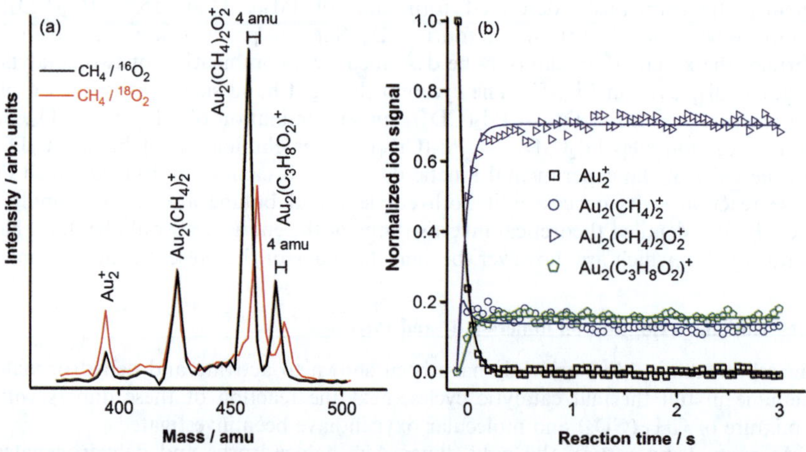

Fig. 6 (a) Ion mass distributions of the reaction between Au_2^+ and a mixture of CH_4 and $^{16}O_2$ (solid black curve) as well as a mixture of CH_4 and $^{18}O_2$ (solid red curve) obtained at 210 K and $t_R = 0.1$ s.[8] (b) Corresponding kinetic data for the reaction $Au_2^+ + CH_4 + {}^{16}O_2$ (0.03 Pa CD$_4$; 0.05 Pa O$_2$; 0.98 Pa He). The open symbols represent experimental data, normalized to the total ion concentration in the ion trap while the solid lines are obtained by fitting the integrated rate equations of a proposed reaction mechanism (with $Au_2(C_3H_8O_2)^+$ as part of a catalytic cycle) to the experimental data.

    This journal is © The Royal Society of Chemistry 2011

$Pd_2(C_3D_{10})^+$, and $Pd_2(C_4D_{12})^+$ are observed as in the case of pure methane-d_4 exposure (*cf.* Fig. 2b and 4). Besides, two low intensity oxygen containing products appear: $Pd_2O_2^+$ and $Pd_2O_4^+$. At around 245 K a first product containing both methane and oxygen, $Pd_2(CD_4)_2O_4^+$, is detected. Further methane–oxygen coadsorption complexes, $Pd_2(CD_4)O_4^+$, $Pd_2(CD_4)_3O_2^+$, and $Pd_2(CD_4)_4O_2^+$, appear in very small intensities (which are not visible in Fig. 6) and/or longer reaction times. From these ion mass distributions it is apparent that molecular oxygen only coadsorbs on non-dehydrogenated Pd_2^+–methane complexes while products containing both activated O_2 and dehydrogenated methane are not observed. Thus, there is no experimental evidence for a successful C–O coupling reaction on Pd_2^+.

Due to the complex ion mass distribution of Pd_2^+ it was not possible to obtain a reaction mechanism that results in a reliable fit to the corresponding kinetic data. However, based on the ion mass distributions in the presence of CD_4 as well as CD_4/O_2 it can be assumed that the catalytic reaction mechanism displayed in Fig. 5b also proceeds in the presence of molecular oxygen, while at the same time CD_4 and O_2 compete for adsorption sites.

4. Discussion

The aim of the following discussion is to compare the activation and dehydrogenation as well as the partial oxidation of methane on small gold and palladium clusters. In this respect, three striking features pertaining to the gas phase cluster reactions will be highlighted: (1) the cooperative action of multiple adsorbed molecules, (2) the temperature dependence and the activation barriers, and (3) the product selectivity.

1. Cooperative molecular coadsorption

The dehydrogenation of methane is observed to proceed on all investigated small palladium clusters Pd_x^+ ($x = 2$–4) while only the gold dimer is found to successfully mediate this reaction. Yet, there seems to be a 'critical number' of adsorbed methane molecules whose cooperative action is necessary to enable the dehydrogenation process.

Au_2^+ reacts with a first CH_4 yielding $Au_2(CH_4)^+$ without any experimental evidence for dehydrogenation. The mere molecular adsorption of the first methane molecule onto Au_2^+ is supported by the fact that a reversible (equilibrium) reaction of Au_2^+ and CH_4 has to be assumed to model the measured kinetic data correctly (*cf.* Fig. 5a).

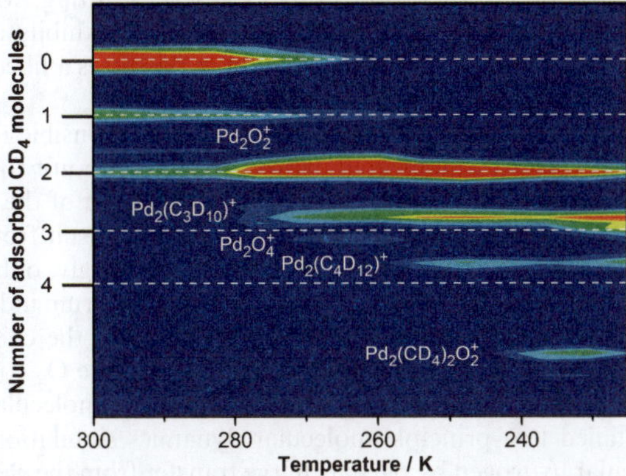

Fig. 7 2D contour plot of the temperature dependent product ion mass distributions after trapping the palladium dimer Pd_2^+ for 0.1 s in the presence of methane CD_4 and molecular oxygen O_2. The relative ion intensity is color coded ranging from blue (0%) to red (100%).

Furthermore, recent first principles computations showed the direct bonding of non-dissociated methane with two hydrogen atoms lying in the plane of the gold cluster and oriented toward it and the other two hydrogen atoms lying in the plane normal to the cluster.[9,13]

Fig. 2a and 3 illustrate that the adsorption of a second CH_4 then enables the activation and dehydrogenation of both methane molecules under formation of ethylene and $2H_2$. This dehydrogenation reaction is only possible due to the cooperative action of both molecules. First principles density functional theory simulations revealed that the adsorption of the second CH_4 can lead to the activation of both CH_4 molecules yielding a structure in which an H atom bridging each of the adsorbed molecules to the nearest Au atom.[9]

Further experimental evidence for a cooperative effect of several molecules is represented by the product $Au_2(C_2H_4)(CH_4)^+$ that is detected as part of the catalytic cycle in conjunction with the cycle closure under re-formation of $Au_2(CH_4)^+$ instead of Au_2^+ (cf. Fig. 3 and 5a). Theoretical studies showed the direct desorption of ethylene from the complex $Au_2(C_2H_4)^+$ to be non-feasible (requiring a large activation energy),[9] instead, the cooperative action of a third adsorbed CH_4 in the intermediate $Au_2(C_2H_4)(CH_4)^+$ energetically facilitates the liberation of ethylene. Thus, on Au_2^+ both the C–H bond activation and subsequent ethylene release is only possible to proceed due to the cooperative action of two and three CH_4 molecules, respectively.

Palladium clusters form the smallest dehydrogenated products $Pd_2(C_3D_{10})^+$, $Pd_3(C_4D_{10})^+$, and $Pd_4(C_4D_{14})^+$ suggesting that on Pd_2^+ the adsorption of a third CD_4 immediately leads to the dehydrogenation, while on Pd_3^+ and Pd_4^+ four methane molecules are necessary for this reaction. As a consequence of the reversible reaction between Pd_2^+ with CD_4 as well as of $Pd_2(CD_4)^+$ with CD_4 (cf. Fig. 5b) it can be concluded that the cooperative action of three CD_4 molecules is mandatory for the dissociation of methane on Pd_2^+. However, for Pd_3^+ and Pd_4^+, due to abovementioned complications in deducing a reaction mechanism and due to the lack of theoretical simulations, it is difficult to draw detailed conclusions concerning the coverage dependent dissociation of methane, the dehydrogenation, or the liberation of reaction products.

For comparison, it is known from previous studies on platinum cations[37,38] and cationic platinum clusters[38,39] under both single collision and multi-collision conditions that these clusters easily activate and dehydrogenate a first adsorbed methane molecule mainly yielding $Pt_xCH_2^+$. The especially high reactivity of platinum clusters compared to gold and palladium might be attributed to the electronic configurations of these metals. The platinum atom exhibits both, a partially filled 5d and 6s orbital (electronic configuration $5d^9\,6s^1$) with a small s–d level splitting[4] which makes the atom and the clusters highly reactive. In contrast, gold exhibits a filled 5d and a singly occupied 6s orbital ($5d^{10}\,s^1$) while palladium possesses a filled 4d and empty 5s orbital ($4d^{10}\,5s^0$) making such metal clusters less reactive.

The cooperative action of multiple molecules is also responsible for the catalytic formation of $Au_2(C_3H_8O_2)^+$. Since gold cations generally do not form stable reaction products with molecular oxygen alone,[8,10] the formation of the experimentally observed products $Au_2(CH_4)_2O_2^+$ and $Au_2(C_3H_8O_2)^+$ must result from the cooperative action of pre-adsorbed methane on Au_2^+. Most interestingly, only $Au_2(CH_4)_2O_2^+$ is detected while $Au_2(CH_4)O_2^+$ is not present in the mass spectrum indicating that the pre-adsorption of two methane molecules is mandatory for the cooperative coadsorption of molecular oxygen. A similar effect of cooperative O_2 coadsorption has previously[8] been observed to occur after pre-adsorption of molecular hydrogen on even x Au_x^+. Detailed first-principle molecular dynamics simulations revealed the binding of molecular hydrogen by partial charge transfer from the electron donating H_2 to the cationic cluster that facilitates the subsequent charge transfer from the hydrogen-cluster complex to O_2 that is necessary for binding and activating O_2.[8] Since CH_4 also acts as an electron donor toward cationic clusters a similar charge transfer processes can be supposed to be responsible for the formation of methane-cluster complexes and the subsequent cooperative oxygen coadsorption. Thus, $Au_2(CH_4)_2O_2^+$ might be assumed to already comprise activated O_2 that can then further react under the cooperative action of a third CH_4 yielding $Au_2(C_3H_8O_2)^+$. Detailed theoretical investigations will be required to reveal further mechanistic and structural details of this reaction.[8]

Consequently, for all investigated reactions (with methane only as well as a methane/oxygen mixture) resulting in the formation of dehydrogenated products on both, Au_2^+ and palladium cations Pd_x^+ ($x = 2$–4) the cooperative action of multiple molecules is indispensable for the product formation.

2. Energetic considerations

A further striking feature is the observed temperature dependence of the formation of dehydrogenated product complexes on Au_2^+ and Pd_x^+. On Au_2^+ ethylene is already detected at room temperature and its formation is hampered with decreasing temperature while on Pd_x^+ the dehydrogenated products appear at lower temperature and the number of products increases with decreasing temperature on Pd_3^+ and Pd_4^+.

Fig. 2a demonstrates that the intensity of the dehydrogenated product $Au_2(C_2H_4)^+$ decreases with decreasing temperature while the intensity of $Au_2(CH_4)_2^+$ containing non-dehydrogenated methane increases indicating an energy barrier for ethylene formation (indicated by ΔE_1 in Fig. 5a) that can be less easily surpassed with decreasing temperature. A second barrier must be involved in the release of C_2H_4 from $Au_2(C_2H_4)(CH_4)^+$ (ΔE_2 in Fig. 5a) since this intermediate is not observed at room temperature implying that the barrier is easily overcome at elevated temperature and that the respective reaction step has become faster than the time scale of our experiment. Detailed molecular dynamics simulations indeed revealed several local potential-energy minima connected by energy barriers corresponding to the formation and elimination of $2H_2$ as well as the formation and release of ethylene.[9]

In contrast to Au_2^+, which dehydrogenates methane already at room temperature, the dehydrogenated palladium dimer product $Pd_2(C_3D_{10})^+$ only appears below 270 K. This difference can be explained in terms of the Lindeman energy transfer model for association reactions (*cf.* Section 2.B.) in conjunction with the increased 'critical number' of methane molecules necessary for dehydrogenation. After adsorption of a neutral molecule on a metal cation the formed complex must be stabilized by a third body collision according to eqn (2a) and (2b). This stabilization is more effective the lower the temperature. Consequently, the adsorption of a third molecule occurs at lower temperature than a second molecule as can be seen from Fig. 2. Since methane dehydrogenation on Pd_2^+ requires the cooperative action of three CD_4 compared to two methanes on Au_2^+, the dehydrogenated product $Pd_2(C_3D_{10})^+$ occurs at lower temperature than $Au_2(C_2H_4)^+$. Similarly, the products $Pd_2(C_4D_{12})^+$, $Pd_3(C_4D_{10})^+$, $Pd_3(C_4D_{14})^+$, $Pd_3(C_5D_{12})^+$, $Pd_3(C_5D_{16})^+$, $Pd_4(C_4D_{14})^+$, $Pd_4(C_5D_{16})^+$, and $Pd_4(C_5D_{18})^+$ appear at even lower temperature than $Pd_2(C_3D_{10})^+$.

Apart from the dehydrogenated products also complexes $Pd_x(CD_4)_y^+$ ($x = 2$–4, $y = 3 - 5$) containing non-dehydrogenated methane are detected at decreasing temperature (*cf.* Fig. 2) indicating activation barriers for methane dehydrogenation (labeled ΔE_1 and ΔE_3 in Fig. 5b). Furthermore, the kinetics obtained at 260 K and 225 K (*cf.* Fig. 4d and f) reveal high relative ion concentrations of $Pd_2(C_3D_{10})^+$ and $Pd_2(C_4D_{12})^+$ suggesting additional activation barriers that are connected with the release of the products C_2D_6, C_2D_4, and/or C_3D_8 from Pd_2^+ (ΔE_2 and ΔE_4 in Fig. 5). Thus, it can be concluded that the observed temperature dependence of methane dehydrogenation on Pd_x^+ most sensitively depends on the adsorption and stabilization of a 'critical number' of CD_4 molecules.

3. Product selectivity

The reaction of Au_2^+ with methane revealed one single dehydrogenated product $Au_2(C_2H_4)^+$ and the elimination of ethylene is facilitated upon adsorption of a further methane molecule in a thermal catalytic reaction mechanism. In contrast, the palladium dimer Pd_2^+ forms two dehydrogenated products $Pd_2(C_3D_{10})^+$ and $Pd_2(C_4D_{12})^+$ in two competing catalytic cycles yielding the neutral molecules ethane, ethylene, and/or butane. The larger palladium clusters Pd_3^+ and Pd_4^+ even exhibit up to four and three different dehydrogenated products, respectively. This demonstrates the outstanding efficiency of Au_2^+ to catalyze the formation of ethylene in a highly product selective manner. The capability of Au_2^+ as a catalytic model system is corroborated by the above presented studies of the dimers in the presence of CH_4 (CD_4) and O_2. While Pd_2^+ does not form any oxygen containing products with dehydrogenated methanes, Au_2^+ exhibits again only one dehydrogenated product containing oxygen, $Au_2(C_3H_8O_2)^+$.

Although the structure of this product cannot be deduced in the present experimental approach, it can be reasoned from the kinetic data that a neutral oxidation product is liberated from the Au_2^+–complex in a catalytic reaction mechanism with a product selectivity of 100% at low temperatures. At this point theoretical simulations are mandatory to suggest possible applicable reaction mechanisms and product complex structures. The results of the combined theoretical investigation and experimental analysis will be communicated in a separate contribution.[8] An additional very remarkable result of the investigations with Au_2^+ is the observation that the product selectivity is tunable between ethylene and the oxidation product by the mere change of the reaction temperature.

The formation of methane oxidation products on free ions has been previously investigated on the cationic platinum atom Pt^+ yielding a mixture of the neutral products methanol (10%), formaldehyde (25%), and "CH_2O_2"(65%) with the latter being either formic acid, CO_2/H_2, or CO/H_2O.[40] Though Pt^+ has been demonstrated to act as an efficient catalytic model system, the product selectivity is rather poor due to the high reactivity of the platinum atom. This drawback might be circumvented by employing the more selective gold, palladium, or even binary gold–palladium clusters.

Acknowledgements

This research was partially supported by the Deutsche Forschungsgemeinschaft and the Fonds der Chemischen Industrie (FCI). In particular, SML acknowledges a Kekulé fellowship of the FCI. The authors would like to thank Uzi Landman for many very fruitful discussions and the suggestion of several experimental studies.

References

1 G. A. Olah, A. Goepper and G. K. S. Prakash, *Beyond Oil and Gas: The Methanol Economy*, Wiley-VCH, Weinheim, 2006.

2 K. Otsuka and Y. Wang, *Appl. Catal., A*, 2001, **222**, 145.

3 D. Schröder and H. Schwarz, *Top. Organomet. Chem.*, 2007, **22**, 1.

4 A. A. Herzing, C. J. Kiely, A. F. Carley, P. Landon and G. J. Hutchings, *Science*, 2008, **321**, 1331.

5 B. Yoon, H. Häkkinen, U. Landman, A. S. Wörz, J.-M. Antonietti, S. Abbet, K. Judai and U. Heiz, *Science*, 2005, **307**, 403; A. Sanchez, S. Abbet, U. Heiz, W.-D. Schneider, H. Häkkinen, R. N. Barnett and U. Landman, *J. Phys. Chem. A*, 1999, **103**, 9573.

6 D. K. Böhme and H. Schwarz, *Angew. Chem. Int. Ed.*, 2005, **44**, 2336.

7 S. M. Lang, D. M. Popolan and T. M. Bernhardt, in *Atomic Clusters: From Gas Phase to Deposited*, ed. P. D. Woodruff, Elsevier, Amsterdam, 2007, vol. 12, pp. 53; S. M. Lang and T. M. Bernhardt, *J. Chem. Phys.*, 2009, **131**, 024310; L. D. Socaciu, J. Hagen, T. M. Bernhardt, L. Wöste, U. Heiz, H. Häkkinen and U. Landman, *J. Am. Chem. Soc.*, 2003, **125**, 10437.

8 S. M. Lang, T. M. Bernhardt, R. N. Barnett, B. Yoon and U. Landman, *J. Am. Chem. Soc.*, 2009, **131**, 8939.
9 S. M. Lang, T. M. Bernhardt, R. N. Barnett and U. Landman, *Angew. Chem. Int. Ed.*, 2010, **49**, 980.
10 S. M. Lang, T. M. Bernhardt, R. N. Barnett and U. Landman, *J. Phys. Chem. C*, 2011, **115**, 6788.
11 K. K. Irikura and J. L. Beauchamp, *J. Am. Chem. Soc.*, 1991, **113**, 2769; K. K. Irikura and J. L. Beauchamp, *J. Phys. Chem.*, 1991, **95**, 8344; M. Schlangen and H. Schwarz, *Angew. Chem. Int. Ed.*, 2007, **46**, 5614.
12 D. M. Cox, R. Brickman, K. Creegan and A. Kaldor, *Mat. Res. Soc. Symp. Proc.*, 1991, **206**, 43; D. M. Cox, R. Brickman, K. Creegan and A. Kaldor, *Z. Phys. D: At., Mol. Clusters*, 1991, **19**, 353.
13 S. M. Lang, T. M. Bernhardt, R. N. Barnett and U. Landman, *ChemPhysChem*, 2010, **11**, 1570.
14 P. Fayet, A. Kaldor and D. M. Cox, *J. Chem. Phys.*, 1990, **92**, 254.
15 A. Božović, S. Feil, G. K. Koyanagi, A. A. Viggiano, X. Zhang, M. Schlangen, H. Schwarz and D. K. Bohme, *Chem.–Eur. J.*, 2010, **16**, 11605.
16 T. M. Bernhardt, *Int. J. Mass Spectrom.*, 2005, **243**, 3.
17 R. Keller, F. Nöhmeier, P. Spädtke and M. H. Schönenberg, *Vacuum*, 1984, **34**, 31.
18 G. A. Miller, *J. Phys. Chem.*, 1963, **67**, 1359; R. C. Bell, K. A. Zemski, D. R. Justes and A. W. Castleman Jr, *J. Chem. Phys.*, 2001, 114.
19 T. M. Bernhardt, J. Hagen, S. M. Lang, D. M. Popolan, L. D. Socaciu-Siebert and L. Wöste, *J. Phys. Chem. A*, 2009, **113**, 2724.
20 E. Schuhmacher, *DETMECH - Chemical Reaction Kinetics Software*, (2003) University of Bern, Chemistry Department.
21 J. I. Steinfeld, J. S. Francisco and W. L. Hase, *Chemical Kinetics and Dynamics*, 2nd edn, Prentice Hall, Upper Saddle River, 1999.
22 P. M. Langevin, *Ann. Chim. Phys.*, 1905, **5**, 245.
23 R. A. Marcus, *J. Chem. Phys.*, 1952, **20**, 359.
24 K. A. Holbrook, M. J. Pillings and S. H. Robertson, *Unimolecular Reactions*, 2nd edn, Wiley, Chichester, 1996.
25 L. Drahos and K. Vékey, *J. Mass Spectrom.*, 2001, **36**, 237.
26 X. Ding, J. Yang, J. G. Hou and Q. Zhu, *THEOCHEM*, 2005, **755**, 9; J. Ho, M. L. Polak, K. M. Ervin and W. C. Lineberger, *J. Chem. Phys.*, 1993, **99**, 8542; J. Ho, K. M. Ervin, M. L. Polak, M. K. Gilles and W. C. Lineberger, *J. Chem. Phys.*, 1991, **95**, 4845; V. A. Spasov and K. M. Ervin, *J. Chem. Phys.*, 1998, **109**, 5344.
27 M. F. Jarrold and J. E. Bower, *J. Chem. Phys.*, 1987, **87**, 5728.
28 T. Shimanouchi, *Tables of Molecular Vibrational Frequencies Consolidated Volume I*, National Bureau of Standards, 1972.
29 W. E. Billups, M. M. Konarski, R. H. Hauge and J. L. Margrave, *J. Am. Chem. Soc.*, 1980, **102**, 7393.
30 S. M. Wetterer, D. J. Lavrich, T. Cummings, S. L. Bernasek and G. Scoles, *J. Phys. Chem. B*, 1998, **102**, 9266.
31 J. F. Weaver, C. Hakanoglu, J. M. Hawkins and A. Asthagiri, *J. Chem. Phys.*, 2010, **132**, 024709.
32 C.-L. Kao and R. J. Madix, *J. Phys. Chem. B*, 2002, **106**, 8248.
33 C.-Q. Lv, K.-C. Ling and G.-C. Wang, *J. Chem. Phys.*, 2009, **131**, 144704.
34 M. R. A. Blomberg, P. E. M. Siegbahn and M. Svensson, *J. Phys. Chem.*, 1994, **98**, 2062.
35 M. R. A. Blomberg, P. E. M. Siegbahn and M. Svensson, *J. Am. Chem. Soc.*, 1992, **114**, 6095.
36 E. Broclawik, J. Haber, A. Endou, A. Stirling, R. Yamauchi, M. Kubo and A. Miyamoto, *J. Mol. Catal. A: Chem.*, 1997, **119**, 35; E. Broclawik, R. Yamauchi, A. Endou, M. Kubo and A. Miyamoto, *Int. J. Quantum Chem.*, 1997, **61**, 673.
37 C. Heinemann, R. Wesendrup and H. Schwarz, *Chem. Phys. Lett.*, 1995, **239**, 75; A. Kaldor and D. M. Cox, *J. Chem. Soc., Faraday Trans.*, 1990, **86**, 2459.
38 U. Achatz, C. Berg, S. Joos, B. S. Fox, M. K. Beyer, G. Niedner-Schattenburg and V. E. Bondybey, *Chem. Phys. Lett.*, 2000, **320**, 52; K. Koszinowski, D. Schröder and H. Schwarz, *J. Phys. Chem. A*, 2003, 4999.
39 G. Kummerlöwe, J. Balteanu, Z. Sun, O. P. Balaj, V. E. Bondybey and M. K. Beyer, *Int. J. Mass Spectrom.*, 2006, **254**, 183.
40 R. Wesendrup, D. Schröder and H. Schwarz, *Angew. Chem. Int. Ed. Engl.*, 1994, **33**, 1174.

Gold catalyzed liquid phase oxidation of alcohol: the issue of selectivity

L. Prati,[*a] A. Villa,[a] C. E. Chan-Thaw,[a] R. Arrigo,[b] D. Wang[c] and D. S. Su[b]

Received 11th February 2011, Accepted 2nd March 2011
DOI: 10.1039/c1fd00016k

Commercial carbon nanotubes (CNTs) and carbon nanofibers (CNFs) modified in various ways at the surface have been used as supports for gold nanoparticles (AuNPs) in order to study their influence on the activity/selectivity of catalysts in the aqueous oxidation of alcohol. Particularly oxidative treatment was used to introduce carboxylic functionalities, whereas subsequent treatment with NH_3 at different temperatures (473 K, 673 K and 873 K) produced N-containing groups leading to an enhancement of basic properties as the NH_3 treatment temperature was increased. The nature of the N-containing groups changed as the temperature increased, leading to an increase in the hydrophobicity of the support surface. Similar Au particle size and similar textural properties of the supports allowed the role of chemical surface groups in both the activity and the selectivity of the reaction of glycerol oxidation to be highlighted. An increase of basic functionalities produced a consistent increase in the activity of the catalyst, which was correlated to the promoting effect of the basic support in the alcoholate formation and the subsequent C–H bond cleavage. The selectivity towards primary oxidation products (C3 compounds) was the highest for the catalysts treated with NH_3 at 873 K, which presented the most hydrophobic surface. The same trend in the catalyst activity has been obtained in the aqueous benzyl alcohol base-free oxidation. As in the case of glycerol, the increasing of basicity and/or hydrophobicity increased the consecutive reactions.

Introduction

The success of gold as a catalyst for selective oxidation in the liquid phase is principally due to the peculiar characteristics of this metal compared to more classical Pd or Pt based catalysts when O_2 is used as the oxidant. In fact, it has been shown that a Au catalyst is less prone to deactivation in the presence of O_2 compared to Pd or Pt.[1-3] Moreover, gold catalysts show high selectivity towards the oxidation of primary alcohol with respect to secondary alcohol.[1,2] Since the first observation on ethylene glycol, a lot of papers have dealt with the application of gold catalysts to this important class of reaction, mainly with two aims. The first aim is to try to impreve the activity and the second aim rs to avoid parallel or consecutive reactions, thus enhancing the selectivity to the desired product. This latter aspect is very important from an application point of view and most recent studies focused on it. Moreover, the mechanistic understanding of the factors ruling selectivity become a priority task for optimizing catalyst design.

[a]Dipartimento di Chimica Inorganica Metallorganica e Analitica L. Malatesta, Università degli Studi di Milano, Via Venezian 21, 20133 Milano, Italy. E-mail: Laura.Prati@unimi.it
[b]Fritz Haber Institute of the Max Planck Society, Faradayweg 4-6, D-14195 Berlin, Germany
[c]Institut für Nanotechnologie, Forschungszentrum Karlsruhe in der Helmholtz-Gemeinschaft, Hermann-von-Helmholtz-Platz 1, 76344 Eggenstein-Leopoldshafen, Germany

One of the most studied substrates is benzyl alcohol, which can be selectively oxidized to benzaldehyde. Dehydrogenation on gold-supported nanoparticles to produce aldehyde can be competitive with the disproportionation of benzyl alcohol to benzaldehyde and toluene,[4] but dehydrogenation prevails when small gold nanoparticles (< 4 nm) are present. This is due to the increasing of the unsaturated coordination sites, which promotes the β C–H bond cleavage, which represents the rate determining step. This trend was confirmed also by periodic density functional theory (DFT) calculations[5] addressing the higher activity of small nanoparticles to the roughness of the surface. High selectivity to benzaldehyde (> 99%) has been reported depending on reaction conditions, *i.e.* solvent, temperature, but principally on the support employed (MgO and hydrotalcite).[4] The main by-products are represented by benzoic acid, benzyl benzoate (from consecutive reactions) or toluene and benzene (from parallel reactions). The selectivity of the reaction can be strongly affected by the presence of a base, whose function is still under investigation. It is believed that the base can cleave the O–H bond of the alcohol to form an alkoxide intermediate.[6] In this process, the role of the support (especially at the interface with the Au nanoparticles) and also the role of coadsorbed oxygen atoms on Au surface could be relevant.[4,7]

Glycerol represents another well-studied substrate, in view of its importance as a cheap material derived from biomass, on which a new chemical platform can be based.[8,9] Glycerol presents some fundamental differences compared to benzyl alcohol: the non-activating nature of the OH groups ($pK_a = 14$), the presence of primary beside secondary OH groups and the chelating nature (also of the oxidation products), which can strongly affect the catalyst life by providing irreversible adsorption and/or enhancing the leaching of metal. Moreover, another important difference lies in the aqueous solubility and the cheap availability of glycerol only in aqueous solution. Therefore, all the catalytic tests should be carried out in water instead of in organic solvent (toluene, xylene, cyclohexane) or solventless, as in the case of benzyl alcohol.

Since the first studies on glycerol oxidation,[1,2] it appeared evident that basic conditions enhanced both activity and selectivity positively. In fact, the presence of a base enhanced the reaction rate by facilitating the alkoxide formation[10,11] and by favoring

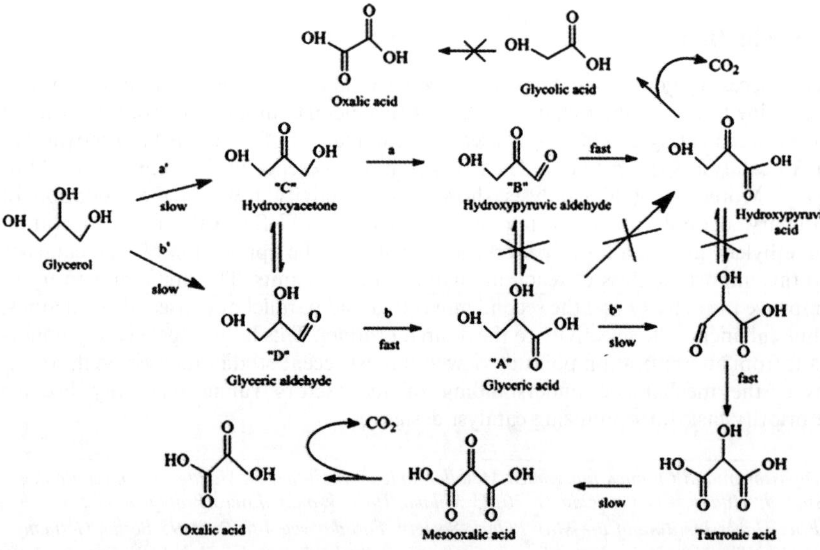

Scheme 1 Reaction scheme for glycerol oxidation under basic conditions.

the desorption of the highly-chelating hydroxyacid, thus decreasing irreversible adsorption on metal active sites.[12] In addition, basic conditions also have a beneficial effect on the selectivity towards the primary alcohol oxidation products (*i.e.* glycerate) (Scheme 1) through keto-enolic equilibrium and through accelerating the decomposition rate of H_2O_2 formed during the reaction, which has been recognized as responsible for the majority of the C–C bond cleavage products (*i.e.* glycolate, oxalate, formate).[10,13]

More recently, efforts in oxidizing glycerol in a base-free medium revealed the specific role of the support, but also showed evidence of the limitation in using monometallic gold based catalysts.[14] In fact, even under stronger conditions (373 K) than those used when the base was present (323–333 K), the reaction rates were very low and only the addition of a second metal (typically Pt) yielded a discrete reaction rate. Selectivity appeared to be determined mainly by the support. According to the model suggested by Davis, on the basis of DFT calculations and labeling tests,[11] the role of the support becomes fundamental in the second elementary step of the reaction, *i.e.* the hydration of aldehyde to acetal with the subsequent formation of the corresponding carboxylic acid. Thus, in this study, we investigated the role of the support in the selective glycerol oxidation further, mainly focusing our attention on the selectivity of the process.

Experimental

Materials

The gold sponge, 99.9999% purity, was purchased from Fluka. Commercial CNFs PR24-PS from Applied Science (average diameter of 88 ± 30 nm and a specific surface area of 43 m^2g^{-1}) and CNTs Baytubes (average diameter of 10 ± 2 nm and a specific surface area of 288 m^2g^{-1}) from Bayer were employed. The functionalization of the CNFs and the CNTs was performed according to the procedure report in ref. 15. The oxygen-containing nanocarbons were obtained by treating the pristine support with HNO_3 according to the following procedure: a solution of CNTs in HNO_3 concentrate (20 g of CNT per liter of HNO_3) was kept at 373 K for 2 h under continuous stirring, then rinsed with distilled water, and finally dried at 343 K for several hours. N-containing CNFs were obtained from the pre-oxidized CNFs by thermal treatment (10 g for each batch) with NH_3 in the temperature range 473–873 K for 4 h. $NaBH_4$ of purity > 96% purchased from Fluka and polyvinylalcohol (PVA) (M 10,000) from Aldrich were used. NaOH of the highest purity available was also from Fluka. Gaseous oxygen from SIAD (99.99% pure) was used. Glycerol (87% wt solution), glyceric acid and all the intermediates were purchased from Fluka. Benzyl alcohol, benzaldehyde, toluene and all the intermediates were also from Fluka.

Catalyst preparation

a) **PVA-protected gold sol preparation.** Solid $NaAuCl_4 \cdot 2H_2O$ (0.043 mmol) and PVA (2% wt) solution (1.64 mL) were added to 130 mL H_2O. After 3 min, $NaBH_4$ (0.1 M) solution (1.3 mL) was added to the yellow solution under vigorous magnetic stirring. The ruby red Au(0) sol was formed immediately.[16]

b) **Immobilization step.** Within a few minutes of sol generation, the sol was immobilised by adding the support under vigorous stirring. The amount of support was calculated to have a final gold loading of 1% wt. After 2 h, the slurry was filtered and the catalyst washed thoroughly with distilled water; it was then used in the wet form. ICP analyses were performed on the filtrate using a Jobin Yvon JV24 to verify the gold loading on the support.

Catalyst characterisation

a) Titration procedure. Potentiometric titration of the basic sites was carried out on the samples using a Mettler Toledo Titrator. Approximately 0.2 g of sample was suspended in 50 mL of KCl 10^{-3} M and then sonicated and equilibrated for several hours. Prior to each measurement, the suspension was continuously saturated with argon to eliminate the influence of CO_2, until the pH was constant. Volumetric standards of HCl (0.01M) or NaOH (0.10M) were used as titrant, starting from the initial pH of the CNFs suspension.

b) Metal loading. The gold content for the sol prepared catalysts was checked by ICP analysis of the filtrate on a Jobin Yvon JY24. The water content was determined by drying a sample at 423 K in air for 5 h. A check of Au loading was also performed directly on the catalysts, confirming the quantitative adsorption of Au NPs of the sol. The samples were weighed and loaded into a home-made NaCl crucible, then annealed at 973 K for two hours in air to burn off the carbon support. The residue was then dissolved in 5 mL of freshly made aqua-regia (3 : 1 hydrochloric/nitric acid – Omnipure grade – EMD). The sample was diluted to 50 mL with deionized water. ICP Standards were prepared by serial dilution of an Alfa-Aesar Au ICP standard. The dissolved NaCl had no effect on the ICP data.

c) Morphology and microstructure of the catalysts were characterized by TEM. The powder samples of the catalysts were ultrasonically dispersed in ethanol and mounted onto copper grids covered with carbon film. A Philips CM200 FEG electron microscope, operating at 200 kV and equipped with a Gatan Tridiem imaging filter, was used for TEM observation. EDX analysis was performed in the same microscope using a DX4 analyzer system (EDAX).

Oxidation experiments

The reactions were carried out in a thermostatted glass reactor (30 mL) provided with an electronically controlled magnetic stirrer connected to a large reservoir (5000 mL) containing oxygen at 300 kPa.

Glycerol oxidation: 0.3 M Glycerol solution, NaOH (NaOH/glycerol = 4 mol mol^{-1}) and the Au catalyst (glycerol/metal = 1000 mol mol^{-1}) were added (total volume 10 mL). The reactor was pressurised at 300 kPa of O_2 and thermostatted at 323 K. After an equilibration time of 5 min, the reaction was initiated by stirring and samples were taken every 15 min and analysed by HPLC on a Varian 9010 HPLC equipped with a Varian 9050 UV (210 nm) and a Waters R.I. detector in series. An Alltech OA-1000 column (300 mm × 6.5 mm) was used with aqueous H_3PO_4 0.1% wt/wt M (0.5 mL min^{-1}) as the eluent. Samples of the reaction mixture (0.5 mL) were diluted (5 mL) using the eluent. Products were assigned by comparison with authentic samples.

Base-free reactions were carried out under the same conditions except for the temperature, which was increased to 373 K and glycerol/metal ratio (500 mol mol^{-1})

Benzyl alcohol oxidation: 0.3 M benzyl alcohol and the catalyst (substrate/metal = 500 mol mol^{-1}) were mixed in distilled water (total volume 10 mL). The reactor was pressurized at the desired pressure of O_2 and thermostatted at 333 K. The reaction does not start during the heating to temperature, as verified by sampling when the reaction temperature was reached without stirring ($t = 0$). The reaction was initiated by stirring. After the end of reaction, the catalyst was filtered off and the product mixture was extracted with cyclohexane. Recoveries were always 98% ± 3 with this procedure. For the identification and analysis of the products, a GC-MS and GC (a Dani 86.10 HT Gas Chromatograph equipped with a capillary column, BP21 30m × 0.53mm, 0.5 μm Film, made by SGE) were used. Comparison with authentic samples was used. For the quantification of the reactant-products, the calibration method using an external standard was employed.

Test for hydrogen peroxide degradation

A 1 mM solution of H_2O_2 (15 mL) was stirred at the appropriate temperature under N_2 atmosphere. The pH of the solution was adjusted by adding H_2SO_4 6 M (pH2). The amount of the catalyst added was the same as in the glycerol oxidation (about 20–30 mg). Hydrogen peroxide was quantified time by time by permanganate titration. The following procedure has been used: a sample (5 mL) of the filtered reacting solution was titrated with a 0.01 N $KMnO_4$ solution at constant pH of 2.0 ± 0.1 (by adding concentrated H_2SO_4). The detection limit of H_2O_2 was 0.01 mM.

Results and discussion

The selectivity of the reaction, as discussed above, can be strongly affected by the presence of a base. However, a relationship between particle size and selectivity was found.[1,2] Larger gold particles (> 10 nm) were more selective towards C_3-oxidation products than smaller ones, which produced a relevant amount of C–C bond cleavage products. However, more recent studies revealed that selectivity could be ruled by a more complex platform of factors.[17–19] Particularly the textural and chemical properties of the supporting material appeared to be fundamental.[17] In fact, it was reported that selectivity in the glycerol oxidation of Au on $MgAl_2O_4$ is tuned by the Al/Mg ratio at the surface besides the particle size and the exposure of the metal. Therefore, for studying the pure support effect in more detail, any other differences such as particle size, surface area and porosity of the supporting materials should be avoided. The selection of the materials was thus performed, taking into account the surface area (as high as possible to increase the metal dispersion) and porosity (avoiding as much as possible micropores for diffusional problems). Moreover, a surface which is easy to modify should be considered. Carbon nanotubes (CNTs) appeared to be good candidates, as they show regular textural properties with surface groups which can be modified by chemical/physical treatments without damaging the overall structure.

Two different commercial samples were used as starting materials. The characterizations have been recently reported:[20] one from Bayer Material Science A.G. (Baytubes, Carbon Nanotubes CNT) and one from Pyrograf Products Inc. (PR24, Carbon Nanofiber CNF). The physical properties of these two materials are reported in Table 1. An oxidative treatment (HNO_3 65% wt, 2 h, 373 K) increased the functionalization on the outermost walls, but did not significantly affect either the grain shape and the size, or the structure. An important effect of the treatment was, in both cases, to remove residual amorphous carbon from the surface. This procedure is known to increase the presence of carboxylic acids.[20] The acidic site titration confirmed an increase of the amount of acidic groups (Table 1). A basic functionalization was carried out by the subsequent treatment with NH_3 at different temperatures, following the procedure reported elsewhere.[21] Table 2 reports the basic site titration data that show the increase of the basic groups with increasing post-treatment temperature (473 K–673 K–873 K). It should be noted that, even at the highest temperature (873 K), the overall structure of CNFs was maintained.[21]

Table 1 Physical data for CNTs and CNFs

	AS ($m^2 \, g^{-1}$)	Micropore area ($m^2 \, g^{-1}$)	diameter (nm)	pH	mol acid sites/g
CNT	288^a	40^a	10 ± 2	10	—
CNT ox	321^a	38^a	10 ± 2	4.35	2.6×10^{-3}
CNF	43^a	0^a	88 ± 30	6.5	—
CNF ox	37^a	5^a	88 ± 30	4.52	4.5×10^{-3}

[a] From ref. [20,21].

Table 2 Titration data for N-CNFs

Sample	mol basic sites/g	Generated pH
N-CNF 473 K	$4.5*10^{-4}$	8.0
N-CNF 673 K	$6*10^{-4}$	8.2
N-CNF 873 K	$1*10^{-3}$	8.5

Table 3 Au particle size on CNTs and CNFs

	Au sol	Au/CNT pristine	Au/ CNT ox	Au/CNF pristine	Au/ CNF ox	Au/N CNF 473 K	Au/ N-CNF 673 K	Au/ N-CNF 873 K
Statistical median (nm)	2.45	4.61	3.83	3.80	3.65	3.15	3.11	2.94
Standard deviation σ	0.27	1.32	0.89	0.95	0.78	0.57	0.64	0.51

CNTs and CNFs presented different surface areas, but the contribution of micropores could be considered negligible in both cases (Table 1).

We prepared thus Au supported nanoparticles (AuNPs) on CNTs and CNFs through the sol immobilization technique which represents the most suitable method for obtaining similar particle size distribution, regardless of the supporting material employed.[16] The Au particle size distributions obtained are reported in Table 3. During the immobilisation step, we noticed an increase in the size of the nanoparticles, probably due to coalescence that increases with decreasing interaction strength between the AuNPs and the support surface. As preformed AuNPs are identically generated in all the cases, we can address the difference in diameter of supported AuNPs to the different surface chemical groups and in particular to their density. Indeed, we observed similar particle diameter for AuNPs immobilized on all the treated supports (about 3–4 nm) with a decreasing of the discrepancy between diameter in the sol and on the support (*i.e.* before and after the deposition) by increasing the functional group density.

The acidic groups, introduced by oxidative treatment, appeared less efficient in stabilizing the nanoparticle diameter than the basic groups. In fact Au/CNT ox and Au/CNF ox showed the most remarkable increase in the Au particle diameter (3.83 and 3.65 nm *vs.* 2.94 nm for a comparable density of functional groups – Table 3). Pristine CNTs and CNFs lead to poor metallic dispersion, probably also due to the presence of amorphous carbon, which preferentially adsorbs the AuNPs from the sol. Some representative TEM pictures for Au/CNF and Au/CNT are shown (Fig. 1 and 2), demonstrating the increase in metal dispersion obtained after functionalization.

Regarding the activity of the catalysts, there are a lot of examples in the literature confirming that the activity of the catalysts in the liquid phase is related to the surface exposure of the active sites that should increase by decreasing the particle size.[17,18] In the present case, the external functionalization of the support surface, the negligible presence of micropores and the methodology chosen for the deposition of the gold nanoparticles should avoid the limitation due to diffusion problems. We should thus expect a trend of activities which follows the particle sizes. However, looking at the reaction profiles (Fig. 3 and 4) it can be seen clearly that this expected trend was not followed. In fact, Au/CNT ox and Au/CNF ox, even showing a very similar particle size distribution (mean size 3.83 nm *vs.* 3.65 nm – Table 3), presented

TOFs of 742 h^{-1} and 186 h^{-1}, respectively (Table 4 – Fig. 3). Also, Au/N-CNF 473 K and Au/N-CNF 673 K with almost the same particle size (3.15 nm and 3.11 nm, respectively) showed TOFs of 112 and 597 h^{-1}, respectively (Table 4 – Fig. 4). This definitely supports the idea that the supporting material plays an active role in determining the rate of glycerol oxidation. This is in agreement with the

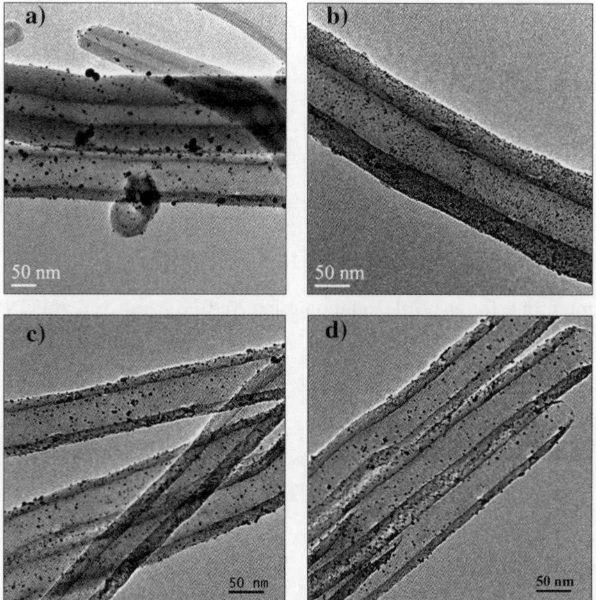

Fig. 1 TEM images of a) Au/CNF, b) Au/CNF ox, c) Au/CNT, d) Au/CNT ox.

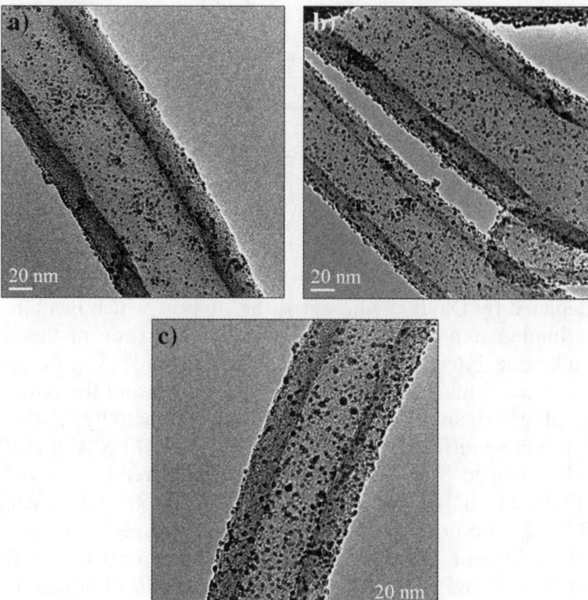

Fig. 2 TEM images of a) Au/N-CNF 473 K, b) Au/N-CNF 673 K, c) Au/N-CNF 873 K.

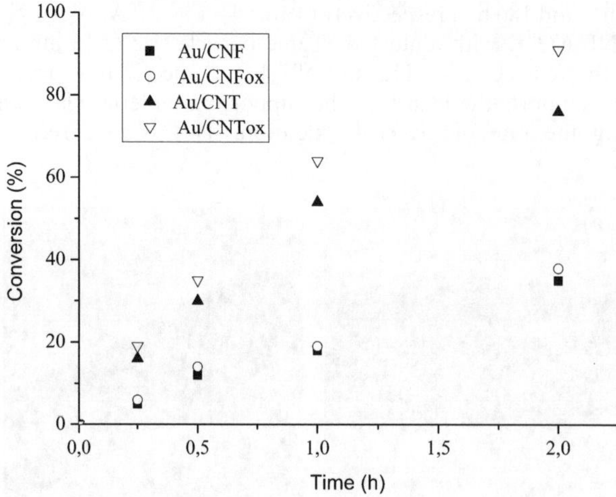

Fig. 3   Reaction profile for glycerol oxidation using Au/CNF, Au/CNT, Au/CNF ox and Au/CNT ox.

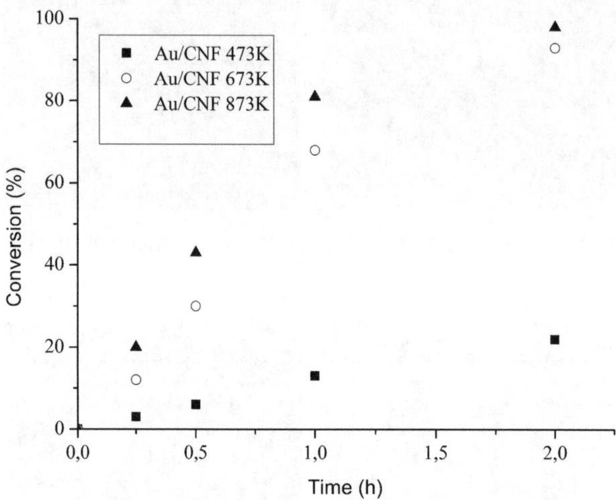

Fig. 4   Reaction profile for glycerol oxidation using Au/N-CNF 473 K, Au/N-CNF 673 K, Au/N-CNF 873 K.

mechanism depicted by Davis,[11] where it is the support which mediates the second and rate determining step of the reaction, *i.e.* the activation of the C–H bond of the ensuing alkoxide intermediate to form the aldehyde. Considering the NH_3 treated CNFs, it was thus not surprising that by increasing the basic site amount (Table 2) we observed an increasing of the catalyst activity in the same order (Fig. 4) even the huge difference between Au/N-CNF 473 K and Au/N-CNF 673 K was not fully justified by simply considering the increase of basic sites [Au/N-CNF 473 K (TOF 112 h^{-1}) < Au/N-CNF 673 K (TOF 597 h^{-1}) < Au/N-CNF 873 K (TOF 853 h^{-1})]. The positive effect on the catalytic performance in the liquid phase of NH_3 treatment has also been observed in the case of Pd supported NPs.[22] It should be noted, that the different functionalization temperature determined the presence of different surface chemical groups in the samples: at lower temperature (473 K) prevalent amide-like groups were observed, which were transformed at higher temperature into pyridine-like groups.[21] In the sample treated at higher temperature (873 K), microcalorimetric measurements also revealed the presence of strong oxygen basic sites similar to those observed in hydrotalcite[23] and in γ-Al$_2$O$_3$,[24] along with a majority of weaker basic sites. The amination process produces a gradual increase of the hydrophobic character of the surface by increasing the temperature of the treatment.

Concerning the selectivity of the glycerol oxidation reaction (Scheme 1), Table 4 showed some unexpected results. On the one hand, using catalysts with similar Au particle size, we were expecting similar selectivity.[1,2] This did not occur, clearly indicating that some other factors influence the selectivity of the process. Pristine CNTs and CNFs showed different selectivity, CNTs being more selective toward C2 products (glycolate and oxalate) than CNFs (Table 4). In both cases, the oxidative treatment increased the selectivity towards C3 products (glycerate and tartronate) and did not alter the activity of the catalysts very much (Table 4 and Fig. 3). In the case of CNF, we observed a breakdown of selectivity for NH$_3$ treated samples, but only when the temperature of treatment was over 673 K, *i.e.* when the catalysts become more active. Indeed, selectivity did not change very much for Au/CNF, Au/CNF ox and Au/N-CNF 473 K, with 55–65% selectivity to glycerate, none to tartronate, 20–27% selectivity to glycolate and 10–16% selectivity to formate (Table 4— all the selectivities reported are at 90% conversion). On the contrary, in the case of Au/N-CNF 673 K and Au/N-CNF 873 K, a decrease of C2–C1 products (glycolate and formate) to 13% selectivity and 5–6% selectivity, respectively, was detected. Meanwhile an increase in C3 products (glycerate and tartronate) to 79–82% selectivity was observed (Table 4—all the selectivities reported are at 90% conversion). This behavior could be possibly explained by the surface properties related to the presence of different chemical groups. Accepting the H$_2$O$_2$ detected during the reaction as responsible for the C–C bond cleavage[10,13] and accepting that H$_2$O$_2$ derives from the O$_2$ reduction by H$_2$O adsorbed on the catalyst surface,[11] we could expect H$_2$O$_2$ formation to be related to the water adsorption capacity of the support, *i.e.* its hydrophilicity. The functionalization with HNO$_3$ and the subsequent treatment with NH$_3$ at different temperatures has been shown to increase the hydrophobicity of the surface progressively by increasing the temperature. Thus in our case, the increase of the basicity in Au/N-CNF 873 K should be expected to increase the activity by promoting the alcoholate formation and C–H bond cleavage, whereas the increase of surface hydrophobicity could lead to an higher selectivity to C3 by decreasing the H$_2$O$_2$ formation. However, during the tests, no significant differences in H$_2$O$_2$ concentration have been revealed. This was not surprising as the basic media could influence the degradation rate of the peroxide. Therefore, to better

Table 4 Glycerol oxidation with Au/CNTs and Au/CNFs under basic conditions[a]

Catalyst	TOF[b] (h^{-1})	Selectivity[c]				
		Glycerate	Tartronate	Glycolate	Oxalate	Formate
Au/CNTs	648	26	7	14	53	<1
Au/CNT ox	742	59	11	17	13	<1
Au/CNF	182	55	1	27	0	16
Au/CNF ox	186	65	<1	20	0	10
Au/N-CNF 473 K	112	64	<1	22	0	12
Au/N-CNF 673 K	597	66	13	13	0	6
Au/N-CNF 873 K	853	68	14	13	0	5

[a] Reaction conditions: in water, glycerol/metal: 1000 mol mol^{-1}, T = 323 K, pO$_2$ = 3 atm., 4eq NaOH [b] TOF calculated at 30 min on the basis of total metal loaded [c] Selectivity at 90% conversion

Table 5 Base free oxidation of glycerol and H_2O_2 detected.[a]

Catalyst	TOF[b] (h^{-1})	Selectivity[c]					H_2O_2[d] mmol L^{-1}
		Glycerate	Hydroxy pyruvic	Glycolate	Oxalate	Formate	
Au/CNF	0.5	41	17	12	0	30	1.5
Au/CNF ox	0.6	33	26	19	0	22	2.0
Au/N-CNF 473 K	1.3	30	20	22	3	23	2.0
Au/N-CNF 673 K	3.8	66	7	10	0	15	0.5
Au/N-CNF 873 K	4.7	62	8	13	-	16	0.3

[a] Reaction conditions: in water, glycerol/metal: 500 mol mol^{-1}, $T = 373$ K, $pO_2 = 3$ atm. [b] TOF calculated at 30 min on the basis of total metal loaded. [c] Selectivity at 20% conversion. [d] H_2O_2 from titration at isoconversion (20%).

clarify this point, we carried out some experiments under base-free conditions, as H_2O_2 is more stable at neutral-acidic pH. Tests under base-free conditions (373 K without any addition of base) showed a higher (even low) activity for Au/N-CNF 873 K, the most basic and hydrophobic catalyst. As under non-basic conditions we expected H_2O_2 to degrade more slowly than under basic conditions,[13] we can use the amount of C–C cleavage as a probe of H_2O_2 formation. Au/N-CNF 873 K, also under these conditions, produced C–C bond cleavage products (glycolate and formate) in a smaller amount than Au/CNF 473 K (Table 5). H_2O_2 quantification by titration confirmed a lesser concentration in the case of Au/N-CNF 873 K (0.3 mmol L^{-1} for Au/N-CNF 873 K *vs* 2.0 mmol L^{-1} for Au/CNFox – Table 5). Moreover blank experiments highlighted the same degradation rate for H_2O_2 in the presence of both catalysts (Fig. 5). These results seem to confirm our hypothesis that the hydrophobic nature of N-CNF 873 K contributed to the decrease in peroxide formation *i.e.* an increase in C3-product selectivity. More accurate kinetic investigations are being carried out in order to confirm this finding.

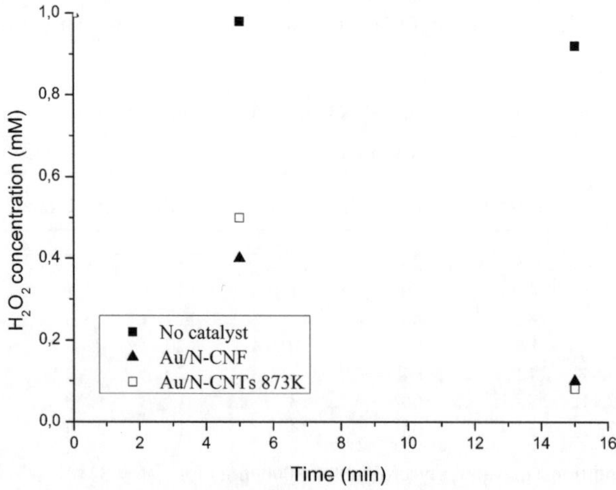

Fig. 5 H_2O_2 degradation profile for Au/CNF and Au/CNF 873 K under acidic conditions (pH 2.5).

Table 6 Benzyl alcohol oxidation[a]

Catalyst	TOF[b] (h^{-1})	Selectivity[c]			
		Benzaldehyde	Benzoic acid	Benzyl benzoate	Toluene
Au/CNF	7	92*	—	—	8
Au/N-CNF 473 K	65	93	2	—	5
Au/N-CNF 673 K	88	88	5	3	3
Au/N-CNF 873 K	102	85	8	3	3

[a] Reaction conditions: in water, alcohol/metal: 500/1, $T = 333$ K, $pO_2 = 3$ atm. [b] TOF calculated at 30 min on the basis of total metal loaded. [c] Selectivity at 90% conversion. *selectivity at 40% conversion.

Moreover, under base-free experimental conditions, it was shown that secondary alcohol can also be oxidized and hydroxypyruvic acid formed, highlighting for these two catalysts a different selectivity towards primary and secondary OH. Au/CNF 473 K appeared the most reactive and, considering that selectivity to glyceric acid is quite low, we could conclude that the oxidation of secondary alcohol is consecutive and most probably depends on a different adsorption mode.

Additional tests have also been performed using benzyl alcohol. The reaction was carried out in water without the addition of a base at 333 K. The increase of activity with the increase of basicity of the support could be expected, as well as the increase of benzoic acid, the formation of which is known to be favored by a basic environment[23] (Table 6). It should be noted, however, that, as in the case of glycerol, the consecutive reaction producing benzoic acid from benzaldehyde is favored by an increase in hydrophobicity of the support *i.e.* N-CNF 673 K and N-CNF 873 K

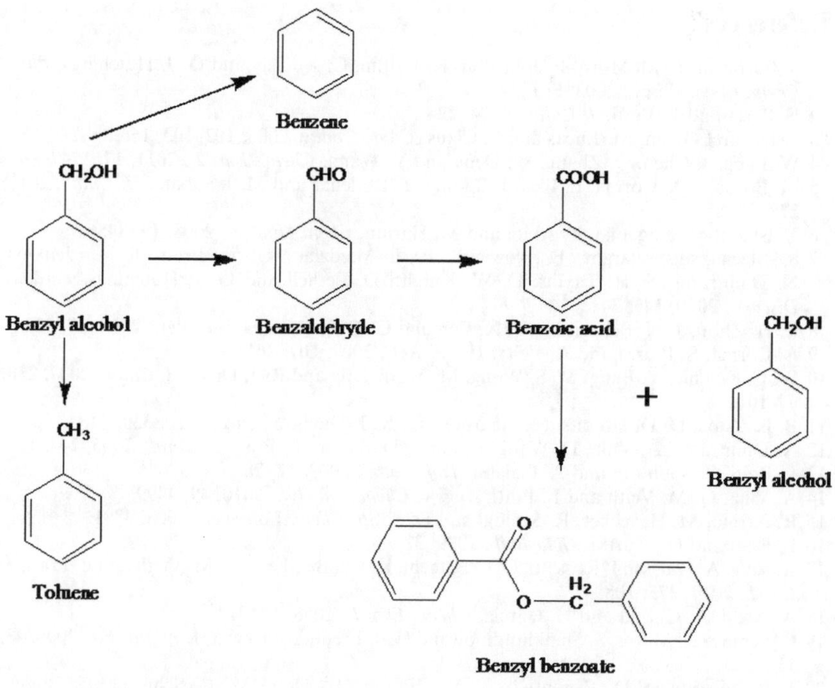

Scheme 2 Reaction scheme for benzyl alcohol oxidation.

(Scheme 2 – Table 6). We also observed a different distribution in the minor products of the reaction, namely benzyl benzoate and toluene. Benzyl benzoate could be considered formed through a consecutive reaction,[7] but the formation of toluene follows a different pathway, probably through dehydrogenation[5] (Scheme 2).

Conclusions

Alcohol oxidation in the liquid phase has been investigated, paying particular attention to the selectivity of the process and particularly to the role that the support can play. Reactions performed in water are of interest, because of the application to water soluble alcohol and particularly to glycerol, an important renewable feedstock. Commercially available CNTs and CNFs have been used as starting materials, because they did not present a relevant microporosity that could lead to diffusional limitation during liquid phase oxidation. Functionalization of CNTs and CNFs has been performed using the procedure set up by Arrigo et al.,[21] which produced fully characterized supports. Preformed gold sol has been immobilized on the supports, yielding Au catalysts with almost the same particle size, but showing different catalytic activity/selectivity as a function of the supporting material. In the glycerol selective oxidation it has been shown that the activity increased with the basicity of CNFs, whereas the selectivity appeared most related to the type of surface groups. Indeed, basic and hydrophobic surfaces enhanced the selectivity to C3 products, whereas more hydrophilic surfaces increased the C–C bond cleavage products. This behavior is possibly correlated to a higher concentration of native H_2O_2 detected in the presence of hydrophilic support.

Acknowledgements

Fondazione Cariplo is gratefully acknowledged for financial support

References

1 S. Carrettin, P. McMorn, P. Johnston, K. Griffin, C. J. Kiely and G. J. Hutchings, *Phys. Chem. Chem. Phys.*, 2003, **5**, 1329.
2 F. Porta and L. Prati, *J. Catal.*, 2004, **224**, 397.
3 S. Demirel-Gulen, M. Lucas and P. Claus, *Catal. Today*, 2005, **102–103**, 166.
4 W. Fang, J. Chen, Q. Zhang, W. Deng and Y. Wang, *Chem.–Eur. J.*, 2011, **17**, 1247.
5 M. Boronat, A. Corma, F. Illas, J. Radilla, T. Ródenas and M. J. Sabater, *J. Catal.*, 2011, **278**, 50.
6 T. Ishida, M. Nagaoka, T. Akita and M. Haruta, *Chem.–Eur. J.*, 2008, **19**, 8456.
7 S. Meenakshisundaram, E. Nowicka, P. J. Miedziak, G. L. Brett, R. L. Jenkins, N. Dimitratos, S. H. Taylor, D. W. Knight, D. Bethell and G. J. Hutchings, *Faraday Discuss.*, 2010, **145**, 341.
8 C. H. Zhou, J. N. Beltramini, Y. X. Fan and G. Q. Lu, *Chem. Soc. Rev.*, 2008, **37**, 527.
9 A. Corma, S. Iborra and A. Velty, *Chem. Rev.*, 2007, **107**, 2411.
10 W. C. Ketchie, Y. Fang, M. S. Wong, M. Murayama and R. J. Davis, *J. Catal.*, 2007, **250**, 94–101.
11 B. N. Zope, D. D. Hibbits, M. Neurock and R. J. Davis, *Science*, 2010, **330**, 74.
12 N. Dimitratos, A. Villa, D. Wang, F. Porta, D. Su and L. Prati, *J. Catal.*, 2006, **244**, 113.
13 L. Prati, P. Spontoni and A. Gaiassi, *Top. Catal.*, 2009, **52**, 288.
14 A. Villa, G. M. Veith and L. Prati, *Angew. Chem., Int. Ed.*, 2010, **49**, 4499.
15 R. Arrigo, M. Hävecker, R. Schlögl and D. S. Su, *Chem. Commun.*, 2008, 4891.
16 L. Prati and G. Martra, *Gold Bull.*, 1999, **32**, 96.
17 A. Villa, A. Gaiassi, I. Rossetti, C. L. Bianchi, K. van Benthem, G. M. Veith and L. Prati, *J. Catal.*, 2010, **275**, 108.
18 A. Abad, A. Corma and H. Garcia, *Chem.–Eur. J.*, 2008, **14**, 212.
19 C. Lemire, R. Meyer, S. Shaikhutdinov and H.-J. Freund, *Angew. Chem., Int. Ed.*, 2004, **43**, 118.
20 J. P. Tessonnier, D. Rosenthal, T. W. Hansen, C. Hess, M. E. Schuster, R. Blume, F. Girgsdies, N. Pfaender, O. Timpe, D. S. Su and R. Schloegl, *Carbon*, 2009, **47**, 1779.

21 R. Arrigo, M. Haevecker, S. Wrabetz, R. Blume, M. Lerch, J. McGregor, E. P. J. Parrott, J. A. Zeitler, L. F. Gladden, A. Knop-Gericke, R. Schloegl and D. Su, *J. Am. Chem. Soc.*, 2010, **132**, 9616.
22 A. Villa, D. Wang, P. Spontoni, R. Arrigo, D. Su and L. Prati, *Catal. Today*, 2010, **157**, 89.
23 W. Winter, X. Xia, B. P. C Hereijgers, J. H Bitter, A. J. van Dillen, M. Muhler and K. P. de Jong, *J. Phys. Chem. B*, 2006, **110**(18), 9211.
24 J. Shen, R. D. Cortright, Y. Chen and J. A. Dumesic, *Catal. Lett.*, 1994, **26**(3–4), 247.

Enhanced performance of the catalytic conversion of allyl alcohol to 3-hydroxypropionic acid using bimetallic gold catalysts

Ermelinda Falletta,[a] Cristina Della Pina,[*a] Michele Rossi,[a] Qian He,[b] Christopher J. Kiely[b] and Graham J. Hutchings[c]

Received 8th April 2011, Accepted 6th May 2011
DOI: 10.1039/c1fd00063b

One of the strategic building blocks in organic synthesis is 3-hydroxypropionic acid, which is particularly important for the manufacture of high performance polymers. However, to date, despite many attempts using both biological and chemical routes, no large scale effective process for manufacturing 3-hydroxypropionic acid has been developed. One potentially useful starting point is from allyl alcohol, as this can be obtained in principle from the dehydration of glycerol, thereby presenting a bio-renewable green pathway to this important building block. The catalytic transformation of allyl alcohol to 3-hydroxypropionic acid presents interesting challenges in catalyst design, particularly with respect to the control of selectivity among the products that can be expected, as acrylic acid, acrolein and glyceric acid can also be formed. In this paper, we present a novel eco-sustainable catalytic pathway leading to 3-hydroxypropionic acid, which highlights the outstanding potential of gold-based and bimetallic catalysts in the aerobic oxidation of allyl alcohol.

Introduction

The critical challenge in developing eco-friendly technologies for chemical processes is to reach a balance between economic and environmental aspects. The availability of inexpensive, renewable feedstocks that derive from bio-renewable resources and the employment of specific processes represent precious tools for establishing a "green" manufacturing industry.

In this context, the bio-refining industry will be crucial in providing widespread availability of chemical intermediates thereby by-passing fossil resources. An important platform based on environmentally friendly materials should include 2-hydroxypropionic acid (lactic acid) and butanedioic acid (succinic acid), advantageously produced by fermentation,[1] along with 3-hydroxypropionic acid (3-HP) which occupies the third position in the classification of the main strategic building blocks from biological resources, as emphasized by Werpy et al.[1]

With respect to its chemical characteristics, 3-HP is a useful starting material for cyclization and polymerization reactions to produce propiolactone, polyesters and oligomers. A further important market arising from 3-hydroxypropionic acid is represented by malonic acid and its derivatives, which can be obtained by selective oxidation of the alcoholic functional group.[2] The bi-functionality of 3-hydroxypropionic

[a] Dipartimento di Chimica Inorganica, Metallorganica e Analitica "L. Malatesta", Università degli Studi di Milano, CNR-ISTM, via Venezian, 21, 20133 Milano, Italy. E-mail: cristina.dellapina@unimi.it; Fax: +39 02 50314405; Tel: +39 02 50314397
[b] Department of Material Science and Engineering, Lehigh University, USA
[c] Cardiff Catalysis Institute, School of Chemistry, Cardiff University, UK

acid allows its conversion into a broad range of valuable products by simple chemical transformations (Scheme 1).[2] In particular, the self-condensation of 3-HP leading to poly(3-hydroxypropionic acid) is a relatively novel technology which can compete with traditional polymers owing to its excellent properties, biocompatibility and biodegradability.[3,4]

Among many different chemicals that can be derived from 3-HP, acrylic acid is, perhaps, the most attractive from an economic viewpoint, being the starting material for the preparation of widely employed polymers. Thus, the synthesis of 3-HP *via* an eco-friendly route is highly desirable since this is presently not possible.

To date, no large scale chemical process is available for the production of 3-HP despite major research efforts, while a biological process is still waiting to be identified. Besides traditional stoichiometric reactions, a number of catalytic methods have been reported mainly in the patent literature[5–8] but, to our knowledge, none is yet industrially exploited owing to unfavourable economics of the starting materials, (*e.g.* 1,3-propanediol, acrylic acid) as well as severe environmental constraints.

As a novel application of the exceptional performance of gold nanoparticles in the aerobic oxidation of different hydroxylated molecules,[9] we recently explored allyl alcohol oxidation using a carbon-supported gold catalyst (0.3%Au/C), under alkaline conditions. Most importantly, besides small amounts of the expected sodium glycerate and sodium acrylate products, 3-hydroxypropionate was also detected.[10] The pathways leading to this valuable intermediate for organic synthesis are shown in Scheme 2.

At present, allyl alcohol is manufactured from propene *via* allyl chloride, propeneoxide, acrolein, or allyl acetate intermediates, following chemical processes which are considered to be environmentally troublesome.[11] The prospect of deriving allyl alcohol from glycerol, however, opens up the attractive new possibility of making 3-HP from a green feedstock. As a consequence, we optimized the experimental conditions in order to improve 3-HP production. We observed the following

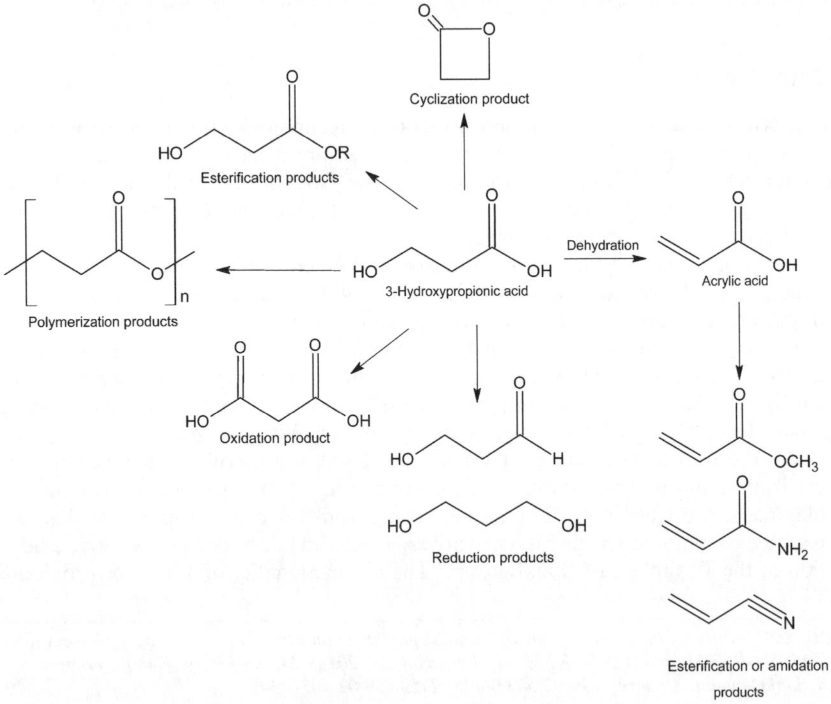

Scheme 1 Possible reactions of 3-hydroxypropionic chemistry.

  This journal is © The Royal Society of Chemistry 2011

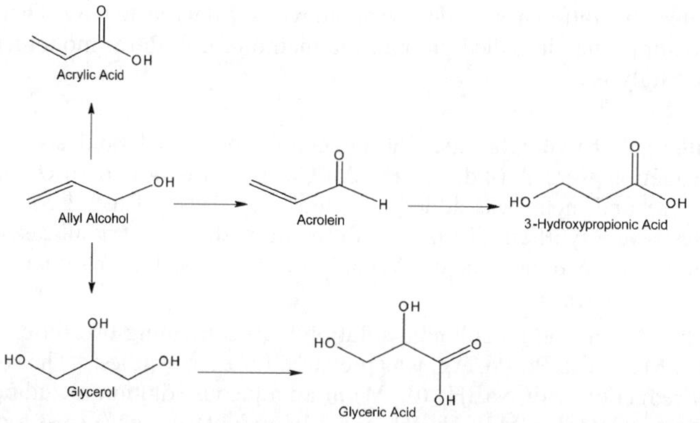

Scheme 2 Possible reaction pathways occurring during allyl alcohol oxidation.

trends:[10] a) the 3-HP yield increased with the temperature from 19% at 25 °C to 42% at 50 °C, and b) by increasing the NaOH/allyl alcohol molar ratio from 1 to 3 at 50 °C, the yield increased from 42% up to 79%. Unfortunately, the Au/C catalyst was not stable and the selectivity to 3-HP was decreased after the second reaction cycle, thereby shedding doubt on the viability of this new process as it is a crucial requirement that catalysts can be re-used over many reaction cycles.

The limited stability prompted us to improve this crucial aspect of the catalyst lifetime. To overcome the problem of selectivity loss, we have investigated a series of bimetallic Au-M nanoparticles (M = Pt, Pd, Cu or Ag) supported on carbon. In a number of previous studies[11–24] the catalytic performance of gold nanoparticles has been enhanced markedly by alloying with a second metal. In particular, alloying Au with Pd has been found to be effective for a number of reactions including the direct synthesis of hydrogen peroxide, the oxidation of alcohols and recently the oxidation of toluene.[16–18] In this paper, we report on the effect of these bimetallic formulations on controlling reaction selectivity with sequential use of the catalyst.

Experimental

Materials

$HAuCl_4$ (Au concentration = 10 mg/ml) prepared by dissolution of gold in *aqua regia*, carbon X40S (A_s = 1300 m^2 g^{-1}, V_p = 0.37 cm^3 g^{-1}) was supplied by Camel Chemicals, $NaBH_4$, $CuCl_2$, $KPtCl_4$, $PdCl_2$, $AgNO_3$, NaOH, LiOH, KOH, Ca $(OH)_2$, H_3PO_4, HCl were purchased from Fluka and used without purification, 3-hydroxypropionic acid (30% aqueous solution) was supplied by TCI Europe.

Catalyst preparation

Monometallic gold catalyst. 0.3%Au/C was synthesized as described elsewhere[10] by colloidal deposition, according to the following protocol: an aqueous solution of $HAuCl_4$ (25 mg l^{-1}) was prepared, glucose was added as a protecting agent (glucose/metal = 50:1 molar ratio), and a freshly prepared solution of $NaBH_4$ (0.1 M, $NaBH_4$/metal = 1:1 by weight) was rapidly added with stirring to obtain a dark yellow colloidal dispersion. The pH of the gold colloid was decreased to pH = 3 by addition of HCl. Carbon powder (X40S) was then added to the sol and left to stand for *ca.* 15 min, and the resulting solid was filtered, washed, dried, and calcined at 400 °C under H_2 (50 ml/min) for 2 h. The gold nanoparticles had a mean diameter of 3.0 nm, as determined by X-ray powder diffraction (XRDP) and confirmed by transmission electron microscopy (TEM). We repeated the preparation of the catalyst but avoiding the calcination step at 400°C under H2 and we did not detect any difference in the morphology and catalytic activity. Therefore, we decided to adopt this simplified preparation method for both the monometallic and bimetallic catalysts.

Bimetallic gold-based catalysts. The bimetallic gold-based catalysts selected for this investigation were Au-Pd, Au-Pt, Au-Cu and Au-Ag supported on carbon X40S. The second metal was added as 1 mol% of the total metal concentration. The catalysts were synthesized *via* three different methods, but in all cases the total metal loading was chosen to be 0.3 weight% of the supporting material.

Method 1. An individual colloidal solution (sol) containing two different metals (M1 = Au, M2 = Cu, Pt, Pd, Ag) was prepared for each synthesis. The sol was obtained by reduction with $NaBH_4$ (0.1M) in an aqueous solution including the two metal precursors ((M1 + M2): $NaBH_4$ = 1:1 by weight) and glucose as a protecting agent ((M1 + M2): glucose = 1:50 molar ratio). The resulting sol was adsorbed on carbon (X40S) and the pH was adjusted at 3 by HCl 0.1M addition. After 30 min, the catalyst was filtered, washed with water several times and dried in air at room temperature overnight.

Method 2. A gold sol was prepared by the reduction of an aqueous solution of $HAuCl_4$ with $NaBH_4$ (0.1M, Au: $NaBH_4$ = 1:1 by weight), using glucose as a protecting agent (Au: glucose = 1:50 molar ratio). The resulting sol was adsorbed on carbon (X40S) and the pH was adjusted at 3 by adding HCl 0.1M. After 30 min, the catalyst was filtered and washed several times with water.

Another sol was synthesized in the same manner but using the second metal precursor ($CuCl_2$, $KPtCl_4$, $PdCl_2$, $AgNO_3$, metal concentration = 0.1 mg ml^{-1}). This second sol was rapidly mixed with the previously prepared monometallic gold catalyst (Au: second metal = 99:1 molar ratio) for 30 min with stirring and then filtered, washed with water and dried in air at room temperature overnight.

Method 3. A gold sol was prepared by the reduction of an aqueous solution of $HAuCl_4$ with $NaBH_4$ (0.1M, Au: $NaBH_4$ = 1:1 by weight), using glucose as a protecting agent (Au: glucose = 1:50 molar ratio). The resulting sol was adsorbed on carbon (X40S) and the pH was adjusted to pH = 3 with HCl (0.1M) addition. After 30 min the catalyst was filtered and washed with water. A salt of the second metal ($CuCl_2$, $KPtCl_4$, $PdCl_2$, $AgNO_3$, metal concentration = 0.1 mg ml^{-1}) was added to an aqueous solution containing the gold catalyst and the mixture was reduced under H_2 flow (50 ml min^{-1}) for 2h. Then, the catalyst was filtered, washed with water and dried in air at room temperature overnight.

Characterization

X-ray powder diffraction (XRPD) was performed on the sol immobilized catalysts using a Rigaku DIII-MAX Horizontal-scan powder diffractometer operating with Cu K_α radiation). The Scherrer equation was used in order to determine, the average metal particle size based on peak width measurements. Samples for examination by scanning transmission electron microscopy (STEM) were prepared by dispersing the dry catalyst powder onto a holey carbon film supported by a 300 mesh copper TEM grid. STEM high angle annular dark field (HAADF) images of the bimetallic particles were obtained using an aberration corrected JEOL 2200FS TEM operating at 200kV.

Catalyst testing

The oxidation of allyl alcohol was carried out in the 25–50 °C temperature range, using a glass reactor (50 ml) connected to a large reservoir containing oxygen. Allyl alcohol (1g, 17.24 mmol) and the catalyst (reactant/metal = 2000–4000 molar ratio) were mixed in distilled water (17.24 ml) in the presence of different amounts of NaOH or other bases (LiOH, KOH, Ca(OH)$_2$), with a molar ratio of base/allyl alcohol of 1;3. The reactor was pressurized to 3 bar oxygen, and the mixture was stirred for 24h. The reaction products were analyzed by HPLC (Shimadzu LC-10) equipped with a refractive index detector RID-10 and Varian MetaCarb H Plus column using aqueous 0.01 M H$_3$PO$_4$ as the eluent. 3-Hydroxypropionic acid (30 % aqueous solution) was used as a standard reference material for ^{1}H NMR and ^{13}C NMR spectroscopy, as well as ^{1}H-^{1}H COSY and [1]H-^{13}C HMQC techniques and HPLC analysis. Analytical data were reproducible within a 3% deviation.

Results and discussion

Allyl alcohol aerobic oxidation using monometallic supported gold catalysts

Before deciding to address our strategy of testing bimetallic systems, we first tried to improve the performance of the monometallic gold catalyst (0.3% Au/C) which suffered from a short life-time, as shown by the data in Figure 1.

A drawback in recycling the monometallic gold catalyst was, in fact, the selectivity loss observed after the second reaction cycle as the 3-HP yield decreased to 26% in favor of glyceric acid (34%), while the formation of acrylic acid remained almost constant. Further use of the catalyst in subsequent reaction cycles confirmed such a trend (Figure 1). The first strategy attempted to overcome this loss of catalyst performance was to change the nature of the basic reaction medium, leaving the catalyst unaltered, by testing a series of alkaline compounds as an alternative to NaOH (KOH, LiOH, Ca(OH)$_2$). The results are reported in Table 1.

In the first reaction, using NaOH, KOH and LiOH, 100% conversion was reached with similar selectivity (79, 83, 75% in 3-HP respectively), while with Ca(OH)$_2$ the conversion was much lower (34%). In each case, however, the selectivity dramatically decreased during the third run, whereas with LiOH the catalyst was totally inactive on first re-use.

Allyl alcohol aerobic oxidation using bimetallic supported gold catalysts

Doping noble metals with one or more different elements often promotes an improvement in, and stabilization of, the catalytic activity. This has been previously observed in the case of the Pt-Pd/C and Pt-Pd-Bi/C catalysts, when they were used for the oxidation of 1,3-propanediol[5,12] and for the Ir-Re catalyst used in the

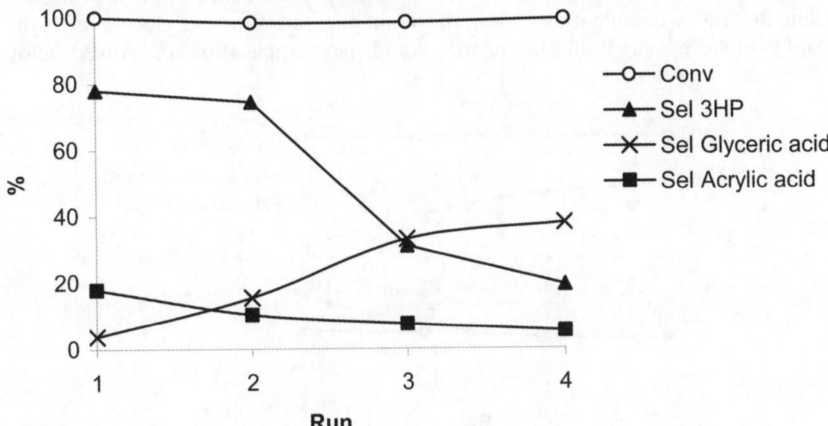

Fig. 1 Loss of selectivity of 0.3% Au/C catalyst on recycling. Reaction conditions: [allyl alcohol] = 1M, pO$_2$ = 3 bar, allyl alcohol/metal = 4000 (molar ratio), T = 50 °C; NaOH/allyl alc. 3, t = 24 h.

Table 1 The effect of different basic compounds on the activity and selectivity to 3-HP during the gold catalyzed oxidation of allyl alcohol.[a]

Run	Base	Conv%	Sel% (3-HP)
1	NaOH	100	79
	KOH	100	83
	LiOH	100	75
	Ca(OH)$_2$	34	5
2	NaOH	98	76
	KOH	99	61
3	NaOH	98	31
	KOH	97	21
4	NaOH	99	19
	KOH	47	5

[a] Reaction conditions: T = 50 °C, PO$_2$ = 3bar; base:allyl alcohol molar ratio = 3; 0.3%Au/C: allyl alcohol = 1:2000 (molar ratio); t = 24 h.

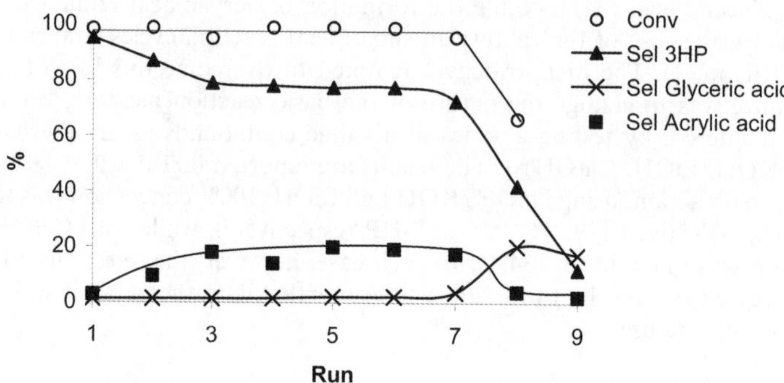

Fig. 2 Activity and selectivity of 0.3% Au$_{99}$-Cu$_1$/C, synthesized with Method 1, on recycling. T = 50 °C, PO$_2$ = 3bar; base:allyl alcohol = 3 (molar ratio); catalyst:allyl alcohol = 1:2000 (molar ratio); t = 24 h.

hydrogenolysis of glycerol.[13] The benefits in applying gold-based bimetallic catalyst formulations have been also pointed out in the gas phase oxidation of benzyl alcohol using the Au-Cu combination,[14] or in the liquid phase oxidation of glucose using the Au-Pt mixture,[15] as well as in the brand new application of Au-Pd alloy

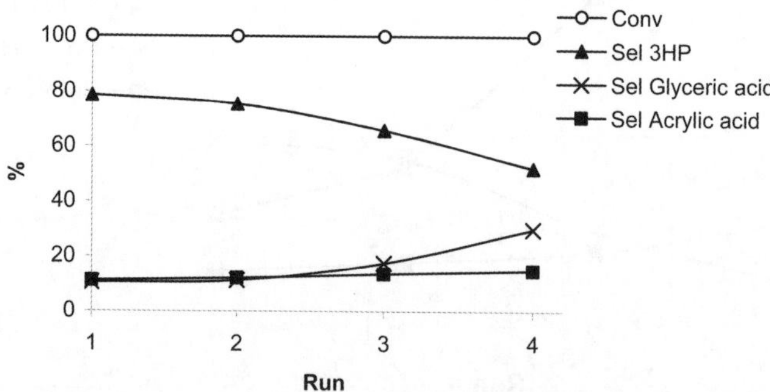

Fig. 3 Activity and selectivity of 0.3% Au$_{99}$-Cu$_1$/C, synthesized with Method 2, on recycling. T = 50 °C, PO$_2$ = 3bar; base:allyl alcohol = 3 (molar ratio); catalyst:allyl alcohol =1:2000 (molar ratio); t = 24 h.

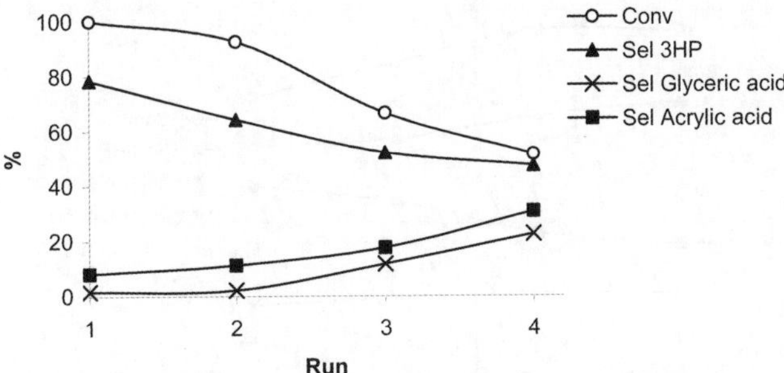

Fig. 4 Activity and selectivity of 0.3% Au$_{99}$-Cu$_1$/C, synthesized with Method 3, on recycling. T = 50 °C, PO$_2$ = 3bar; base:allyl alcohol = 3 (molar ratio); catalyst:allyl alcohol = 1:2000 (molar ratio); t = 24 h.

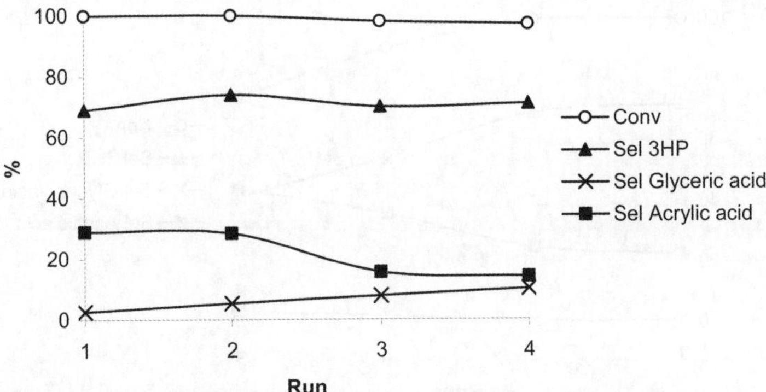

Fig. 5 Activity and selectivity of 0.3% Au$_{99}$-Pt$_1$/C, synthesized with Method 1, on recycling. T = 50 °C, PO$_2$ = 3bar; base:allyl alcohol = 3 (molar ratio); catalyst:allyl alcohol = 1:2000 (molar ratio); t = 24 h.

nanoparticles in the solvent-free oxidation of primary carbon-hydrogen bonds in toluene.[16] Hence, a series of bimetallic Au-M on carbon catalysts (M = Pt, Pd, Cu and Ag) were prepared using three methods, as described previously, and tested for the aerobic oxidation of allyl alcohol. Initial experiments showed that using relatively large amounts of the second metal (M:Au = 2–0.5 molar ratio) inhibited the reaction rate with respect to the performance of the monometallic gold catalyst. However, following careful experimentation we found that a rather minor amount

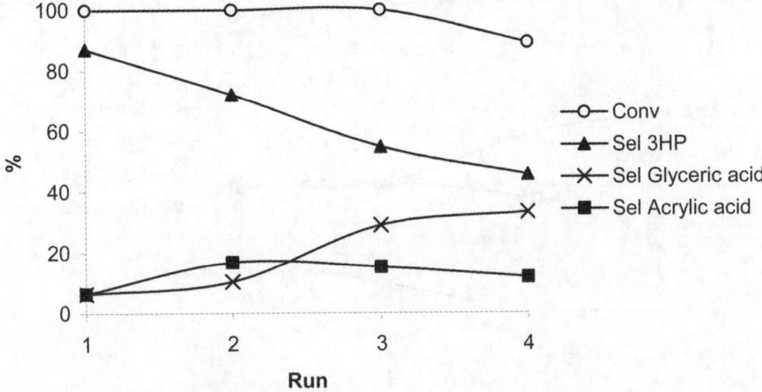

Fig. 6 Activity and selectivity of 0.3% Au$_{99}$-Pt$_1$/C, synthesized with Method 2, on recycling. T = 50 °C, PO$_2$ = 3bar; base:allyl alcohol = 3 (molar ratio); catalyst:allyl alcohol = 1:2000 (molar ratio); t = 24 h.

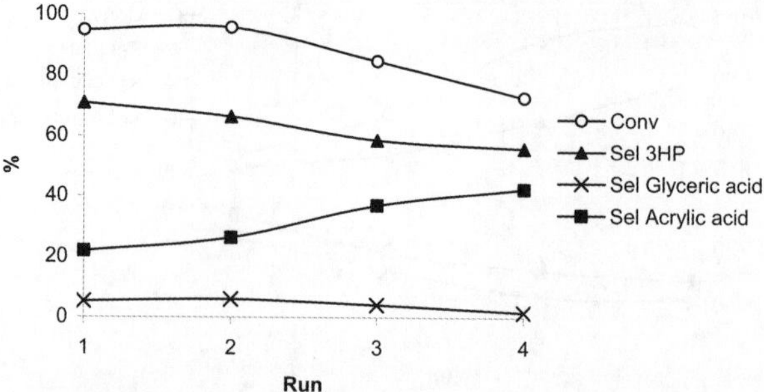

Fig. 7 Activity and selectivity of 0.3% Au$_{99}$-Pt$_1$/C, synthesized with Method 3, on recycling. T = 50 °C, PO$_2$ = 3bar; base:allyl alcohol = 3 (molar ratio); catalyst:allyl alcohol = 1:2000 (molar ratio); t = 24 h.

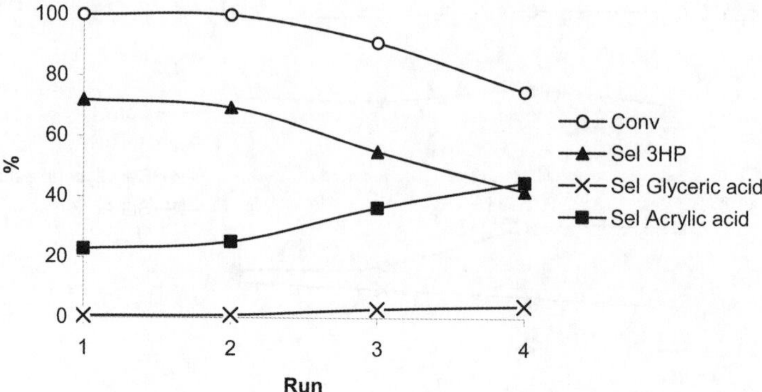

Fig. 8 Activity and selectivity of 0.3% Au$_{99}$-Pd$_1$/C, synthesized with Method 1, on recycling. T = 50 °C, PO$_2$ = 3bar; base:allyl alcohol = 3 (molar ratio); catalyst:allyl alcohol =1:2000 (molar ratio); t = 24 h.

of the second metal (only 1% mol with respect to the total amount of the two metals) induced the desired benefit with most of the alloying metals we investigated (Figures 2–11).

Regarding 0.3%Au-Cu/C system prepared following the three different protocols, the tests reported in Figures 2–4 highlight how the first preparation method

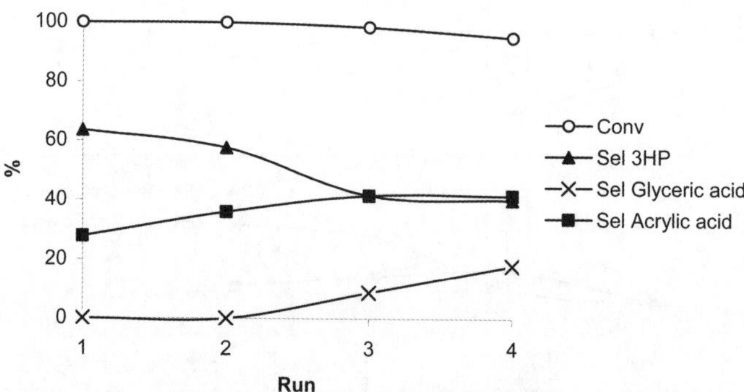

Fig. 9 Activity and selectivity of 0.3% Au$_{99}$-Pd$_1$/C, synthesized with Method 2, on recycling. T = 50 °C, PO$_2$ = 3bar; base:allyl alcohol = 3 (molar ratio); catalyst:allyl alcohol = 1:2000 (molar ratio); t = 24 h.

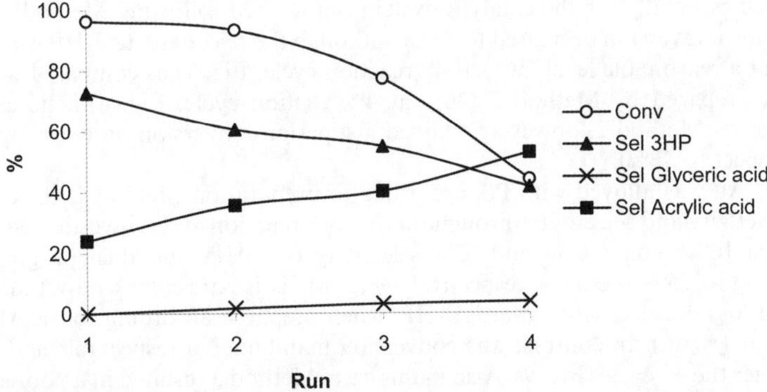

Fig. 10 Activity and selectivity of 0.3% Au$_{99}$-Pd$_1$/C, synthesized with Method 3, on recycling. T = 50 °C, PO$_2$ = 3bar; base:allyl alcohol = 3 (molar ratio); catalyst:allyl alcohol = 1:2000 (molar ratio); t = 24 h.

(Method 1) is observed to be the most effective, with the conversion remaining high after 7 reaction cycles. Selectivity toward 3-HP was also notable, reaching the value of 95% during the first reaction cycle and retaining a high level up to the 7th cycle, but then, unfortunately, a marked decrease in both activity and selectivity was observed (Fig. 2).

As to Method 2 used in the catalyst preparation, even though the total conversion was always achieved during recycling, this catalyst sustained a loss in selectivity towards 3-HP (from 78% at the 1st run to 52% at the 4th run) and glyceric acid production was observed (Fig. 3).

The poorest performance was observed for the catalyst synthesized according to Method 3, the conversion being halved after the 4th reaction cycle. The selectivity to 3-HP, however, was similar to that obtained with catalyst prepared following the second protocol, but an increase in acrylic acid production was observed during the 4th run (Fig. 4).

Also the 0.3%Au-Pt/C catalytic system, prepared following the three protocols, was found to be most effective when synthesized according to Method 1. The performance in terms of conversion was maintained at a high level after 4 reaction cycles, whereas a constant selectivity towards 3-HP (around 70%) was achieved throughout the four reaction cycles (Fig. 5).

Preparation of the 0.3% Au-Pt/C catalyst according to the Methods 2 and 3 (Figs. 6 and 7 respectively) gave a poor performance on recycling. In fact, conversion

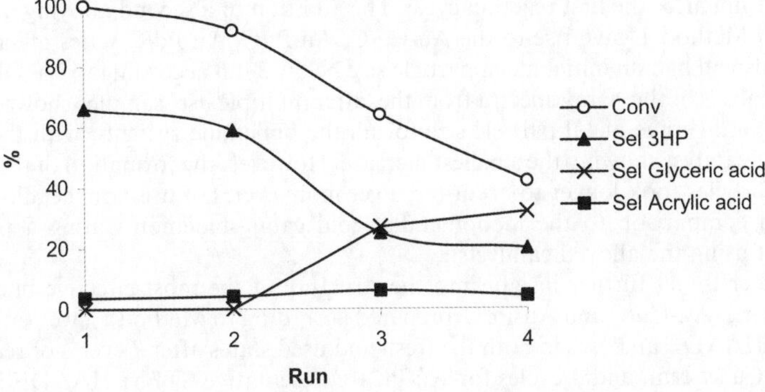

Fig. 11 Activity and selectivity of 0.3% Au$_{99}$-Ag$_1$/C, synthesized with Method 1, on recycling. T = 50 °C, PO$_2$ = 3bar; base:allyl alcohol = 3 (molar ratio); catalyst:allyl alcohol = 1:2000 (molar ratio); t = 24 h.

decreased especially for the catalytic system synthesized following Method 3 (after the 4th run, conversion decreased to 73%), although the selectivity to 3-HP was maintained at a reasonable level (56% at 4th reaction cycle, Fig. 7) as compared with the catalyst prepared by Method 2 (46% at 4th reaction cycle, Fig. 6). The catalyst prepared by Method 2, however, retained a superior conversion on recycling (90% with respect to 73%).

When Au was alloyed with Pd, according to preparation Method 1, the catalyst lost its activity and selectivity throughout the four reaction cycles investigated, starting from 100% conversion and 72% selectivity to 3-HP, and then progressively decreasing to 75% and 42%, respectively (Fig. 8). This particular catalyst also sustained a loss in selectivity towards 3-HP, when prepared according to the Method 2 (63% at 1st run). In contrast, the conversion maintained a respectable and stable value after the 4th cycle (Fig. 9). Also in this case, Method 3 resulted in a poorer catalyst, in terms of both conversion and selectivity to 3-HP (Fig. 10). In the present study we did not synthesize the 0.3% Au-Ag/C catalyst *via* Methods 2 and 3, as this particular bimetallic combination was not effective even when prepared by the most promising protocol (Method 1, Fig. 11).

The data reported here clearly demonstrate that Method 1 was the most promising for preparing the bimetallic catalysts. Regarding the three main catalytic systems (Au-Cu/C, Au-Pt/C and Au-Pd/C), the performance in terms of conversion was maintained at a high level even after a significant number of cycles: 7 for the Au-Cu system, 4 for Au-Pt, and 3 in the case of Au-Pd, while the Au-Ag alloy exhibited a sharp decrease in activity even after the second reaction cycle. As far as the selectivity toward 3-HP is concerned, the Au-Cu/C and Au-Pt/C catalysts displayed the best results. In particular, the Au-Cu catalyst reached 80% selectivity to 3-HP, maintaining this high level up to the 7th cycle, but then, unfortunately, displayed a marked decrease in both activity and selectivity. Similarly, the Au-Pt system maintained a respectable performance for four reaction cycles. As for the second method of the catalyst preparation, despite retaining high values of conversion during recycling, these catalysts sustained a loss in selectivity which followed the trend: Au-Pd>Au-Pt>Au-Cu. Generally, with these catalysts the selectivity towards 3-HP decreased and an increase in acrylic acid production was observed. A similar diminishing trend in catalytic performance was particularly apparent when the catalysts were prepared according to Method 3.

Catalyst characterization

The monometallic catalyst 0.3%Au/C synthesized according to Method 1 was characterized by the XRPD technique both before and after catalytic testing. This investigation showed that the average gold particle diameter sharply increased from 3 nm up to 11 nm after the first reaction cycle. The addition of a second alloying metal to gold *via* Method 1 gave rise to the Au-Cu/C, Au-Pt/C, Au-Pd/C series of samples, which also all had an initial mean particle size below 3 nm according to XRDP analysis. Analysis of the X-ray spectra from the 'after multiple use' samples showed some increase in average metal particle size for all the bimetallic systems, with the 0.3% Au-Pt/C catalyst showing the smallest increase. However, the formation of the larger metal particles took longer to occur (*i.e.* over more cycles of use) for the alloy catalysts, in comparison to the monometallic gold catalyst demonstrating a positive effect of using the alloyed catalysts.

In order to get further insight into the structure of the most effective bimetallic systems (*i.e.* Au-Cu/C and Au-Pt/C, prepared according to Method 1), we performed STEM-HAADF analyses in both the fresh and used states after 7 cycles of reaction, for Au-Cu system, and 4 cycles for Au-Pt. Representative STEM-HAADF images and corresponding particle size distributions of the 'fresh' and 'used' Au$_{99}$Pt$_1$/C andAu$_{99}$Cu$_1$/C catalysts are shown in Figures 12 and 13 respectively. Both samples showed a very similar initial average particle size: *i.e.* 1.4 nm for Au$_{99}$Pt$_1$/C and

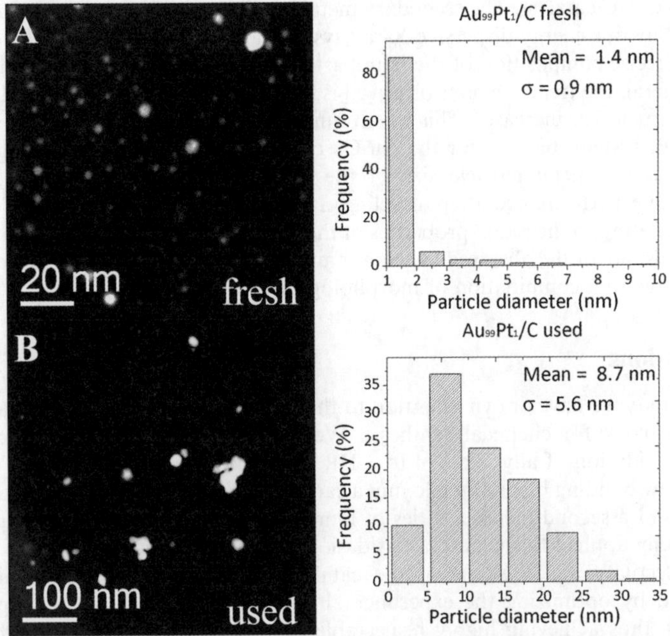

Fig. 12 STEM-HAADF images of the fresh and used (4 cycles) $Au_{99}Pt_1$/C catalysts and their corresponding particle size distributions.

1.6 nm for $Au_{99}Cu_1$/C. In both cases, the metal particles showed definite signs of sintering after reaction. The average particle size showed a modest increase to 8.7 nm for the $Au_{99}Pt_1$/C catalyst, but was more substantial (up to 18.7 nm) for

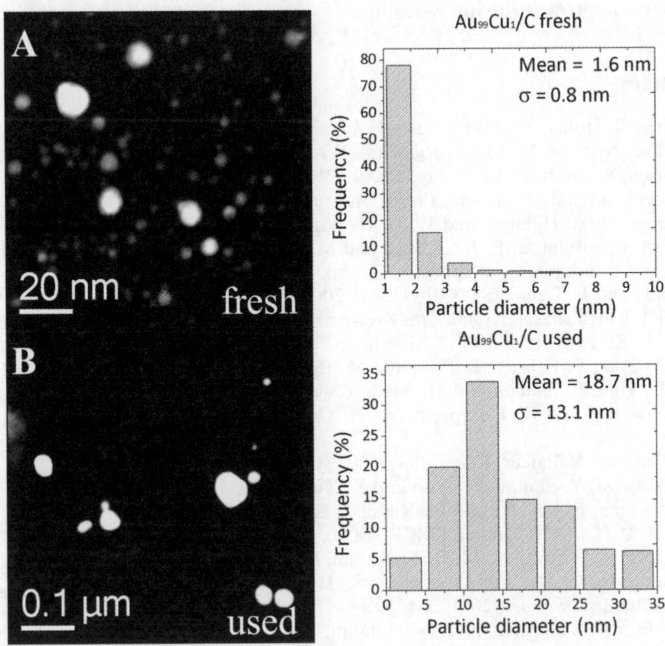

Fig. 13 STEM-HAADF images of the fresh and used (7 cycles) $Au_{99}Cu_1$/C catalysts and their corresponding particle size distributions.

the $Au_{99}Cu_1$/C catalyst. The secondary metal loading was too low in each case to be detected by our energy dispersive X-ray system, and therefore we cannot comment on the precise composition of the nanoparticles from our microscopy studies.

In general, the trend we have observed is that the catalytic performance decreases as the particle size increases. This was confirmed by our STEM-HAADF results for the Au-Pt system, but not for the Au-Cu catalyst. In this latter case, a significant growth in the metal particle size of the recycled catalyst occurred, despite its outstanding performance after recycling. In this case, we consider that electronic factors relating to the redox properties of the alloying metal added to the gold could be responsible for the observed behaviour and so the observed catalytic performance may be due to a combination of morphological and electronic factors.

Conclusions

In this study we have drawn attention to the possible importance of 3-HP as a new building block for chemical synthesis. We have demonstrated that the catalytic aerobic oxidation of allyl alcohol to 3-HP under mild conditions is a useful route to this new building block. By alloying a carbon supported gold catalyst with minor amounts of a second metal, a series of bimetallic catalysts has been generated and successfully applied to the aerobic oxidation of allyl alcohol to 3-HP. The conversion, selectivity and life-time of the catalysts could be tuned and significantly improved by optimizing the experimental conditions, as well as the preparation methods, thus achieving highly respectable performances, especially for the 0.3% $Au_{99}Cu_1$/C bimetallic system. The stabilization of the catalytic activity due to the presence of the dilute second metal, most likely through alloying, as evidenced in the case of Au-Pt,[15] could be a consequence of electronic modifications in the metal nanoparticle. A key factor in controlling the catalytic performance is often the metal particle size which, however, seems not to be of fundamental importance in the case of the Au-Cu/C system. In this case the low redox potential of copper could be implicated in its sustained high performance, thus outweighing any limitations imposed by the particle growth during recycling.

References

1 T. Werpy, J. Holladay, J. White, *Top Value Added Chemicals From Biomass*. PNNL-14808, 2004. Pac. Northw. Nat. Lab., Richland, WA. http://environment.pnl.gov/staff/staff_info.
2 A. Corma, S. Iborra and A. Velty, *Chem. Rev.*, 2007, **107**, 2411.
3 M. Mochizuki and M. Hirami, *Polym. Adv. Technol.*, 1997, **8**, 203.
4 D. Zhang, M. A. Hillmyer and W. B. Tolman, *Macromolecules*, 2004, **37**, 8198.
5 A. Behr, A. Botulinski, F. J. Carduck and M. Schneider, 1992, DE 4107987, to Henkel.
6 Ullmann's Encyclopedia of Industrial Chemistry 6th Edition, vol 17, p. 322.
7 H. Ishida and E. Ueno, JP 2000159724 A 20000613 to Asahi Chemical Industry Co.
8 See URL 9362554/asia-acrylates-may-set-new-highs-as-supply-heads-to-us-europe.html.
9 (a) A. S. K. Hashmi and G. J. Hutchings, *Angew. Chem., Int. Ed.*, 2006, **45**, 7896; (b) C. Della Pina, E. Falletta, L. Prati and M. Rossi, *Chem. Soc. Rev.*, 2008, **37**, 2077.
10 C. Della Pina, E. Falletta and M. Rossi, *ChemSusChem*, 2009, **2**, 57.
11 K. Weissermel and H. J. Arpe, *Industrial Organic Synthesis*, VCH, Weinheim, 1997, 3rd edition.
12 T. Mallat and A. Baiker, *Chem. Rev.*, 2004, **104**, 3037.
13 Y. Nakagawa, Y. Shinmi, S. Koso and K. Tomishige, *J. Catal.*, 2010, **272**, 191.
14 C. Della Pina, E. Falletta and M. Rossi, *J. Catal.*, 2008, **260**, 384.
15 M. Comotti, C. Della Pina and M. Rossi, *J. Mol. Catal. A*, 2006, **251**, 89.
16 L. Kesavan, R. Tiruvalam, M. H. Ab Rahim, M. I. bin Saiman, D. I. Enache, R. L. Jenkins, N. Dimitratos, J. A. Lopez-Sanchez, S. H. Taylor, D. W. Knight, C. J. Kiely and G. J. Hutchings, *Science*, 2011, **331**, 195.
17 J. K. Edwards, B. Solsona, N. Edwin Ntainjua, A. F. Carley, A. A. Herzing, C. J. Kiely and G. J. Hutchings, *Science*, 2009, **323**, 1037.
18 D. I. Enache, J. K. Edwards, P. Landon, B. Solsona-Espriu, A. F. Carley, A. A. Herzing, M. Watanabe, C. J. Kiely, D. W. Knight and G. J. Hutchings, *Science*, 2006, **311**, 362.

19 M. C. Daniel and D. Astruc, *Chem. Rev.*, 2004, **104**, 293.
20 Y. Mizukoshi, T. Fujimoto, Y. Nagata, R. Oshima and Y. Maeda, *J. Phys. Chem. B*, 2000, **104**, 6028.
21 Y. Mizukoshi, K. Okitsu, Y. Maeda, T. A. Yamamoto, R. Oshima and Y. Nagata, *J. Phys. Chem. B*, 1997, **101**, 7033.
22 H. Tada, F. Suzuki, S. Ito, T. Akita, K. Tanaka, T. Kawahara and H. Kobayashi, *J. Phys. Chem. B*, 2002, **106**, 8714.
23 C. Mihut, C. Descorme, D. Duprez and M. D. Amiridis, *J. Catal.*, 2002, **212**, 125.
24 C. L. Bianchi, P. Canton, N. Dimitratos, F. Porta and L. Prati, *Catal. Today*, 2005, **102**, 203.

Preparation of ultra low loaded Au catalysts for oxidation reactions

Adrian Thomas,[a] Qian He[b] and Jennifer K. Edwards[*a]

Received 15th February 2011, Accepted 10th March 2011
DOI: 10.1039/c1fd00021g

Cyanide leaching was used to obtain Au/SiO_2 catalysts with very low gold loadings. A number of catalysts with a nominal 5 wt% target loading were prepared using impregnation and deposition precipitation techniques and these were found to be active catalysts for the solvent-free aerobic oxidation of benzyl alcohol. Exposure of these catalysts to a basic solution of NaCN for 10 min leached gold from the materials to give very pale pink catalysts which were found to contain just 0.06 wt% Au. The concentration of Au removed from the catalyst was constant regardless of the length of NaCN exposure. When the cyanide leached materials were employed for benzyl alcohol oxidation under the same conditions as the unleached parent catalysts, the conversions were identical. However, when the catalytic activity was normalised to the Au content (determined by ICP analysis) the TOFs were much higher for the NaCN treated catalysts ($>400,000$ h^{-1}). These results clearly demonstrate that NaCN leaching is an effective route to the development of catalysts containing very low gold content, whilst maintaining high activity. The leached materials were found to comprise metallic Au nanoparticles. The successful utilisation of ultra low loaded Au catalysts for selective oxidation, where the majority of the Au present is active and 98% of the spectator Au is removed, makes the industrial application of such materials more economically viable. Information obtained through the identification of these active structures using state of the art techniques may provide useful insights into how the reaction proceeds on the Au surface.

Introduction

Selective oxidation is a very important process in the synthesis of fine chemicals and intermediates and the selective oxidation of primary alcohols to aldehydes provides a direct route to clean high value perfumery chemicals.[1] Today the concept of green chemistry and sustainability is a key consideration and processes are required that can effectively utilise raw materials, reduce waste and avoid the use of toxic intermediates under mild reaction conditions. In order to satisfy the 12 principles of green chemistry,[2] the toxic, stoichiometric oxidants traditionally employed for this transformation (chromate, permanganate) should be avoided and much simpler oxidants used instead.[1] There is substantial interest in the development of highly active and selective heterogeneous catalysts for the oxidation of alcohols in a simple one pot method. The major advantage of this approach is that heterogeneous metal catalysts are recoverable and reusable, with no metal loss and therefore no metal waste, and when oxygen or hydrogen peroxide are employed as oxidants, the sole by-product of

[a]Cardiff Catalysis Institute, Cardiff University, Main Building Park Place, Cardiff, CF10 3AT. E-mail: edwardsjk@cf.ac.uk
[b]Department of Materials Science and Engineering, Lehigh University, 5 East Packer Avenue, Bethlehem, PA, 10185-3195, USA

the reaction is water.[3] Thus, catalysts based on heteropolyacids,[4,5] hydrotalcites,[6] molecular sieves,[7,8] mixed oxides,[9] and Au, Pd, Ru and Pt[10–16] supported catalysts have been used for the oxidation of alcohols[7] and recently it has been shown that bimetallic catalysts based on Au and Pd are highly effective in the oxidation of alcohols and polyols. Kaneda *et al.*[17] showed that Pd nanoparticles supported on hydroxyapetite gave turnover frequencies (TOF, moles benzyl alcohol reacted per mole Au present per hour) of 9800 h^{-1} for the oxidation of phenylethanol and benzyl alcohol under mild conditions.

The initial discovery by Haruta[18] that finely dispersed supported Au nanoparticles are exceptionally active for CO oxidation at sub-ambient temperatures was followed by the successful utilisation of Au catalysts for a wide range of highly selective, clean oxidation reactions which operate under very mild conditions. One of the most significant advances in the field of alcohol oxidation has been the studies by Corma and co-workers[15,19] showing that an Au/CeO_2 catalyst is active for the selective oxidation of alcohols to aldehydes and ketones, and the oxidation of aldehydes to acids. In these reactions the catalysts are active at relatively mild conditions, without the addition of a solvent, and use O_2 as the oxidant, without the need for NaOH addition to achieve high activity. The results were shown to be comparable to, or higher than, the highest activities that had been previously observed by Kaneda with supported Pd catalysts.[17] The catalytic activity was ascribed to the Au/CeO_2 catalyst stabilising a reactive peroxy intermediate from O_2. Subsequently, Enache *et al.*[20] showed that alloying Pd with the Au in supported Au/TiO_2 catalysts enhanced the activity for alcohol oxidation under solvent free conditions by a factor of >25. The extensive work that has been carried out on supported Au (and Au–Pd) nanoparticles as heterogeneous catalysis typically uses high Au(Pd) concentrations. Prati[21–23] and Hutchings[24] have demonstrated that Au catalysts can be prepared using a sol-immobilisation technique to give materials which have 1 wt% total AuPd loading, although there are no current preparation methodologies able to produce active reusable catalysts with <0.5 wt% total metal loading. The catalyst prepared by Kaneda[17] comprises 0.2 wt% Pd, the highly active nanocrystalline Au/CeO_2 catalyst prepared by Corma[16] contains 2.8 wt% Au and the bimetallic $AuPd/TiO_2$ catalysts prepared by Enache[25] has a 5 wt% total metal content. Whilst these materials are extremely active for the selective oxidation of alcohol, it is unlikely that all of the metal present is in the active form required for catalysis. Detailed characterisation of the 2.5 wt% $AuPd/TiO_2$ catalyst showed a bimodal particle size distribution of AuPd nanoparticles ranging from 5–60 nm and thus, this catalyst will have a number of different sites on the surface.[26] A recent study by Hutchings *et al.*[27] elegantly demonstrated that only a small percentage of gold in an Au/Fe_2O_3 catalyst is active for CO oxidation. The imaging of the Au/Fe_2O_3 catalyst with aberration corrected high annular dark field scanning transmission electron microscopy (HAADF-STEM) showed a number of Au structures present on an active and inactive catalyst (atoms, bilayers and clusters). The bilayer gold structures on the Fe_2O_3 surface were only found in the samples which were active for CO oxidation. These bilayer species (obtained through controlled thermal treatment) are made up of around 10 Au atoms, and were found to comprise just 1% of the surface Au atoms, suggesting that just 0.05 wt% of the metal on the catalyst is active. This discovery should come as no surprise as in 2003, Flytzani-Stephanopoulos and co-workers[28] demonstrated that cationic gold was an important element in obtaining a high activity water gas shift catalyst. These catalysts were prepared *via* the deposition precipitation of Au onto nano-crystalline 10% La-doped CeO_2 and were subsequently leached with 2% NaCN removing 90% of the gold. After NaCN leaching no Au particles were detected by XRD however the catalytic activity was not only retained, it was significantly enhanced, on a catalyst that was found to contain just 0.001 wt% Au.

Whilst highly active low loaded catalysts have been successfully prepared and applied for the water gas shift reaction, these materials have not been used in liquid phase transformations, nor has the active site of the catalyst been clearly determined. Here we demonstrate for the first time that ultra low loaded catalysts can be sucessfully applied for the oxidation of benzyl alcohol.

Experimental

Catalyst preparation

Impregnation An aqueous solution of gold was prepared (2.5 g_{Au} in 75 ml). For 1 g of 5 wt% Au/SiO$_2$ catalyst 1.5 ml of Au solution was stirred (300 rpm) at 50 °C. To this 0.95 g SiO$_2$ (Grace, 50 g, pore volume 1.5 ml g^{-1}) was added and stirring continued until the paste was homogenous in colour (pale yellow). The paste was dried at 80 °C for 16 h and then calcined under air to give the final catalyst (500 °C, 15 min, ramp rate 10 °C min^{-1}) Catalyst denoted Au-IMP.

Deposition precipitation (DP) with NaOH For 5 g 5 wt% Au/SiO$_2$, 4.75 g of SiO$_2$ was suspended in deionised water (200 ml). NaOH (0.1 M) was added to this suspension until a steady pH of 8 was achieved. The Au (18.25 ml, 0.25 g_{Au}) was added drop-wise with stirring to the slurry while the pH was maintained at 8 by further addition of NaOH. After complete addition of the Au^{3+} the yellow suspension was stirred overnight at pH 8. Once stirring was stopped the solution was filtered and the catalyst washed with deionised water (1 L). The catalyst was then dried (16 h) in an oven at 110 °C to give a lilac powder which was calcined in air to give the pale red catalyst (500 °C, 15 min, ramp rate 10 °C min^{-1}). Catalyst denoted Au-DPH.

Deposition precipitation with Na$_2$CO$_3$ SiO$_2$ (4.75 g) was suspended in deionised water (100 ml). Na$_2$CO$_3$ (2 M) was added to the suspension until a steady pH of 8 was achieved. Once this was reached Au (18.25 ml, 0.25 g_{Au}) was added drop-wise with stirring to the slurry while the pH was maintained at 8 by addition of Na$_2$CO$_3$. After gold addition the yellow suspension was stirred at pH 8 overnight. Once stirring was stopped the solution was filtered and the catalyst washed on the filter with deionised water for 8 h. The retentate was then dried in an oven overnight at 110 °C (burgundy) and calcined in air to give the final catalyst (500 °C, 15 min, ramp rate 10 °C min^{-1}). Catalyst denoted Au-DPC.

Cyanide leaching 0.5 g of catalyst was placed in a round bottom flask to which a small amount of NaOH (1 M) was added to ensure no acid was present on addition of NaCN. 2% NaCN (50 mg) was then added to the suspension and O$_2$ bubbled through the solution. The solution was stirred for a set time (5 min, 10 min, 30 min or 1 h (denoted Au-IMP$_5$ for Au-IMP catalyst exposed for 5 min)) at pH 12, after which the catalyst was recovered by filtration, washed with plenty of water to remove any cyanide residue and dried overnight at 110 °C.

Catalyst testing

Benzyl alcohol oxidation was carried out in a stirred reactor (100 ml, Parr Instruments). The autoclave was charged with benzyl alcohol (40 ml) and catalyst (0.05 g) and then purged 5 times with oxygen leaving the vessel at 10 barg. The stirrer was set to 1500 rpm, the reaction mixture was heated to the required temperature and the reaction time was started as soon as the required reaction temperature was reached. Samples from the reactor were taken periodically, *via* a sampling system and analysed using GC (Varian Star 3400 cx with a 30 m CP-Wax 52 CB column).

Catalyst characterisation

X-ray photoelectron spectroscopy (XPS) was carried out on a Kratos Axis Ultra DLD spectrometer employing a monochromatic Al K$_\alpha$ X-ray source (75–150 W) and analyser pass energies of 160 eV (for survey scans) or 40 eV (for detailed scans).

Samples were mounted using double-sided adhesive tape and binding energies referenced to the C(1s) binding energy of adventitious carbon contamination which was taken to be 284.7 eV.

Scanning electron microscopy (SEM) and energy dispersive X-ray (EDX) analyses were performed on a Carl Zeiss EVO-40 fitted with backscatter detector and Oxford Instruments Si(Li) detector. Samples were mounted on aluminium stubs using adhesive carbon discs and they were analysed uncoated.

Samples were prepared for transmission electron microscopy (TEM) and scanning transmission electron microscopy (STEM) by dry dispersing the catalyst powder onto a holey carbon TEM grid. High resolution electron microscopy (HREM) and high-angle annular dark field (HAADF) imaging experiments were carried out using a 200 kV JEOL 2200FS transmission electron microscope equipped with a CEOS aberration corrector. All the STEM-HAADF images were treated with a light low pass filter using a 3×3 kernel to decrease the high frequency noise. The JEOL 2200FS TEM/STEM was equipped with a Si(Li) detector Thermo Scientific Inc Si(Li) detector for X-ray energy dispersive spectroscopy (XEDS) analysis. Point spectra were acquired with a total acquisition time of 120 s. Spectrum images were acquired using a pixel dwell time of 800 ms. Multi-variate statistical analysis (MSA) of the XEDS data cubes was carried out utilizing the MSA plug-in for Digital Micrograph.

Samples for ICP analysis were prepared by digesting them in HF prior to analysis using a JY Horiba Ultima system.

Results and discussion

Effect of cyanide treatment on catalyst activity

Initially three catalysts were prepared on SiO_2 using deposition precipitation with NaOH (Au-DPH), deposition precipitation with Na_2CO_3 (Au-DPC) and by impregnation (Au-IMP). These catalysts had a target metal loading of 5 wt% Au, and were calcined in air at 500 °C. Each of these parent unleached catalysts were investigated for the oxidation of benzyl alcohol at 140 °C with 10 barg O_2 in the absence of solvent. The results (Fig. 1) demonstrate that each of the catalysts showed some activity for this reaction, with the catalysts prepared by deposition precipitation showing 14% conversion at 3h compared to 6% for the impregnated sample.

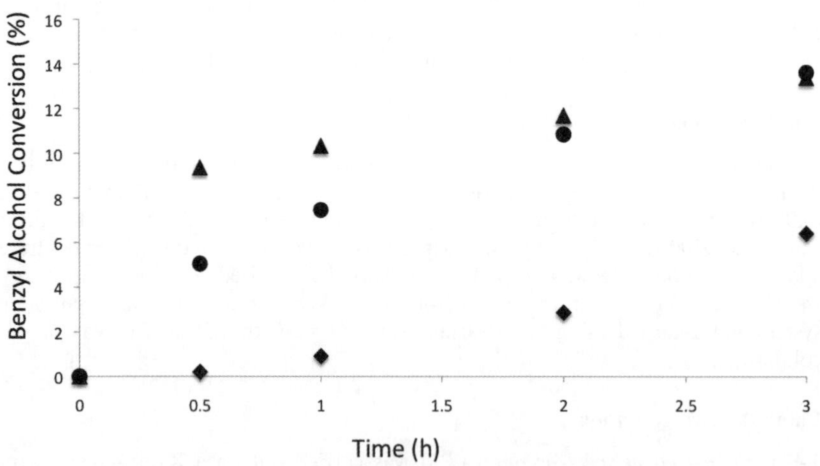

Fig. 1 Benzyl alcohol oxidation conversion using Au-DPH (▲); Au-IMP (◆); and Au-DPC (●). Reaction conditions: 50 mg catalyst, 40 ml benzyl alcohol, O_2 (10 barg), 140 °C, 1500 rpm.

   This journal is © The Royal Society of Chemistry 2011

Following these results, a small portion (50 mg) of each parent catalyst was leached with NaCN for 1 h, following the method of Stephanopolous[28] (Au-IMP$_{60}$, Au-DPH$_{60}$ and AuDPC$_{60}$). It was obvious from a visual inspection of these leached samples that removal of gold had occurred. The Au-IMP and Au-DCP catalysts were white in colour, whilst the Au-DPH catalyst retained a pale pink colour. No gold was observed in the any of the leached catalysts when analysed by atomic absorption and no Au^0 reflections were found in the XRD pattern of these catalysts.

In an attempt to control the amount of Au remaining on the catalysts after NaCN leaching, small portions of the most active parent catalyst (Au-DPH) were subjected to NaCN washing for 5 min, 10 min, 30 min and 1 h. The results showed very similar conversions for each of the catalysts after 3 h (Fig. 2).

ICP analysis of the 3 parent catalysts and their cyanide leached (10 min) counterparts was undertaken to accurately quantify the amount of Au remaining on the catalyst after leaching (Au-DPH$_{10}$, AuDPC$_{10}$ and Au-IMP$_{10}$). The results (Table 1) show that the parent Au/SiO$_2$ catalyst prepared by impregnation contains more Au (4.38 wt%) compared to the parent catalysts prepared by deposition precipitation (2.35–2.69 wt%). This is to be expected as the impregnation method essentially comprises the evaporation of a predetermined Au concentration onto a support, followed by drying and calcination. Although the final loading of the deposition

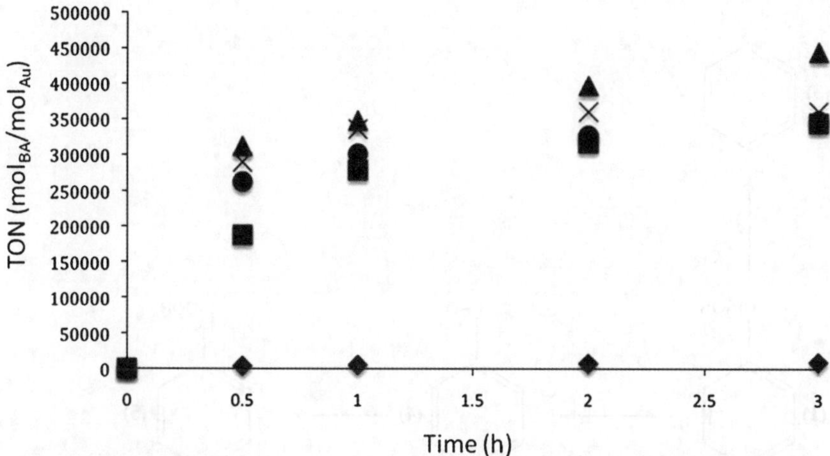

Fig. 2 Benzyl alcohol oxidation conversion over Au-DPH; untreated (◆); leached with 50 ml NaCN for 1 h (●); 30 min (×); 10 min (▲); and 5 min (■). Reaction conditions: 50 mg catalyst, 40 ml benzyl alcohol, O$_2$ (10 barg), 140 °C, 1500 rpm.

Table 1 Metal loading of the Au/SiO$_2$ catalysts determined by ICP analysis

Catalyst	Preparation method	Leaching time (min)	Assay (wt% Au)
Au-IMP	Impregnation	0	4.38
Au-IMP$_{10}$	Impregnation	10	0.06
Au-DPH	Deposition precipitation with NaOH	0	2.35
Au-DPH$_{10}$	Deposition precipitation with NaOH	10	0.06
Au-DPH$_{30}$	Deposition precipitation with NaOH	30	0.06
Au-DPH$_{60}$	Deposition precipitation with NaOH	60	0.05
Au-DPC	Deposition precipitation with Na$_2$CO$_3$	0	2.69
Au-DPC$_{10}$	Deposition precipitation with Na$_2$CO$_3$	10	<0.05

precipitated catalysts was approximately half the targeted amount, this is a surprisingly good yield. Generally, SiO_2 is not considered a good support in the deposition precipitation process as the highly negatively charged surface does not allow the adsorption of negatively charged Au complexes formed during the precipitation onto the support surface, which is necessary for the formation and stabilization of small gold particles[29,30] These results for the NaCN treated catalysts show that approximately 0.06 wt% Au remains on the catalyst, irrespective of the parent materials preparative route or length of time the parent is exposed to NaCN.

There are a number of products that can be formed from benzyl alcohol oxidation (Scheme 1), and a representative product distribution for the Au-DPH catalyst is given in Table 2. The selectivity profile for each of the unleached catalysts was similar regardless of the preparation technique, with 90% selectivity towards benzaldehyde after 1 h, which decreases to 85% after 3 h. The reduction in selectivity corresponds to the increasing concentration of the secondary oxidation product (benzoic acid) with time. The benzyl benzoate concentration remains constant over 3 h, indicating the condensation reaction only occurs during the first hour and then stops. The NaCN leached catalysts all showed much lower selectivity towards benzaldehyde, around 80% over 3 h. However, whilst the toluene yield (formed by a hydrogen transfer reaction) is small over the parent catalysts, NaCN leaching increases the toluene formation to 11% after 3 h reaction time. This significant difference could be due to the exposure of new sites on the Au catalysts by the NaCN pre-treatment.

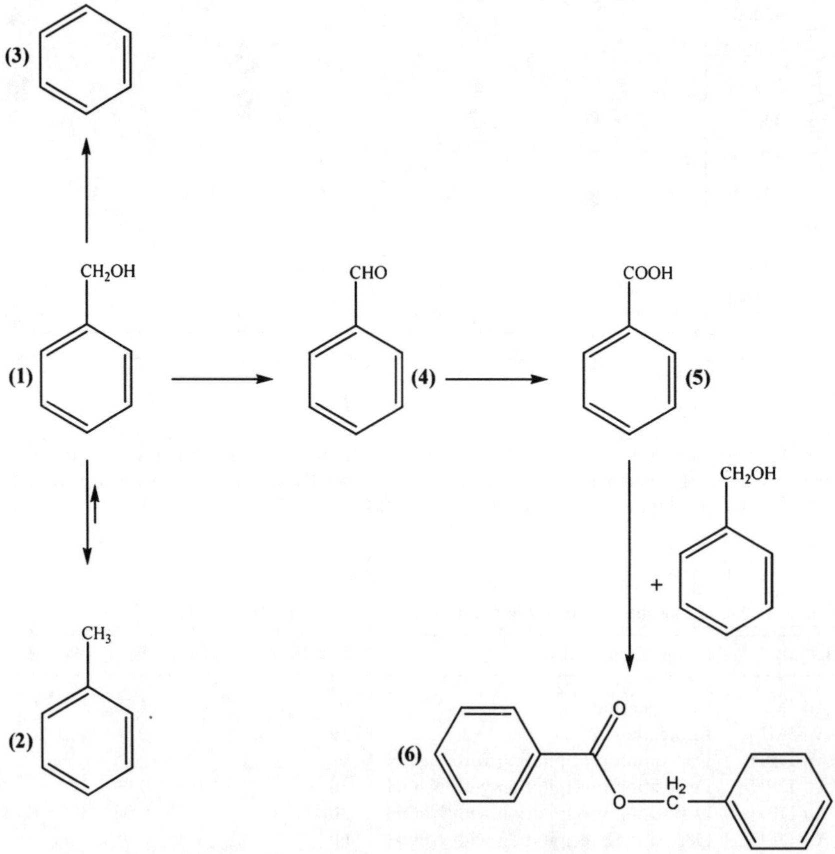

Scheme 1 The oxidation of benzyl alcohol; (1) benzyl alcohol, (2) toluene, (3) benzene, (4) benzaldehyde, (5) benzoic acid, (6) benzyl benzoate.

Table 2 Product distribution for benzyl alcohol oxidation[a] over catalysts prepared by deposition precipitation with NaOH (Au-DPH) and NaCN leached (Au-DPH$_{10}$)

	Selectivity at time (%, h)		
Product	1	2	3
Au-DPH			
Benzene	0.1	0.2	0.2
Toluene	0.5	0.6	0.6
Benzaldehyde	90.2	86.4	84.3
Benzoic acid	4.9	8.3	10.6
Benzyl benzoate	4.3	4.5	4.3
Au-DPH$_{10}$			
Benzene	3.7	2.8	2.6
Toluene	13.4	12	8.1
Benzaldehyde	79.2	80	82
Benzoic acid	2.2	3.4	5.9
Benzyl benzoate	1.5	1.8	1.4

[a] Reaction conditions: 50 mg catalyst, 40 ml benzyl alcohol, O_2 (10 barg), 140 °C, 1500 rpm.

The data from ICP analysis was used in conjunction with the benzyl alcohol oxidation data to calculate turn over numbers for each of the leached catalysts (TON = mol$_{BA}$/mol$_{Au}$). The results (Fig. 2) clearly show that the leached catalysts exhibit superior activity to their unleached counterparts. At 1 h the TON of the Au-DPH$_{10}$ catalyst is 350000, compared to 5100 for the Au-DPH parent catalyst. Usually high turnovers are achieved with good selectivity when Au–Pd catalysts are employed. Enache *et al.* reported TOFs of 86500 h^{-1} over a AuPd/TiO$_2$ catalyst for benzyl alcohol oxidation at 160 °C under similar conditions, whereas the Au/TiO$_2$ catalyst was an order of magnitude lower. Corma reports a TOF of 12500 h^{-1} for 1-phenylethanol oxidation using Au/CeO$_2$. The activity of our NaCN leached materials is much higher.

A number of experiments were performed to verify the high activity of these catalysts. The blank reaction with no catalyst at 140 °C and 10 barg O_2 showed a very low conversion (0.5%) over 3 h. Before each test run the autoclave reactor was thoroughly cleaned with aqua regia and a blank reaction was performed to ensure no background activity. Following the reaction the autoclave was again cleaned with aqua regia and a further blank performed to ensure the reactor was clean. In each case, the TONs were the same as those reported in Fig. 2. Whilst unlikely, in order to rule out any positive effect the NaCN may have on the activity of the catalyst, SiO$_2$ and NaCN treated SiO$_2$ were evaluated under the same experimental conditions. The blank reaction with SiO$_2$ present was identical to that without any catalyst

Table 3 Blank reactions for the oxidation of benzyl alcohol at 140 °C

Benzyl Alcohol conversion at time[a] (h)					
	0	0.5	1	2	3
Blank	0	0.00	0.01	0.30	0.50
SiO$_2$	0	0.00	0.12	0.40	0.51
SiO$_2$ NaCN	0	0.00	0.09	0.33	0.47

[a] Conditions: 40 ml benzyl alcohol, 10 barg O_2 1500 rpm. 40 mg SiO$_2$.

or SiO_2, and there was no observed difference over SiO_2 with or without NaCN treatment. This clearly indicates that the small percentage of Au remaining on the catalyst after NaCN leaching is responsible for the activity. The results for all blanks are given in Table 3.

Catalyst characterisation

In an attempt to determine the properties of the supported Au nanoparticles, and to identify the nature of the active site, detailed structural and chemical characterisation was carried out using SEM and XPS.

To evaluate the surface concentration of Au (expressed as the Au/Si surface atom ratio) and the Au oxidation state the fresh and NaCN leached catalysts were analysed by XPS (Fig. 3). Only metallic gold was observed on the fresh and leached catalysts, the latter exhibiting a major decrease in surface gold. Table 4 summarizes and compares the ICP assays with the XPS-derived Au/Si values.

The ICP assays show the sample prepared by impregnation to have a significantly higher Au loading than the two DP samples, which is not unexpected since the DP preparation includes a washing step and because of the low propensity for Au deposition on SiO_2. In contrast, the surface Au/Si ratio for the Au-IMP sample is much lower than that for the DP samples and this is clear also from Fig. 3. The Au(4f) signal from Au nanoparticles is affected by particle size: **for a specific Au loading** the photoemission signal from a set of large particles is reduced compared with that from small particles, due to the small sampling depth of the technique. This suggests that the mean Au particle size for Au-DPC is smaller than for Au-DPH, consistent with the fact that a larger fraction of the gold is removed from Au-DPC_{10} compared with Au-DPH_{10}. Moreover, supported catalysts prepared by impregnation are known to exhibit significantly larger particle sizes than materials prepared by deposition precipitation, which could explain the low Au/Si ratio observed for Au-IMP (Table 4). The fraction of gold removed for Au-IMP is much lower than for the DP samples, consistent with a much larger Au particle

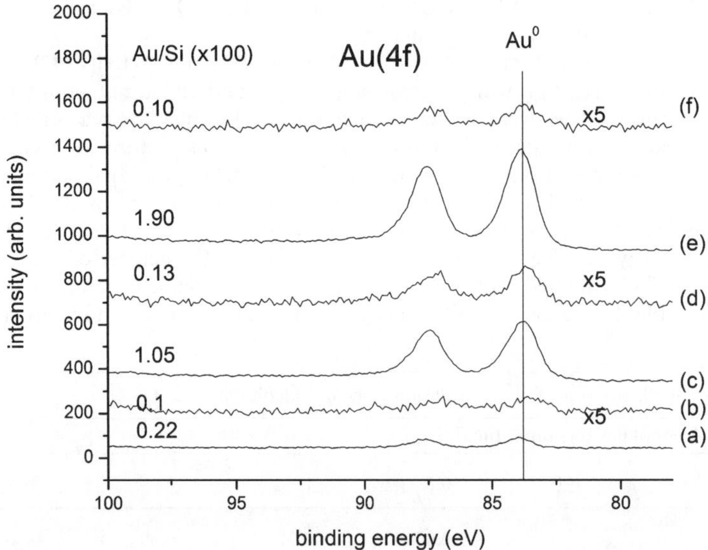

Fig. 3 Au(4f) photoemission spectra for fresh and cyanide leached (10 mins) catalysts Au-IMP (a); Au-IMP_{10} (b); Au-DPH (c); Au-DPH_{10} (d); Au-DPC (e); and Au-DPC_{10} (f). Note that the spectra for the NaCN leached catalysts have been scaled by a factor of 5. The atom ratios Au/Si ($\times$100) derived from the integrated Au(4f) and Si(2p) intensities are indicated on the figure.

   This journal is © The Royal Society of Chemistry 2011

Table 4 Comparison of ICP assays and XPS-derived Au/Si atom ratios for the 3 parent catalysts and leached counterparts as shown in Fig. 3

	Catalyst	Fresh/leached (10 min)	Assay (wt% Au)	Au/Si atom ratio ($\times 100$)
(a)	Au-IMP	Fresh	4.38	0.22
(b)	$Au-IMP_{10}$	Leached	0.06	0.10
(c)	Au-DPH	Fresh	2.35	1.05
(d)	$Au-DPH_{10}$	Leached	0.06	0.13
(e)	Au-DPC	Fresh	2.69	1.90
(f)	$Au-DPC_{10}$	Leached	<0.05	0.10

size. For all catalysts, leached and unleached, the Au was found to be metallic in nature.

Scanning electron microspcopy (SEM) was used to image the catalysts and to get a qualitative idea of the metal particle size. Some representative images of the parent and 10 min leached catalyst are given in Fig. 4. The micrographs of the unleached catalysts clearly show that large gold metal particles are obtained on SiO_2 regardless of the preparative route.

In stark contrast, the SEM images of the CN leached samples show much fewer large particles. The catalyst prepared by impregnation and washed for 10 min

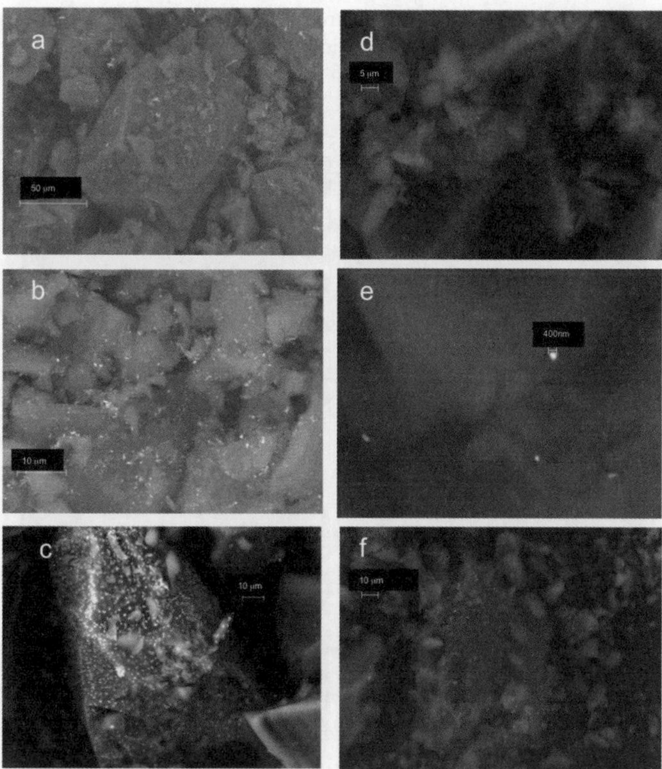

Fig. 4 Micrographs of the parent catalysts Au-DPH (a); Au-DPC (b); and Au-IMP (c); with corresponding images for the NaCN leached counterparts $Au-DPH_{10}$ (d); $Au-DPC_{10}$ (e); and $Au-IMP_{10}$ (f).

(Fig. 4f) shows more large Au particles on the surface when compared to the washed deposition precipitation catalysts. The unleached parent catalyst also shows larger particles than the samples prepared by deposition precipitation, which is consistent with the XPS data. Whilst useful for identifying larger Au particles, small nanoparticles (<100 nm) will not be observed using SEM and thus whilst we may comment on the frequency of large particles on the leached catalysts, it is unlikely that these particles are the ones responsible for catalytic activity. A number of reports have shown that usually the active particles in selective oxidation are 2–3 nm,[16,31–33] and some studies have suggested the active site is even smaller.[27] It is worth noting here that the Au/SiO_2 catalysts are very mobile when subjected to the electron beam, thus making it very difficult to obtain a useful image. Prolonged focusing on the large metal particles on the unleached parent catalysts caused significant damage to the catalysts, resulting in what appeared to be an explosion of the gold cluster.

The use of HAADF-STEM has proven to be a very useful tool in the analysis of supported Au and AuPd particles. With the use of aberration corrected microscopy, a electron beam 1 Å in diameter is able to image individual atoms. In an attempt to successfully image the small nanoparticles on the NaCN leached catalysts, the most active sample (Au-DPH$_{10}$) was analysed by HAADF-STEM. No Au clusters (large or small) were observed (Fig. 5). This technique has been successfully used to image 0.5 nm bilayer structures comprising 10 Au atoms on a Au/Fe_2O_3 catalyst and thus identify them as the active species for CO oxidation. However, to observe these small gold clusters very small areas of the support must be imaged (<100 nm). In this study of ultra low loaded catalysts which contain very few gold particles on a high surface area support the active Au clusters will be widely spaced and thus not easy to observe for the sampling size of the STEM, although this may in part be due to the erratic behaviour of these catalysts in the electron beam. A more thorough study may provide a solution. Whilst accurate detection of the Au species has not been successful using microscopy, other supports such as TiO_2 or activate carbon may be more

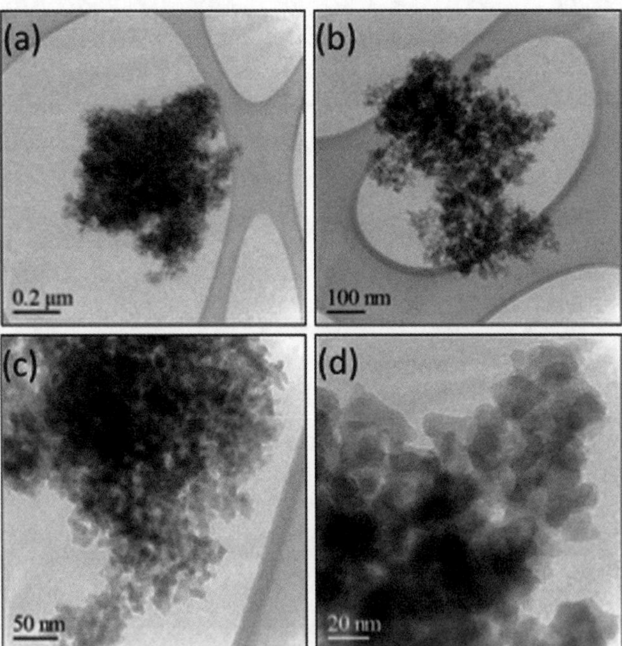

Fig. 5 TEM images of Au-DPH$_{10}$ at 200 nm (a); 100 nm (b); 50 nm (c); and 10 nm (d) magnification. No Au was detected.

    This journal is © The Royal Society of Chemistry 2011

suitable candidates for leaching as we have previously identified AuPd nanostructures on these materials.[34,35]

Conclusions

Au/SiO$_2$ catalysts prepared by impregnation and deposition precipitation are active for the selective oxidation of benzyl alcohol. The activity of these materials is maintained following the removal of ~98% of the Au by NaCN leaching. This small fraction of Au is responsible for the catalytic activity, and is able to convert 400,000 moles of benzyl alcohol per mole of active Au present per hour. Analysis of the leached materials by XPS indicates the particles are metallic in nature and are smaller on the catalysts prepared by deposition precipitation. SEM and STEM analyses have been unable thus far to locate and identify the nanoclusters on the catalyst surface indicating that they are very well spread over the high surface area support.

The successful utilisation of ultra low loaded Au catalysts for selective oxidation, where the majority of the Au present is active and 98% of the spectator Au is removed, makes the industrial application of such materials more economically viable. Information obtained through the identification of these active structures using state of the art techniques may provide useful insights into how the reaction proceeds on the Au surface.

Acknowledgements

I would like to thank Dr Albert Carley for his useful discussions of the XPS data. I would also like to thank Professor Graham Hutchings and Professor Chris Kiely for their continued support and input. Dr Peter Ellis, Dr Emma Moxham and Dr David Thompsett for their contributions to the discussions. A special thanks to Johnson Matthey for financial support for Adrian Thomas, and the Leverhulme Institute for their financial support of Dr Jennifer Edwards.

References

1 R. A. Sheldon and J. K. Kochi, *Metal Catalysed Oxidations of Organic Compounds*, Academic Press, New York, 1981.
2 http://www.epa.gov/gcc/pubs/principles.html.
3 R. A. Sheldon, I. Arends and A. Dijksman, *Catal. Today*, 2000, **57**, 157–166.
4 R. Neumann and M. Levin, *J. Org. Chem.*, 1991, **56**, 5707–5710.
5 S. Fujibayashi, K. Nakayama, M. Hamamoto, S. Sakaguchi, Y. Nishiyama and Y. Ishii, *J. Mol. Catal. A: Chem.*, 1996, **110**, 105–117.
6 B. M. Choudary, M. L. Kantam, A. Rahman, C. V. Reddy and K. K. Rao, *Angew. Chem., Int. Ed.*, 2001, **40**, 763–766.
7 M. G. Clerici, *Top. Catal.*, 2000, **13**, 373–386.
8 J. D. Chen, J. Dakka, E. Neeleman and R. A. Sheldon, *J. Chem. Soc., Chem. Commun.*, 1993, 1379–1380.
9 M. Musawir, P. N. Davey, G. Kelly and I. V. Kozhevnikov, *Chem. Commun.*, 2003, 1414–1415.
10 P. A. Shapley, N. J. Zhang, J. L. Allen, D. H. Pool and H. C. Liang, *J. Am. Chem. Soc.*, 2000, **122**, 1079–1091.
11 C. Keresszegi, T. Burgi, T. Mallat and A. Baiker, *J. Catal.*, 2002, **211**, 244–251.
12 A. Abad, A. Corma and H. Garcia, *Chem.–Eur. J.*, 2008, **14**, 212–222.
13 M. S. Kwon, N. Kim, C. M. Park, J. S. Lee, K. Y. Kang and J. Park, *Org. Lett.*, 2005, **7**, 1077–1079.
14 K. Ebitani, H. B. Ji, T. Mizugaki and K. Kaneda, *J. Mol. Catal. A: Chem.*, 2004, **212**, 161–170.
15 A. Abad, P. Concepcion, A. Corma and H. Garcia, *Angew. Chem., Int. Ed.*, 2005, **44**, 4066–4069.
16 S. Carrettin, P. Concepcion, A. Corma, J. M. Lopez Nieto and V. F. Puntes, *Angew. Chem., Int. Ed.*, 2004, **43**, 2538–2540.
17 K. Mori, T. Hara, T. Mizugaki, K. Ebitani and K. Kaneda, *J. Am. Chem. Soc.*, 2004, **126**, 10657–10666.

18 M. Haruta, T. Kobayashi, H. Sano and N. Yamada, *Chem. Lett.*, 1987, 405–408.
19 A. Corma and M. E. Domine, *Chem. Commun.*, 2005, 4042–4044.
20 D. I. Enache, J. K. Edwards, P. Landon, B. Solsona-Espriu, A. F. Carley, A. A. Herzing, M. Watanabe, C. J. Kiely, D. W. Knight and G. J. Hutchings, *Science*, 2006, **311**, 362–365.
21 L. Prati and F. Porta, *Appl. Catal., A*, 2005, **291**, 199–203.
22 C. L. Bianchi, S. Biella, A. Gervasini, L. Prati and M. Rossi, *Catal. Lett.*, 2003, **85**, 91–96.
23 L. Prati and G. Martra, *Gold Bulletin (London)*, 1999, **32**, 96–101.
24 J. A. Lopez-Sanchez, N. Dimitratos, P. Miedziak, E. Ntainjua, J. K. Edwards, D. Morgan, A. F. Carley, R. Tiruvalam, C. J. Kiely and G. J. Hutchings, *Phys. Chem. Chem. Phys.*, 2008, **10**, 1921–1930.
25 D. I. Enache, J. K. Edwards, P. Landon, B. Solsona-Espriu, A. F. Carley, A. A. Herzing, M. Watanabe, C. J. Kiely, D. W. Knight and G. J. Hutchings, *Science*, 2006, **311**, 362–365.
26 J. K. Edwards, B. E. Solsona, P. Landon, A. F. Carley, A. Herzing, C. J. Kiely and G. J. Hutchings, *J. Catal.*, 2005, **236**, 69–79.
27 A. Herzing Andrew, J. Kiely Christopher, F. Carley Albert, P. Landon and J. Hutchings Graham, *Science*, 2008, **321**, 1331–1335.
28 Q. Fu, H. Saltsburg and M. Flytzani-Stephanopoulos, *Science*, 2003, **301**, 935–938.
29 A. Wolf and F. Schth, *Appl. Catal., A*, 2002, **226**, 1–13.
30 R. Zanella, L. Delannoy and C. Louis, *Appl. Catal., A*, 2005, **291**, 62–72.
31 J. K. Edwards, B. Solsona, N. E. Ntainjua, A. F. Carley, A. A. Herzing, C. J. Kiely and G. J. Hutchings, *Science*, 2009, **323**, 1037–1041.
32 S. Biella and M. Rossi, *Chem. Commun.*, 2003, 378–379.
33 M. Haruta, T. Kobayashi, H. Sano and N. Yamada, *Chem. Lett.*, 1987, 405–408.
34 J. K. Edwards, B. E. Solsona, P. Landon, A. F. Carley, A. Herzing, C. J. Kiely and G. J. Hutchings, *J. Catal.*, 2005, **236**, 69–79.
35 J. K. Edwards, E. Ntainjua, A. F. Carley, A. A. Herzing, C. J. Kiely and G. J. Hutchings, *Angew. Chem., Int. Ed.*, 2009, **48**, 8512–8515, S8512/8511-S8512/8518.

General discussion

Professor Meyerstein opened the discussion of the paper by Professor Friend: In the proposed mechanism always a H atom is abstracted from an alcoholic O–H bond though in all H atom abstractions from alcohols by radicals the abstracted H atom stems from an α C–H bond. I wonder whether this indicates that the first step involves a proton transfer and is therefore not a redox process? Alternatively it could be suggested that the mechanism is more complex and that an Au–O bond is coherently formed and this bond is considerably stronger than the Au–C bond.

Dr Xu said: It appears that a finite but small surface coverage of atomic oxygen is key to high selectivity for coupling products, whereas a significant coverage of atomic oxygen would facilitate combustion. How might atomic oxygen be generated and its coverage properly controlled on typical supported Au catalysts? Would H_2O_2 be a more practical oxidant?

Professor Friend responded: I agree with your statement that a low steady-state coverage of atomic O is key for selective oxidation. Possible methods for controlling the O coverage involve both materials control and process conditions. Materials control: The gold could be modified so that more O can be supplied to the surface. This could potentially be done by: (1) Design of the support and catalyst particle to facilitate O migration. (It is possible that small particles on titania, for example, enable O migration (also called "reverse spillover".)) (2) Use of alloys where the second metal more readily dissociates O_2. The case of nanoporous Au, that was discussed by Dr Wittstock, is an example of AgAu alloys. There are also examples in the literature of supported alloy catalysts, including the beautiful work of Prof. Hutchings on AuPd alloys. In the case of alloys, the right composition is needed so that the metal alloyed with gold does not dominate the chemistry. This could lead to either too high an O coverage or chemistry dominated by the alloying metal. Process control: (1) Use of a stronger oxidant, as per your suggestion of H_2O_2. (We have not been able to induce dissociation of hydrogen peroxide on our gold surface; however, it always comes with water and it seems to be strongly hydrogen bound to the water. Therefore, it desorbs at low temperature.) Free hydrogen peroxide or ozone are stronger oxidants than O_2 and could be used to form adsorbed O. Ozone is probably not practical to use, though, because of the explosion hazard. (2) Adjustment of the O_2: alcohol ratio can also be used to control selectivity. This approach was used to obtain very high selectivity for methanol coupling catalyzed by nanoporous Au in the presence of O_2 at atmospheric pressure.[1] The idea is to use relative pressures to overcome the differences in the rates of the key steps in the process.

1 A. Wittstock, V. Zielasek, J. Biener, C. M. Friend, and M. Bäumer, *Science*, 2010, **327**, 319-322.

Professor Hutchings asked: Concerning the use of alternative oxidants to ozone, it may be possible to use H_2O_2 that is extracted into another solvent *e.g.* ether, and therefore concentrated and decreases the contamination with water. H_2O_2 forms adducts with some compounds, *e.g.* urea, and this may be another approach.

Professor Friend replied: This is an interesting suggestion and one worth pursuing since H_2O_2 may be an effective oxidant even in the vapor phase.

Professor Poliakoff asked: How do you obtain pure ozone (as opposed to a mixture of $O_3 + O_2$) and where does the water that you observe originate?

Professor Friend responded: A mixture of ozone and O_2 are exposed to our surface. Independent experiments show that O_2 does not react within our detection ability.

The water formed in our experiments is due to activation of O–H and C–H bonds in our reactants to form transient OH, which disproportionates to water and adsorbed O.

Dr Willock remarked: You use adsorbed O atoms to generate methyl formate. Does the amount of methyl formate generated balance with the amount of oxygen adsorbed? Do you have an idea of the mobility of O on the Au surface and can this explain why the atoms do not desorb as dioxygen under your experimental conditions?

Professor Friend replied: There is a balance between methyl formate production and combustion in methanol oxidation that depends on O coverage. In our experiments, all adsorbed O is consumed during the reaction. We have also studied the mobility of atomic O on the surface. Mobile Au–O complexes are present by 400 K, based on STM experiments. The mobility leads to an ordered phase of O on Au that produces a 2-D surface "oxide" with rectangular symmetry. Dioxygen forms at higher temperature—above 500 K—and the rate of its formation is determined by the decomposition of the ordered layer. These results suggest that there is a barrier to formation of O_2 that is higher than the barrier for forming the ordered phase.

Professor Fortunelli remarked: You showed that H_2O or H_2 are not indispensable for catalytic CO oxidation by Au systems. We found the same for Au_3 cluster supported on a simple ionic oxide (MgO): facile CO oxidation by Au_3 supported on the regular surface without any hydroxilic groups. In our case a crucial role is played by the presence of the electrostatic field due to the ionic (charge-separated) surface that stabilizes the dipole moment of oxidized (and hence charge-separated) Au (or Ag–Au) clusters. Can we make an analogy with your case, in which image charges due to the underlying Au extended system may have a similar stabilizing effect on oxidized (and hence charge-separated) Au structures?

Professor Friend responded: You raise a fascinating point about the possible role of dipolar interactions in the reaction. We did not investigate the role of charge explicitly. I do note that we are working on an extended metal surface with small particles on top of that so any charge separation would be highly screened by the metal electrons. We cannot rule out the role of dipole effects, especially at the edges of our particles. The image dipole may also play a role. I am not sure how this could be investigated, though, since these are subtle effects. Please let me know if you have a suggestion.

Professor Bowker commented: There is always residual oxygen in these experiments. Do you ever see a hydrogen desorption peak (200 K) on Au? For instance—can you completely titrate oxygen with formaldehyde to form the dioxymethylene intermediate? On Ag this decomposes to produce formate and H_2 gas, with a very nice decomposition-limited H_2 peak in TPD at $\sim$200 K.

Professor Friend answered: We do not see H_2 formation from the reactions we have studied, which is one significant difference between Au and Ag. This could be because H is very weakly bound on Au, as suggested by DFT studies, and that it reacts with neighboring species such as adsorbed O, methoxy, carboylates, *etc.* It also may be the case that Au itself does not activate C–H bonds. My view is that C–H activation is assisted by attack of an electron-rich species, such as adsorbed O, methoxy, *etc.*; however, this is a point of discussion in our lab and we are planning to study atomic H on the surface.

Professor Bowker enquired: Are these studies always carried out with oxygen dosing at low temperatures? Is it possible that these oxygen species are only present at low temperature and that different reactive species are present for higher temperature dosing?

Professor Friend replied: The dosing temperature definitely effects the distribution of oxygen species on the surface. In addition, the rate of oxidation and the oxygen coverage affect the distribution of O on the surface. We used STM, vibrational spectroscopy (HREELS), and XPS to study this effect.[1] In summary, high O coverages and high temperatures produce a more ordered, 2-D oxide structure that is less reactive overall and less selective. The lower activity of more ordered structures is demonstrated for the CO oxidation.[2] The lower selectivity is demonstrated for olefin oxidation, *e.g.* propene[3] and styrene.[4]

1 T. A. Baker, X. Liu, and C. M. Friend, *Phys. Chem. Chem. Phys.*, 2011, **13**, 34–46, and references therein for a complete description.
2 B. K. Min, A. R. Alemozafar, D. Pinnaduwage, X. Deng, and C. M. Friend, *J. Phys. Chem. B*, 2006, **110**(40), 19833–19838.
3 B. K. Min, X. Deng, X. Liu, C. M. Friend and A. R. Alemozafar, *Chem. Cat. Chem.*, 2009, **1**, 116–121.
4 R. G. Quiller, X. Liu, and C. M. Friend, *Chem.–Asian J.*, 2010, **5**, 78–86.

Dr Cadete Santos Aires said: Is the periodic contrast in your STM image a remnant of the "herringbone" reconstruction on Au(111)?

The sense of my question is: do you have the same surface you have at low coverage (where the "herringbone" reconstruction is still present meaning that your Au surface is not or only slightly modified) and at higher coverages when the Au surface is much more modified?

Professor Friend responded: There is cluster nucleation at the herringbone elbow sites at low coverage and low temperature. This is a kinetic effect since the gold atoms at the herringbone elbows are somewhat more weakly bound than others on the surface and, therefore, they are extracted first so that clusters containing O and Au tend to nucleate there. The exact structure depends on the temperature and rate of oxidation and the total O coverage, as described in my reply to the previous question number from Professor Bowker. It is the case that at higher coverage the surface is more modified and a 2-D ordered "oxide" is formed. [See response to previous question for several references.]

Professor Campbell remarked: It seems to me that at low O coverage you should be able to get a condition where there is no residual oxygen after the primary reaction, but you say not. Maybe this is due to the fact that the O is in islands and only the perimeter O atoms can react in the primary reaction.

Professor Friend answered: We have thus far not been able to obtain a condition in ultrahigh vacuum where there is no residual oxygen. One factor is that OH adsorbed on Au is unstable with respect to disproportionation; therefore, when a proton is transferred to adsorbed O, two of them will react even at low temperature to form water and adsorbed O. [Note that my statement about OH refers to vapor phase conditions.] As water desorbs, around 170 K, adsorbed O is left behind. Your point about islanding playing a role is a good one and could be part of the reason that we have residual O under our reaction conditions. We are currently trying to study the distribution of reactions involving O adsorbed on Au using STM to address this point. Islanding may also contribute to the facile disproportionation of the OH. There is certainly precedent for such effects due to islanding, as you know, so this may be a factor.

Professor Campbell commented: Thanks for a great presentation.

Professor Hutchings asked: Can you comment on the role of water in your system as you indicated low levels are present at the start of an experiment. Water is weakly bound on Au111, but H_2O is important for some reactions in particular CO oxidation. In your case is CO oxidation going *via* a different mechanism rather than that proposed to involve a surface OOH species. Have you looked at activation energies for CO oxidation in your system and do these compare with those that Professor Bond quoted in this discussion. It might be interesting to look at different surfaces of Au and conditions and see if a similar compensation effect is observed.

Professor Friend answered: As you note, we do not need water to be present for efficient CO oxidation. On the other hand, water is known to increase the rate of CO oxidation, in particular, for supported catalysts. We have not performed a detailed analysis of the kinetics of CO oxidation under our experiments. Your suggestion is an excellent one and we will try to make the comparison.

Professor Golunski opened the discussion of the paper by Dr Nijhuis: How much of the on-line deactivation that you observe is due to reversible site-blocking and how much to irreversible structural changes? What is the nature of the site-blocking species?

Dr Nijhuis responded: The deactivation is caused by the formation of strongly adsorbed consecutive oxidation species on the catalyst, like carboxylates. These species are adsorbed on titanium sites. This deactivation is completely reversible, if we heat the catalyst to 300 °C in 20% oxygen in helium, we can completely restore its activity. We do not see deactivation by for example sintering of the gold particles.

Professor Hutchings remarked: Is it possible to fine tune your chemistry by the addition of a second metal? Perhaps this could interrupt the coupling process?

Dr Nijhuis replied: It is indeed possible to fine tune the chemistry/the catalyst by adding a second metal. It has been reported in literature that adding promotors can improve the hydrogen efficiency of the catalyst. However, until this moment we have decided not the add promotors to our catalysts. We would like to first get a good understanding on what is happening at the catalyst and to design the gold–Ti catalyst system in an optimal manner. Once we have achieved that goal, we will be adding promotors.

Professor Haruta queried: In the reaction of propylene with O_2 and H_2, we have found that a series of separate double bed catalysts, Au/SiO_2 in the inlet and titanium-based metal oxides in the outlet, can also produce PO but with yields about half of the $Au/Ti\text{-}SiO_2$ catalysts. This means that H_2O_2 *in-situ* formed can move in gas phase to react with propylene. In your mechanism, do you assume gas phase transformation of H_2O_2?

Dr Nijhuis answered: In our mechanism we do not make any assumptions in how the peroxo or hydrogen peroxide species is transferred from the gold to the titanium sites. In the slides that I have shown, we assume that it is a transfer via the catalyst surface, the model that I presented, however, does not assume that and will also be able to describe experiments well in which transport of hydrogen peroxide occurs via the gas phase. Your observations with double catalyst beds are a very strong indication of gas phase transport. We should also perform such experiments as I think it is of value to fit our model to these experiments and to see if any of the rate parameters are changing.

Professor Haruta asked: Selectivity to PO varies depending on reaction conditions and reactant concentrations. When you changed reactant compositions, did you observe appreciable change in selectivity? Can you define the optimum conditions for higher PO selectivity?

Dr Nijhuis answered: The experiments that I presented today, were all performed at the same temperature with a catalyst that produces negligible amounts of propane. For these experiments, we did not see significant changes in the PO selectivity. If we go up in temperature, however, we do see that the PO selectivity goes down. We also did similar experiments with other catalysts. If we take low gold loaded catalysts with very small gold particles, we see significant propane formation. For such a catalyst we see the propene hydrogenation increase as the hydrogen concentration increases and therefore the selectivity goes down.

For the higher gold loaded catalyst that I presented, the main difference is in the activity and the hydrogen efficiency as a function of the gas composition. Since a high hydrogen efficiency is essential for a viable process, this means that one should operate at low hydrogen concentrations and high propene concentrations. Our experiments and calculations showed that even with this non-optimized catalyst a hydrogen efficiency of 50% is possible at the reaction temperature we used in this study. However, in that case the price to pay is a process running at very low conversions. For this reason further catalyst optimization is needed, which is what we are working on.

Professor Hutchings said: In your case do you think it would be better to have a catalyst that is good at making hydrogen peroxide as this is potentially a key intermediate in the process? Of course Pd would not be a good choice as it will be active for propene hydrogenation, but there may be other candidates.

Dr Nijhuis responded: A catalyst which is very active in the hydrogen peroxide production, but inactive in the propene hydrogenation, is indeed a very good candidate to become a good propene epoxidation catalyst, once it is combined with a proper Ti-containing support. As you mentioned, attempts to work with bimetallic Au–Pd catalysts, highly active in the peroxide generation, were unsuccessful because of their high propene hydrogenation activity. If you are aware of alternatives, I am very interested in trying these.

Professor Campbell remarked: I think you said that the –OOH intermediate resides on the TiO_2 surface, not the Au? Is this based on calculations or what evidence?

Dr Nijhuis replied: We think that the OOH intermediate is produced on the Au and used on the Ti for the epoxidation. This is the common view in literature for this system. It is based on the fact that Au can produce hydrogen peroxide, and Ti-based catalysts are highly effective epoxidation catalysts using hydrogen peroxide. In addition, using spectroscopy the presence of OOH on Ti has been observed during the epoxidation using this type of catalysts. Using DFT calculations the formation of OOH on gold out of H_2 and O_2 has been confirmed. Direct evidence for this mechanism is now available however.

Professor Campbell asked: It seems that your results could be explained by the mechanism we proposed in our paper in this proceedings, whereby H_2 dissociates (perhaps starting on the oxide) to make 2 H(ad) on the Au particles, and then O_2 inserts into the Au–H bond to make Au–OOH. This –OOH then oxidizes propene to make its epoxide and leave Au–OH behind. As Prof. Friend reported here, 2 OH(ad) on Au surfaces rapidly react to make H_2O and O(ad). This O(ad) then reacts with 2 H(ad) to make H_2O. This mechanism explains your observed strong correlation of the epoxide and water product amounts. This mechanism would set a theoretical maximum limit of 1 propene epoxide produced for every 1 H_2O side product. That could be a nice goal for future research. You see more water than that by $2\times$ at best, if I understood correctly. Your extra water could come from the side reaction of H_2 oxidation somewhere on this material. Is such a mechanism consistent with your kinetics? If so, I think it would be very interesting to measure the rate of the H_2+O_2 reaction (without propene but otherwise at the same conditions and after pretreating the catalyst by running the propene reaction). If the H_2O production rate (in absence of propene) gives the correct amount of extra water (*i.e.*, beyond the 1 : 1 ratio of my proposed epoxidaion mechanism), it would be strong support for my mechanism.

Dr Nijhuis answered: The theoretical limit for the least amount of water produced per propene oxide is indeed a ratio of 1 : 1. Apart from water produced as co-product of PO, we see that the water produced is in 3 ways. Direct water formation—not linked to PO formation. This is probably at gold particles not neighboring Ti. We also see water formation by hydrogenation of the OOH oxidizing species, this can be reduced by lowering the hydrogen concentration or increasing the propene concentration. We also see water produced by decomposition of the OOH species, we can also reduce this by increasing the amount of propene. In the absence of propene, the water formation is generally significantly higher than in the presence of propene, however, this information cannot be used easily to conclude something on the OOH utilization, as the adsorption of propene on gold affects the water formation rate. We are working to reach the goal of a 1 : 1 water–PO ratio. In my presentation I showed that by changing the concentrations of the reactants a 1 : 2 ratio is possible. We think that we can approach the 1 : 1 ratio by a more controlled catalyst synthesis—making sure gold is always near Ti—by using promotors, and other reaction conditions.

Professor Bowker said: Since the selectivity to propene oxide is only high at extremely low conversions is this reaction feasible commercially? In a real system you would need a huge recycle ratio—such a process would be very expensive.

Dr Nijhuis responded: The performance of the current catalysts is indeed insufficient for a commercial process. However, also the ethylene epoxidation process runs at relatively low conversions. The targets that we need to reach for a commercially attractive process are a conversion of about 10% and a hydrogen efficiency of 50%. The propene epoxidation selectivity is not a problem, we easily reach over 90%. More active catalysts, using for example promotors, have already been reported in literature, for example by Haruta. We are currently working to improve our mechanistic insight and use that to develop a better catalyst and select the optimal process conditions. A process would indeed need a big recycle, however, the separation of propylene oxide and the co-product water from the feed gases should be relatively simple.

Mr Piccinini said: Since it is known that the addition of halides can decrease the contribution of subsequent H_2O_2 decomposition reactions, I was wondering if you have ever tried to dope your catalyst with Br or Cl? In this way, I believe you might have some insights on the nature of the active sites.

Dr Nijhuis answered: We have not added halides to the catalysts. In contrast, we have taken special care that after catalyst synthesis our catalysts are completely chloride free. The presence of halides makes the gold nanoparticle sinter easily and their presence is therefore negatively effecting the catalyst stability. In addition, we have decided not to work with promotors yet, as this complicates the catalyst system. Once we have gained sufficient mechanistic insight, we will start adding promotors. I do think, however, that your suggestion is worth trying.

Dr Huang asked: Did the authors investigate the kinetic study of propylene with O_2 and H_2 mixture over other Au catalysts such as Au/TS-1? Were similar results obtained?

Dr Nijhuis responded: Gold on TS-1 is also a very active epoxidation catalyst. We are working with that catalyst as well and the performance is more or less the same as that of the gold on Ti dispersed on amorphous silica support that I showed in our paper. In general, small gold nanoparticles on a Ti containing support with Ti in a tetrahedral coordination will work best.

Dr Huang asked: 1) How is water produced? 2) As shown in Scheme 1 in the paper, hydrogen peroxide may decompose directly to water. However, the authors propose that at high H_2 concentration, water is produced dominantly by the hydrogenation of H_2O_2, based on the work of Edwards *et al.* (ref. 40 in the paper). However, in both cases reaction conditions (temperature and pressure) are very different. Therefore, the possibility that water is formed mainly by the direct decomposition of H_2O_2 at 403 K cannot be excluded. Could the authors give some comments about this?

Dr Nijhuis replied: 1. Water is produced in three ways: as a side product of the epoxidation, by hydrogenation of the peroxo species produced on the gold and by decomposition of the peroxo species into water and oxygen.
2. In our work, we managed to distinguish between water produced by peroxide decomposition and water produced by peroxide hydrogenation. In our paper, we show these relative contributions in Figure 4 as a function of the feed gas compositions. At low hydrogen concentrations the decomposition is dominant, at hydrogen concentrations above approximately 5 vol%, for our catalysts at our reaction conditions, the hydrogenation of the peroxo species is the most important route toward the undesired direct water formation.

Dr Huang questioned: Can oxygen be dissociated directly over Au/Ti–SiO_2 in the absence of H_2?

Dr Nijhuis answered: I am not sure if oxygen can be dissociated directly over our Au/Ti–SiO_2 catalysts in the absence of hydrogen. In the absence of hydrogen the epoxidation activity of our catalysts is extremely low. In any case oxygen can be activated in the absence of hydrogen, as our catalysts are also highly active in the CO oxidation.

Dr Xu said: Could you please define the measure of H_2 efficiency, and clarify what the highest H_2 efficiency achieved in your study was?

Dr Nijhuis answered: The hydrogen efficiency is defined as the amount of propene oxide produced divided by the amount of hydrogen consumed—in most cases it can also be calculated by taking the amount of propene oxide produced divided by the amount of water produced. For our catalysts the hydrogen efficiency is typically at best 30%. What I showed in our paper, is that for the catalyst used in this study a hydrogen efficiency of 50% should be possible by operating with a low hydrogen concentration and a high propene concentration in the feed, however, this will be at the expense of the productivity. 50% is approximately the minimum that is needed for an industrially attractive process. Apart from optimizing the reaction conditions, as I have shown here, we will also be able to improve the hydrogen efficiency by improving the gold–titanium interaction—making sure that all gold particles are located near a titanium site—or by adding promotors to the catalyst as has been reported in literature. I think that by combining these approaches we will be able to reach the required hydrogen efficiency at a sufficiently high productivity.

Professor Bowker opened the discussion of the paper by Professor Bernhardt: You propose that the methane is molecularly adsorbed on these small clusters, but how do you know they are not dissociated. The binding energy seems very high, and you cannot tell from the mass spectra whether CH_4 is dissociated or not.

Professor Bernhardt responded: From the mass spectra alone it is indeed not possible to decide whether the methane is bound molecularly or dissociatively to the gold clusters. A first indication that the methane bound to the gold dimer is not dissociated comes from the kinetic data. In order to properly fit a reaction mechanism to the measured kinetics, it is mandatory to include an equilibrium reaction step in the adsorption of the first methane molecule (Figure 5a in our paper, see also ref 1). However, the re-desorption of the CH_4 molecule would be very difficult, if it would be bound dissociatively. Further support for the non-dissociated bonding of methane to the gold clusters comes from the investigation of larger cluster sizes. In contrast to the gold dimer, where we detect dehydrogenated product, the loss of hydrogen was never observed for larger cluster sizes.[2] Finally, concurrent theoretical simulations show no indication for the dissociation of the first adsorbed methane molecule.[1,2] Only with the co-adsorption of a second CH_4 methane can be activated.

1 S. M. Lang, T. M. Bernhardt, R. N. Barnett, U. Landman, *Angew. Chem., Int. Ed.*, 2010, **49**, 980.
2 S. M. Lang, T. M. Bernhardt, R. N. Barnett, U. Landman, *ChemPhysChem*, 2010, **11**, 1570.

Dr Lalik asked: In the mechanism you propose for the methane activation, the CH_4 or O_2 molecule (L) and the metal cluster Mx+ form an energized intermediate (MxL+)*. Once formed, the latter has to lose its energy and it does so by colliding with a helium atom. As the result, the energy of the cluster is transferred to the He atom to form what you call the energized helium atom, He*. I wander whether the metastable He atoms can be formed in such collision. If this is the case than using other noble gases apart from He may expectedly lead to different results since the excitation energies for noble gases differ considerably from one another and they decrease on going down the periodic table, namely: 20.61 eV for He,[1] 16.62 eV for Ne,[1] 11.72 eV for Ar[2] and 10.56 eV for Kr.[2] Did you try to use other than He noble gases in your experiments, and are the result different from those obtained with helium?

1 L. P. Shishatskaya, *J. Appl. Spectr.*, 1972, **16**, 187.
2 C. K. Fagerquist, M. K. Hellerstein, D. Fauber and M. J. Bertrand, *J. Am. Soc. Mass Spectrom.*, 2001, **12**, 754.

Professor Bernhardt answered: In the Lindemann energy transfer mechanism for gas phase association reactions that has to be assumed to be operative at our experimental conditions the excess energy is primarily converted into kinetic energy of the colliding helium atoms. Considering the thermal conditions under which the reactions take place in the ion trap the formation of metastable He atoms might be rather unlikely. Nevertheless, in previous experiments we also investigated the influence of different buffer gases, like Ar, on the metal cluster reactions. We found that a change in the buffer gas does not influence the observed reaction mechanism, but it does change the measured pseudo first order rate constants due to the different energy transfer efficiency.

Dr Molina commented: In the adsorption of methane by Au clusters, is it mandatory to have positively charged gold clusters? That is, is there a possibility of methane adsorption on the neutral species?

Professor Bernhardt responded: For methane adsorption and C–H bond activation interactions with electrophilic metal centres are important. Thus, positively

   This journal is © The Royal Society of Chemistry 2011

charged metal clusters are ideal. Moreover, in the case of gold clusters, previous experiments indicate that methane does not bind to negatively charged clusters at all.[1] Also stable reaction products of neutral gold clusters with methane have not yet been reported to our knowledge.

1 D. M. Cox, R. O. Brickman, K. Creegan, A. Kaldor, *Mat. Res. Soc. Symp. Proc.*, 1991, **206**, 43; D. M. Cox, R. Brickman, K. Creegan, A. Kaldor, *Z. Phys. D*, 1991, **19**, 353.

Professor Hutchings asked: Concerning the activation of oxygen on the gold dimers, it is clear that this requires methane to be adsorbed. Have you tried a larger hydrocarbon with the possibility of an improved inductive effect that could enhance O_2 activation? Isobutane would be a logical choice or maybe propane. This effect could be similar to the aspects we were discussing previously with Professor Madix concerning the effects of carbon containing surface molecules affecting oxygen activation.

Professor Bernhardt replied: Thank you very much for this interesting proposal. We did not yet investigate the interaction of larger saturated hydrocarbon molecules with the small free gold clusters. This could certainly provide valuable insight into the cooperative oxygen activation mechanism and we should definitely look into this in the future because the experiments are not difficult and indeed would be a logical extension of our work performed so far. We investigated, however, the effect of hydrogen pre-adsorption on the activation of molecular oxygen and saw similar promoting effects as in the case of methane.[1] We also studied the interaction of unsaturated hydrocarbons, namely propylene, with gold cluster cations and found a very strong bonding (as expected) and saturated coverage that excluded the co-adsorption of other ligand molecules.[2]

1 S. M. Lang, T. M. Bernhardt, R. N. Barnett, B. Yoon, U. Landman, *J. Am. Chem. Soc.*, 2009, **131**, 8939.
2 S. M. Lang, T. M. Bernhardt, *Eur. Phys. J. D*, 2009, **52**, 139.

Dr Willock remarked: You show that the C–H bond cleavage on a Au cation dimer is much lower when two methane molecules are adsorbed. In your calculations do both methane molecules undergo similtaneous C–H bond activation producing H_2 directly?

Professor Bernhardt answered: From the experimental mass spectrometric and kinetic data we can conclude that C–H bond activation resulting in dehydrogenation only occurs, if two methane molecules are co-adsorbed on the gold dimer cation. The concurrent theoretical simulations by Uzi Landman and Robert Barnett clearly indicate that the C–H bond cleavage is sequential in the way that two methyl radicals are formed sequentially at the two gold atoms, respectively, through hydrogen atom elimination. The two hydrogen atoms stay attached to the gold dimer and recombine to H_2 which can readily desorb. The formation of the second H_2 molecule occurs after hydrogen atom elimination in the course of the generation of ethylene on the gold dimer.[1]

1 S. M. Lang, T. M. Bernhardt, R. N. Barnett, U. Landman, *Angew. Chem., Int. Ed.*, 2010, **49**, 980.

Dr Willock asked: What level of theory was used for calculating the binding energy of molecular methane to the Au clusters? In particular are dispersion interactions accounted for?

Professor Bernhardt replied: Thank you very much for your question. However, our contribution is only concerned with the experimental ion trap studies. The theoretical simulations concerning the gold cluster systems that we investigated were performed by Professor Uzi Landman and his co-workers with the use of density functional theory calculations. In particular, they employed the Born–Oppenheimer–spin density functional–molecular dynamics method with norm-conserving soft pseudopotentials (including a scalar relativistic pseudopotential for Au) and the generalized gradient approximation for electronic exchange and correlations. For details and references we would like to refer to the respective publications.[1–3]

1 S. M. Lang, T. M. Bernhardt, R. N. Barnett, U. Landman, *Angew. Chem., Int. Ed.*, 2010, **49**, 980.
2 S. M. Lang, T. M. Bernhardt, R. N. Barnett, U. Landman, *ChemPhysChem*, 2010, **11**, 1570.
3 S. M. Lang, T. M. Bernhardt, R. N. Barnett, U. Landman, *J. Phys. Chem. C*, 2011, **115**, 6788.

Professor Friend asked: You show very nice data for the coordination of species to your clusters; however, you do not truly know if there is a reaction or not for one methane molecule—you only know the total mass. Your interpretation relies heavily on the theory and as we have heard these last few days, weak interactions may be important for Au and possibly not accounted for in the theory. My questions are (1) have you done any vibrational spectroscopy or other experiments that provide information about structure and bonding that can be compared to predictions from theory? and (2) is there is a way you can see the final products of your reaction as shown in the catalytic cycle?

Professor Bernhardt responded: Thank you very much for raising these important questions. (1) So far we do not have spectroscopic data of our reaction products. Nevertheless, these experiments are certainly on our agenda. However, the ion trap, in which the catalytic reactions take place, always contains a mixture of different reaction products. Thus, in order to be able to properly characterize individual product complexes, we are currently setting up a second ion trap which will serve to isolate and store intermediate reaction products. Once isolated these intermediate product complexes will be characterized by different spectroscopic techniques such as vibrational spectroscopy. (2) The final products of the catalytic reactions like carbon dioxide or ethylene are neutral species. Because our experiment relies on mass spectrometric techniques we are not able to detect these molecules in a straight forward manner. Nevertheless, it is, in principle, possible to ionize these neutral reaction products, *e.g.*, by laser irradiation and to detect the ionized species. Such experiments have been performed for different reaction systems.[1]

1 S. Li, A. Mirabal, J. Demuth, L. Wöste, T. Siebert, *J. Am. Chem. Soc.*, 2008, **130**, 16832.

Dr Xu asked: The reaction energy profile calculated using density functional theory that was shown during the presentation appeared to indicate that closing the catalytic loop requires a highly endothermic step at the end. Could you please comment on how this amount of energy might be supplied to the gas-phase Au dimer at reaction temperatures less than or equal to 300 K?

Professor Bernhardt replied: This is a very important issue and it is related to the reaction conditions in the ion trap. In contrast to comparable studies under ultra-high vacuum conditions, our ion trap is filled with about 1 Pa of helium buffer gas and small partial pressures of the reactant gases. These pressure conditions ensure a multi-collision environment and a complete thermalisation of the clusters and the reaction products on the timescale of our experiments. Under these conditions it is therefore possible to surmount even comparably high activation energy barriers due to the statistical action of the high energy tail of the Maxwell-Boltzmann distribution. As a consequence, of course, the corresponding rate constants are comparably small. Nevertheless, the rate constants deduced from our kinetic measurements compare very favourably with the relative energy barriers determined by first principle calculations of Uzi Landman.[1,2]

1 S. M. Lang, T. M. Bernhardt, R. N. Barnett, U. Landman, *Angew. Chem. Int. Ed.*, 2010, **49**, 980.
2 S. M. Lang, T. M. Bernhardt, R. N. Barnett, U. Landman, *J. Phys. Chem. C*, 2011, **115**, 6788.

Professor Bowker continued the discussion of Professor Friend's paper: When you adsorb O_2, you extract Au metal atoms from the surface. Would the same happen for other surfaces of Au? Can you also tell us in what sense are these findings of atom extraction by oxygen general for the IB/11 metals?

Professor Friend answered: I think the abstraction of metal atoms from the surfaces of group IB metals is very general. I do not think this is specific to the (111) surface of Au, although we have not studied other surfaces. It is well known that O leads to chains of M–O–M on Cu(110) and Ag(110) surfaces and that an ordered O phase forms on Ag(111) involving added metal atoms.

Professor Madix commented: On all three metals, copper, silver and gold, oxygen adsorbs and incorporates metal atoms into island structures beginning at low oxygen coverages. Thus with predosed oxygen, it is difficult to achieve a dispersion of isolated oxygen atoms on the metal surface and render them all accessible to reactants subsequently dosed.

Professor Hutchings remarked: Have you considered substituent effects or regioselectivity. Are secondary alcohols reactive. With our AuPd catalysts we have found that while 1-octanol is reactive, 2-octanol is completely inactive because it adsorbs preferentially. It might be interesting to consider longer chain alcohols.

Professor Friend replied: We have looked at a few secondary alcohols: cyclohexanol and 2-cyclohexenol.[1] In these cases, there is facile formation of ketones. We have also studied longer-chain primary alcohols and the effect of phenyl rings on the coupling reactions and there are effects on selectivity. There are two main factors determining selectivity for straight-chain alcohols, as described in our paper: (1) the ease of β-H elimination from the alkoxide, and (2) the binding strength of the alkoxide to the surface. The rate of β-H elimination and the strength of bonding both increase with alkyl chain length. As a result, there is a considerable amount of aldehyde formation for, *e.g.*, 1-butanol oxidation. The alcohols containing aromatic rings, *e.g.* phenol and benzyl alcohol, tend to decompose on the surface. (Unpublished results.) It appears that O adsorbed on Au is very active for activation of the ring, probably because on an extended surface the phenyl ring tilts toward the surface to allow for bonding to the metal *via* the ring π-system.

1 X. Liu and C.M. Friend, *Langmuir*, 2010, 16552–16557.

Professor Hutchings asked: Is it possible to fine tune your chemistry by the addition of a second metal? Perhaps this could interrupt the coupling process?

Professor Friend responded: I do think it is possible to tune the chemistry *via* a second metal. We are now beginning to study planar alloys. Based on DFT studies of methanol coupling, the relatively weak binding of intermediates, *e.g.* the alkoxides, to the Au facilitates coupling because the intermediates can readily migrate and be in sufficient proximity to couple.[1] This will most likely be different when another metal is introduced. Generally, the second metal will probably more strongly bind the intermediates and may even decompose them before they can migrate to couple. A second metal could, however, increase the supply of O to the Au, which is necessary for reaction to occur. I think that this is a very fruitful avenue for investigation.

1 B. Xu, E. Kaxiras, C. M. Friend, *J. Phys. Chem. C*, 2011, **115**(9) 3703–3708

Professor L. M. Rossi opened the discussion of the paper by Professor Prati: As pointed out by Prof. Bond in paper 23, the accurate measurement of reaction rate under specified conditions is very important, as well as the representation of such results, as for example, as specific rate or turnover frequency. In Table 4 in your paper we can find the TOF determined at a fixed reaction time. Why do you not measure specific rate or TOF at a fixed conversion, as usual?

Professor Prati replied: Turnover Frequency (TOF) is defined[1] as number of revolution of the catalytic cycle per unit time. It is a chemical reaction rate and like all the catalytic rate, it is hard to measure. First of all the main limitation we encountered is the determination of active site number. This measurement is actually under debate in the case of gold and one should also consider that all the main techniques able to provide reliable numbers are carried out in the gas-phase. Therefore, as we applied our catalyst for liquid-phase reactions, no methodologies can be presently use to obtain this number and this is why we calculated TOF with respect to the total metal. I would also like to note that in the present paper we carried out a comparative study, *i.e.* compare values of catalyst with a very similar size distribution of AuNPs. We can than argue that reasonably the active site number should be the same in all the catalyst. Thus, even relative, it should be equivalent for the calculation of TOF values to measure conversion at fixed time or time at fixed conversion. We preferred to fix the time as short as possible to avoid any problems due to deactivation that can occur during the reaction thus providing a more reliable comparison among the catalysts. The time we used (30 min) appeared the shortest possible time for sampling the reaction in our system for obtaining a measurement of the initial rate.

1 M. Boudart, G. Djega-Mariadassou, *Kinetics of heterogeneous catalytic reactions*, Princeton University Press, Princeton, NJ, 1984.

Professor L. M. Rossi remarked: In your experiments you have used 4 eq of base. Did you change this parameter? We can find in the literature reaction without base, with base in excess and also with sub-stoichiometric amounts of base. Can you comment on the role of base in the reaction mechanism?

Professor Prati responded: The use of basic environment in gold catalyzed liquid phase oxidation was envisaged since the earliest reports on this topic. After almost twenty year, although the detailed mechanism is still under debate, it is now generally accepted that the mechanism of aerobic alcohol oxidation should involve four steps namely i) the adsorption of reactants; ii) and iii) the OH and αC–H bond cleavage; iv) removal of H by O_2 or its derivatives. As when acidic support as TiO_2 is used, Au on TiO_2 appeared a very poor catalyst for alcohol oxidation in the absence of a base but it was reported that TiO_2 is able to adsorb and activate O_2 even at low temperature in CO oxidation, we can argue that the base is mainly involved in the ii) or iii) steps. It was also demonstrated the αC–H bond cleavage is the rate determining step. Conversely the direct use of basic supports allows an efficient alcohol oxidation even in the absence of a base. Therefore it can be figured out that the use and the relative amount of a base can be regulated by some factors as for example the support but also the pK_a of the alcohol.

Mr Pritchard asked: When describing the pre-treatment/functionalisation of the CNT/CNF material with concentrated nitric acid, has a post-TPD (Temperature

     This journal is © The Royal Society of Chemistry 2011

Programmed Desorption) experiment been performed on either material to confirm what functionalities have been introduced and whether these functionalities are carboxylic acid groups exclusively?

Professor Prati answered: The HNO_3 treatment of CNT/CNF is a well-known procedure to introduce oxygenated groups [see for example Ref. 1], the characterization of which is not always clear. TPD-MS revealed CO_2/CO desorption in the case of our oxidized CNFs confirming the introduction of functionalities. Moreover an XPS study particularly in the region of O1s (530–536 eV) was performed on these samples highlighting the presence of carboxylic (the most abundant) and conjugated carbonyl groups. For more details on these characterizations see R. Arrigo, *et al.*,[2] where it can be also found the detailed characterisation of NH3-treated samples at different temperatures.

1 A. J. Plomp, D. S. Su, K. P. de Jong, J. H. Bitter, *J. Phys. Chem. C*, 2009, **113**, 9865.
2 R. Arrigo, *et al.*, *J. Am. Chem. Soc.*, 2010, **132**, 9616.

Professor Bond asked: In view of the sensitivity of the reaction to the acid/base character of the support, I wonder if it is possible that your reaction is bifunctional, *i.e.* if some part of the reaction cycle, perhaps an isomerisation or other acid–base catalysed step, may be taking place on it. Do you think this is likely?

Professor Prati replied: For sure an active role of the support can be envisaged. Particularly in the alcohol oxidation the first step is the formation of an alcoholate ion which adsorbs on the catalytic surface and undergoes to the subsequent hydride abstraction. Moreover quite recently Figueiredo *et. al.*, comparing Au supported on active carbon and carbon nitride (an oxygen free support), have shown that the support oxygen sites are accounted for the molecular oxygen activation. The role of the oxygen containing groups was also confirmed by Orfao *et al.*[2]

1 Figueiredo *et al.*, *J. Catal.*, 2010, **274**, 207.
2 Orfao *et al.*, *J. Catal.*, DOI:101016/j.jcat.2011.04.008.

Professor Hutchings said: In the ^{18}O labelling experiments that have been carried out in the studies for alcohol oxidation, the ^{18}O does not appear in the products. However, what happens under non basic aqueous conditions. Is there a different mechanism?

Professor Prati replied: The labeling experiments have shown that under basic conditions with gold catalysts as well as with Pt or Pd based catalysts the oxygen atoms introduced in the reaction products are from water and not from O_2. This is also true under base-free condition in the presence of Pt catalyst.[1] This finding seems to support the same mechanism under basic or base-free conditions. It is reasonable to extend this conclusion also for gold catalyzed alcohol oxidation. In fact, as predicted by recent experimental/theoretical data, the role of the basic environment (added base or basic support) in gold catalyzed alcohol oxidation is mainly involved in the rate determining step, *i.e.* the OH and αC–H bond cleavage. The addition of a base apparently just facilitate the reaction but does not modify the mechanism. The selectivity on the contrary can be greatly affected. I would like to emphasize that what is discussed above is true when water is used as the solvent. Changing the solvent could lead to modify the mechanism.

1 B. N. Zope *et al.*, *Science*, 2010, **330**, 74.

Dr Louis commented: How stable are your catalysts upon reaction? Did you check whether the acid–base properties of the catalysts have changed after reaction?

Professor Prati responded: Actually we have not performed recycling experiments. However we check the used catalysts and no variation of basic propertied was revealed.

Professor L. M. Rossi opened the discussion of the paper by Dr Della Pina: In paper 26, Dr Della Pina shows that the particle size increases and the catalytic activity remains when a second metal is added to gold. Is it possible that the particle size of the Au core plays a minor role in this kind of bimetallic nanoparticle catalysts? Does this apply to other catalytic reactions? Please comment.

Dr Della Pina answered: Yes, the particle size of the gold core could play a minor role in this kind of bimetallic catalysts as the experimental data and STEM analyses seem to underpin, in particular for the system Au–Cu/C prepared according to method 1: it is a fact that the gold size growth detected on recycling did not prevent the catalyst from prolonging its high performance for 7 reaction cycles, that is why we must suppose that other factors could outweigh the limit of sintering. To date, we have applied this kind of bimetallic system (Au99-M1/C by sol deposition) only to the aerobic oxidation of allyl alcohol in liquid phase, but in principle it could be used also for other kinds of reaction. We successfully employed a Au–Cu/SiO2 catalyst (1% Au–Cu/SiO2, Au : Cu = 4 : 1 w/w by incipient wetness) in the gas phase aerobic oxidation of benzyl alcohol to benzaldehyde[1] but this is indeed a different bimetallic system, either for the preparation method (*via* incipient wetness instead of sol deposition) and supporting material (SiO_2 instead of C) or for the metals amount.

1 C. Della Pina *et al.*, *J. Catal.*, 2008, **260**, 384.

Professor L. M. Rossi remarked: In this paper you have shown that the addition of 1 wt% of a second metal on gold changes significantly the reaction rates. Can you please discuss a little more about the reasons for this enhancement? Can it be related to changes in the rate of O_2 dissociation? Please comment.

Dr Della Pina replied: First a specification: the addition of the second metal is 1 mol%, not 1 wt%.

We suppose that electronic factors, related to the low redox potential of copper, could act as enhancers of the catalytic performance. Copper, as also mentioned by Professor Friend, is more oxophilic than gold and this could affect the rate of O_2 dissociation as you interestingly suggest. Unfortunately, the very low amount of the second metal makes it difficult to verify the nature of gold–copper interaction, but we are trying to find a technique, or a concert of techniques, which can help us to figure it out.

Dr Whiston asked: What is the change in morphology between the three catalyst preparation methods used in the paper? And why are these obtained?

Dr Della Pina responded: In this work we limited the characterization by STEM-HAADF and XRDP techniques to the most promising catalysts prepared according to method 1 only. In particular, we performed XRDP analysis on the fresh and used mono- and bimetallic catalysts (Au/C, Au–Cu/C, Au–Pt/C and Au–Pd/C), thus registering a metal particle growth on recycling for all of them, while STEM-HAADF analysis was used for picturing only the two best catalytic systems, fresh and used Au–Cu/C and Au–Pt/C, which confirmed the sintering after re-use.

Dr Manzoli said: In your paper, you compared the catalytic activity in the aerobic oxidation of allyl alcohol of a monometallic gold catalyst with a series of bimetallic gold-based catalysts (Au–Pd, Au–Pt, Au–Cu and Au–Ag). After preparation, the monometallic gold catalyst was treated at 400 °C in H_2 for 2h, while the bimetallic

    This journal is © The Royal Society of Chemistry 2011

catalysts were all simply dried in air at room temperature overnight. Why did you contrast samples pre-treated at different temperature and in different atmosphere?

As for the stability of the metallic particles, to how many catalytic cycles the samples were undergone?

Finally, have you direct experimental evidence of an interaction between the two metals? Being the amount of the second metal 1 mol% added to the 0.3% Au/C catalyst, which kind of conformation of the metal particles is obtained? Is it an alloy? Or is it something similar to a coreshell organisation, where the gold is the core and the second metal is decorating it?

Dr Della Pina replied: It is a proper consideration, thank you for underlining it. We started preparing the catalyst by including the calcination step under H_2—as we did for the monometallic gold catalyst in our previous paper[1] which we referred to—but we then simplified the protocol by deleting this step, limiting to dry the catalyst in air overnight, and we detected no difference in both morphology and catalytic activity. As a consequence, we decided to follow such a simplified preparation method for all the catalysts, thus saving time and energy. Thereby, the tests have been carried out by comparing catalysts prepared according to the same procedure. Since we actually did not specify it in the text, we have added this in the revised proofs of the paper. As indicated in the abscissa of the 13 figures, each catalyst (included the monometallic one, but excluded Au99Cu1/C prepared according to method 1) underwent 4 reaction cycles, whilst the Au99Cu1/C prepared following method 1 was tested for 9 catalytic cycles as it prolonged much more its high catalytic performance. STEM-HAADF images were taken of the 4 cycle-reused Au-Pt/C and 7 cycle-reused Au-Cu/C catalysts.

Unfortunately up to now, the very low amount of the second metal has prevented us from verifying the precise nature of the gold–copper interaction. STEM-XEDS analysis of the Au-containing particles has been attempted, but unluckily the amount of second metal is just below the detection limit of the technique. We are sure, however, from our analysis that a second population of the pure monometallic Cu particles does not co-exist with the Au-containing particles. We are trying to find a technique, or a concert of techniques, which can help us to understand how the two metals alloy (as already found in the case of 1% Au–Pt/C (Au : Pt = 1 : 1 w/w)[2]). It is unlikely that a 'core -shell' like structure will be formed in this particular case as the amount of second metal present is insufficient to give complete coverage of the Au core. However this does not preclude the second metal preferentially decorating sites on the Au particle surface in a more dilute fashion.

1 C. Della Pina *et al.*, *ChemSusChem*, 2009, **2**, 57.
2 C. Della Pina *et al.*, *J. Mol. Catal.*, 2006, **251**, 89.

Dr Louis commented: 1. You observed sintering of your bimetallic particles during catalytic reaction, do you also observe sintering of monometallic gold particles?

2. Maybe a way to determine whether your particles are bimetallic is to used DRIFTS coupled to CO adsorption on your samples. For instance CO vibrates at different frequency when it is adsorbed on Au0 and Pd0 ; moreover under CO, Pd segregates on particle surface, leading to a decrease of the band intensity of CO adsorbed on Au and an increase of that on Pd.[1] We observed Pd migration on Au particles in the case of bimetallic with a Au/Pd atomic ration of 20.[2] This should also work for Au–Pt. For Au–Cu, the difficulty is that the frequency of CO adsorbed on Au is close to that of CO adsorbed on Cu.

1 T. Wei, J. Wang, D. W. Goodman, *J. Phys. Chem. C*, 2007, **111**, 8781; K. Luo, T. Wei, C.-W. Yi, S. Axnanda, D. W. Goodman, *J. Phys. Chem. B*, 2005, **109**, 23517.
2 A. Hugon, L. Delannoy, J.-M. Krafft, C. Louis, *J. Phys. Chem. C*, 2010, **114**, 10823.

Dr Della Pina answered: 1. Yes, we also observed sintering for monometallic gold nanoparticles: XRDP analysis registered a growth of the average nanoparticle diameter from $d < 3$nm (fresh sample) up to $d = 11$ nm (used sample). 2. We thank you for your interesting suggestions, supported by many references. We surely will consider them for shedding light on the nature of our bimetallic systems. My only concern regards the use of CO adsorption when carbon is utilized as the supporting material, since I know this could be troublesome and behaves differently from the other kinds of catalytic supports (*e.g.* alumina, titania, zirconia, ceria and silica) mentioned in these papers.

Professor Friend said: Comment: XPS is subject to final state effects, especially in small particles, therefore, you should be careful about using XPS to evaluate electronic effects. I suggest using X-ray absorption focusing on the near edge of the metal absorptions to probe for possible electronic effects.

Question: Given your very small concentrations of the second metal, it is hard to imagine electronic effects if the alloy has a uniform distribution of the second metal because they would be relatively long range. Is it not possible that the second metal segregates to the surface, especially under oxidizing conditions? Copper, for example, is more oxophilic than Au and, therefore, may segregate to the surface under reaction conditions. In this case, the alloy would not be uniform and the surface would be enriched in the second, minority metal. Under these circumstances, electron effects on the surface as well as simple functional additivity may be more likely. Have you tested for this or see any evidence in the microscopy studies?

Dr Della Pina replied: We acknowledge you for the useful suggestions, especially for underlining the limit of XPS technique to evaluate electronic effects as, you are right, final state effects can mask particle size ones. These will inform our future research, aiming to disentangle this intriguing issue. Despite STEM-HAADF imaging analysis showing metal particles sintering on recycling, the experimental data registered a prolonged high catalytic performance of Au–Cu/C prepared according to method 1. Thus, other factors seem to outweigh the consequences of particle size growth. The hypothesis that electronic factors, related to the low redox potential of copper, could play a fundamental role in enhancing the catalytic performance of this bimetallic system is quite plausible. Owing to the extremely low amount of the second metal, it was not possible to detect copper by energy dispersive X-ray spectroscopy (XEDS). In addition, STEM-HAADF imaging of individual lower mass Cu atoms against the heavier Au support particles is not technically feasible. We therefore could not get much insight into the elemental distribution within the Au–Cu alloy nanoparticles using aberration corrected STEM methods.

Your suggestion that copper could segregate to the surface, thus favouring electronic effects, should be taken into account and we will focus our future efforts on solving this problem.

Professor Madix asked: You might find it useful to consult the papers on the oxidation of allyl alcohol (as well as other alcohols and aldehydes) on silver and propylene on gold single surfaces in order to construct a viable mechanism for your reaction on gold.[1] There appear to be elements that fit nicely the products you are seeing.

1 J. Solomon and R. J. Madix, *J. Chem. Phys.*, 1987, **91**, 6241.

Dr Della Pina responded: Thank you for drawing out attention to this reference. A mechanism of the possible reaction pathways occurring during allyl alcohol oxidation has already been suggested in our previous paper[1] and again mentioned in this new manuscript. Since it fits the products we detected reasonably well, it presently seems to be the most viable route we can propose. Even though we could not directly detect acrolein—the suggested transient species—its formation is in line with what it is expected during gold catalyzed aerobic oxidation of alcohols in the liquid phase: the oxidative dehydrogenation to the corresponding aldehyde, acrolein in this case, which can undergo a Michael-type addition of water in alkali conditions. In order to reinforce such a hypothesis, we also used acrolein as the reagent in gradual increasing amounts and we actually observed 3-hydroxypropionate. Regarding the two other minor products, acrylate and glycerate, the oxidation of the primary alcoholic group in aqueous alkali is widely documented, as well as the ability of supported gold to catalyze the epoxidation of olefins.

1 C. Della Pina *et al.*, *ChemSusChem*, 2009, **2**, 57.

Professor Bond opened the discussion of the paper by Dr Edwards: Can you say what steps you have to take to recover Au from the cyanide solution, and how you safely dispose of what remains? I ask because although the method is obviously viable on a small scale, I can envisage costly effluent disposal problems if it were desired to scale it up. For this reason I also wonder if it is really necessary to start with such a high Au concentration. May it not be that starting with 1 or 0.5% Au would still lead to enough small particles remaining after leaching?

Why do you think that all the Au does not dissolve in the NaCN solution? What is it about the smallest particles that prevents their dissolution? Could it be due to the rapid decrease in the redox potential as size decreases?[1]

1 L. D. Burke, A. J. Ahern and A. P. O'Mullane, *Gold Bulletin*, 2002, **35**, 3.

Dr Edwards answered: The dissolution of Au from the catalysts described in the papers follows the equation:

$$4Au + 8NaCN + O_2 + 2H_2O \rightarrow 4Na[Au(CN)_2] + 4NaOH$$

This process is widely used for Au extraction in mining on a large scale.[1] Following cyanidation, the aurocyanide ion can be extracted using a variety of processes. The carbon in pulp methodology, which is the cheapest and simplest process for Au extraction, involves the adsorption of the aurocyanide ion onto activated carbon. Once loaded, the Au is removed by desorption at high temperature and pH.[2] The procedure employed to safely dispose the aurocyanide generated in our experiments follows this methodology.

At the present time only catalysts with 5 wt% loading have been leached with cyanide. It is unclear from the analysis of the leached materials why the same concentration of Au remains on the SiO_2 despite rather harsh leaching conditions, with a huge excess of CN present. One would expect that all metallic Au present on the SiO_2 should be removed using the methodology employed. Certainly this may imply that the remaining Au is not all metallic (despite XPS analysis showing metallic Au), and this is an avenue of research we are currently exploring.

1 A. Rubo, R. Kellens, J. Reddy, N. Steier and W. Hasenpusch, in *Ullmann's Encyclopedia of Industrial Chemistry*, Wiley-VCH Verlag GmbH and Co, KGaA, 2000.
2 P. F. Sorensen, *Hydrometallurgy*, 1988, **21**, 235–241.

Dr Raj asked: In your work, Au loss in the preparation with cyanide leaching is very high (98%) compared to the other preparation techniques reported in the literature. Though the activity is due to the 0.06 wt% metal in the catalyst, what is the state of the remaining Au. Is there any work done on Au recovery or to avoid the loss.

Dr Edwards replied: The Au remaining on the CN leached materials appears to be metallic (XPS). Due to the sensitivity of the SiO_2 in the electron beam we have been unable to ascertain a particle size distribution.

I have commented on the recovery of Au in the previous question posed by Professor Bond

Dr Lalik commented: It seems that your procedure resembles somehow that of preparation of homeopathic medicines. In the homeopathic preparation, it starts from preparing a solution of active ingredient which is then repeatedly diluted up to the point that the ingredient is barely detectable, but the medicine remains (supposedly) effective. In your procedure you start from obtaining the supported Au catalyst and then you attempt to remove as much of gold from the support as possible, by prolonged leaching, until reaching an extremely low Au content, and yet the resulting material remains as active as the parent catalyst. Is it possible to leach it until it becomes totally inactive?

Dr Edwards responded: We were very surprised to observe that the same proportion of Au remained on the SiO_2 regardless of the length of exposure to CN. The CN is in major excess, and one would expect all metallic Au to be removed using the procedure outlined in the paper. I have expanded more on this issue in my response to the earlier question from Professor Bond. To date we have been unable to prepare a completely inactive catalyst, with no Au remaining after leaching.

Dr Lalik asked: In a table that you have shown in your paper (Table 1) the amount of Au in certain cases appears to be as low as 0.05%. Is it still possible to "see" such minute amounts of Au, that is, to detect them reliably with the characterisation techniques that you have applied in your research of these materials?

Dr Edwards replied: Typically the solutions prepared from the 0.06 wt% Au/SiO_2 catalysts which are analysed by ICP-MS contain 1500 µg L^{-1} Au. The detection limit for the ICP instrument used for Au analysis is 0.6 µg L^{-1} which is more than adequate to detect the low Au loadings reported in the paper. (See: www.horiba. com/uk/scientific/products/atomic-emission-spectroscopy/downloads)

Dr Jupille queried: My concern is the estimate of the gold content of the silica-supported gold catalyst. By photoemission (XPS), the Au/Si atomic ratio of the fresh catalyst ranges between 0.22×10^{-2} and 1.9×10^{-2} (Figure 3 and Table 4 in the paper). The gold loading of *ca.* 1 to 6 wt% that can be derived from this approach is in qualitative agreement with the values given by bulk analysis (Table 4). After leaching, the Au/Si atomic ratio determined by XPS decreases to $0.1–0.13\times10^{-2}$, which corresponds to about 0.3–0.4 Wt% Au. It is then counterintuitive to find bulk loading of 0.05–0.06 wt% Au. Upon leaching, the more surprising case is that of the Au-IMP of which surface atomic ratio is only decreased by a factor two while the bulk loading is divided by a factor 70 (Table 3). Photoemission determines averaged values of surface concentrations. Then, it is fully relevant to compare these to estimated bulk loadings of the catalyst. In addition, leaching should enhance the depletion in gold concentration at the surface of the catalyst which is analyzed by photoemission, so that the gold content of the bulk is expected to exceed that of the surface. Could you comment on the huge discrepancies between the two series of measurements?

Dr Edwards responded: The discrepancy between the bulk and XPS-derived loadings is likely to arise from differences in the population of gold atoms in the pores of the carbon (invisible to XPS) compared with the concentration of gold on the surface . A smaller effect is due to differences in Au particle size.

Professor L. M. Rossi asked: Why did you use such different substrate to catalyst mole ratio in the experiments with the gold catalyst before and after leaching of gold? Can you please inform the conversion obtained in the experiments in Fig 2 in the paper? Is it relevant for practical applications?

Dr Edwards answered: The reason for choosing to perform these experiments in neat benzyl alcohol, in the absence of any solvent is an environmental one. I refer to point 5 of the 12 principles of green chemistry:
"The use of auxiliary substances (*e.g.*, solvents, separation agents, *etc.*) should be made unnecessary wherever possible and innocuous when used."[1] The conversion for the catalysts in Fig. 2 after 3h for the 5 min, 30 min and 1 h CN leached catalyst is 15%, the 10min sample is 17.5%. The conversion for the unleached parent catalyst is 13%.

I have developed a number of exceptionally active AuPd bimetallic catalysts for the solvent free oxidation of benzyl alcohol.[2] It is generally known that monometallic Au catalysts are much less active than the bimetallic counterparts and as such my Au/SiO_2 catalysts which have been CN leached have lower conversions than $AuPd/TiO_2$. However, my catalysts contain far less Au and it is this discovery which forms the basis of this paper, and the work I am doing at the moment.

1 P. T. Anastas, J. C. Warner, *Green Chemistry: Theory and Practice*, Oxford University Press, New York, 1998, p.30.
2 J. K. Edwards *et. al.*, *Science*, 2006, **311**, 362.

Mr Forde remarked: Why have you only used silica? Does this method of cyanide leaching work with other oxide supports to produce the same Au loadings?

Dr Edwards responded: The use of other supports for this methodology is work in progress. Silica was chosen as the alcohol conversion for the unleached parent catalyst was particularly high.

Mr Forde questioned: Seeing that the Au loading is very low and your Au particles are very small have you tried using other techniques, such as chemical vapour deposition, to get similar loading or very small Au particle size?

Dr Edwards answered: I have developed a direct "wet chemical" route to these ultra low loaded materials. CVD has not been explored.

Mr Forde commented: In some work I have done we saw the trace impurities of iron in silica was responsible for the catalytic activity of the support. It is said that all catalysis is due to metal impurities in materials. What is the form of the Au left after leaching? Is it like a trace impurity and if so have you tried other materials which may contain trace impurities of Au such as carbon nano-fibres synthesised with Au as a template metal or synthesising them in this way?

Dr Edwards replied: As indicated in the paper, the bare SiO_2 is not active for the oxidation of benzyl alcohol. The Au could be described as a trace impurity on the SiO_2, indeed it has been described as homeopathy during these discussions (see the earlier question from Dr Lalik). Unfortunately, the exact nature of these Au nanoparticles is not known at the present time. Other materials perhaps containing Au impurities have not been investigated.

Dr Cadete Santos Aires said: Indeed HAADF imaging is a very rewarding technique to detect very small particles (even isolated atoms) since its visual. However since this is not possible due to the effect of the beam that rendered the support unstable, spectroscopy in the STEM (although not a visual technique) such as EELS could be very useful to detect small amounts of metals on the support (especially if you use very small probes). Another solution, although not a spatially resolved one, to have information on the Au on the support is XPS. Since you have performed XPS you can try to quantify the Au on the support by monitoring accurately the Au/Si ratio. Indeed knowing the surface specific area of silica, mean size of particles, real amount of Au by ICP you can make a layer by layer model that can give you information on the way Au is dispersed on the surface. Can you please comment on that?

(Comment: of course if you try to extract from XPS the total amount of Au it is normal that you have higher values that ICP since XPS will probe all Au but only part of silica (up to ~5–10 nm from the surface) thus overestimating Au whereas ICP will probe all Au taking into account the total weight (thus volume) of silica.)

Dr Edwards replied: Such modelling, though possible in theory, is fraught with practical difficulties and not as feasible as the question implies. I think a program called TOPOL was developed by a catalyst company some years ago to do this kind of modelling but it was specific to very well defined samples and made several key assumptions. We have therefore avoided this approach as it would be difficult to validate the results.

Dr Cadete Santos Aires enquired: Is there a particular reason for not trying to make directly 0.05% Au samples instead of leaching a more Au-charged sample?

Dr Edwards responded: These catalysts were prepared and evaluated for benzyl alcohol oxidation following the submission of this paper to the Faraday Discussions. The preparation of 0.06 wt% Au/SiO_2 by deposition precipitation and impregnation afforded a material with no activity for benzyl alcohol oxidation. Increasing the metal loading to 1 wt% using impregnation and deposition precipitation afforded material with approximately one quarter the activity of the leached materials after 3 h. This demonstrates that the CN leaching provides a unique route to these highly active Au/SiO_2 catalysts.

Dr Huang said: Did the authors analyze the content of sodium (Na) in the leached Au/SiO_2? It is well known that alkaline NaOH or alkaline sodium salts can greatly enhance the catalytic activity of Au catalysts in alcohol oxidation with molecular oxygen.

Dr Edwards responded: XPS analysis of the CN leached materials indicated no Na on the surface, within detection limits. The CN leaching of the catalyst is followed by a lengthy washing procedure (>1 L) which would be expected to remove all Na from the catalyst.

Dr Groszek asked: Why did you select sodium cyanide for the extraction of gold from the catalyst? I am asking this question because in the industrial practice potassium cyanide is used to recover gold from active carbons, which is used in the industrial processes.

Could it be that sodium cyanide is less effective in the extraction of gold?

Dr Edwards responded: Both KCN and NaCN are widely used in the mining industry for the extraction of Au from ore. We have not studied other CN salts for Au leaching, so the effect of the counter ion isn't known. NaCN was chosen as has shown to be very effective for removing Au in Au/CeO_2 catalysts, providing highly active materials for the water gas shift reaction.[1]

1 Q. Fu, H. Saltsburg and M. Flytzani-Stephanopoulos, *Science*, 2003, **301**, 935–938.

Professor Haruta commented: You leached metallic Au particles by using aqueous NaCN solution but it would cause the formation of $NaAu(CN)_2$ (ref. 1), which might be active for benzyl alcohol oxidation. If you leach Au particles by using iodine tincture, the catalyst samples are free from cyanide. Have you tried to replace aqueous NaCN solution with iodine tincture?

1 S. T. Oyama *et. al.*, *ChemCatChem*, 2010, **2**, 1582

Professor Bowker asked: Have you just tried soaking it in cyanide and then thermally decomposing it? Then wash off the Na? In this way you might get a better overall yield by getting a larger number of the very small particles which you get after cyanide washing/leaching.

Dr Edwards answered: This approach has not been investigated. The surprising results highlighted in the paper indicated that only a very small fraction of Au in the 5 wt% Au/SiO2 catalyst is actually active for the oxidation of benzyl alcohol. Whilst a re-dispersion of the 5 wt% Au/SiO2 catalyst using CN may result in a much better dispersion of smaller Au nanoparticles, the focus of this research at the present time is in the preparation of catalysts with very small Au loadings, where most (or all) of the Au is present as the active site and participates in the catalytic reaction. However, this redispersion approach may prove useful on catalysts with low Au loading prepared by conventional impregnation methodologies.

Dr Nijhuis remarked: In our propene epoxidation work we have observed a similar remarkable high activity for low gold loaded catalysts. A similar observation for the propene epoxidation has also been reported by the Delgass group for the propene epoxidation, in their case for even lower gold loadings. In both their and our case, these catalysts were prepared by just depositing such a small amount of gold. You mentioned that when you tried to prepare these low loaded catalysts by just depositing a small amount, these catalysts were not active. Can you explain what is the difference between these catalysts and those prepared by leaching when you characterize them? When you tried to prepare the low loaded catalysts, did it simply not end up on the support?

Dr Edwards responded: When the very low loaded catalysts (0.06 wt%) were prepared by impregnation and deposition precipitation Au was found on the support (ICP-MS). Unfortunately these materials suffer the same instability under the electron beam and no further analysis is available at the present time.

Concluding remarks: from match to flamethrower

Martyn Poliakoff*

Received 15th August 2011, Accepted 15th August 2011
DOI: 10.1039/c1fd00106j

These remarks give some impressions of Faraday Discussion 152 from the point of view of an outsider and suggest a number of actions which might help bind the Gold Catalysis Community more strongly and increase the long term impact of their science.

Introduction

I came to this Discussion as a complete outsider. I am a Green Chemist[1,2] with a background in Inorganic Chemistry.[3,4] Of course, like the rest of the participants, I had a pile of preprints to read but I have never published any papers involving Gold and have been an author on only two publications[5,6] that could possibly be construed as 'serious catalysis'.

So you can imagine that I was filled with increasing apprehension as the date of the Discussion approached and my presentation drew nearer. What could I say to fill the 30 min so generously allocated to me? Then I had an inspiration. I would e-mail Professor D. D. Eley, legendary pioneer of catalysis,[7] who had recently 'gone on-line' for the first time at the age of 95, and seek his advice. He replied quickly:

"Dear Martyn, thank you for your email enquiry. In my catalysis work on Pd alloys, I chose PdAu and not PdAg (periodic table!), because Au was a non-catalyst, at least in those days, 1950, 61 years ago! I do hope this might be of some help for your Faraday closing remarks. With best wishes for the success of the Faraday Discussion on Gold catalysis! Dan Eley"

I then found that several authors in this Discussion were using Pd/Au as a catalyst! In an attempt to find out why things had changed since 1950, I turned to Paper 23, by another pioneer of catalysis, Geoff Bond. It quickly became clear that the difference was connected with particle size. Gold is only catalytic when the particles are small, possibly < 3nm. Bond suggests that this size may be associated with a conductor/insulator transition. However, I began wondering whether some of the difference between gold and the other noble metal oxidation catalysts (*e.g.* Pd or Pt) might be due to differences in their reactivity towards O_2. Bulk gold does not react with O_2 but might Au atoms or nanoparticles react with O_2?

At this point I began feeling rather like Agatha Christie's fictional character Miss Marple who solved murders in exotic locations by comparing them to events in her English village many years before.[8] Earlier in my career, I was active in the field of matrix isolation where reactive species are trapped in a solid noble gas (the matrix) at low temperature *e.g.* 4 K)[9,10] and are then studied at leisure using IR spectroscopy. I remembered that Lester Andrews had pioneered the reactions of laser-ablated metal atoms with O_2 and other gases. Had he studied Au with O_2? A quick search found the reaction[11] in Scheme 1.

The School of Chemistry, University of Nottingham, University Park, Nottingham, NG7 2RD, England. E-mail: Martyn.Poliakoff@nottingham.ac.uk; Tel: +44 115 951 3520

$$\text{Au} + \text{X} \xrightarrow{\quad\text{Ar matrix, 10 K}\quad} \text{HOAuOH}$$

$$\text{X} = H_2O_2 \text{ or } H_2 + O_2$$

Scheme 1 Reaction of Au atoms isolated in a matrix.[11]

The product, $Au(OH)_2$, appears to be pertinent to Au catalysis of the formation of H_2O_2 from H_2. Furthermore, its structure, Fig. 1, has an O–Au–O group similar to that featured in the structure of oxidized Au in Paper 8. Additional searching revealed more papers by Andrews[12,13] with Au reactions of possible relevance to FD152.

Thus armed, I arrived in Cardiff and listened to the presentations. I was struck that many of the participants appeared to be expecting a single explanation for all the experimental observation in Au catalysis. This seemed unrealistic on two grounds. Firstly, there is a well known scientific cliché that any theory that explains all the facts is bound to be wrong because some of the facts will be wrong. Secondly, and more personally, I was involved in a study of a single reaction of $Ni(CO)_3N_2$ which switched between two totally different mechanisms with only a modest rise in temperature.[14] I was also struck by two statements; one by Ye Xu during Paper 16 *"Variability of Au catalysts obfuscates understanding"* and the other by Cynthia Friend in the discussion following Paper 19 *"Can you not do more experiments to be more convincing?"* These and other remarks prompt me to suggest the following points for consideration by researchers in this field.

1. For experimentalists

There seems to be a need for more data that can be compared between one experiment and another. Together these data would form a solid *corpus* which could provide a solid basis for theoretical work and, at the same time, would link researchers together as a true community. After the Meeting, Geoff Bond wrote to me: *"In an attempt to obtain 'more good data', I persuaded the World Gold Council some years ago to have made batches of 'standard' Au catalysts, which in fact have been quite widely and helpfully used."* (These standards are available for people to use for benchmarking their own catalysts). Geoff also cautioned *"Catalysis is essentially a kinetic phenomenon, and to start with we need to know the rate of our reaction, and where relevant the product selectivities, under a range of fully specified conditions. Care has to be taken to allow for deactivation and for the onset of diffusion limitation. Identification of factors bearing on catalytic performance necessitates comparison of work performed in different laboratories, and this requires it to be conducted carefully and reported accurately."*

2. For theoreticians

Much of the theoretical work presented at this Meeting was aimed at explaining what had already been observed. There seemed to be a real reluctance to make

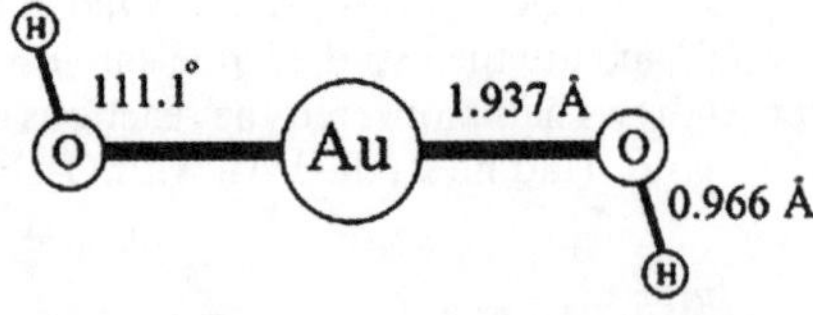

Fig. 1 Calculated B3LYP structure of $Au(OH)_2$ consistent with IR spectra in low temperature matrices.[11] Reproduced with permission - copyright The Royal Society of Chemistry.

predictions of what might be found in new experiments. Without predictions there cannot be a dialogue between Theory and Experiment. There is no need to be afraid; incorrect predictions are just as valuable as those that are found to be correct and may even attract more citations. All predictions help to advance science. Conversely, experimental advances are unlikely without theoretical predictions. Furthermore Cynthia Friend commented to me afterwards: *"Firstly, gold is a very subtle material with weak interactions playing important roles in its behavior, yet the theoretical treatment du jour (DFT) does not capture these effects. Somehow the methodology needs to be improved. Furthermore, it is very important that these caveats be noted when theoretical treatments are being presented. As of now, the theory is often stated as proof of something instead of noting the possible inaccuracies. [We could, of course, make the same statement about experiments having limitation.] Secondly, catalysis is about kinetics, but DFT probes thermodynamics. The 'nudged elastic band' method does try to probe the energy of the transition state; however, molecular dynamic simulations and/or construction of a microkinetic model is critical to understand complex reactions where selectivity is an issue."*

3   The role of the interface

The nature of the interface between oxide support and Au nanoparticle seems to be a crucial question which has to be answered before Au catalysis can be even poorly understood. More experimental data are required before theoreticians can even begin to address the question. During the Meeting, Sir John Meurig Thomas sent me an e-mail: *"After numerous studies using a multiplicity of techniques it is still not clear whether it is the interface between a minute Au particle and its support or, non-metallic Au [as suggested by Flytzani-Stephanopoulos[15]] where there are ionic-covalent bonds between the Au and its support e.g. Au–O–Ce that constitutes the locus of the exceptional catalytic activity of gold nanoparticles and nanoclusters."* After the Meeting, Cynthia Friend also commented to me: *"We, as a field, should devise experiments to test the postulated effects, which include electronic modification at the interface, reaction in part on the oxide support, and migration between the Au and the oxide. This will improve our understanding of the interface and guide the design of effective catalysts."*

4   The role of water

Several of the speakers stressed that the presence of H_2O was essential for successful catalytic oxidation over Au. Professor Haruta did discuss this briefly in his Opening Address and I hope that it will be covered in his printed paper. However, almost none of the other Papers gave even a vague suggestion of why H_2O is needed. One exception was Stephen Hashmi who did report, in Paper 11, some modelling of H_2O as a proton shuttle in a *homogeneous* gold catalyst. Again there is a clear need for more data. Perhaps D_2O might give some sort of isotope effect similar to that mentioned in Paper 22?

5   Fluxionality

In Paper 9 it was suggested that Au clusters might not adopt the simplest shape (e.g. an icosahedron for Au_{13}). This gave me another "Miss Marple moment" because it reminded me how our understanding of co-ordinatively unsaturated metal carbonyls was transformed[16] when it was realised that these species did not necessarily adopt the most symmetrical structures[17] and that these structures were fluxional.[9] Could fluxionality be influencing Au catalysis? As Mike Bowker said during the discussion of Paper 9, *"Fluxionality is not good for catalysis"*. There were also suggestions that Au atoms may be mobile on surfaces. In short, there is a need to answer Cynthia Friend's rather blunt question *"What's on the surface?"*

6 "Communality of Group 1 reactivity"

This was a phrase used by Mike Bowker in the discussion of Paper 14. What he was asking was *"What lessons can be drawn from Cu and Ag to illuminate Au catalysis?"* As an Inorganic chemist, it seems quite natural to me to use the Periodic Table in this way. After my lecture, I was surprised to learn that Bob Madix has recently published three papers/reviews on this topic[18–20] but, until now, they seem to have been largely overlooked by the Au Catalysis Community. This point is amplified in his answer to one of the questions about Paper 14.

7 The role of the solvent

Relatively few of the Papers concerned liquid phase reactions but those that did seemed largely to ignore the role of the solvent. As mentioned under point 2, weak interactions are likely to be very important in Au catalysis. This means that catalyst-solvent interactions are almost certain to play a role. The challenge to both experimentalists and theoreticians alike is how to identify that role.

8 Engagement with industry

Au catalysis is over 25 years old. It is newer than Pt or Pd catalysis but cannot really be regarded as new. Au is relatively abundant and much cheaper than other noble metals. So there is an opportunity for a real step change in industrial catalysis if Au-based processes can be commercialized. So I was amazed to find only two industrial participants at the Meeting. Furthermore, hardly any of the speakers addressed even the simplest questions needed for industrial exploitation, for example catalyst lifetime, stability and degradation. So my final challenge to the Au Catalysis Community is for them to generate real industrial impact. I threw down this challenge by analogy, with a simple practical demonstration. I struck a match. If you do nothing, the match will either go out or burn your fingers. But if you use the match to light something else, you can have a flamethrower. (Fortunately, my match went out!)

I ended my remarks by showing a picture of the UK's first ATM gold-vending machine that had just been installed at the Westfield shopping centre in West London.[21] This machine underlines the current high price of gold and, as Graham Hutchings, organizer of FD152, commented *"Gold is now ca. £1000/ounce and this is a factor that has been a problem. It means we need continually to strive for very low levels of gold i.e. 0.1–0.3% by weight."*

Postscript

A chance conversation with two of the delegates in the taxi as we departed for the station indicated that there is far more industrial activity in this area than suggested at the Meeting. There are also strong rumours of a few processes already on-stream involving gold catalysts. In addition, there were occasions when speakers at FD152 hid behind industrial confidentiality during the discussion of their papers. While fully appreciating the need to protect intellectual property and industrial knowhow, I feel that such levels of secrecy are disappointing and cannot be in the best long term interests of either the science or its exploitation.

Acknowledgements

I thank the participants, especially Keith Whiston, for welcoming me to this Discussion, and Dan Eley, Lester Andrews, Geoff Bond, Bob Madix, John Meurig Thomas, Cynthia Friend and Graham Hutchings for sending me comments and papers.

References

1 M. Poliakoff, J. M. Fitzpatrick, T. R. Farren and P. T. Anastas, *Science*, 2002, **297**, 807–810.
2 M. Poliakoff and P. Licence, *Nature*, 2007, **450**, 810–812.
3 P. M. Hodges, S. A. Jackson, J. Jacke, M. Poliakoff, J. J. Turner and F. W. Grevels, *J. Am. Chem. Soc.*, 1990, **112**, 1234–1244.
4 J. A. Darr and M. Poliakoff, *Chem. Rev.*, 1999, **99**, 495–541.
5 G. J. Hutchings, J. A. Lopez-Sanchez, J. K. Bartley, J. M. Webster, A. Burrows, C. J. Kiely, A. F. Carley, C. Rhodes, M. Havecker, A. Knop-Gericke, R. W. Mayer, R. Schlogl, J. C. Volta and M. Poliakoff, *J. Catal.*, 2002, **208**, 197–210.
6 G. J. Hutchings, J. K. Bartley, J. M. Webster, J. A. Lopez-Sanchez, D. J. Gilbert, C. J. Kiely, A. F. Carley, S. M. Howdle, S. Sajip, S. Caldarelli, C. Rhodes, J. C. Volta and M. Poliakoff, *J. Catal.*, 2001, **197**, 232–235.
7 E. K. Rideal and D. D. Eley, *Discuss. Faraday Soc.*, 1950, **8**, 96–104.
8 A. Christie, *A Caribbean Mystery - A Jane Marple Murder Mystery*, Pocket Books, New York, 1966.
9 M. Poliakoff, *Inorg. Chem.*, 1976, **15**, 2892–2897.
10 B. Davies, A. McNeish, M. Poliakoff and J. J. Turner, *J. Am. Chem. Soc.*, 1977, **99**, 7573–7579.
11 X. F. Wang and L. Andrews, *Chem. Commun.*, 2005, 4001–4003.
12 L. Andrews and X. F. Wang, *J. Am. Chem. Soc.*, 2003, **125**, 11751–11760.
13 L. Andrews, X. F. Wang, L. Manceron and K. Balasubramanian, *J. Phys. Chem. A*, 2004, **108**, 2936–2940.
14 J. J. Turner, M. B. Simpson, M. Poliakoff and W. B. Maier, II, *J. Am. Chem. Soc.*, 1983, **105**, 3898–3904.
15 Z. Zhou, S. Kooi, M. Flytzani-Stephanopoulos and H. Saltsburg, *Adv. Funct. Mater.*, 2008, **18**, 2801–2807.
16 J. K. Burdett, *J. Chem. Soc., Faraday Trans. 2*, 1974, **70**, 1599–1613.
17 M. Poliakoff and J. J. Turner, *J. Chem. Soc., Dalton Trans.*, 1974, 2276–2285.
18 R. J. Madix, C. M. Friend and X. Y. Liu, *J. Catal.*, 2008, **258**, 410–413.
19 X. Y. Liu, R. J. Madix and C. M. Friend, *Chem. Soc. Rev.*, 2008, **37**, 2243–2261.
20 C. G. Freyschlag and R. J. Madix, *Mater. Today*, 2011, **14**, 134–142.
21 http://www.guardian.co.uk/money/2011/jul/01/au-atm-gold-vending-machine, Accessed 13 August, 2011.

Poster titles

Electrochemistry at gold electrodes, **P. Rodriguez, N. Garcia-Araez, A. Koverga, A. Yanson and M. T. M. Koper**, *Leiden University, The Netherlands*

Molecular clusters to nanoparticles: entrapment and catalytic activity of gold nanoparticles generated from an Au_{13} precursor on oxide supports and zeolite matrix, **R. Sarip, G. Hogarth and G. Sankar**, *University College London, UK*

Hydrogen interaction with gold nanoparticles and clusters supported on different oxides. A FTIR study, **M. Manzoli, A. Chiorino, F. Vindigni and F. Boccuzzi**, *University of Torino, Italy*

Role of the size of gold particles for selective oxidations with molecular O_2, **F. Boccuzzi, A. Chiorino, M. Manzoli, F. Menegazzo, F. Pinna, M. Signoretto, V. Trevisan and F. Vindignia**, *University of Torino, Italy*

Investigaing the evolution of catalytic surfaces under realistic conditions: CO adsorption on Au (110), **M. A. Languille, Y. Jugnet, E. Ehret, J. C. Bertolini and F. J. Cadete Santos Aires**, *CNRS/Université Lyon I, France*

Dynamic electro-chemo-mechanical analysis during cyclic voltammetry, **M. Smetanin, Q. Deng and J. Weissmüller**, *Karlsruhe Institute of Technology, Germany*

Pd-Promoted gold catalysts for selective hydrogenation of 1,3-butadiene and *p*-chloronitrobenzene, **A. Hugon, L. Delannoy, C. Louis, F. Cárdenas-Lizana, S. Gómez-Quero and M. A. Keane**, *Université Pierre et Marie Curie – CNRS, France*

Au and Cu electrocatalysts for CO_2 reduction, **D. Plana, M. Montes de Oca and D. J. Fermín**, *University of Bristol, UK*

Design of heterogeneous catalysts within a toolbox approach: functionalization of nanoporous gold, **A. Wichmann, A. Wittstock, L. Mädler, K. Frank, A. Rosenauer and M. Bäumer**, *University Bremen, Germany*

A reactive global optimization (RGO) approach to heterogeneous catalysis (oxidation) by $(Ag-Au)_3$ nanoclusters, **F. Negreiros, E. Apra and A. Fortunelli**, *IPCF-CNR, Italy*

Van der Waals interactions between nanotubes and nanoparticles for controlled assembly of composite nanostructures, **G. A. Rance, D. H. Marsh, S. J. Bourne, T. J. Reade and A. N. Khlobystov**, *University of Nottingham, UK*

Supported alkylthiol-derivatised gold nanoparticles in catalysis, **W. A. Solomonsz, G. A. Rance, A. La Torre, S. Miners, P. D. Brown, and A. N. Khlobystov**, *University of Nottingham, UK*

On the lifetime of the transients $(NP)-(CH_3)_n$ ($NP = Ag^0$, Au^0 nanoparticles) formed in the reactions between methyl radicals and nanoparticles suspended in aqueous solution, **R. Bar-Ziv, I. Zilbermann, T. Zidki, G. Yardani, H. Cohen and D. Meyerstein**, *Ariel University Center of Samaria, Israel*

Preparation of a highly active core-shell bimetallic AuPd catalyst with low palladium loading, **L. M. Rossi, T. A. Silva, and E. Teixeira Neto**, *Universidade de São Paulo, Brazil*

Carbon monoxide adsorption on doped gold clusters, **H. T. Le, J. De Haeck, P. Claes, S. M. Lang, S. Bhattacharyya, E. Janssens and P. Lievens**, *KU Leuven, Belgium*

Photocatalytic reductions under visible light using supported gold catalysts, **X. Ke, J. Zhao, X. Zhang, S. Sarina, J. Chang and H. Zhu**, *Queensland University of Technology, Australia*

A QM-MM study on the self-assembly of thiol monolayers on gold surfaces, **S. A. Serapian, M. J. Bearpark and F. Bresme**, *Imperial College London, UK*

Electrocatalytic properties of Au-Pd core-shell nanostructures towards CO and formic acid oxidation, **M. Montes de Oca, D. Plana and D. J. Fermín**, *University of Bristol, UK*

Theoretical study of structure and catalytic properties of gold and bimetallic clusters, **D. F. Mukhamedzyanova, D. A. Pichugina, A. V. Beletskaya, M. S. Askerka, A. F. Shestakov and N. E. Kuz'menko**, *Moscow State University, Russia*

Evaluation of methods to predict reactivity of gold nanoparticles, **T. C. Allison, Y. Ye and J. Tong**, *National Institute of Standards and Technology, USA*

Glycerol oxidation over gold-supported mesoporous metal oxides, **P. R. Murthy, A. Alagarasi, B. Viswanathan and P. Selvam**, *National Centre for Catalysis Research, India*

Abnormally high heat evolutions generated by the interaction of oxygen with hydrogen chemisorbed on gold, **A. J. Groszek and E. Lalik**, *Microscal Ltd, UK*

Catalytic activity of Au clusters on "inert" h-BN support, **M. Gao, A. Lyalin and T. Taketsugu**, *Hokkaido University, Japan*

STEM-HAADF characteristics of Au/ZnO catalysts for CO oxidation, **Q. He, A. Thomas, J. K. Edwards, A. F. Carley, G. J. Hutchings and C. J. Kiely**, *Lehigh University, USA*

Size-selected Au cluster interactions with supporting graphite substrates: a DFT approach, **A. J. Logsdail, R. L. Johnston and J. Akola**, *University of Birmingham, UK*

Identifying cluster geomites from HAADF-STEM images: a kinematic model coupled with structural searches, **A. J. Logsdail, D. S. He, Z. Y. Li and R. L. Johnston**, *University of Birmingham, UK*

An optimum size of gold for CO oxidation over Au/CeO_2 at room temperature, **J. Huang, T. Takeia, H. Ohashib and M. Haruta**, *Tokyo Metropolitan University, Japan*

Active oxygen species on Au nanoparticles supported on $TiO_2(110)$ investigated using *in situ* XPS, **K. Dumbuya, G. Cabailh, R. Lazzari, J. Jupille, L. Ringel, M. Pistor, O. Lytken, H. P. Steinrück and J. M. Gottfried**, *Université Pierre et Marie Curie, France*

Selective synthesis of secondary amines by *N*-alkylation of primary amines with alochols over gold catalysts, **R. Takamura, T. Ishida, T. Takei, T. Akita and M. Haruta**, *Tokyo Metropolitan University, Japan*

Deposition of gold nanoparticles on carbon nanohorn and their catalytic activity for glucose oxidation, **J. Yamada, T. Ishida, T. Takei, R. Yuge, T. Yoshitake and M. Haruta**, *Tokyo Metropolitan University, Japan*

Deposition of Au11 clusters on ZrO_2 by oxygen plasma irradiation, **Y. Yu, T. Ishida and M. Haruta**, *Tokyo Metropolitan University, Japan*

Preparing highly active CO oxidation catalysts by simple variations in the co-precipitation procedure of Au/FeO_x, **S. J. Freakley, Q. He, Y. Mineo, J. K. Edwards, A. F. Carley, T. Takei, C. J. Kiely, M Haruta and G. J. Hutchings**, *Cardiff University, UK*

Not so loosely bound rare gas atoms: vibrational fingerprints of neutral gold cluster complexes, **L. M. Ghiringhelli, P. Gruene, J. T. Lyon, G. Meijer, A. Fielicke and M. Scheffler**, *Fritz-Haber-Institut der Max-Planck-Gesellschaft, Germany*

The Skinner Prize for the best poster was awarded to Miss Maria Montes de Oca of University of Bristol, UK, for her poster on Electrocatalytic properties of Au-Pd core-shell nanostructures towards CO and formic acid oxidation.

List of participants

Dr T Allison, *NIST, U.S.A.*
Prof Dr M Bäumer, *University Bremen, Fachbereich 2, Germany*
Prof Dr T Bernhardt, *University of Ulm, Germany*
Prof Dr S Bhargava, *RMIT University, Australia*
Prof Dr F Boccuzzi, *University of Torino, Inorganic, Physical & Materials Chemistry, Italy*
Professor G Bond, *Brunel University., United Kingdom*
Professor M Bowker, *Cardiff University, United Kingdom*
Miss G Brett, *Cardiff University, United Kingdom*
Dr F Cadete Santos Aires, *IRCELYON UMR5256 - CNRS/Univ. Lyon, France*
Professor C Campbell, *University Of Washington, U.S.A.*
Mr J Chen, *Eindhoven University of Technology, The Netherlands*
Dr E Corbacho Beret, *Fritz-Haber-Institut der MPG, Germany*
Mr J Counsell, *Royal Society of Chemistry, United Kingdom*
Dr C Della Pina, *Universita degli Studi di Milano-Dip.Chimica Inorganica, Metallorganica, Analtica, Italy*
Dr J Edwards, *Cardiff Catalysis Institute, United Kingdom*
Professor P Edwards, *University of Oxford, United Kingdom*
Mrs B Fetene Shawil, *Addis Ababa University, Ethiopia*
Mr M Forde, *Cardiff Catalysis Institute, United Kingdom*
Mr A Fortunelli, *IPCF-CNR, Italy*
Mr S Freakley, *Cardiff University, United Kingdom*
Professor C Friend, *Harvard University, U.S.A.*
Dr L Ghiringhelli, *Fritz-Haber-Institut der Max-Planck-Gesellschaft, Germany*
Professor S Golunski, *Cardiff Catalysis Institute, United Kingdom*
Dr A Groszek, *United Kingdom*
C Hammond, *Cardiff University, United Kingdom*
Mr N Harby, *Rand Refinery Ltd, South Africa*
Professor M Haruta, *Tokyo Metropolitan University, Japan*
Professor S Hashmi, *University of Heidelberg, Germany*
Mr Q He, *Lehigh University, U.S.A.*
Miss A Hefer, *Rand Refinery Ltd, South Africa*
Professor P Hu, *Queens University Belfast, United Kingdom*
Dr J Huang, *Faculty of Urban Environmental Science, Tokyo Metropolitan University, Japan*
Professor G Hutchings, *Cardiff University, United Kingdom*
Dr C Jia, *Max Planck Institut Fur Kohlenforschung, Germany*
Dr J Jupille, *CNRS, France*
Dr X Ke, *QUT, Australia*
Dr T Keel, *World Gold Council, United Kingdom*
Prof Dr C Kiely, *Lehigh University, U.S.A.*
Dr E Lalik, *Institute of Catalysis and Surface Chemisrtry, Poland*
Dr S Lang, *Ulm University, Germany*
Miss H Le, *K.U.Leuven, Belgium*
Mr A Logsdail, *University of Birmingham, United Kingdom*
Dr C Louis, *Laboratoire de Reactivite de Surface, CNRS-UPMC, France*
Dr A Lyalin, *Hokkaido University, Japan*
Professor R Madix, *Harvard University, U.S.A.*
Dr M Manzoli, *University of Torino, Department of Inorganic, Physical & Materials Chemistry, Italy*

Professor D Meyerstein, *Ben Gurion University, Israel*
Dr L Molina, *University of Valladolid, Spain*
Miss M Montes-de Oca Yemha, *University of Bristol, United Kingdom*
Ms D Mukhamedzyanova, *M.V. Lomonosov Moscow State University, Russia*
Dr A Nijhuis, *Eindhoven University of Technology, The Netherlands*
Dr T Ntho, *Mintek, South Africa*
Dr A OMullane, *RMIT University, Australia*
Mr M Piccinini, *Cardiff University, United Kingdom*
Prof Dr F Pinna, *University of Venice, Italy*
Dr D Plana, *University of Bristol, United Kingdom*
Professor M Poliakoff, *University of Nottingham, United Kingdom*
Professor L Prati, *Universita degli Studi di Milano, Italy*
Mr J Pritchard, *Cardiff University, United Kingdom*
Professor P Pyykkö, *University of Helsinki, Finland*
Dr A Raj, *Johnson Matthey, United Kingdom*
Dr G Rance, *University of Nottingham, United Kingdom*
Dr P Rodriguez, *Leiden Institute of Chemistry, Germany*
Prof Dr L Rossi, *University of Sao Paulo, Brazil*
Professor V Rotello, *University of Massachusetts-Amherst, U.S.A.*
Dr M Saint-Lager, *Institut Nel - CNRS, France*
Miss R Sarip, *University College London, United Kingdom*
Professor P Selvam, *Indian Institute of Technology Madras, India*
Mr S Serapian, *Imperial College London, United Kingdom*
Dr M Smetanin, *Technical University Hamburg-Harburg, Germany*
Mr W Solomonsz, *University of Nottingham, United Kingdom*
Miss R Takamura, *Tokyo Metropolitan University, Japan*
Mrs R Thompson, *Royal Society of Chemistry, United Kingdom*
Dr B Timmins, *Materion, Republic of Ireland*
Professor R Van Santen, *Eindhoven University of Technology, The Netherlands*
Dr A Watson, *Royal Society of Chemistry, United Kingdom*
Dr K Whiston, *Invista UK Ltd, United Kingdom*
Mr D Widmann, *Ulm University/Institute of Surface Chemistry and Catalysis, Germany*
Dr D Willock, *Cardiff University, United Kingdom*
Dr A Wittstock, *Lawrence Livermore National Laboratory, U.S.A.*
Dr Y Xu, *Oak Ridge National Laboratory, U.S.A.*
Mr J Yamada, *Tokyo Metropolitan University, Japan*
Miss Y Yu, *Faculty of Urban Environmental Science, Tokyo Metropolitan University, Japan*

Index of contributors*

* The page numbers in **bold** type indicate papers submitted for discussions.